Green Energy and Technology

For further volumes:
http://www.springer.com/series/8059

Marta S. Basualdo · Diego Feroldi
Rachid Outbib
Editors

PEM Fuel Cells with Bio-Ethanol Processor Systems

A Multidisciplinary Study of Modelling, Simulation, Fault Diagnosis and Advanced Control

Springer

Dr. Marta S. Basualdo
CIFASIS-(CONICET-UNR-UPCAM)
-UTN-FRRo
French-Argentine International Center
of Information and Systems Sciences
27 de Febrero 210 bis
S2000EZP Rosario
Argentina
e-mail: basualdo@cifasis-conicet.gov.ar

Dr. Rachid Outbib
LSIS-Domaine Universitaire
de Saint-Jérôme
Laboratoire des Sciences
de l'Information et des Sytèmes
Avenue Escadrille Normandie-Niemen
13397 MARSEILLE CEDEX 20
France
e-mail: rachid.outbib@lsis.org

Dr. Diego Feroldi
CIFASIS-(CONICET-UNR-UPCAM)
French-Argentine International Center
of Information and Systems Sciences
and LCC-FCEIyA-UNR
27 de Febrero 210 bis
S2000EZP Rosario
Argentina
e-mail: feroldi@cifasis-conicet.gov.ar

ISSN 1865-3529
e-ISSN 1865-3537
ISBN 978-1-4471-2687-4
ISBN 978-1-84996-184-4 (eBook)
DOI 10.1007/978-1-84996-184-4
Springer London Dordrecht Heidelberg New York

British Library Cataloguing in Publication Data
A catalogue record for this book is available from the British Library

Cover design: eStudio Calamar, Berlin/Figueres

Printed on acid-free paper

Springer is part of Springer Science+Business Media (www.springer.com)

To Pablo, Sebastian, Javier and Hernan, to the lovely memory of my mother

MB

To my beloved family

DF

To my family

RO

Preface

The control of PEM fuel cell-based systems has been the subject of several papers during the last years. However, until now only few books have been devoted to analyze the various aspects of this matter. Moreover, the involved technology is quite extensive and more dedicated to specific areas. In fact, there are also many disciplines involved in this subject which have to work in a complementary way to give efficient solutions to the most common problems arising from PEM fuel cells-based systems.

The goal of this book is to present some useful tools for studying the entire system from hydrogen production through bio-ethanol reforming to power generation in the fuel-cell stack. In this context, the book tries to present a deep discussion about several aspects concerning some typical problems in PEM fuel cells-based systems and to propose alternative ways of operation through conventional and advanced control techniques in a safe manner. Aspects such as efficient energy management strategies are performed too in order to command the power split in a hybrid system such as the fuel cell stack coupled with a supercapacitor bank.

We know that an apparently appropriate control scheme for the PEM fuel cell may actually lead to an in-operable plant when it is connected to other unit operations in a process with recycle streams and energy integration. Our objective is to design a control system that provides basic regulatory control of the hydrogen production process; i.e., the plant will be at the wanted operating point despite disturbances. Above this regulatory structure we can then build fault diagnosis systems to improve plant safety performance. Although one of the methods discussed here is heuristic, we certainly recommend the use of a systematic approach that can aid the sensor network together with the plantwide control designs. However, the primary mathematical tool employed in this book is a rigorous, nonlinear model of the entire plant. This model can faithfully capture the nonlinearity and the constraints encountered in the hydrogen production process under consideration. It is considered that the control scheme for the overall system must be tested on this type of model because linear, unconstrained models are not adequate to predict many of the important plant-wide phenomena.

So mathematical modeling and simulation are vital tools in the solution of the optimal sensor location, multi-variable control, and fault diagnosis problems.

Another significant element in this book is the important place that we give to the computer-aided engineering software tools and computer horsepower that permit us to assemble a flowsheet, perform the steady-state analysis (mass and energy balances, engineering economics, and optimization), and then evaluate the dynamic performance of the plant. The commercial software packages used here, able to combine steady-state and dynamic realistic models, give us a major breakthrough in the tools available to support our studies. The concepts presented in this book can be applied at different levels such as in the conceptual development of this new process, as plant revamps and in the operation of an existing process. However, the emphasis is on the new plant design because this is the level at which the effect of considering these tools can have the most significant impact on profitability. The cost of modifying the process at the design stage is usually fairly low and the effect of these modifications on the dynamic controllability can be enormous. It is remarked here because several dynamic operability problems cannot be seen in a steady-state flowsheet.

This book is intended for use by students in senior courses of chemical, electrical-electronic and mechanical engineering in which dynamics and control are incorporated with the traditional steady-state coverage of flowsheet synthesis, engineering economics, and optimization. Practicing engineers and researchers could find in this book new proposals which can be useful too. Anyway, we assume that this book can be followed by a wide range of readers, as only basic concepts of modeling, simulation and control are necessary to have. Besides, each issue in the book is presented starting from its basic concepts, providing to the readers a guidance to follow the subject and an important review of the available literature for the different topics addressed here.

The main subjects of this book are dedicated to these new power generation systems, based on fuel-cell technology, involving diverse disciplines to be well-handled. In selecting the authors and the topics, the emphasis has been on giving the most adequate tools to achieve accurate, practical, and useful solutions. Accordingly, they are supported by several fields such as control theory, computational tools, artificial intelligence, expert systems, and experimental results, among others. Therefore, the book will give to the readers, from several areas, opportunity of learning how to handle these kind of problems and to propose new solutions too.

The book is divided into two main parts, the first one mainly is dedicated to fuel-cells systems and the second one to fuel processor systems from renewable energy sources such as bio-ethanol for hydrogen production. In both parts, modeling, simulation, advanced control, and fault diagnosis aspects are deeply analyzed. Part I includes the following issues: an introduction in Chap. 1 where the reader could have a picture about the main contents of the book and which are the reasons that justify this selection. Then, a thorough description of PEM Fuel Cells System is provided Chap. 2. In Chap. 3, two advanced control strategies based on Model Predictive Control to control the oxygen level in the cathode of a

PEM fuel-cell system are presented with the objectives of achieving high efficiency and maintaining the necessary level of the oxygen in the cathode to prevent short circuit and membrane damage. In Chap. 4, an alternative modeling approach (using partial differential equations) is presented for analysis and simulation of the thermal behavior of PEMFC, then a strategy to control the thermal behavior is proposed. Chapter 5 deals with the problem of fault diagnosis for PEM fuel-cell, presenting two strategies for fault diagnosis that are based on electrical equivalent technical and statistical approach, respectively. In Chap. 6, the fault diagnosis of PEM Fuel Cell Systems is addressed, using a model-based methodology. Besides, in this chapter it is addressed the Fault-Tolerant control based on Model Predictive Control. In Chap. 7, it is approached the study of fuel-cell hybrid systems with some energy storage, showing the important advantages of hybridization in Fuel Cell Hybrid Vehicles. In Chap. 8, three energy management strategies, based on the fuel-cell efficiency map, are presented and validated through an experimental setup to control the energy flow between the fuel cell, the energy storage system, and the electrical load in Fuel Cell Hybrid Vehicles.

Part II is dedicated to Fuel Processors Systems with Bio-Ethanol to produce the necessary hydrogen to supply a PEM fuel-cell system. This part includes the following chapters. Chapter 9 deals with the synthesis of an integrated bio-ethanol processor for PEM fuel-cell and a first proposal of control structure based on more heuristic concepts. Chapter 10 approaches the control-oriented dynamic model of the bio-ethanol processor system useful to test different strategies of control and fault diagnosis. Chapter 11 contains several details about the three software packages used to construct the nonlinear pseudo-dynamic model presented in Chap. 10. Hence, it addresses the model implementation issue for simulation purposes and is intended for those readers who have some previous knowledge on programing issues. Chapter 12 covers the development of a systematic approach for solving the optimal sensor location and plant-wide control for the fuel processor system with PEM fuel-cell. Finally, Chaps. 13 and 14 are dedicated to give insight about new tools for improving systems for fault diagnosis through the use of principal component analysis. Hence, Chap. 13 approaches the fault detectability index for optimal monitoring system design and, Chap. 14, considers the use of delay adjustment are addressed. Chapters 13, 12 and 14 need to use genetic algorithms for solving the combinatorial problem that arise from this kind of large-scale system. It allows testing several potential solutions and avoiding adopt decisions based on heuristic considerations.

Rosario, Argentina, July 2011 M. S. Basualdo
Rosario, Argentina, July 2011 D. Feroldi
Marselle, France, July 2011 R. Outbib

Acknowledgments

The authors would like to thank a number of people and institutions who contributed in several ways to make this book possible. First, we want to express our gratitude to the French Argentine International Center for Information and Systems Sciences (CIFASIS) (where Marta Basualdo and Diego Feroldi developed their work) and to the Laboratoire des Sciences de L'Information et des Systèmes (LSIS) (where Rachid Outbib is working). D. Feroldi also wants to express his gratitude to FCEIA-UNR and M. Basualdo to FRRo-UTN.

We are very grateful to Maria Serra and Jordi Riera from the Institut de Robòtica i Informàtica Industrial (IRI-UPC-CSIC) for their contributions in Chaps. 3, 6, 7, and 8. Our thanks also to Joseba Quevedo, Viçenc Puig, Teresa Escobet and Salvador de Lira from the Advanced Control Systems group (SAC) of the Universitat Politècnica de Catalunya for their contribution in Chap. 6.

Our thanks to the National Agency for Scientific and Technological Promotion (ANPCyT) for the financial support through the projects PRH 2007-24, PICT-2006-02226, and PICT-2009-0017. Our thanks also to the Spanish Ministry of Science and Innovation who contributed through the CICYT projects DPI2007-62966 and DPI2004-06871-C02-01.

Finally, a very special thanks to our families for the support and understanding during the long process of writing this book.

Rosario, February 2011 — Marta Basualdo and Diego Feroldi
Marseille, February 2011 — Rachid Outbib

Contents

Part I PEM Fuel Cells: Modeling, Simulation, Advanced Control, and Diagnosis

Introduction 3
Diego Feroldi, Marta Basualdo and Rachid Outbib

Description of PEM Fuel Cells System 49
Diego Feroldi and Marta Basualdo

Advanced Control Strategies for the Oxygen in the Cathode 73
Diego Feroldi, David Zumoffen and Marta Basualdo

Simulation and Control of PEMFCS Thermal Behavior 117
Abdelkrim Salah, Rachid Outbib and Jaafar Gaber

Fault Diagnosis of PEM Fuel Cell 151
Andres Hernandez, Rachid Outbib and Daniel Hissel

Fault Diagnosis and Fault Tolerant Control of PEM Fuel Cell Systems 185
Diego Feroldi

Fuel Cell Hybrid Systems 207
Diego Feroldi

Energy Management Strategies for Fuel Cell Hybrid Systems 233
Diego Feroldi

Part II PEM Fuel Cells in the Context of the Fuel Processor System with Bio-Ethanol

Design and Control of an Integrated Bio-Ethanol Processor with PEMFC 261
L. Nieto Degliuomini, S. Biset and M. Basualdo

Control Oriented Dynamic Model of the Bio-Ethanol Processor System 299
L. Nieto Degliuomini, S. Biset, P. Luppi and M. Basualdo

Mathematical Model Implementation for Simulation Purposes 329
L. Nieto Degliuomini, S. Biset, D. Feroldi and M. Basualdo

Plant-Wide Control for Fuel Processor System with PEMFC: Control Structure Selection and Optimal Sensor Location 361
L. Nieto Degliuomini, D. Zumoffen and M. Basualdo

Fault Detectability Index for Optimal Monitoring System Design 405
D. Zumoffen, L. Nieto Degliuomini and M. Basualdo

Improved Principal Component Analysis Models for Fault Detection Using Delay Adjustment 431
E. Musulin and M. Basualdo

Index 453

Contributors

Dr. Marta Basualdo CAPEG-CIFASIS-(CONICET-UNR-UPCAM), 27 de Febrero 210 bis, S2000EZP Rosario, Argentina; UTN-FRRo, Zeballos 1341, S2000BQA Rosario, Argentina, e-mail: basualdo@cifasis-conicet.gov.ar

Sebastian Biset CAPEG-Chemical Eng. Dpt. FRRo-UTN, Zeballos 1341, S2000BQA, Rosario, Argentina, e-mail: sbiset@frro.utn.edu.ar

Lucas Nieto Degliuomini Chemical Eng. CAPEG-CIFASIS-(CONICET-UNR-UPCAM), 27 de Febrero 210 bis, S2000EZP, Rosario, Argentina, e-mail: nieto@cifasis-conicet.gov.ar

Dr. Diego Feroldi CAPEG-CIFASIS-(CONICET-UNR-UPCAM), 27 de Febrero 210 bis, S2000EZP Rosario, Argentina; DCC-FCEIA-UNR, Pellegrini 250, S2000BTP Rosario, Argentina, e-mail: feroldi@cifasis-conicet.gov.ar

Dr. Jafaar Gaber SET-UTBM, 90010 Belfort cedex, France. e-mail: jafaar.gaber@utbm.fr

Dr. Andres Hernandez Escuela Colombiana de Ingeniería Julio Garavito, Bogotá, Colombia, e-mail: andres.hernandez@escuelaing.edu.co

Dr. Daniel Hissel FEMTO-FClab, rue Thierry MIEG, 90010 Belfort cedex, France. e-mail: daniel.hissel@utbm.fr

Patricio Luppi Electronic Eng. CAPEG-CIFASIS-(CONICET-UNR-UPCAM), 27 de Febrero 210 bis, S2000EZP, Rosario, Argentina, DCC-FCEIA-UNR, Pellegrini 250, S2000BTP, Rosario, Argentina, e-mail: luppi@cifasis-conicet.gov.ar

Dr. Estanislao Musulin CAPEG-CIFASIS-(CONICET-UNR-UPCAM), 27 de Febrero 210 bis, S2000EZP Rosario, Argentina; DCC-FCEIA-UNR, Pellegrini 250, S2000BTP Rosario, Argentina. e-mail: musulin@cifasis-conicet.gov.ar

Dr. Rachid Outbib LSIS, Aix-Marseille University, av. Escadrille Normandie Niemen, 13397 Marseilles, France. e-mail: rachid.outbib@lsis.org

Dr. Abdelkrim Salah LSIS, Aix-Marseille University, av. Escadrille Normandie Niemen, 13397 Marseilles, France. e-mail: abdelkrim.salah@lsis.org

Dr. David Zumoffen CAPEG-CIFASIS-(CONICET-UNR-UPCAM), 27 de Febrero 210 bis, S2000EZP Rosario, Argentina; FRRo-UTN, Zeballos 1341, S2000BQA Rosario, Argentina. e-mail: zumoffen@cifasis-conicet.gov.ar

Acronyms

AFC	Alkaline fuel cells
APC	Adaptive predictive control
APCWRF	Adaptive predictive control with robust filter
APRFC	Adaptive predictive robust filter control
BPS	Bio-ethanol processor system
CM	Compressor motor
CO-PrOx	Preferential oxidation of carbon monoxide
CPO	Catalytic partial oxidation
CPV	Cumulative percent variance
CSD	Control structure design
CSS	Control structure selection
CSTR	Continuous stirred tank reactors
CV	Controlled variable
DAPCA	Delay adjusted PCA
DC	Direct current
DMC	Dynamic matrix control
DMFC	Direct methanol fuel cell
DPCA	Dynamic PCA
ESR	Ethanol steam reforming
ESS	Energy storage system
EWMA	Exponential weighted moving average
FB	Feedback
FC	Fuel cell
FCHS	Fuel cell hybrid system
FCHV	Fuel cell hybrid vehicle
FCS	Fuel cell system
FD	Fault detection
FF	Feed forward
FIR	Finite impulse response
FTP	Federal test procedure
GA	Genetic algorithm

GA-DAPCA	Genetic algorithm based delay adjusted PCA
H_2-FCV	Hydrogen fuel cell vehicle
HHV	Higher heating value
HR	Hardware requirements
HTS	High Temperature water gas shift
HWFET	Highway fuel economy test
IAC	Indirect adaptive control
IAE	Integral absolute error
ICE	Internal combustion engine
ICV	Indirect controlled variable
IMC	Internal model control
LHV	Lower heating value
LTS	Low temperature water gas shift
MCFC	Molten carbonate fuel cells
MEA	Membrane electrode assembly
MFM	Minimal fault magnitude
MIMO	Multiple-input/multiple-output
MPC	Model predictive control
MSD	Minimum square deviations
MV	Manipulated variable
NEDC	New European driving cycle
NLE	Net load evaluation
OMSD	Optimal monitoring system design
OSL	Optimal sensor location
PAFC	Phosphoric acid fuel cell
PC	Principal component
PCA	Principal component analysis
PEM	Proton exchange membrane
PEMFC	Proton exchange membrane fuel cell
PID	Proportional-integral-derivative controller
PMDC	Permanent magnet direct current
PrOx	Preferential oxidation
RDGA	Relative disturbance gain array
RGA	Relative gain array
RI	Reliability index
RLS	Recursive least square
RNGA	Relative normalized gain array
SC	Super capacitor
SEDC	Separately excited direct current
SI	System identification
SISO	Single-input/single-output
SMA	Smoothed moving average
SoC	State of charge
SoE	State of energy
SOFC	Solid oxide fuel cells

SPE	Square predictive error
SSD	Sum of square deviations
SVD	Singular value decomposition
TCS	Traction control system
TFM	Transfer function matrix
UDDS	Urban dynamometer driving schedule
UV	Uncontrolled variable
WGS	Water gas shift

Symbols

A	Principal component retained
A	Number of selected latent dimensions (Used in Chap. 14)
$\mathbf{A}$	Penalty matrix
A_f	Vehicle frontal area
A_t	Cathode output valve area
$\mathbf{b}$	Mean vector
$\mathbf{B}$	Penalty matrix
c_i	ith gen parameter
c_i^*	ith cost parameter
cp	Mean specific heat
cp_{gases}	Mean specific heat of hot gases
$\mathbf{c}$	Cost vector
C_d	Drag Coefficient
C_p	Specific heat capacity of air
C_R	Supercapacitor capacitance
$Cons\ H_2$	Cumulative consumption of hydrogen
$\mathbf{C}_i$	ith chromosome
$\mathbf{C}_{full}$	Full selection
$\mathbf{C}_{op}$	Optimal selection
$\mathbf{C}_{unf}$	Unfeasible selection
$\mathbf{C}_{NLE}$	Parameterized chromosome for NLE
$CPV(A)$	CPV to A
$\mathbf{DV}$	Delay vector
d	Process delay
$\mathbf{d}_*(s)$	Disturbance vector
$d_*^p(i)$	Vector with unitary entry in the position i
D_0	Nominal controller denominator
$d_{\max}$	Maximum delay considered
D_t	Tubular reactor diameter

$\mathbf{D}(s)$	Complete disturbance TFM model (Laplace domain)
$\mathbf{D}_A$	First A eigenvalues matrix
$\mathbf{D}_\lambda$	Full eigenvalues matrix
$\mathbf{D}_r(s)$	Remaining disturbance TFM model (Laplace domain)
$\mathbf{D}_s(s)$	Selected disturbance TFM model (Laplace domain)
D^*	Controller denominator
$\mathbf{D}_*$	UD-factorization
e	Scalar tracking error
$\mathbf{e}_d(i)$	Deviation due to disturbance no. i
$\mathbf{e}_{set}\ (i)$	Deviation due to reference no. i
emf	Electromotive force
emf_{ref}	Electromotive force reference
$\tilde{e}$	A-priori prediction error
E	Energy
E^0	Single cell open circuit voltage
E_{cap}	Maximum storage energy of supercapacitors
E_j	Activation energy
E_{rev}	Reversible open circuit voltage
ECN	Net load evaluation index
$\mathbf{E}$	Residual matrix
f	Output free response
f_r	Rolling friction coefficient
$\mathbf{f}_i$	Fault components vector
$\Vert f_j \Vert_{MFM}^{i}$	MFM for i statistic and fault j
F	Faraday number
F_{gases}	Feed flow to hot gases chamber
$\mathbf{F}(s)$	Low-pass filter TFM
FA	Water flow to the reformer
FE	Ethanol flow to the reformer
FEC	Ethanol flow to the burner
Fr	Exit flow in each slice
Fr_0	Nominal flow in each slice
g	Gravity
$\hat{g}(i)$	Estimated FIR component
g_i	Dynamic matrix coefficients
$G(z^{-1})$	Plant FIR model
$\hat{G}(z^{-1})$	Estimated FIR model
$G_0(z^{-1})$	Nominal FIR model
$\mathbf{G}$	Dynamic matrix in *DMC* controller
$\mathbf{G}(s)$	Complete process TFM model (Laplace domain)
$\mathbf{G}_c(s)$	Controller TFM based on IMC (Laplace domain)
$G_c^*(s)$	Controller TFM based on unitary output feedback (Laplace domain)

$\mathbf{G}_r(s)$	Remaining process TFM model (Laplace domain)
$\mathbf{G}_s(s)$	Selected process TFM model (Laplace domain)
$\tilde{G}_{s\Gamma}(s)$	Approximate process model of $\mathbf{G}_s(s)$
$\tilde{G}^{-}_{s\Gamma}(s)$	Invertible part of $\tilde{G}_{s\Gamma}(s)$
$\tilde{G}^{+}_{s\Gamma}(s)$	Non invertible part of $\tilde{G}_{s\Gamma}(s)$
$\mathbf{G}_{fir}$	Dynamic matrix of prediction
h	Specific heat
$h_1 \cdot D$	Heat transmission coefficient
h_{gases}	Accumulated moles in hot gases chamber
h_m	Control horizon
h_p	Prediction horizon
H_p	End of the prediction horizon
H_u	Control horizon
H_w	Beginning of the prediction horizon
HD	Hybridization degree
i	Current density
i_{cell}	Load current cell
I	Current
I_a	Armature current
$I_{a,\ ref}$	Armature current reference
I_e	Excitation circuit current
$I_{e,\ ref}$	Excitation current reference
I_{st}	Stack current
J_{cp}	Compressor inertia
J	Abnormal events
$\mathbf{J}$	Cost function
k_c	PI controller proportional gain
k_j	Frequency factor
k_{SC}	Constant depending on the particular SC
k_{mix}	Gaseous mix conductivity
k_*	Evaluation sample
K_g	Static compensation
Keq	Equilibrium constant
L	Wavelet decomposition level
L_{ind}	Individual length
m	Data matrix samples
m	Number of samples (Used in Chap. 14)
mH_2	Hydrogen mass (anode)
mN_2	Nitrogen mass (cathode)
mO_2	Oxygen mass (cathode)
m_{sm}	Air mass (supply manifold)
m_{veh}	Vehicle mass

$m_{w,\ an}$	Water mass (anode)
$m_{w,\ ca}$	Water mass (cathode)
m_T	Vehicle total mass
MH_2	Hydrogen molar mass
MN_2	Nitrogen molar mass
M_{O2}	Oxygen molar mass
M_v	Water vapor molar mass
$\mathbf{M}$	Combined statistic matrix
N	FIR order
N_c	Chromosome length
N_G	Number of generations
N_g	Total generations
N_i	Initial population
N_{ind}	Number of individuals in the generation
N_m	Moving window dimension
n	Data matrix variables
n	Number of variables (Used in Chap. 14)
n_c	Number of cells in a stack
n_t	Accumulated moles in each slice
N_*	Evaluation window
$\mathbf{P}$	Loading matrix
$\tilde{P}$	Discarded principal components, residual subspace
P_m	Mutation probability
p	Pressure
P_c	Crossover probability
$\hat{p}_o$	Operating point parameter
pH_2	Hydrogen partial pressure
pO_2	Oxygen partial pressure
p_{rm}	Return manifold pressure
p_{sm}	Supply manifold pressure
$\mathbf{p}_i$	Eigenvector *i*
P	Power
P_{cm}	Power compressor motor
P_{cell}	Actual electrical power generated
P_{co}	Crossover probability
P_{cp}	Compressor power load
P_m	Mutation probability
P_{net}	Net power
P_{net}^{max}	Maximum net power
P_{system}	Net power produced by the fuel cell system
P/E	Storage system *Power/Energy* ratio
$\mathbf{P}$	PCA model
Q	SPE statistic

$\mathbf{Q}$	Weight matrix in *DMC*
Q_{PEM}	Heat produced by the cell
r	Resistance
$(-r_j)$	Reaction speed
R_0	Nominal noncausal polynomial
R_a	Air gas constant
R_d	Internal resistance (DC) of *ESS*
R_g	Universal gas constant
$R_{w/e}$	Water/ethanol molar ratio
$\mathbf{R}$	Weight matrix in *DMC*
$\mathbf{R}_c$	correlation matrix
$R(z)$	Noncausal polynomial
Re	Reynolds for flow through packed bed
s_1	Supervision indicator
$\mathbf{s}$	Variance vector
$\mathbf{S}_d$	UVs-disturbances gain matrix
$\mathbf{S}_{fir}$	Dynamic matrix of regression
$\mathbf{S}_{sp}$	UVs-references gain matrix
SRa	Air stoichiometry
SSD_{yr}	Sum of square deviations of $\mathbf{y}r$
T	Temperature
$T_{cp,\ out}$	Compressor exit air temperature
T_l	Load torque
T_g	Hot gases chamber temperature
T_{rm}	Return manifold gas temperature
T_s	Sample time
T_{sm}	Supply manifold gas temperature
$T\#i$	Temperature of stream number i
T_2	Hotelling statistic
$\mathbf{T}$	latent matrix
$\mathbf{T}_i$	Transformation matrix
$\mathbf{t}$	latent vector
t	Time
$\mathbf{t}$	Latent variable
t_c	Duration of the driving cycle
u	Scalar manipulated variable
$\hat{u}$	Estimated manipulated variable
$u00$	Operation point
u_i	Input i
U	Manipulated variables
U_a	Armature voltage
U_e	Excitation circuit voltage
$\mathbf{u}$	Manipulated variable vector
$\mathbf{u}(s)$	MVs vector
$\mathbf{U}$	Eigenvector matrix

$\mathbf{U}_*$	UD-factorization
v	Velocity
v_{cm}	Compressor voltage
v_{cm}^{opt}	Optimal compressor voltage
v_{ref}	Voltage reference for DC converter
V_{cell}	Actual cell voltage
V_{cm}	Compressor voltage
V_{fcs}	Fuel cell voltage
V_{rm}	Return manifold volume
V_{sl}	Volume of each slice
V_{sm}	Manifold volume
w_{cat}	Catalyzer mass
W	Mass flow rate
W_{cp}^{opt}	Optimal compressor air flow
W_{comp}	Work of the compressor
$\bar{\mathbf{X}}$	Normalized data matrix
x	Signal
x_f	Filtered signal
$\mathbf{X}$	Process data matrix
$\bar{\mathbf{x}}$	Normalized process sample
$\bar{\mathbf{x}}_*$	Normalized process sample model
$\bar{\mathbf{x}}_0$	Normal behavior case
$\mathbf{x}$	Normalized sample vector
$\mathbf{X}_{adj}$	Delay adjusted data matrix
$\bar{\mathbf{X}}$	Normalized process data matrix
XE	Ethanol molar fraction
y	Scalar plant output
$y(k+1\|k)$	Prediction of y_i at time $k + 1$ done in time k
y_{00}	Operation point
y_i	Molar fraction of component
y_r	Scalar reference signal
$\hat{\mathbf{y}}$	Scalar estimated plant output
$\mathbf{y}$	Plant outputs vector
$\mathbf{y}_r$	Reference signal vector
$\hat{\mathbf{y}}$	Estimated plant outputs vector
$\mathbf{y}(s)$	Process outputs vector
$\mathbf{y}_r(s)$	UVs vector
$\mathbf{y}_s(s)$	CVs vector
$y_s^{net}(s)$	Net load effect vector (Laplace domain)
$y_s^{sp}(s)$	Reference vector
$\mathbf{y}_{set}(i)$	Vector with unitary entry in the position i
$\mathbf{y}_r(s)$	UVs vector
$y_s^{net}(s)$	Net load effect vector (Laplace domain)
$y_s^{sp}(s)$	Reference vector

$\mathbf{y}_{set}(i)$	Vector with unitary entry in the position i
Y	Controlled process variables
z	Combined statistic
z_n	Supervision indicator
Z	Performance variables

Greek Letters

α	Road slope
α^j	Reaction order
α_i	Penalty coefficients
α_r	Filter smoothing coefficient
β_i	Penalty coefficients
δ_i	Confidence limit i
α^2_{SPE}	Control limit for SPE
α^2_T	Control limit for T2
$\Delta(.)$	Variation from the nominal value
$\hat{\Delta}G$	Process-model mismatch estimation
ΔH_j	Reaction heat
ΔH_{HHV}	Higher heating value of hydrogen
ΔH_{LHV}	Lower heating value of hydrogen
Δu	Variation in the control action
ε	Prediction error
γ	Ratio of the specific heats of air
γ_{ij}	ij component of Γ
Γ	Parametrization matrix
F	Fault directions matrix
η	Efficiency
η_{fc}	Efficiency of a single fuel cell
η_{fcs}	Efficiency of the fuel cell system
θ_{ij}	Dead time between output i and input j
Θ_j	j Fault subspace
Θ_1/Ξ_1	Disturbance weight matrix
Θ_2/Ξ_2	Output relative weight matrix
$\hat{\eta}$	Disturbance estimation
$\hat{\theta}$	Estimation parameters
$\hat{\theta}^{LS}_N$	LS estimation parameters
κ_i	ii Filter order from $\mathbf{F}(s)$
λ	Forgetting factor
λ	Correlation matrix eigenvalues, PC variance (Used in Chap. 14)
λ^{ij}_1	ij component of $\Lambda 1$
Λ_1/Δ_1	References weight matrix
Λ_2/Δ_2	Output relative weight matrix
λ_i	Eigenvalue i

λ_{O2} Oxygen excess ratio
$\lambda_{O_2}^{opt}$ Optimal oxygen excess ratio
μi Mean value for i signal
μ_{min} Minimum improvement
v Stoichiometric coefficient
ρ_a Air density
σ_i Variance value for i signal
τ Torque
τ_i^c PI controller integral time constant
τ_{ij}^m ij Component time constant of $\tilde{G}_s(s)$
τ_{fi} ii filter time constant of $\mathbf{F}(s)$
Φ Fitness function
ϕ Combined index
ϕ_{Ca} Relative humidity at the cathode
φ_i Robustness parameter for loop i
ψ Regressor vector
ω_{cp} Compressor speed
ω_{ref} Shaft speed reference
ω_s Shaft DC motor speed

Subscripts and Superscripts

an Anode
atm Atmospheric
aux Auxiliary
Burn Burner, Post-combustion reactor
ca Cathode
cap Capacity
Cell Fuel cell stack
cm Compressor motor
cp Compressor
el Electrical
ess Energy storage system
Et Ethanol
fc Fuel cell
fcs Fuel cell system
gen Generated
H_2 Hydrogen
i Index
i Principal components (Used in Chap. 14)
in Inlet
j Index
j Process variables (Used in Chap. 14)
k Individuals in the population

l	Wavelet decomposition level
N_2	Nitrogen
O_2	Oxygen
out	Outlet
rct	Reacted
rm	Return manifold
Ref	Reformer reactor
Rev	Reversible
sm	Supply manifold
sys	System
t	Throttle
T	Transposed
v	Water vapor
w	Water

Part I
PEM Fuel Cells: Modeling, Simulation, Advanced Control, and Diagnosis

Introduction

Diego Feroldi, Marta Basualdo and Rachid Outbib

Abstract The use of fuel cell-based systems is advantageous because of the high efficiency in the energy conversion and null emissions of these electrochemical devices. In this chapter, a general overview of fuel cell systems is provided. Special emphasis is placed on proton exchange membrane fuel cell (PEMFC) systems because of the important benefits of technology. First, the main concepts about different fuel cell technologies are introduced, focusing on PEMFC. Important issues such as operation, efficiency and working temperature of each type of fuel cells are covered. Then, an introduction to the control and diagnosis of PEMFC systems is provided. Later, the fuel cell hybrid systems are introduced and described, showing the important advantages of hybridization and energy management. Since the PEMFC needs to be fed with high-quality hydrogen the problematic related to hydrogen production for an onboard implementation is addressed at the end of the chapter. Therefore, topics such as design, modeling, simulation, control and fault diagnosis are briefly introduced here and covered in detail in the rest of the chapters of this book.

D. Feroldi (✉) · M. Basualdo
CAPEG-CIFASIS-(CONICET-UNR-UPCAM), 27 de Febrero 210 bis,
S2000EZP Rosario, Argentina
e-mail: feroldi@cifasis-conicet.gov.ar

D. Feroldi
DCC-FCEIA-UNR, Pellegrini 250,
S2000BTP Rosario, Argentina

M. Basualdo
UTN-FRRo, Zeballos 1341,
S2000BQA Rosario, Argentina
e-mail: basualdo@cifasis-conicet.gov.ar

R. Outbib
LSIS, Aix-Marseille University,
av. Escadrille Normandie Niemen,
13397 Marseilles, France
e-mail: rachid.outbib@lsis.org

M. S. Basualdo et al. (eds.), *PEM Fuel Cells with Bio-Ethanol Processor Systems*,
Green Energy and Technology, DOI: 10.1007/978-1-84996-184-4_1,

1 Fuel Cells

Hydrogen is an energy vector with great potential and it is expected to take a leading role in the medium term in a wide spectrum of applications. The main advantage of hydrogen is the high energy density, although having difficulty of storage. Hydrogen can be used both in heat engines and in fuel cells. Fuel cells are devices for electrochemical energy conversion into electricity and heat, which function as an electrochemical battery with the only difference that it generates electricity as long as fuel is supplied.

Fuel cells have very different applications from low power such as portable electronics to high power such as electric vehicles or stand-alone residential applications. In fact, fuel cells are an important technology for a potentially wide variety of applications including micropower, auxiliary power, transportation power, stationary power for buildings and other distributed generation applications, and central power. For this reason, it is expected that these applications will be present in a large number of industries in the short term.

With regard to efficiency, fuel cells are more efficient than heat engines in the order of two to three times [1]. Besides, fuel cells use preferably the chemical energy of hydrogen, which is the most common fuel in fuel cell systems to produce electricity and heat without combustion process. In spite of the frequently mentioned fact that hydrogen is a fuel (and hence the origin of the name fuel cells), this is not completely correct. Hydrogen, in reality, is an energy vector and not an energy source [2, 3]. Hydrogen may be obtained from fossil sources (such as methane) or water electrolysis. In this book, the production of hydrogen is discussed in Sect. 4.

However, the attractive potential of hydrogen is countered by uncertainties about the development and the economics of the implied key enabling technologies, such as renewable energy sources, advanced production processes, fuel cells, storage technologies, safety and infrastructure [4].

1.1 Fuel Cell Description

Fuel cells are devices that convert the chemical energy stored in a fuel like hydrogen through an electrochemical reaction directly into electrical energy [5]. This reaction does not imply any type of combustion and thus the reaction does not have any harmful by-product. The basic physical structure of a fuel cell consists of an electrolyte layer in contact with a porous anode and cathode on either side. A schematic representation of a generic fuel cell with the reactant/product gases and the ion conduction flow directions through the cell is shown in Fig. 1.

In a typical fuel cell, gaseous fuels are fed continuously to the anode (negative electrode) compartment and an oxidant (i.e., oxygen from air) is fed continuously to the cathode (positive electrode) compartment. The electrochemical reactions

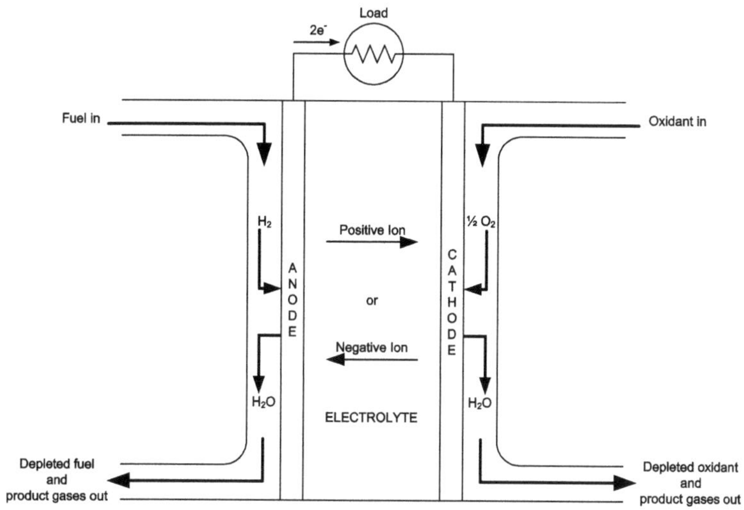

Fig. 1 Schematic of a generic single fuel cell

take place at the electrodes to produce an electric current, which flows through the load [5].

This is the general principle of operation of a generic fuel cell. However, there are different types of fuel cells, which differ mainly in the type of electrolyte, the fuel and the operating temperature. Therefore, the principle of operation of each type has its own characteristics. In Sect. 1.3, a review of the different fuel cell types is done.

1.2 Advantages of Fuel Cell Systems

The advantages of fuel cell systems depend on each type of fuel cell. However, there are common advantages for all types. Some of these common advantages are the following [2]:

- *Efficiency* The efficiency of fuel cells is generally higher than combustion engines (this issue will be further studied). Besides, the efficiency of fuel cells do not depend on the the size of the system, unlike engines, which is very valuable in small applications.
- *Simplicity* Fuel cells are inherently simple and the only moving parts are in the auxiliary equipment, which positively affects the life span of the system.
- *Low emissions* Fuel cells are of low emission and, when hydrogen is used, there are strictly zero emissions. This issue is very important especially in vehicle applications because it can reduce significantly the pollution in cities. However, the matter must be analyzed globally taking into account the hydrogen production. The problematic of hydrogen production is studied in more detail in Sect. 4.

- *Silence* Fuel cell systems are very quiet even considering the auxiliary equipment (compressor, fan, fuel processor, etc.).

1.3 Different Types of Fuel Cell Technologies

There are different fuel cell technologies that have been tried. From the point of view of its components, the main difference between different types of fuel cells is the electrolyte used. From the point of view of the operation, the different types of fuel cells differ in the operating temperature, from ambient to 1,000°C, and the type of fuel that can be used, in addition to hydrogen, such as natural gas and methanol. The following is a list of the main fuel cell technologies:

- Proton Exchange Membrane Fuel Cell (PEMFC)
- Direct Methanol Fuel Cell (DMFC)
- Alkaline Fuel Cell (AFC)
- Phosphoric Acid Fuel Cell (PAFC)
- Molten Carbonate Fuel Cell (MCFC)
- Solid Oxide Fuel Cell (SOFC)

1.3.1 Proton Exchange Membrane Fuel Cells

PEMFC have a thin polymer membrane (usually perfluorosulfonated acid polymer [6]) as electrolyte. The main feature of this membrane is that it allows the passage of protons but, in contrast, prevents the passage of electrons. On the other hand, the catalyst is typically platinum supported on carbon. The more attractive feature of this type of fuel cell is the low operating temperature between 60°C and 80°C. This feature allows its use in many very different applications, among which can be highlighted the automotive applications, the stationary power generation and the portable electronics applications. The efficiency of these types of fuel cells is between 30 and 45%, although new PEMFCs can reach 60%.

The principle of operation of the fuel cell is quite simple and the first development was by the scientist William Grove in 1839 [2]. The basic principle of operation is as follows. Hydrogen flows through the feeding channels of the anode, spreads through the diffusion layer and reaches the catalytic layer where it is oxidized releasing electrons and protons, according to

$$2H_2 \longrightarrow 4H^+ + 4e^-. \qquad (1)$$

The released electrons are led through the catalytic metal and the granulated coal of the catalytic layer of the anode, arriving at the cathode via the external circuit, whereas the protons are transported through the membrane to the catalytic layer of the cathode. At the same time, oxygen is injected into the feeding channels of the cathode and spreads through the diffusion layer toward the catalytic layer, where it reacts with protons and electrons, generating water, according to

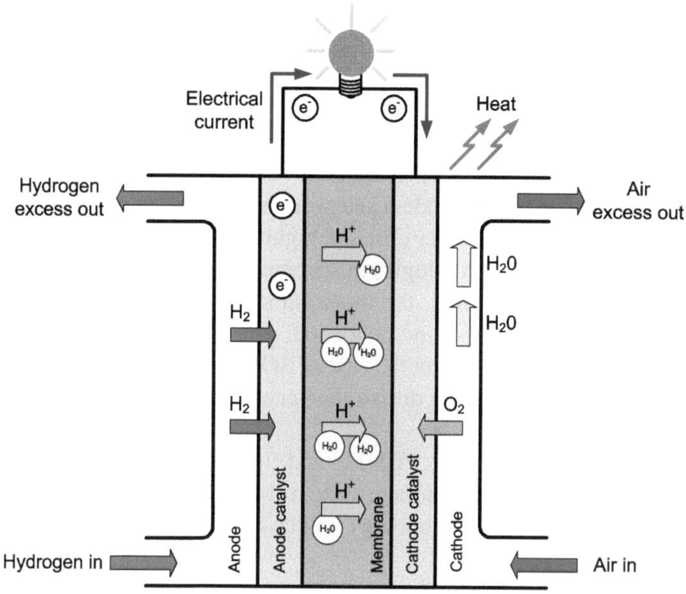

Fig. 2 Schematic diagram showing the PEM fuel cell principle of working

$$O_2 + 4H^+ + 4e^- \longrightarrow 2H_2O. \tag{2}$$

Therefore, the global reaction in the fuel cell is

$$2H_2 + O_2 \longrightarrow 2H_2O. \tag{3}$$

The reaction on the cathode is exothermic: the heat release is dependent upon the voltage, which is directly related to its efficiency. The previous explanation is schematically shown in Fig. 2, where the circulation of gases and electrons, and the passage of ions through the membrane are shown.

The water management in a cell is critical for efficient operation. As mentioned previously, a requirement for these types of cells is to maintain a high water content in the electrolyte to assure high protonic conductivity [5]. The protonic conductivity of the electrolyte is high when the membrane is completely saturated of water, and thus offers minimum resistance to the ion passage, increasing the efficiency of the cell especially at high current densities [7].

The water content in the cathode and anode is determined by the water balance in the respective volumes, the balance being a result of the inputs and outputs of water. There is normally an input of water in the incoming gas and there is water transport through the membrane. The water transport during the operation is a function of the cell current and the characteristics of the membrane and the electrodes.

The main mechanism of water transport through the membrane is the drag of water molecules by the protons: each proton drags between 1 and 2.5 water molecules

[8]. If more water leaves the cell than is produced, it is important to humidify the incoming gas in the anode and/or in the cathode. Nevertheless, if there is too much humidification, the diffusion layers are flooded, which causes problems in the gas diffusion.

PEMFC is one of the most advanced fuel cell systems along with SOFC. However, there are still important areas such as material development (especially relating to electrolyte and catalysts for electrodes) and process control that will further advance the technology [9] permitting lower cost and higher reliability.

Diverse features in both technologies make potential applications also differ considerably. In the case of PEMFC, the more suitable applications are in vehicles, portable electronics and stand-alone residential power generation, because of the low operating temperature. While in the case of SOFC, the most suitable application is in large-scale power generation due to the very high efficiencies achieved when cogeneration is also applied.

In this book, we devote to the PEMFC and, in particular, the fuel cell hybrid systems (FCHS) with an energy storage system (ESS) acting as a secondary power source. Therefore, the PEMFC technology will be further explained in the following chapter.

1.3.2 Direct Methanol Fuel Cells

DMFC use the same type of electrolyte as PEMFC but they use methanol directly as a fuel instead of hydrogen. For this reason, DMFC is not usually considered in the category of fuel cells. In addition, the handling of these fuel cells is very different, thus DMFC deserves to be in a different category. The main advantages of this type of fuel cell are the high net energy density (18.9 MJ kg^{-1} [2]), the storage facility and the low production cost. These advantages make the DMFC an excellent option for portable applications.

However, the most important problem associated with DMFC is that the fuel anode reactions are much slower than with hydrogen. The result is that the performance of the DMFC is significantly worse than other fuel cell types. The overall reaction in the DMFC is represented by the following equation:

$$CH_3OH + 1\frac{1}{2}O_2 \longrightarrow 2H_2O + CO_2. \tag{4}$$

The anode and cathode reactions depend on the electrolyte used. Methanol can be used in a fuel cell with any of the standard electrolytes. However, to exploit the advantages of a liquid fuel system, the only possible electrolytes are the low-temperature electrolytes: alkaline and PEM. Therefore, there are two main types of DMFCs: the Alkaline Direct Methanol fuel cells and the PEM Direct Methanol fuel cell.

Alkaline Direct Methanol FC

The anode reaction of the alkaline DMFC is the following:

$$CH_3OH + 6OH^- \longrightarrow CO_2 + 5H_2O + 6e^-. \quad (5)$$

Note the six electrons produced for each molecule of methanol, which is one of the main advantages of this fuel cell type, in addition to the advantage that the fuel is liquid. However, a major drawback is that for this reaction to occur the reactants must interface with the alkaline electrolyte. Thus, the carbon dioxide reacts with the hydroxide, forming carbonate, which is a salt of carbonic acid:

$$2OH^- + CO_2 \longrightarrow CO_3^{2-} + H_2O. \quad (6)$$

The result is that the electrolyte loses its alkalinity and, therefore, the life of this fuel cell type is not very long unless the electrolyte is frequently replaced. For this reason, the conclusion is that alkaline DMFC is not practical, although the cost and performance are better, in comparison with PEM-based systems [2].

PEM Direct Methanol FC

Due to the previously explained problem of alkaline electrolytes, the best option for Direct Methanol FC is the use of PEM electrolyte. In this case, the overall anode reaction is the following:

$$CH_3OH + H_2O \longrightarrow 6H^+ + 6e^- + CO_2. \quad (7)$$

The H^+ ions move through the electrolyte and the electrons move around the external circuit. Note that water is required at the anode, though it is produced more rapidly at the cathode via the following reaction:

$$O_2 + 6H^+ + 6e^- \longrightarrow 3H_2O. \quad (8)$$

The result is the same as for the alkaline DMFC: six electrons per molecule of methanol fuel. However, the reaction of (7) takes place in a more complex manner [2].

1.3.3 Alkaline Fuel Cells

AFC is one of the pioneers of modern fuel cells to be developed in 1960 and its first application was to provide onboard electric power for the Apollo space vehicle [5]. Thus, AFC is amongst the most mature fuel cell technologies.

In this type of fuel cell, the operating temperature is between 50 and 200°C and the fuel is also hydrogen while the efficiency of this fuel cells is about 60%. In fact, advanced AFCs have the highest electrical efficiency (at nearly 70%) and have shown excellent reliability when run at low current densities [10].

The AFC converts gaseous hydrogen and gaseous oxygen into electricity using a direct, low-temperature, electrochemical reaction. The hydrogen is usually compressed and the oxygen is taken from the air. It uses a circulating liquid alkaline electrolyte that is also an effective heat transfer and water management medium

[11]. In this type of fuel cell, the electrolyte, as the name implies, needs to be an alkaline solution, for example, sodium hydroxide or potassium hydroxide solution (KOH).

The reaction at the anode of the AFC is the following [2]:

$$2H_2 + 4OH^- \longrightarrow 4H_2O + 4e^-. \tag{9}$$

While, the electrons released in this reaction pass around the external circuit, reaching the cathode, where they react, forming new OH^- ions:

$$O_2 + 4e^- + 2H_2O \longrightarrow 4OH^-. \tag{10}$$

The overall reaction becomes

$$2H_2 + O_2 \longrightarrow 4H_2O + \text{electrical energy} + \text{heat}, \tag{11}$$

where it can be seen that through the combination of oxygen and hydrogen, water is produced at the anode and electrical energy and heat are generated by the fuel cell.

Although AFCs used in space applications are complex and costly, the terrestrial versions, operating as hybrid systems with a buffer battery, are not. Furthermore, the AFC does not rely on high volume manufacture to reduce the cost to an affordable level since the materials are not very expensive like other types of fuel cells and the processing is straightforward [11]. It is possible to significantly reduce the cost of the stack since AFCs are able to use non-noble catalysts [10].

1.3.4 Medium and High Temperature Fuel Cells

The fuel cell types discussed so far correspond to low temperatures. However, operating at high temperature has important advantages. These advantages are the following [2]:

- The electrochemical reactions proceed more quickly, which results in lower activation voltage losses. Also, noble metal catalysts are often not needed.
- The high temperature of the cell allows the extraction of hydrogen from other more readily available fuels, such as natural gas.
- The high-temperature exit gases and cooling fluids can be exploited to heat buildings or other processes near the fuel cell.
- The high-temperature exit gases and cooling fluids can be used to drive turbines that can drive generators, producing further electricity.

Now, we address the different types of fuel cells that operate at higher temperatures.

1.3.5 Phosphoric Acid Fuel Cell

PAFC is a mature technology and is particularly suitable for power applications. PAFCs were the first fuel cells to cross the commercial threshold. At the moment, conventional PAFCs are mainly developed for stationary power generation because they are based on an acid electrolyte and are not suitable for portable and transport applications due to limited thermal cyclability and life issues related to the leaching of the acid [12].

The operating temperature (~220°C [2]) places this type of fuel cell in an intermediate situation between the low temperature fuel cell types (PEMFC, DMFC, and AFC) and the very high temperature fuel cell types (SOFC and MCFC). The efficiency of these fuel cells is between 40% and 45%.

The electrochemical reactions occurring in PAFCs are as follows [5]:

$$H_2 \longrightarrow 2H^+ + 2e^- \tag{12}$$

at the anode, and

$$\frac{1}{2}O_2 + 2H^+ + 2e^- \longrightarrow H_2O \tag{13}$$

at the cathode. The overall cell reaction is

$$\frac{1}{2}O_2 + H_2 \longrightarrow H_2O. \tag{14}$$

PAFC is a type of fuel cell that uses liquid phosphoric acid as an electrolyte, which can be a disadvantage in some applications. Another important disadvantage is that the electrodes are made of carbon paper coated with a finely dispersed platinum catalyst, which make them expensive to manufacture. An important difference from other fuel cell types is that PAFC are not affected by carbon monoxide impurities in the hydrogen. On the contrary, PAFC has the disadvantage that the phosphoric acid solidifies at a temperature of 40°C, which makes difficult the startup and thus PAFC applications are restricted to continuous operations.

However, at an operating range of 150 to 200°C, the expelled water can be converted to steam for air and water heating. Therefore, PAFCs are used for stationary applications with a combined heat and power efficiency of about 80% [8].

1.3.6 Molten Carbonate Fuel Cells

MCFC operates at approximately 650°C, which places this type of fuel cell within the category of very high temperature. The high operating temperature is needed to achieve sufficient conductivity of its carbonate electrolyte and, on the other hand, allows the use of low cost metal as catalyst instead of noble metal. The fuels for these types of fuel cell are methane and coal gas directly, without an external reformer. The efficiency of these fuel cells is between 45 and 60%.

The electrolyte of the molten carbonate fuel cell is a molten mixture of alkali metal carbonates (usually a binary mixture of lithium and potassium) or lithium and sodium carbonates, which is retained in a ceramic matrix of $LiAlO_2$ [2]. Unlike other fuel cells, carbon dioxide needs to be supplied to the cathode as well as oxygen.

The half-cell electrochemical reaction at the anode is as follows:

$$H_2 + CO_3^{=} \longrightarrow H_2O + CO_2 + 2e^-, \qquad (15)$$

whereas the half-cell electrochemical reaction at the cathode is as follows:

$$\frac{1}{2}O_2 + CO_2 + 2e^- \longrightarrow CO_3^{=}. \qquad (16)$$

The overall reaction of MCFC is

$$H_2 + \frac{1}{2}O_2 + CO_2(\text{Cathode}) \longrightarrow H_2O + CO_2(\text{Anode}). \qquad (17)$$

It is usual to recycle the CO_2 generated at the anode to the cathode where it is consumed.

MCFC is not poisoned by carbon monoxide or carbon dioxide. In fact, MCFC can even use carbon dioxide as fuel, making it more attractive for fueling with gases made from coal. In addition, it is believed that MCFC could even be capable of coal internal reforming coal, since it can be made resistant to impurities such as sulfur and other particles that result from converting coal into hydrogen [5].

On the other hand, the main disadvantage of the current MCFC technology is durability. The high operating temperatures and the corrosive electrolyte decrease the fuel cell life. However, new materials and designs are being investigated to improve the fuel cell life without losing the performance.

1.3.7 Solid Oxide Fuel Cell

SOFC is a very high temperature fuel cell technology. In this type of fuel cell, there is no liquid electrolyte, thus corrosion and other problems regarding electrolyte management are avoided. The operating temperature above 800°C allows a very active electrochemical reaction with nonprecious materials, and, besides, produces high quality by-product heat for cogeneration, similar to MCFC [5]. In fact, the efficiency of this fuel cell can be increased from 60 to 85% when used in cogeneration. This fuel cell type is used in high power applications, with values in the order of 100 kW.

The electrochemical reactions associated with hydrogen fuel are as follows [5]:

$$H_2 + O^{=} \longrightarrow H_2O + 2e^- \qquad (18)$$

at the anode, and

$$\frac{1}{2}O_2 + 2e^- \longrightarrow O^{=} \qquad (19)$$

at the cathode. The overall cell reaction is

$$H_2 + \frac{1}{2}O_2 \longrightarrow H_2O. \tag{20}$$

The use of high temperatures in these types of fuel cells allows the development of internal reforming processes. The recent trends in the research of solid oxide fuel cells concern the use of available hydrocarbon fuels, such as natural gas [13]. On the other hand, the high temperature increases the kinetic of the electrochemical reactions without the use of precious metals, and generates heat with the appropriate characteristics to be used in cogeneration.

In these types of fuel cells, it is also possible to use carbon monoxide and methane as fuels. The hydrogen is generated through the reactions of both compounds with water at high temperature and then oxidizes in the anode. The reactions are as follows:

$$CO + H_2O \longrightarrow H_2 + CO_2, \tag{21}$$

$$CH_4 + H_2O \longrightarrow 3H_2 + CO. \tag{22}$$

The high temperature of SOFC, however, requires more sophisticated materials and, therefore, is more expensive. The development of suitable low cost materials and the low cost fabrication of ceramic structures are the key technical challenges of this technology [14]. Besides, research is currently underway to improve the fuel flexibility of SOFCs.

1.4 Fuel Cell Applications

The important advantages of fuel cells mean that there are many applications from a few watts to megawatts, both developing and proven. In addition, the number of applications is steadily growing. The following is an extensive list of the possible applications:

- Stationary power generation
- Automobiles
- Scooters
- Golf carts
- Utility vehicles
- Backup power
- Portable power
- Space
- Airplanes
- Locomotive
- Boats
- Underwater vehicles
- Telecommunications
- Wastewater treatment plants
- Forklifts

- Portable electronics
- Autonomous robots
- Battery charger

Fuel cell applications may be classified in different ways. One possible classification is as mobile or stationary applications. The mobile applications include transportation systems and portable electronic equipment while stationary applications include combined heat and power (CHP) systems for both residential and commercial utilizations [1].

It is important to note that the possible applications depend on the particular fuel cell type. For example, some PEMFC applications are vehicles, mobile applications and lower power CHP systems. On the other hand, AFC technology is used in space vehicles, e.g. Apollo, Shuttle. DMFC technology is suitable for low power portable electronic systems. PAFC is used in CHP systems larger than PEMFC. MCFC is suitable for medium- to large-scale CHP systems, up to systems in the MW order of magnitude. SOFC is suitable for all sizes of CHP systems, from 2 kW up to several MW [2].

Fuel cell systems are particularly suited for stationary generation, both in non-grid-connected and grid-connected applications [15]. The fuel cell stack generates direct current electric power. However, the stack is only a part of the power system: a large number of ancillary devices such as compressors, fans, valves, humidifiers, sensors, power converters, etc. are necessary to supply the load. Different technologies are used for stationary generation: phosphoric acid fuel cell, molten carbonate fuel cell, solid oxide fuel cell, proton-exchange membrane or polymer electrolyte membrane fuel cell and alkaline fuel cell. Molten carbonate and solid oxide fuel cells are high-temperature fuel cells more suitable for stationary applications [16].

The main application of fuel cells is the generation of electricity. However, due to inevitable losses, a part of the energy is generated in the form of heat. Fuel cell systems thus can be used for the combined generation of heat and power even on a small scale in individual buildings [17]. CHP generation allows a more efficient use of fuel than the traditional way of generating, for example, electricity in centralized power plants or heat in individual boilers. Besides, the application of heat cogeneration to obtain more electricity is also possible [18]. In the case of CHP for buildings, industrial facilities or stand-by generators, since the efficiency of fuel cell power systems is nearly unaffected by size, the stationary plant development has focused on the smaller, several hundred kilowatt to low megawatt capacity plants using PEMFC, AFC, PAFC, MCFC, SOFC technology, where the plants are fueled primarily with natural gas [5].

On the other hand, the automotive industry has been a pioneer in using the technology and all over the world leading car manufacturers have designed at least one prototype vehicle using fuel cells. Examples of fuel cell vehicles running with hydrogen stored onboard are the Class B Mercedes Benz, the Honda FCX, or the recent FCHV-adv of Toyota [19]. Fuel cell technology running on hydrogen is the most attractive long-term option for passenger cars [20] because it can eliminate emissions on the tank-to-wheel [1] path if the hydrogen is produced from renewable sources, providing very high average efficiencies. The latter is particularly based on the fact that the fuel cell reaches the highest efficiency at part load unlike internal combustion engines. Hydrogen production, storage, and distribution are the biggest challenges to be fully introduced in the market.

Buses in which the internal combustion engine is replaced by fuel cells are very attractive because the advantages of high efficiency and elimination of pollution in cities. There are demonstration projects with fuel cell buses in several cities with significant success but some challenges remain to be overcome. The most significant difficulties at present are the cost of the vehicles and the lack of an infrastructure for the supply of the necessary fuel (hydrogen) [21].

Another interesting field of application is in portable electronics. The objective of the application of fuel cells to portable electronics is to provide an uninterrupted power source with a high energy density to replace conventional batteries. In fact, the theoretical energy densities of the energy of fuel cells are much higher compared with different battery technologies [22]. The fuel cell systems suitable for operating portable or micro fuel cells are as follows: pure hydrogen, pure liquid hydrocarbons (e.g., methanol, ethanol, formic acid or ethylene glycol), and onboard hydrogen derived from reformed hydrocarbons such as methanol [23].

On the other hand, fuel cells are a good alternative for uninterruptible and back-up power, especially in critical areas where the negative aspects of batteries and generators make conventional technologies unsuitable. Besides, fuel cells can provide a robust alternative to severe weather conditions [24]. For example, wireless networks is an important niche market for fuel cells applied to uninterruptible and back-up power with great development prospects.

Auxiliary power units (APU) is another interesting niche market for fuel cells, including applications such as passenger cars, trucks, buses and marine units. The independent operation of APU with respect to the main engine, allows the shutdown of the main engine when the automobile is stopped, thereby avoiding the idling of the engine. During this period, APU delivers power and heat, which means an improvement in the system efficiency [25].

2 Control and Diagnosis of PEM Fuel Cell Systems

The transient behavior of the electrochemical reaction that takes place in the fuel cell is very fast. Nevertheless, the behavior of the generated power in a fuel cell system depends on the hydrogen and air flows, the regulation of the pressure and the management of water and heat [7]. When a load is connected to the fuel cell,

---

[1] Tank-to-wheel emission refers to the emissions produced in the path from the tank to the wheel, that is the total emissions produced in the vehicle. This does not include the emissions produced in the process to generate the fuel (hydrogen in this case), which is known as well-to-tank emissions. The sum of both emissions are known as well-to-wheel emissions, comprising the whole path from the primary source to the vehicle wheels.

the control system must control the temperature, the membrane hydration and the partial pressure of gases on both sides of the membrane to avoid voltage degradation and, therefore, a reduction in the efficiency [26]. Besides, it is important to control efficiently the mentioned variables to assure the durability of the cells. These critical variables must be controlled for a wide range of power, with a series of actuators such as valves, pumps, compressors, expanders, fans, humidifiers and condensers [27].

On the other hand, the energy generation systems based on fuel cells are complex since they involve thermal, fluidic and electrochemical phenomena. Moreover, they need a set of auxiliary elements such as valves, compressor, sensors, regulators, etc., to make the fuel cell work at the pre-established optimal operating point. For these reasons, they are vulnerable to faults that can cause the stop or the permanent damage of the fuel cell. Therefore, it is useful to use systematic techniques, like the recent methods of fault-tolerant control (FTC), to guarantee the safe operation of the fuel cell systems [28, 29].

2.1 Control of PEM Fuel Cell Systems

Fuel cell systems are devices that produce electric energy through an electrochemical reaction and, in order to operate efficiently, all the subsystems must work properly. Air supply, which is one of these subsystems, has a great influence on the system efficiency. In fact, one of the most important challenges in fuel cell control is to assure sufficient amount of oxygen in the cathode when current is abruptly drawn from the fuel cell stack. The air supply subsystem is generally composed of an air compressor, which supplies air to the cathode. On the other hand, the hydrogen supply subsystem relies on a hydrogen-pressurized tank and the hydrogen inlet flow rate is regulated by an independent control loop to maintain the working pressure in the anode close to the pressure in the cathode.

The main challenges related with fuel cell control are as follows:

1. Regulation of the stack power.
2. Control of the *oxygen excess ratio* in the cathode.
3. Control of the hydrogen supply.
4. Controllability of the system.
5. Improvement of the system efficiency.

2.2 Regulation of the Stack Power

One of the main control objectives is to control the PEMFC so that the power output meets the power load. Among the possible applications of PEMFC are electrical vehicles propulsion. In automotive applications, the load fluctuates strongly according to the driver requirements. Indeed, the power required by a car depends heavily on the driving profile and traffic conditions. There are different standard driving cycles, representing urban and suburban driving situations, which are widely used in the automotive literature to compare the performance and emissions in different vehicles. One of these cycles is *new European driving cycle* (*NEDC*). The standard driving cycles will be seen in detail in Sect. 4.3 of Chap. 7.

The control system must match the power demand, manipulating appropriately the control variables. Theses variables include the hydrogen and air inlet flow, the hydrogen and air inlet pressure, the stack temperature, the membrane humidification, the coolant temperature and the coolant flow. There is a degree of uncertainty in the behavior of the fuel cell system (FCS) components due to material degradation and the external temperature and pressure [30]. Thus, the control system must be sufficiently robust.

In the work of Golbert and Lewin [30], model-based predictive control (MPC) is used with the objective to demonstrate that it can be used to obtain a robust control and to improve the system efficiency reducing the hydrogen consumption. Since that the work is focused on applications with vehicles, the objective is to meet a certain speed more than the output power. In practice, the car driver manipulates the accelerator and brake pedals according to the desired speed. This behavior can be simulated using a PID controller together with a simple vehicle model. This model calculates the required power to meet the required speed according to the vehicle characteristics. In the simulations, the FCS controller based on MPC is implemented as a slave of a master controller based on PID control, which provides the power set-point to the MPC controller. Then, this controller manipulates the control variables of FCS to achieve the required FCS power. The MPC-based controller uses a nonlinear reduced order model of the FCS model developed in [31]. In this work it is concluded that it is possible to satisfy the fluctuating power demands and to optimize the system efficiency, with a significant hydrogen consumption reduction.

2.3 Control of Oxygen Excess Ratio

One of the most important problems in FCS control is to guarantee sufficient oxygen supply during abrupt changes in the load demand. This objective is difficult to obtain since oxygen reacts instantaneously to load demands whereas, as mentioned previously, the air supply is limited by the gas feeding dynamics and the air compressor dynamics. Several publications approach this problem [27, 32–34]. When the current demanded increases abruptly, the present oxygen level in the cathode diminishes drastically as a result of the electrochemical reaction. The abrupt oxygen diminution can cause damages to the membrane and degradation of the FCS efficiency. This phenomenon is known in the PEMFC literature as "oxygen starvation". This term describes the operation condition of a fuel cell in sub-stoichiometric reactant feedings and is one of the potential causes of fuel cell failure [35]. A parameter that indicates the excess of oxygen in the cathode, namely *oxygen excess ratio* (λ_{O_2}), is defined [32, 34, 36]:

$$\lambda_{O_2} = \frac{W_{O_2,\mathrm{in}}}{W_{O_2,\mathrm{rct}}}, \tag{23}$$

where $W_{O_2,\mathrm{in}}$ is the cathode inlet oxygen mass flow and $W_{O_2,\mathrm{rct}}$ is the reacting oxygen mass flow in the cathode electrochemical reaction. This flow depends on FCS current I_{fcs} as follows:

$$W_{O_2,\mathrm{rct}} = M_{O_2} \frac{n I_{\mathrm{fcs}}}{4F}, \tag{24}$$

where M_{O_2} is the molar mass of oxygen, n is the stack cell number, and F is the Faraday number ($F = 96485\,\mathrm{C mol^{-1}}$). Therefore, when the FCS current increases, the value of λ_{O_2} diminishes instantaneously. It is necessary to implement a control strategy that allows a fast recovery of the oxygen level. In order to do that the controller must provide more air to the FCS increasing the air compressor voltage.

In the work of Pukrushpan et al. [27], the oxygen level regulation is obtained by means of a PI controller in the compressor. The oxygen transitory response is improved using linear observability techniques. The observability improves when the FCS voltage is included as a measured variable in the feedback controller. It is demonstrated that there is a severe conflict between the oxygen excess control and the net power temporary response. The limitation in the temporal response comes from the fact that all the auxiliary equipment is power-supplied directly by the FCS without any secondary power source. In this publication it is mentioned that with the use of an energy storage system (e.g., batteries or supercapacitors) it is possible to overcome this conflict, although this configuration is not developed.

In the work of Grujicic et al. [32], the optimization of the transitory behavior in a PEMFC is analyzed with strategies of model-based control. The controller in this work has the objective to maintain a necessary level of oxygen partial pressure in the cathode during steep changes in the demanded current. The results obtained indicate that the oxygen level in the cathode can be satisfactorily maintained by means of feedforward control, manipulating the air compressor motor voltage. Nevertheless, the FCS power does not respond adequately during the transitory behavior, which also suggests the necessity of a power management by means of a secondary power source (battery or another storage energy system). In [33], a feedback controller is added to the feedforward controller to reduce the stationary state error. Again it is concluded that it is necessary to implement a hybrid configuration with storage of energy in order to improve the transitory response.

In the work of Vahidi et al. [34], a MPC is proposed to solve the oxygen supply problem during steep changes in the current. In this article a hybrid system composed of FCS and supercapacitors is analyzed. MPC is based on minimization of a cost function of the predicted response of a system over a future horizon. Excellent results in the oxygen level control are obtained, although the transitory behavior of the FCS power is not analyzed.

2.4 Hydrogen Supply Control

The hydrogen supply control objective is to feed the FCS with hydrogen at the same pressure as air independently of the hydrogen consumed, which is proportional to the current load [7]. It is very important to reduce the differential pressure through the membrane to avoid damaging it. When pressurized hydrogen is used, the hydrogen flow can be regulated by a servo valve. Since the valve is fast, it is assumed that the flow can be controlled with a proportional controller based on the feedback of the differential pressure [26]. Commonly, a purge valve is installed at the anode output to remove the water in excess. The purge valve can also be used to reduce the anode pressure quickly in case of necessity.

It is generally accepted that the hydrogen supply control is not critical when the system is fed with pressurized hydrogen. However, the hydrogen supply control problem is critical when the hydrogen production is made using an onboard fuel processor with a relatively slow transitory response [36]. This is the case when reforming a hydrocarbon fuel, for example natural gas, in a rich hydrogen gas [37]. This drawback can originate a phenomenon known as "hydrogen starvation" [36, 37]. This topic will be discussed again in Chap. 9.

2.5 System Controllability

The Efficiency improvement in a fuel cell system is very important, but it is also important to maintain the system controllability. The work of Serra et al. [38] deals with the control of PEM fuel cells, including a linear analysis of the system at different operating points, the comparison and selection of different control structures and the validation of the controlled system by simulation. The work is based on a complex nonlinear model developed in [39], which has been linearised at several operating points. The linear analysis tools are the Morari resiliency index, the condition number and the relative gain array. These techniques are employed to compare the controllability of the system with different control structures and at different operating conditions. According to the results, the most promising control structures are selected and their performance with PI-based diagonal controllers is evaluated through simulations with the complete nonlinear model. The conclusions indicate the good performance of several diagonal linear controllers. However, very few have a wide operability range.

Following this analysis, the efficiency and controllability of the system in two operating points for a determined net power is studied in [40]: one point corresponding to the minimum current necessary to obtain the required power and the other with a greater current. A conflict between controllability and efficiency was found. The operating point corresponding to minimum current has greater efficiency whereas the point with higher current has better controllability. Therefore, it is necessary to obtain a convenient balance between efficiency and controllability.

The analysis of fuel-cell controllability is also studied in the work of Golbert and Lewin [31] focuses on the crucial issue of how to control the fuel cell to ensure acceptable response time for the power demand, while achieving high efficiencies over the entire operating range. In that work, is used a spatial, time-dependent model of a fuel cell. The analysis of the fuel cell controllability is performed relying on the developed model as well as the synthesis of an adaptive controller, intended to account for the observed sign-change in the process static gain. Also, they present an alternative approach, relying on a reduced-order model, and the use of the model in a robust MPC scheme, which satisfies the power demand over a wide range of conditions, demonstrating a performance superior to that of the adaptive scheme. In this book, the oxygen excess ratio control will be addressed in Chap. 3.

2.6 Improvement of the System Efficiency

It is meaningful to study the manner of operating an FCS, especially at low loads. An adequate operation produces important benefits, increasing the system efficiency in terms of hydrogen reduction and allowing a greater peak power. In a direct-hydrogen FCS, the air supply subsystem has a crucial role in the improvement of the performance of the system [41]. In fact, there are two external variables that have greater impact on the polarization curve: the air pressure and the air stoichiometry (SRa).

The air pressure and the SRa control the oxygen partial pressure in the catalytic layer of the cathode, which determines the cathode polarization and, therefore, the efficiency. In [7], it is also stated the importance of the air pressure control to improve the FCS efficiency. The efficiency improvement for a given load is based on a trade-off between an increase in the air pressure and SRa, and an increase in the parasitic compressor power.

In the work of Friedman and Moore [41], it is shown that an FCS can be optimized to obtain high peak power and high efficiency over a broad range of output powers. The key to obtain this objective is to vary the pressure and air flow. Based on this result, it may be concluded that an FCS must be operated to the greater possible pressure and SRa. Nevertheless, if the energy necessary to compress the air is considered, the result is different: for a fixed air flow, the compressor power consumption increases significantly when the pressure is increased. This means that it is possible to find an optimal combination of pressure and air flow.

A similar conclusion is stated in [32]. In this work, it is assumed that the FCS net power P_{net} can be approximately defined as the difference between the power produced by FCS, P_{fcs}, and the consumed compressor power P_{cm}. For each load current, an air flow increment increases the cathode pressure and, therefore, increases the oxygen partial pressure, increasing the FCS voltage. This also leads to an increment in the oxygen excess ratio in the cathode, λ_{O_2}. The initial increase in the oxygen excess is translated into an FCS power increase and a net FCS power. Nevertheless, if a limit λ_{O_2} is exceeded, an excessive compressor power produces so that the net power falls. Thus, for each current there is an optimal value of λ_{O_2}, for which the net power has its maximum.

In [42], a single-objective and a multi-objective optimization problem are used to optimize the operating conditions of an experimental fuel cell system for two different applications: vehicular and stationary. The stack used for the study is based on Palcan PC6-1200, which is a PEM fuel cell system with maximum gross power output of 1,200 W. In both problems, two optimizations are performed with different objectives: the maximization of the net power and the system efficiency. The design variables include the operating temperature, the air stoichiometry and the operating pressure of the fuel cell stack.

In this book, the issue of improvement of the system efficiency introduced here will be further addressed in Chap. 3 through adequate control techniques.

2.7 Diagnosis of Fuel Cell Systems

The Energy generation systems based on fuel cells are complex since they involve thermal, fluidic and electrochemical phenomena. Moreover, they need a set of auxiliary elements (valves, compressor, sensors, regulators, etc.) to make the fuel cell work at the pre-established optimal operating point. For these reasons, they are vulnerable to faults that can cause the stop or the permanent damage of the fuel cell. To guarantee the safe operation of fuel cell systems, it is necessary to use systematic techniques, like the recent methods of FTC, which allows to increase the fault tolerance of this technology described in [28].

The first task to achieve active tolerant control consists of the inclusion of a fault diagnosis system operating in real-time. The diagnosis system should not only allow the fault detection and isolation but also the fault magnitude estimation. The fault detection and isolation methods can be classified into three groups [43]:

- Quantitative model-based methods
- Qualitative model-based methods
- Process history-based methods

Among the most frequently used methods are those based on quantitative models. These methods generate inconsistencies between the actual and expected behavior relying on an explicit model of the monitored plant. These inconsistences are called *residuals* and can be used for detection and isolation purposes. The residuals should be close to zero when no fault occurs but show "significant" values when the monitored system changes. The generation of the diagnostic residuals requires an explicit mathematical model of the system: a model derived analytically using first principles or a black-box model obtained empirically.

On the other hand, statistical methods are often required to decide whether there has been a failure. The residual generation can be performed using diagnostic observers, parity relations, Kalman filters or parameter estimation, among other methods. For more details on quantitative model-based methods and an extensive review on qualitative model-based methods and process history-based methods, see [43–45]. In this book, he diagnosis of PEM Fuel Cell Systems will be addressed in Chap. 6.

3 Fuel Cell Hybrid Systems

The fuel cell hybrid systems are composed of an FCS, which is the primary power source and an ESS, e.g., a battery or supercapacitor bank that contributes to supply the load power demand.

Given a load power $P_{\text{load}}(t)$, this can be supplied with a fraction of the power from the FCS, $P_{\text{fcs}}(t)$, the rest of the power being supplied by ESS, $P_{\text{ess}}(t)$:

$$P_{\text{load}}(t) = P_{\text{fcs}}(t) + P_{\text{ess}}(t) \qquad \forall t. \tag{25}$$

A diagram with the energy and power flows between the elements in the hybrid system is shown in Fig. 3. The $P_{\text{fcs}}(t)$ is produced by the fuel cell stack, which is fed with a hydrogen flow $\dot{m}_{H_2}(t)$ from a H_2-pressurized tank, whereas, the bidirectional power flow $P_{\text{ess}}(t)$ can come from two different sources: the FCS or the load, through a regenerative process in automotive applications. The regenerative braking is covered in Sect. 4 of Chap. 6. In the FCHS of Fig. 3, four energy flows can be distinguished:

- *Fuel cell energy flow*: The FCS, which is the primary power source, supplies energy directly to the load.
- *Charging energy flow*: It is possible to charge the ESS with energy from the FCS to subsequently supply the load.
- *Regenerated energy flow*: When the load power is negative, e.g., during regenerative braking in automotive applications, it is possible to regenerate energy and charge the ESS with energy from the load.
- *Boosting energy flow*: It is possible to boost the FCS supplying energy to the load from the ESS with energy previously charged from the FCS or regenerated from the load.

How to manage efficiently this energy flow is the function of the energy management strategies (EMS), which are covered in Chap. 8.

3.1 Advantages and Disadvantages of Fuel Cell Hybrid Systems

The potential advantages of fuel cell hybrid systems are as follows [46, 47]:

- Improve the hydrogen economy
- Overcome the relatively slow fuel cell system transient response

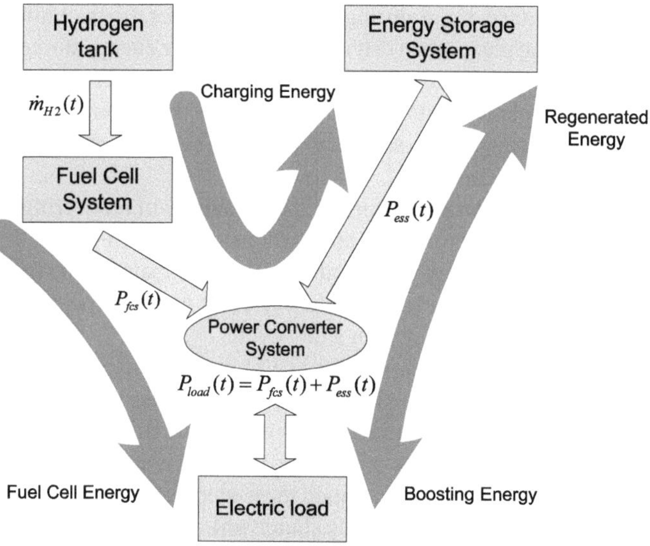

Fig. 3 Diagram showing the energy and power flows in a fuel cell hybrid system

- Reduce the cost and weight of the global system[2]
- Reduce the warm-up time of the fuel cell system to reach full power

The disadvantages of the hybridization are as follows:

- Greater complexity of the vehicle system
- Greater complexity of the control system

The advantages and disadvantages of fuel cell hybrid systems can be better understood by analyzing the roles of the energy storage system in a fuel cell hybrid vehicle, which is the most promising application of hybrid systems. These roles are as follows [48]:

- *Supply traction power during fuel cell start-up*: Under ambient and cold-start conditions the FCS output power is lower than its rated power. Thus, the energy storage system is expected to compensate this drawback until the fuel cell reaches its rate power.
- *Supply power assistance during drive cycles*: The stored energy in the ESS can supply the transitory demands of energy, allowing to assist the fuel cell operation in two ways. First, it is possible to reduce the fuel cell power, supplying only a fraction of the full load. Second, the energy storage can be used to improve the transient response in a system in which the fuel cell alone is not capable of responding quickly enough to an increase in power. The maximum *power* demands from

[2] This is true if the cost per unit power of the storage elements is lower than that corresponding to the fuel cells. Otherwise, it can become a disadvantage. In the same way, the weight per unit power of the storage elements must be lower than that corresponding to the fuel cell.

a vehicle typically take place during maximum accelerations, while the highest *energy* demands take place when driving with slope. The energy storage system can improve significantly the acceleration performance supplying instantaneous power to compensate the relatively slow fuel cell dynamics. However, the contribution to the performance of driving with slope is expected to be minimal since the capability of energy storage is limited.

- *Recapture regenerative braking energy*: The possibility of recovering energy from regenerative braking in automotive applications allows to improve the hydrogen economy recuperating energy which, otherwise, is dissipated at the friction brakes. This amount of energy depends on the driving cycle considered and goes from 5% in a highway cycle to almost 18% in a city cycle.
- *Supply electrical accessory loads*: A fuel cell hybrid vehicle may include several electric loads such as radiator fans, power-assisted steering, electric brake, air conditioning system, etc. These loads vary during the driving cycle but the total net load could be considered as a constant value. It is potentially expected that the energy storage system could maintain these electrical loads for certain periods during the cycle (e.g., if the fuel cell is shutdown).
- *Fuel cell start-up and shutdown*: In fuel cell hybrid vehicles it would be desirable to have start/stop capability. This means that while the vehicle is running, the primary power plant (i.e., the fuel cell) can be shut down and restarted according to an energy management strategy. This type of operation is beyond the scope of this book.

3.2 Applications of Fuel Cell Hybrid Systems

According to the concepts explained above, it follows that FCHS are particularly suitable in applications where the load varies strongly. Therefore, two possible applications for FCHS are as addressed in the following two subsections.

Stand-Alone Residential Application

Fuel cell-based systems are very attractive for stationary energy generation, since they allow the production of electricity and heat in a decentralized, quiet, efficient and environmentally friendly way [49]. However, for stationary applications, an FC stack may not be sufficient to satisfy the load demands, especially during peak demand periods or transient events.

The problem of the slow dynamic response is more pronounced when the fuel cell (FC) stack is fed with hydrogen coming from a reformer instead of a H_2-pressurized tank because of the dynamic response of the reactor. Moreover, in stationary applications the prospective arrangement is composed of a reformer. In addition, it is expected that the load has an aggressive behavior: sudden load switching and load conditions such as active–reactive power and loads with high harmonic distortion [50]. As a result, an auxiliary energy storage system such as a battery or supercapacitor bank to meet all load demands may be necessary.

A dynamic electrochemical simulation model of a grid-independent PEMFC power plant is presented in [51]. The model includes the methanol reformer, the PEM fuel cell stack and the power conditioning unit. In addition to these components, the system is equipped with storage batteries connected in parallel with the DC bus. The model is used to study the transient response of the PEMFC power plant when subjected to rapid changes in a residential load connected to it. The results show the fast response capabilities of the PEMFC following the changes in the load.

In the same way, in [52] and [50], it is concluded based on experimental results that the parallel combination of an FCS and an SC bank exhibit good performance for stand-alone residential applications during the steady-state, load-switching and peak power demand. On the contrary, without the SC bank, the FCS must supply an extra power, thereby increasing the size and cost of the FCS.

In addition, a residential CHP[3] system based on PEMFC is studied in [49]. The system is a co-generative unit, converting natural gas into electricity and heat that can work both under grid-connected and stand-alone configurations. The system considered in this work is also equipped with a set of batteries enabling load tracking and peak loads. Results show that the prototype can satisfy the energy needs of small residential applications.

Fuel Cell Hybrid Vehicles Application

Most major automobile manufacturers are developing prototypes of fuel cell vehicles. During the 1990s, several car companies began to work on developing the PEMFC for passenger car propulsion motivated by the increasing environmental issue. Hydrogen fuel cell vehicles (H_2-FCVs) are a good alternative to conventional internal combustion engine (ICE) vehicles. Actually, fuel cells and hydrogen are considered as the best overall solutions in the long run, although significant technical improvements are necessary [20].

Some of the advantages of H_2-FCVs are as followings:

- The H_2-FCVs have no local contaminant emissions allowing to reduce the urban pollution: the only emissions created during vehicle operation is water vapor. In addition to this advantage, the use of hydrogen for transportation makes it possible to introduce alternative energy sources (e.g., biomass, biofuels, solar power, wind power, wave power, etc.), in order to reduce the dependence on petroleum.
- FCS, in contrast to ICE, has the characteristic that the efficiency does not degrade at part load and, in fact, can be higher. A typical urban vehicle operates mostly at part load conditions. In [53], it is concluded that the hydrogen economy of H_2-FCVs can be 2.5–3 times the fuel economy of conventional ICE vehicles.
- It is possible to improve the hydrogen economy by recovering energy through regenerative braking adding an energy storage system, which constitutes a fuel cell hybrid vehicle (see Chap. 7).
- The specific energy of hydrogen is much larger than batteries, which convert H_2-FCVs into a more competitive solution than pure battery electric vehicles: $1600\,\mathrm{W\,h\,kg^{-1}}$ for hydrogen at 70 MPa contrary to $35\,\mathrm{W\,h\,kg^{-1}}$ for lead-acid

[3] Combined heat and power.

batteries, 70 W h kg^{-1} for nickel-metal-hydride batteries, and 120 W h kg^{-1} for Lithium-ion batteries [20].

Therefore, in this book we concentrate our attention on Fuel Cell Hybrid Vehicles with some kind of Energy Storage System.

3.3 Electrical Topology

One important issue in fuel cell hybrid systems is to determine the proper electric topology. Some configurations for fuel cell vehicles and their control strategies are studied in [54, 55]. Some of these configurations include the use of a propulsion battery. On the other hand, in the work of Santi et al. [56], a case study of a domestic fuel cell-based uninterrupted power supply is presented. In that work, different configurations are also analyzed. Because of the slow dynamics of the fuel cell compared to the electrical system requirement, it is concluded that the use of energy storage elements is necessary. The rate at which the energy from the storage system is required is a significant parameter to choose the correct technology for storage and the use of batteries is decided after an analysis of the different options.

An approach for the design of a power conditioning system (PCS) for fuel cell generation systems is studied in [57] with the help of several PCS examples. From this study, it is noted that full-bridge, half-bridge and push-pull converters and various different topologies are fundamentally used. Also, the comparison of the voltage source and current source topologies has been done, so that a system designer can select the proper topology according to the applications. From this comparison, it is concluded that it is better to connect the energy storage unit to the *DC*-link in parallel through a bidirectional converter. A similar conclusion is found in [58]. This issue will be revisited in Sect. 3 of Chap. 7.

3.4 Hybridization Degree

In the work of Atwood et al. [59], the influence of the hybridization degree on the system efficiency is analyzed, looking for an optimal value. Results for a wide range of combinations between batteries and FC sizes are presented, from a purely electrical vehicle to a pure fuel cell vehicle (without storage batteries). The results of hydrogen economy that are shown demonstrate that hybridization can improve the power efficiency. As expected, the presence of batteries allows the capture of regenerated energy from braking. The main conclusion of this work is that the optimal degree is not independent of the driving cycle considered. The results also show that the control strategy, the power spectrum of the considered cycle and the FCS and battery efficiency interact in a complex form. It is also concluded that the efficiency at low powers depends on the air compressor power and its minimum speed, combined with the hybridization degree.

Similarly, in the work of Friedman [60], the convenience of using a hybrid configuration of FCS and batteries is also investigated. In this work, comparisons between two hybrid systems are made (which differ in the size of the battery bank). A pure FCS system is also considered. Different driving cycles are used to compare the performance. The results are not conclusive about the convenience of the hybrid systems, although the work clarifies that they depend on the technology used. In fact, with the new technologies in batteries and supercapacitors it is possible to achieve better results.

An economic comparison between vehicles with different hybridization degrees is done in [46], considering the initial cost of the vehicle, the fuel cell, the batteries and the hydrogen. The initial cost of a fuel cell vehicle is reduced by means of hybridization when the price of the fuel cell is high. In addition, the operation cost is affected mainly by the hydrogen cost. Since one of the advantages of hybridization is the hydrogen reduction, it is possible to reduce the operative cost of the vehicle. In this work, a hybrid vehicle model was simulated using the *federal urban driving schedule* (FUDS) for different HD, including the case of null hybridization (no batteries). The results show that the best hydrogen economy was obtained for an HD of 33%, meanwhile for an HD of 73% the economy is worse than in the case of pure FC vehicle. Concerning the initial vehicle cost, the results also show that if the fuel cell price is less than 400 US$/kW the hybridization has no benefits.

It is also concluded in [61] that an auxiliary power source can be used in FCV to alleviate the performance limitation, especially when the FC is fed with an onboard fuel reformer. It is shown that for a given vehicle, motor and FCS configuration, there is an optimal choice for the energy capacity for different battery technologies. However, in this work the effect of different driving cycles is not taken into account.

The results of an energy analysis for a load-following fuel cell vehicle (without any energy storage device) versus a battery-hybrid fuel cell vehicle are presented in [62]. The major conclusion from this study is that only for cycles with a large amount of regenerative braking at low to mid power levels (e.g., the FUDS cycle) there are significant advantages in terms of overall hydrogen economy. The fact that the hydrogen economy occurs only in certain driving cycles can be attributed to the hybrid configuration and, fundamentally, the energy recovery from regenerative braking. However, the energy management strategy employed to perform the power split between the energy storage system and the fuel cell is quite elementary. As a result, it is expected to achieve better results implementing a more elaborated strategy exploiting the capabilities of the recovered energy from regenerative braking and operating the FCS in a more advantageous way. Besides, and probably more important, as this study was done with batteries in the energy storage system the use of supercapacitors increases the benefits of the hybrid system as will be shown.

In this book, the analysis of the optimal hybridization degree will be addressed in Sect. 6 of Chap. 7.

3.5 Energy Management Strategies for FCHS

The coordination between several power sources in an FCHS requires a high-level control: the energy management strategy. Different approaches are found in the literature. In the work of Thounthong et al. [63], the strategy implemented is based on *DC* link voltage regulation by controlling the power converters. The fuel cell is operated in almost steady-state conditions in order to minimize the mechanical stresses and ensure a good synchronization between hydrogen flow and load current. Supercapacitors are only functioning during transient energy delivery or transient energy recovery. This strategy addresses the link voltage regulation but not the system efficiency.

In the work of Paganelli et al. [64], a general formulation about the instantaneous power split between the FC and the ESS in an FCHV is presented. The proposed method is based on a control strategy called equivalent consumption minimization strategy (ECMS). This formulation is based on fitting the instantaneous distribution of power between the FC and the ESS to diminish the global hydrogen consumption using an appropriate analytical formulation suited for real-time control. For that, a first step consists in converting the electrical powers that take part in the system into equivalent hydrogen consumptions. Unlike the previous work, the ECMS strategy demonstrates to be robust under a wide range of operating conditions. Nevertheless, the efficiency is sensitive to several crucial parameters: average consumption, the function penalty employed and the limit values of the load range. They conclude that greater investigation is necessary to surpass these problems.

Similarly, in the work of Rodatz et al. [47, 65], is implemented a real-time control to distribute the power between FC and ESS, with the objective of diminishing the hydrogen consumption, maintaining, at the same time, the controllability of the system, using the same concept of equivalent hydrogen consumption. Experimental results are shown in [65] with an FC-supercapacitors hybrid vehicle (experimental Volkswagen HyPower vehicle).

An algorithm is implemented in [47], which resolves an optimization problem at each sampling time and the cost function is constructed from variables considered at the present time. This cost function takes into account the use of the hydrogen energy, since the major objective is to reduce the hydrogen consumption, but also considers the amount of recovered energy. The conclusion of the work is that the efficiency is not significantly improved with the use of this control strategy. This might be due to the small margin for optimization with respect to the case study of the work of Rodatz et al. [65], which is utilized to make the comparisons. Nevertheless, this technique has the advantage that it does not need more adjustment than the horizon time. In addition, ECMS control involves a hydrogen consumption model as a function of the net output power. This feature makes ECMS inherently self-adaptive to achieve full efficiency with any system configuration and operating condition such as temperature.

In the work of Kotz et al. [66], a pilot system on a small scale that represents a hybrid vehicle with FC and supercapacitors is analyzed. The FC–SC hybrid system is tested with the NEDC,[4] demonstrating that supercapacitors are excellent energy storage devices that can be used efficiently in vehicle applications for energy recovering from braking and peak power reinforcing. The recovered energy in different driving cycles for two different car sizes are presented: in an urban driving cycle it is possible to recover until a 25% of the traction energy and the general efficiency of the recovering process is near 70%. The main conclusion of this work is that supercapacitors have a great potential for applications with short duration peak powers, such as in hybrid electrical vehicles. In addition, it is concluded that an elaborated energy management strategy is necessary as presented in [47].

In the work of Caux et al. [67, 68], the energy management between an FCS and supercapacitors is studied in a tramway application. An algorithm that determines two power references (one for the FCS and another for the supercapacitors) is presented with the main objective to fulfill the power demand. Other objectives are as follows: (1) recover the maximum of braking energy, (2) supply the power consumed by the auxiliary systems with the FCS, (3) the FCS must operate preferably at rated power (400 kW), (4) employ the storage elements as little as possible, (5) restrict the FCS power variation, and (6) perform the starting and stopping with the storage elements totally charged. The algorithm computes all that is possible and, with the losses and the efficiency estimation of all components, an effective solution is found for a given profile. The conclusion is that a different number of supercapacitors is required depending on the profile under study.

For the development of a correct energy management strategy, both a good FCS model and an estimation of the energy losses produced in the storage elements and the converters are necessary. The losses in the supercapacitors by Joule effect ($P_{\text{loss}} = r_{\text{sc}} \cdot I_{\text{sc}}^2$) and the losses produced by conduction and commutation in each converter are calculated in [68]. These losses are functions of the commutation frequency (f_c), and the current and voltage on the converter ($P_{\text{loss}} = f(f_c, I_{\text{DC}}, V_{\text{DC}})$).

In the work of Jung et al. [69], a control strategy with two objectives is proposed: one objective is to obtain high efficiency in the hybrid system and the other is to maintain a minimum state of charge in the batteries. If both objectives cannot be fulfilled simultaneously, the priority is given to the battery state of charge. The system efficiency can be improved controlling the power split between the fuel cell and the battery. The value of reference for the battery state of charge can be determined considering the optimal efficiency of charge and discharge. The operation is divided into five different modes of operation: (1) *charge* mode, (2) *discharge* mode, (3) *fuel cell* mode, (4) *regenerative* mode and (5) *electrical vehicle* mode (only batteries). The operation mode is determined according to the battery state of charge and the power required. The effectiveness of the control strategy proposed was evaluated by experimental results, although the obtained effectiveness is not quantified.

In the work of Kim et al. [70], a combined power management/design optimization problem approach for the performance optimization of FCHVs is formulated. Their results indicate that optimality consists in downsizing the fuel cell compressor,

[4] The characteristic load profiles are covered in Sect. 4.3 of Chap. 7.

decreasing the hybridization degree without compromising regenerative braking and employing adequate control strategies.

Schiffer et al. [71] also shows the benefits of using a fuel cell-supercapacitor system for FCHVs. Their results show that it is a good strategy to use the supercapacitors with their maximum power during acceleration and deceleration phases and to control the voltage of the supercapacitors during standstill and constant speed phases to reach an optimum voltage that depends on the vehicle speed. The fuel cell is used as little as possible during acceleration and deceleration phases while it is used to charge the supercapacitors during standstill and constant speed phases if necessary.

4 Problematic Related to the Hydrogen Production

To understand the potential of hydrogen as an energy carrier, it is important to know their basic properties. Hydrogen is the most abundant chemical element in the universe, lightweight and simple, and its predominant isotope being only one proton and one electron. Besides, it is the first element of the periodic table with atomic number 1. The weights are $Z = 1$ and 1.00794 u.m.a, respectively. Under normal temperature and pressure ($0°C = 273.15$ K and 1.013 bar), hydrogen forms a diatomic gas. Hydrogen constitutes 75% of the mass and 90% of the number of atoms in the universe.

However, despite its abundance in nature, hydrogen does not exist in pure form, but is a component of other substances, such as water, hydrocarbons and organic compounds in general. Therefore, certain processes are needed to obtain hydrogen. Till date, steam reforming of hydrocarbons, i.e. natural gas, has been the most commonly used and generally the most economically competitive method for hydrogen production. However, the use of fossil fuels for secondary energy production is non-sustainable [72]. Therefore, there is great interest in the research community to produce hydrogen from renewable sources cleanly and safely. In this section, a review of the different methods for hydrogen production is presented.

4.1 Hydrogen Production

Hydrogen can be produced using diverse, domestic resources including fossil fuels, such as coal (with carbon sequestration) and natural gas; nuclear; and biomass and other renewable energy technologies, such as wind, solar, geothermal and hydroelectric power [73]. Hydrogen has a great potential for diversity of energy supply and hence, is a promising energy carrier. At present, the dominant technology for direct production is steam reforming from hydrocarbons. However, there are other methods to obtain hydrogen, which can be classified in the following three groups:

1. Hydrogen obtained from hydrocarbon

   a. Reforming from natural gas
   b. Reforming from ethanol

c. Coal gasification
d. Partial oxidation of hydrocarbons, etc.

2. Water electrolysis
3. Hydrogen from biomass

From the environmental point of view, the production of hydrogen for use as fuel only makes sense if it is obtained using renewable energy sources, or if the method used to recover the energy content is efficient enough to compensate the energy used in production. Otherwise, the energy balance would be negative. With this in mind, it is more convenient to classify the methods of producing hydrogen according to the source of the raw materials used. Thus, we can establish two groups:

1. Methods based on fossil sources, or production of hydrogen from hydrocarbons.
2. Methods based on renewable sources, or production of hydrogen from water.

Currently, most of the produced hydrogen comes from fossil fuels (around 96% [74]) and is intended mostly for industrial uses. Natural gas contains methane (CH_4) that can be used to produce hydrogen via thermal processes, such as steam-methane reformation and partial oxidation.

4.1.1 Steam-methane Reforming

Steam-methane reforming is a process in which high-temperature steam (700–1,000°C) is used to produce hydrogen from a methane source, such as natural gas. In steam-methane reforming, methane reacts with steam under 3–25 bar pressure in the presence of a catalyst to produce hydrogen, carbon monoxide and a relatively small amount of carbon dioxide. Steam reforming is an endothermic reaction, that is, heat must be supplied to the process for the reaction to proceed:

$$CH_4 + H_2O\,(+\text{heat}) \longrightarrow CO + 3H_2. \tag{26}$$

Then, the carbon monoxide and steam are reacted using a catalyst to produce carbon dioxide and more hydrogen in a water-gas shift reaction, which is an exothermic reaction:

$$CO + H_2O \longrightarrow CO_2 + H_2\,(+\text{small amount of heat}). \tag{27}$$

Finally, the carbon dioxide and other impurities are removed from the gas stream, producing essentially pure hydrogen. This is a method widely used in mass production of hydrogen and the main advantage is the availability of natural gas. A disadvantage is that they require high temperatures (700–1,100°C) and therefore are necessary SOFC or MCFC due to its high operating temperature. Another important disadvantage is that it produces CO_2 (7.5 kg of CO_2 per kg of H_2.) This production method is complicated to implement onboard a vehicle.

Steam reforming can also be used to produce hydrogen from other fuels, such as ethanol, propane or even gasoline. This requires high temperatures (>500°C) but not as high as in the case of natural gas. Below are the steam-reforming reactions corresponding to different fuels [74]:

Propane:

$$C_3H_8 + 3H_2O\,(+\text{heat}) \longrightarrow 3CO + 7H_2 \tag{28}$$

Ethanol:

$$C_2H_5OH + H_2O\,(+\text{heat}) \longrightarrow 2CO + 4H_2 \tag{29}$$

Gasoline[5]:

$$C_8H_{18} + 8H_2O\,(+\text{heat}) \longrightarrow 8CO + 17H_2 \tag{30}$$

$$C_7H_8 + 7H_2O\,(+\text{heat}) \longrightarrow 7CO + 11H_2 \tag{31}$$

4.1.2 Partial Oxidation

On the other hand, in partial oxidation, methane and other hydrocarbons in natural gas are reacted with a limited amount of oxygen (typically, from air) which is not enough to completely oxidize the hydrocarbons to carbon dioxide and water. With less than the stoichiometric amount of oxygen available for the reaction, the reaction products contain primarily hydrogen and carbon monoxide (and nitrogen, if the reaction is carried out with air rather than pure oxygen), and a relatively small amount of carbon dioxide and other compounds. Then, in a water–gas shift reaction, the carbon monoxide reacts with water to form carbon dioxide and more hydrogen.

Partial oxidation is an exothermic process, that is, it releases heat. The process is much faster than steam reforming and requires a smaller reactor vessel. However, this process initially produces less hydrogen per unit of the input fuel than that obtained by steam reforming of the same fuel:

Methane:

$$CH_4 + \frac{1}{2}O_2 \longrightarrow CO + 2H_2\,(+\text{heat}) \tag{32}$$

Propane:

$$C_3H_8 + 1\frac{1}{2}O_2 \longrightarrow 3CO + 4H_2\,(+\text{heat}) \tag{33}$$

Gasoline[6]:

[5] Using iso-octane and toluene as example compounds from the hundred or more compounds present in gasoline.

[6] Ibid.

$$C_8H_{18} + 4O_2 \longrightarrow 8CO + 9H_2\ (+\text{heat}) \quad (34)$$

$$C_7H_8 + 3\frac{1}{2}O_2 \longrightarrow 7CO + 4H_2\ (+\text{heat}) \quad (35)$$

Water–Gas Shift Reaction in partial oxidation method:

$$CO + H_2O \longrightarrow CO_2 + H_2\ (+\text{small amount of heat}) \quad (36)$$

4.1.3 Electrolysis

On the other hand, electrolysis is the process of using electricity to split the water molecules into hydrogen and oxygen with an electrochemical reaction opposite to the fuel cell reaction. This reaction takes place in a unit called an electrolyzer. It is a small, appliance-size equipment well-suited for small-scale distributed hydrogen production. Research is ongoing to examine larger scale electrolysis that could be used directly for renewable electricity production. For example, hydrogen production at a wind farm.

Hydrogen produced via electrolysis can result in zero greenhouse gas emissions, depending on the source of the electricity used (hydro, wind, waste, solar). The source of the required electricity, including its cost and efficiency as well as emissions resulting from electricity generation, must be considered when evaluating the benefits of hydrogen production via electrolysis.

Like fuel cells, electrolyzers consist of an anode and a cathode separated by an electrolyte. The different types of electrolyzers operate similarly with minor differences. In a PEM electrolyzer, the electrolyte is a solid plastic membrane as in a PEM fuel cell (see Sect. 1.3.1). The water reacts at the anode to form oxygen and positively charged hydrogen ions (protons). The electrons flow through the external circuit and the protons move across the membrane to the cathode. At the cathode, the protons combine with electrons from the external circuit to form hydrogen gas. The reactions are as follows:

Anode reaction:

$$2H_2O \longrightarrow O_2 + 4H^+ + 4e^-. \quad (37)$$

Cathode reaction:

$$4H^+ + 4e^- \longrightarrow 2H_2. \quad (38)$$

Different types of electrolytes can be used, similar to the different fuel cell technologies. However, the proton exchange membrane is currently the most common technology used. The hydrogen produced by these types of electrolyzers have high purity since there are no traces of other components as opposed to the reforming process from fuel. The only drawback is the presence of water vapor, but this is countered by using a gas drying prior to storage [2].

A new technology in development that looks very promising is hydrogen electrolyzers at high pressure. Mitsubishi Corporation announced that it has achieved the first production of high-pressure hydrogen gas without a compressor [75]. The maximum pressure level is 350 bar and they plan to develop a model that can operate at 400 bar, which is very promising because this would meet the current standard for hydrogen refueling stations.

4.2 Hydrogen Storage

One of the main problems facing the development of hydrogen technology is the need to find safe and profitable forms of storage. Storage methods can be classified as follows:

1. Gas with pressure vessels.
2. Liquid in cryogenic tanks.
3. Solid as a gas absorbed in porous systems.
4. Using chemical methods.

4.2.1 Hydrogen Storage with Pressure Vessels

The most simple solution is its storage as a compressed gas. It requires the most simple infrastructure, compared with other methods, and is the most commonly developed method currently. The main drawback of this technology is that the volumetric density is low, which complicates the provisioning. However, the volumetric density is improved with the pressure. Actually, the operating pressure is in the range of 50–300 bar, despite systems in the range of 500–600 bar being in development.

Another important advantage is the high hydrogen purity, which prevents contamination and deterioration of the fuel cell. This is particularly important in the case of PEMFC, which is the type of fuel cell that requires higher purity of the fuel.

4.2.2 Hydrogen Storage as a Liquid in Cryogenic Tanks

Another method of hydrogen storage is as liquid in cryogenic tanks. In this case, the hydrogen is stored at atmospheric pressure but the temperature is about 253°C (=20 K). The volumetric density is much higher than in the previous case (approximately three times larger) but the main problem of this method is the high energetic consumption necessary to keep such a low temperature. In fact, it is considered necessary to employ 25% of the stored energy [2]. This issue is even worse in small systems, where the consumption can reach 45%.

4.2.3 Hydrogen Storage using Solid Materials

One new method under development is the storage through physic absorbtion using porous matrices of special materials such as special mixtures of titanium, iron, manganese, nickel, chromium and others that can react with hydrogen to form a metal hydride in a very easily controlled reversible reaction. These materials work like a hydrogen "sponge". This technology greatly improves the volumetric density but not the gravimetric density. For example, the titanium iron hydride $TiFeH_2$ has a gravimetric density of 5.47 kg L^{-1} and the volume to store 1 kg of H_2 is 9.8 L [2]. It is important to note that this value does not include all the extra equipment needed to hold or process the compound. Thus, it is not a fully representative value.

The operation can be explained by the following reaction:

$$M + X/2H_2 \rightleftarrows MH_x + \Delta H, \qquad (39)$$

where ΔH is the enthalpy of the reaction and represents the heat energy that must be removed for the forward reaction (exothermic) and added for the reverse reaction (endothermic). ΔH can vary from 20 to 75 kJ mol^{-1} H_2, depending on the particular material [76].

The main advantages of metal hydride used for hydrogen storage are as follows:

- Safety
- Compact
- Low pressure operation
- Fast kinetics
- Long cycle life
- Compatible with lightweight aluminum or carbon fiber vessel
- Onboard waste heat can be used to produced the reverse reaction

4.2.4 Hydrogen Storage using Chemical Hydrades

Other methods in development use chemical hydrides as "hydrogen carriers". A chemical hydride can be used as a hydrogen carrier or storage medium. The hydrogen in the hydride is released through a reaction with water. Chemical hydride systems are irreversible and require thermal management and regeneration of the carrier to recharge the hydrogen content. One promising chemical hydride for hydrogen storage is commonly referred to as sodium borohydride. However, the proper systematic name for the compound $NaBH_4$ is sodium tetrahydridoborate. The basic hydrogen producing reaction is the following hydrolysis reaction:

$$NaBH_4 + 2H_2O \longrightarrow 4H_2 + NaBO_2 \qquad (\Delta H = -218\,\text{kJ mol}^{-1}). \qquad (40)$$

This reaction does not normally proceed spontaneously and some form of catalyst is needed. Thus, solutions of $NaBH_4$ in water are quite stable. On the other hand, note that some of the hydrogen comes from the water and the system produces four hydrogen molecules for each molecule of sodium borohydride, which results in a very high efficiency as a hydrogen carrier.

The main advantages can be summarized as follows [77]:

- The reaction is highly controllable and can be stopped removing the catalyst.
- The reaction does not require energy and can operate at ambient temperature and pressure.
- It is the simplest and cheapest method of generating hydrogen for small quantities of gas. However, it is still expensive for large quantities.
- No high pressures are produced.
- Neither carbon dioxide nor other caustic vapors are produced.
- If the system is warm, water vapor will be mixed with the hydrogen, which is highly desirable for PEMFC systems.

Another approach using chemical hydride is the lithium hydrade (LiH). The basic reactions is as follows:

$$LiH + H_2O \longrightarrow LiOH + H_2. \tag{41}$$

4.3 Hydrogen Security

Another important problem facing the development of hydrogen technology is the issue of security. Unfortunately, there is a perception in society that hydrogen is unsafe. However, it has a long history of safe use in chemical and aerospace industries. The key to the safe operation lies in the knowledge of its properties, development of security systems and procedures and establishing regulations and standards [74].

Hydrogen must be handled with care like other fuels or energy vectors. However, some specific characteristics of hydrogen make it different from other common fuels like gasoline or natural gas. Hydrogen has a higher leak probability that is potentially dangerous because it has the smallest molecule. However, a natural gas leak would result in a greater energy release than hydrogen because natural gas has energy density per unit volume over 3 times higher than hydrogen [6]. On the other hand, hydrogen disperses faster than other fuels, reducing the hazard levels.

4.4 Hydrogen Uses

Hydrogen is currently used in various industrial processes. Among other applications, it can be distinguished for its importance as a raw material in the chemical industry, both organic and inorganic, and as a reducing agent in metallurgy. It also has some specific use as a means of a cooling medium in electrical equipment and some industrial applications. Currently, pure hydrogen is not used directly for energy production, except in aerospace applications. However, hydrogen can be used as fuel for engines where the mechanical energy obtained from hydrogen is used to move vehicles or in power generation.

A new use is the production of electricity trough an electrochemical reaction using fuel cell, which is one of the most promising technologies. Fuel cells need hydrogen as fuel with a high grade of purity depending on the particular fuel cell properties. Basically, the lower operating temperature and the purity requirements are stronger.

The CO is a poison for PEM fuel cell if it is present in more than 10 ppm, whereas CO is a poison for AFC and PAFC if it is present in more than 0.5%. On the other hand, CO is a fuel in MCFC and SOFC. Actually, CO reacts with H_2O producing H_2 and CO_2 via the shift reaction in (27) [2]. CO_2 and H_2O are poisons for AFC and diluents for the rest of fuel cells, whereas the CH_4 is a diluent for all the fuel cell types. With regard to the S (as H_2S and COS), in general is a poison, especially for PAFC (>50 ppm), MCFC (>0.5 ppm), and SOFC (>1 ppm). In a PEM fuel cell, even small amounts of the H_2S and COS can cause significant performance drop producing an absorbtion on the Pt catalyst [78].

5 Fuel Processor Systems with Bio-Ethanol

The current industrial hydrogen production technology could only partially meet the requirement of small-scale fuel processors, which must be compact, turnkey and of high efficiency by combining component functionalities and eliminating unnecessary components. Although the onboard fuel processor program has been put on hold by the U.S. Department of Energy (DOE), a number of new approaches for the establishment of hydrogen refueling stations, based on some form of a fuel processor, are being pursued [79]. Particularly among all the alcohols, methanol is the most popular fuel for reforming, not only because it requires mild reforming conditions and has potential for attainment of highest possible efficiency, but also because of its possibility to be produced from renewable resources [80, 81].

Another alcohol is ethanol, which is already used as one of the main additives for gasoline. It has high octane number (RON and MON) and low toxicity which explains its gained popularity for its environment friendliness as well as sustainability [82]. Biomass, including bio-ethanol and other high molecular weight materials such as sugar alcohol is considered to be promising raw material in terms of their carbon cycling neutrality. Conventionally, hydrogen can be obtained by high temperature gasification/pyrolysis of biomass followed by catalytic reforming of the gas/liquid product [83].

A good recent review about the different technologies for integrated fuel processors for fuel cell application is given in Qi et al. [84]. They emphasized that process intensification technologies such as engineered catalysts, on-site heat production/removal and product purification can not only allow precise control of reaction and heat/mass transfer rates, but also help to optimize the operation conditions and, consequently, improve overall efficiency and mitigate the requirement for materials and capital investment. On the other side, the various difficulties and risks involved in storage, distribution and refueling of hydrogen makes fuel processors the best way to feed FC.

During the last decades, the need to obtain more efficient processes and improve benefits has been growing exponentially. At the beginning of this millennium, the main focus was set on obtaining improvements in the energy efficiencies. Nowadays, restrictions to avoid the continuous growing of environment deterioration has become a matter of vital importance, mainly due to global warming consequences. The main factors responsible for this fact is the huge amount of CO_2 present, and continuously accumulating, in the atmosphere. One of the contributors to this situation is the automobile industry, a great fossil fuel consumer. In this regard, FC stacks represent an excellent alternative, because they can work with greater efficiencies than internal combustion engines, environmental friendliness and lower emissions, ideally H_2O and heat.

The ideal situation for a fuel cell will be to feed it with pure hydrogen, which carries a lot of implementation issues, from the lack of infrastructure to produce and distribute enough amounts of fuel up to the difficulties and dangers of transportation and storage. To overcome these difficulties, several solutions have been proposed, one of them being the operation of cells capable of being fed directly with gasoline or diesel fuels, but sacrificing the efficiency. The most plausible solution is to process a fuel onboard to produce a hydrogen-rich synthesis gas. Many fuels are being studied such as gasoline, diesel or methane. Although they present good operability and acceptable efficiencies, they do not agree with the premise of substitution of non-renewable raw materials, because they proceed from fossil origins. In this context, bio-ethanol seems to be the best candidate as tomorrow's fuel, since it does not have the toxicity of methanol. Both of them are relatively easy and cheap to produce, although methanol has a more beneficial H/C ratio. Additionally, bio-ethanol has a positive net energetic balance. It can be measured through a relation between the obtained energy during the combustion of ethanol and the required energy to produce it. This refers to the entire cycle of life from raw materials extraction, required inputs, transport, up to the process of transformation into bio-ethanol. The value of this index has been stated by [85] in 8 for the production from sugarcane, and 6 from lignocellulosic residue materials. Obviously, if it is compared with fossil fuels that can achieve values between 10 and 23 [86], it is a poor fuel. But of course, the energy required to recapture CO_2 and turn it into a hydrocarbon again is enormous and would turn the balance into a small fraction.

In order to improve the efficiency of the processor, it is necessary to integrate and reutilize the heat contained in the flows. The flows entering and exiting the system need to be reconditioned in order to keep the process in optimal operating conditions, normally, entering substances must be heated and intermediate products cooled down, also exit gases contain heat that could be used. This heat exchange is not trivial and not always possible, a heat exchanger network capable of doing this should have a large number of operating units.

5.1 Bio-Ethanol Production

The advantages of ethanol, compared to other feedstocks, are the high hydrogen content, the availability, the non-toxicity and the storage and handling safety. Ethanol can be produced renewably by fermentation of biomass sources, such as energy plants, agroindustrial wastes, forestry residue materials and organic fraction of municipal solid waste. Ethanol produced in this way is called bio-ethanol [72]. It is produced by yeast fermentation converting glucose into ethanol and carbon dioxide:

$$C_6H_{12}O_6 \overset{\text{fermentation}}{\longrightarrow} 2CH_3CH_2OH + 2CO_2. \qquad (42)$$

Sugarcane, switchgrass, potatoes, corns and other starch-rich materials can be effectively converted into ethanol by fermentation. This method is mature and easy to control. However, the cost of ethanol production is rather high mainly due to the expensive feedstock plantation. As lignocellulosic biomass accounts for about 50% of biomass in the world, there is increasing interest in producing ethanol from lignocellulosics materials such as wood, agricultural and forest residues [72, 87]. However, production of ethanol from lignocellulosics is more difficult than from sugarcane or starch-rich materials because of its more complex molecular structure, which contains cellulose, hemi-cellulose and lignin.

5.2 Bio-Ethanol Reforming

Ethanol is a very good fuel cell energy source because of its high efficiency and concentration of H_2 that can be obtained, also because of its environmental friendliness. The usual methods of converting bioethanol are catalytic partial oxidation and steam reforming, the latter is considered because high concentrations of hydrogen can be obtained to feed the fuel cell stack. There is considerable research on the reaction pathways and thermodynamics of ethanol steam reforming. Hydrogen production varies significantly with different reaction pathways. In order to maximize hydrogen production, it is crucial to ensure sufficient supply of steam and minimize ethanol dehydration and decomposition [72]. Besides, catalysts play a crucial role in the conversion of ethanol to hydrogen. However, each catalyst induces different pathways and, therefore, the catalyst plays a crucial role in ethanol-steam reforming for hydrogen production.

Active catalysts should maximize hydrogen selectivity and inhibit coke formation as well as CO production. Haryanato et al. [88] analyzed the advantages of obtaining hydrogen by the ethanol-steam reforming reaction, first comparing the current H_2 production processes. They studied the different reaction pathways that can occur during ethanol-steam reforming over metal catalysts and presented the most common groups of catalysts used with their respective highlights. In addition, they selected the best-performing catalysts for the steam reforming and the Water Gas Shift (WGS) reactions. Steam reforming process engineering aspects are well discussed in [89], pointing to thermodynamic considerations and then making an overview on the published literature on catalysts. They made an extensive analysis on the process considerations and operating conditions to favor hydrogen production, such as high temperature, low pressure and high water-to-ethanol ratios.

Ethanol reforming processes for hydrogen production can be generally classified into two groups: steam reforming and autothermal reforming. Compared with autothermal reforming, steam reforming of bio-ethanol has received more attention due to its relatively higher conversion efficiency [72]. The basic bio-ethanol reforming reaction to produce hydrogen is as follows [89]:

$$C_2H_5OH + 3CO_2 \longrightarrow 2CO_2 + 6H_2 \qquad (\Delta H^0_{298} = 174 \text{ kJ mol}^{-1}). \tag{43}$$

6 Control and Diagnosis of the Fuel Processor System

According to the considerations given above the definition of the complete control strategy integrated to the fault diagnosis for the BPS with PEMFC will be thoroughly analyzed in the second part of this book. The main concepts to attack this problem are taken from the plantwide control and statistical methods such as principal component analysis (PCA). Hence, in the following two subsections the main philosophy to be applied on the overall system (BPS with PEMFC) so as to achieve the process objectives, will be introduced.

6.1 Multivariable Control of the Bio-Ethanol Processor System

The traditional approach to developing a new process has been to perform the design and control analysis sequentially. First, the design engineer constructs a steady-state process flowsheet, with particular structure, equipment, design parameters and operating conditions.

The objective is to operate the plant in the face of potentially known and unknown disturbances, production rate changes and transitions from different kind of demands generated by a medium automobile in different standard circuits. Luyben et al. [90] remarked that this staged approach has long been recognized as deficient, although it is defensible from a certain perspective. For example, it would be difficult for the control perspective to specify the instrumentation and the distributed control system (DCS) without knowing exactly what process it was intended for. Similarly, it would make no sense for the process engineers to request a control system design for all those flowsheets that were considered but rejected on the basis of steady-state economics alone. However, this staged approach can result in missed opportunities because of the close connection between process design and controllability. How a process is designed fundamentally determines its inherent controllability, which means qualitatively how well the process rejects disturbances and how easily it moves from one operating condition to another. Based on these last considerations, the final flowsheet of the BPS with PEMFC, presented by Francesconi et al. [91], is deeply studied for analyzing its controllability and proposing a plantwide control strategy based on two different kinds of philosophies. As a result of this stage it could be concluded the necessity to modify or not the proposed flowsheet according to its associated controllability.

Plantwide process control involves the systems and strategies required to control an entire chemical plant consisting of many interconnected unit operations. It means that chemical process control must move beyond the sphere of unit operations into the realm of viewing the plant as a whole system. The chemical industry is mature enough for the development of a systematic plantwide control design procedure. The technology, insight and understanding have reached a state where general guidelines can be presented. Several computer software, needed for plantwide at steady state and for dynamic simulations, are commercially available. While linear methods are very useful to analyze control concepts, the final evaluation of any plantwide control structure requires rigorous nonlinear dynamic simulations, and not only linear transfer function analysis. Given a complex, integrated process and a diverse assortment of equipment, it is necessary to devise a proper logic, instrumentation and strategies to operate the plant safely and achieve its design objectives. In this book, an integral strategy is presented based on the consideration that the field of process control and safety has six parts of a continuous spectrum. Each part is important, can be economically significant and strongly interacts in some manner with the others. It is demonstrated through the application on this open case study of the BPS with PEMFC.

The first part of the spectrum is the design of the process itself. Sometimes the flowsheet structure can be changed, use different design parameters and employ different types of process equipment to produce a plant that can be controlled more easily than other alternatives. This part is widely detailed in the work of Francesconi et al. [91] and reproduced here. It is well recognized that, at this level, a good trade-off between process design and control strategy can potentially have an enormous economic impact. One of the parts must deal with the control hardware and infrastructure required to operate a plant. It is necessary to assemble the proper types of control valves and process measurements (for temperature, flow, pressure, composition, etc.). These are the sensory devices of the plant and are essential for any control system to function. Any control strategy, no matter how clever, will have severe difficulties without the right measurements and valves in the process. They will be closely related with the main objectives of the process. The next part is the determination of the control system structure. Accounting the available measurements and how they should be paired with the proper manipulated variables. The control structure is vitally important because a poor strategy will result in poor performance no matter what type of control algorithm is used or how much effort is applied to tune it.

The following part of the spectrum deals with the controller algorithms, the type of controller to use (proportional, integral, derivative, multivariable, nonlinear, model predictive, etc.) must be decided. The next part involves controller tuning. While this task is often performed by using heuristics and experience, it can sometimes be a nontrivial exercise for certain loops. In this book the methodology based on the internal model control (IMC) theory given in [92] for the control loop, from which controller settings can be efficiently calculated, is used. The final step involved is which is the best measurement selection to ensure the fault detectability when a fault diagnosis system is working online. This is a really important task, linked with the already installed sensors so as to reduce the initial investment and guaranteeing safety of the process. This subject is discussed in detail in the following subsection. There is little information or guidance in the literature or in process control textbooks (both introductory and advanced) about how to develop an effective control structure for an entire complex chemical plant. This is one of the main subjects of this book specifically dedicated to the BPS with PEMFC and automobile requirements. That is, apply a new methodology on this kind of a challenging process. This systematic approach was previously tested with success on other complex and highly integrated benchmark chemical plants (see [93–95]). Therefore, it gives enough confidence about the final results to be obtained for the BPS with PEMFC process.

Part II of this book treats extensively the fuel processor systems with bio-ethanol with PEM fuel cell for automobile applications. In addition, it addresses questions such as: What issues are necessary to consider? What is of essential importance within this immense amount of detail? How does the dynamic behavior of the interconnected plant differ from that of the individual unit operations? What, if anything, is necessary to model or test? How is it more convenient to begin? Therefore, the fundamental ideas of control system synthesis are explained through two different methodologies. In Chap. 9, the one consisting of a general heuristic design procedure that generates an effective plantwide base-level regulatory control structure is detailed. The other, more systematic and integrating optimal sensor location with control for then improving some safety issues for the entire, complex process BPS flowsheet is detailed in Chap. 12.

6.2 Improvements on Fault Diagnosis of the Bio-Ethanol Processor System

In large-scale processes, monitoring systems play an important role in supervising the state of the plant. Hundreds and thousands of variables must be handled in each sample time. The increasing demand for obtaining a more efficient and rentable process generates complex control systems and very interconnected plants. In this context, it is extremely difficult to achieve a suitable monitoring without the support of a decision-maker system [96–98].

PCA-based monitoring systems are widely used in industrial processes as well as in academic research [99–101] due to their excellent properties for handling noise and large data bases. However, there are still some drawbacks that justify to propose some improvements so as to get better results. One of the drawbacks that will be considered here is that PCA does not account for time-delays present in data. It causes an inefficient dimensionality reduction of process variables and the subsequent poor monitoring and disturbance detection performance. The other problem analyzed here is, which is the minimum number of signals to be chosen for developing an efficient PCA monitoring system? This question is usually not addressed because the tendency is to use all the available information sources for developing the PCA model. Therefore, in part II of this book, mainly in Chap. 13 and Chap. 14, two methodologies are developed and applied in the BPS with PEMFC process which offer alternative solutions to the drawbacks mentioned above.

In Chap. 13 the new approach to improve the PCA-based monitoring systems performance consists of its optimal design based on maximizing the abnormal events detection and minimizing the investment costs related to the number of sensors to be bought. For this purpose, the detectability indexes are used for determining which set of variables is enough to achieve the best fault classification. Particularly, here it is proposed to do that based on combined statistic [96–98, 102]. Additionally, due to the great problem dimensionality a good solution can be found by using genetic algorithm (GA). Thus, the objective function within the GA also considers a cost that penalizes the use of new measurement devices that are not in the current control structure. In other words, this approach tries to avoid the use of additional sensors for monitoring purposes so as to prevent increases in the hardware investment cost.

In Chap. 14 the new method named genetic algorithm-based delay adjusted PCA (GA-DAPCA), is proposed to improve the PCA performance in the presence of time delays between process signals. The optimization via GA is performed in two loops. The first loop finds the shift between variables that minimizes the number of principal components to be considered as common cause variance, the second maximizes the variance contained in the previously selected principal component dimensions. In Chap. 13 and Chap. 14, the same faults are considered and tested via dynamic simulation. Through the obtained results it is possible to evaluate the potentiality of the improvements introduced here.

7 Conclusions

The goal of this chapter is to present a picture of the main topics considered in this book and the corresponding justification through an updated review of the literature. It is mainly dedicated to some useful tools for studying the entire system from hydrogen production through bio-ethanol reforming to power generation in the fuel cell stack. In this context, this chapter discusses about some typical problems in PEM fuel cells-based systems and proposes alternative ways of operation through conventional and advanced control techniques in a safe manner. Aspects such as efficient energy management strategies are also performed in order to command the power split in a hybrid system such as the fuel cell stack coupled with a supercapacitor bank.

This chapter serves as an introduction to the study that is carried out throughout the book. It is done having in mind that an apparently appropriate control scheme for the PEM fuel cell may actually lead to an inoperable plant when it is connected to other unit operations in a process with recycle streams and energy integration. The main objective was to present alternative designs of control system that provides basic regulatory behavior of the hydrogen production process; i.e., the plant would be at the wanted operating point despite disturbances. Above this regulatory structure can then be built fault diagnosis systems to improve plant safety performance.

The emphasis is on providing the most adequate tools to achieve accurate, practical and useful solutions for the problems reported by a great number of researchers worldwide. Accordingly, these tools are supported by several fields such as plant-wide process control and fault diagnosis, computer-aided, artificial intelligence, expert systems and experimental results, among others. Hence, the subjects discussed in this chapter represent an invitation to the readers to get more insight into them.

References

1. Holland B, Zhu J, Jamet L (2001) Fuel cell technology and application. J Power Source 100:171–187
2. Larminie J, Dicks A (2003) Fuel cell systems explained, 2nd edn. Wiley and Sons, New York, USA
3. Bose T, Bénard P (2003) Hydrogen as an energy vector. In: Proceedings of the power engineering, 2003 large engineering systems conference on, pp 196–199
4. Conte M (2009) Energy: Hydrogen economy. In: Encyclopedia of electrochemical power sources. Elsevier, Amsterdam, pp 232–254
5. EG & G Technical Services, Inc. Science Applications International Corporation (2002) Fuel Cell Handbook, 6th Edn. US Department of Energy, Department of Energy, Office of Fossil Energy and National Energy Technology Laboratory, Morgantown
6. Barbir F (2005) PEM fuel cells: theory and practice. Elsevier, Burlington, MA, USA
7. Yang W, Bates B, Fletcher N, Pow R (1998) Control challenges and methodologies in fuel cell vehicles development. Fuel cell technology for vehicles, pp 249–256
8. Laughton MA (2002) Fuel cells. Power Eng J 16:37–47
9. Irvine JTS (2004) The Bourner lecture: power sources and the new energy economy. J Power Source 136(2):203–207
10. Lin BYS, Kirk DW, Thorpe SJ (2006) Performance of alkaline fuel cells: a possible future energy system?. J Power Source 161(1):474–483
11. De Geeter E, Mangan M, Spaepen S, Stinissen W, Vennekens G (1999) Alkaline fuel cells for road traction. J Power Source 80(1–2):207–212
12. Lan R, Xu X, Tao S, Irvine JTS (2010) A fuel cell operating between room temperature and 250 c based on a new phosphoric acid based composite electrolyte. J Power Source 195(20):6983–6987
13. Sun C, Stimming U (2007) Recent anode advances in solid oxide fuel cells. J Power Source 171(2):247–260
14. Minh NQ (2005) Ceramic fuel cells. J Am Ceram Soc 76(3):563–588
15. Appleby AJ (2009) Applications-stationary-fuel cells. In: Jürgen Garche, Encyclopedia of electrochemical power sources. Elsevier, Amsterdam, pp 76–107
16. Chu H-S, Tsau F, Yan Y-Y, Hsueh K-L, Chen F-L (2008) The development of a small pemfc combined heat and power system. J Power Sources 176(2):499–514. Selected Papers presented at the10th ULM ElectroChemical Days-W. Tillmetz, J. Lindenmayer, 10th Ulm ElectroChemical Days.

17. Jörissen L (2009) Applications-stationary | residential energy supply: fuel cells. In: Jürgen Garche, Encyclopedia of electrochemical power sources. Elsevier, Amsterdam, pp 108–123
18. Cabot P.-L, Alcaide F, Brillas E (2009) Applications-stationary | cogeneration of energy and chemicals: fuel cells. In: Jürgen Garche, Encyclopedia of electrochemical power sources. Elsevier, Amsterdam, pp 146–156.
19. Campanari S, Manzolini G, Garciadela Iglesia F (2009) Energy analysis of electric vehicles using batteries or fuel cells through well-to-wheel driving cycle simulations. J Power Source 186(2):464–477
20. Helmolt R, Eberle U (2007) Fuel cell vehicles: status 2007. J Power Source 165:833–843
21. Whitehouse N, Whitehouse S, Huss A (2009) Applications-transportation | buses: fuel cells. In: Encyclopedia of electrochemical power sources. Elsevier, Amsterdam, pp 193–202
22. Dyer CK (2002) Fuel cells for portable applications. J Power Source 106(1–2):31–34
23. Kundu A. Jang JH (2009) Applications-portable | portable devices: fuel cells. In: Encyclopedia of electrochemical power sources. Elsevier, Amsterdam, pp 39–45.
24. Benz U, Busche D, Lutterbeck D (2009) Applications-stationary | uninterruptible and back-up power: fuel cells. In: Encyclopedia of electrochemical power sources. Elsevier, Amsterdam, pp 135–145.
25. Henne RH, Friedrich KA (2009) Applications-transportation | auxiliary power units: fuel cells. In: Encyclopedia of electrochemical power sources. Elsevier, Amsterdam, pp 157–173.
26. Pukrushpan J, Peng H, Stefanopoulou A (2004) Control-oriented modelling and analysis for automotive fuel cell systems. J Dyn Syst 126:14–25
27. Pukrushpan J, Stefanopoulou A, Peng H (2004) Control of fuel cell breathing: initial results on the oxygen starvation problem. IEEE Control Syst Mag 24:30–46
28. Blanke M, Kinnaert M, Lunze J, Staroswiecki M (2006) Diagnosis and fault-tolerant control. Springer second edition, Heidelberg, Germany
29. Puig V, Quevedo J, Escobet T, Morcego B, Ocampo C (2004) Control tolerante a fallos (parte II): mecanismos de tolerancia y sistema supervisor. Revista Iberoamericana de Automatica e Informatica Industrial 1(2):5–21
30. Golbert J, Lewin D (2005) Fuel efficient model predictive control of pem fuel cells. In: Proceedings of 16th World Congress, Prague
31. Golbert J, Lewin D (2004) Model-based control of fuel cells: (1) regulatory control. J Power Source 135:135–151
32. Grujicic M, Chittajallu KM, Law EH, Pukrushpan JT (2004) Model-based control strategies in the dynamic interaction of air supply and fuel cell. Proc Inst Mech Eng, Part A: J Power Energy 218(7):487–499
33. Grujicic M, Chittajallu KM, Pukrushpan JT (2004) Control of the transient behaviour of polymer electrolyte membrane fuel cell systems. Proc Inst Mech Eng, Part D: J Automob Eng 218(11):1239–1250
34. Vahidi A, Peng A (2004) Model predictive control for starvation prevention in a hybrid fuel cell system. Am Control Conf. Proc 2004(1):834–839
35. Liu Z, Yang L, Mao Z, Zhuge W, Zhang Y, Wang L (2006) Behavior of PEMFC in starvation. J Power Source 157(1):166–176
36. Pukrushpan J (2003) Modelling and control of fuel cell systems and fuel processors. PhD thesis, University of Michigan
37. Varigonda S, Pukrushpan JT, Stefanopoulou AG (2003) American institute of chemical engineers challenges in fuel cell power plant control: the role of system level dynamic models. American institute of chemical engineers
38. Serra M, Aguado J, Ansade X, Riera J (2005) Controllability analysis of decentralized linear controllers for polymeric fuel cells. J Power Source 151:93–102
39. Pukrushpan JT, Stefanopoulou AG, Peng H (2004) Control of fuel cell power systems: principles, modeling, analysis and feedback design. Springer, Godalming, Surrey, UK

40. Serra M, Husar A, Feroldi D, Riera J (2006) Performance of diagonal control structures at different operating conditions for polymer electrolyte membrane fuel cells. J Power Source 158(2):1317–1323
41. Friedman D Moore R (1998) PEM fuel cell system optimization. In: Proceedings Electrochemical Society, vol 27, pp 407–423
42. Wishart J, Dong Z, Secanell M (2006) Optimization of a PEM fuel cell system based on empirical data and a generalized electrochemical semi-empirical model. J Power Source 161(2):1041–1055
43. Venkatasubramanian V, Rengaswamy R, Yin K, Kavuri SN (2003) A review of process fault detection and diagnosis. Part I: quantitative model-based methods. Comput Chem Eng 27(3):293–311
44. Venkatasubramanian V, Rengaswamy R, Kavuri SN (2003) A review of process fault detection and diagnosis. Part II: qualitative models and search strategies. Comput Chem Eng 27(3):313–326
45. Venkatasubramanian V, Rengaswamy R, Kavuri SN, Yin K (2003) A review of process fault detection and diagnosis. Part III: process history based methods. Comput Chem Eng 27(3):327–346
46. Jeong K, Oh B (2002) Fuel economic and life-cycle cost analysis of a fuel cell hybrid vehicle. J Power Source 105:58–65
47. Rodatz P, Paganelli G, Sciarretta A, Guzzella L (2004) Optimal power management of an experimental fuel cell/supercapacitors-powered hybrid vehicle. Control Eng Pract 13:41–53
48. Markel T, Zolot M, Wipke K, Pesaran A (2003) Energy storage system requirements for hybrid fuel cell vehicles. In: Proceedings of the 3rd international advanced automotive battery conference, Nice, France, June
49. Gigliucci G, Petruzzi L, Cerelli E, Garzisi A, La Mendola A (2004) Demonstration of a residential CHP system based on PEM fuel cells. J Power Source 131(1–2):62–68
50. Uzunoglu M, Onar O, Alam M (2007) Dynamic behavior of PEM FCPPs under various load conditions and voltage stability analysis for stand-alone residential applications. J Power Source 168(1):240–250
51. El-Sharkh M, Rahman A, Alam M, Byrne P, Sakla A, Thomas T (2004) A dynamic model for a stand-alone PEM fuel cell power plant for residential applications. J Power Source 138:199–204
52. Uzunoglu M, Alam MS (2006) Dynamic modeling, design, and simulation of a combined PEM fuel cell and ultracapacitor system for stand-alone residential applications. Energy Convers IEEE Trans 21(3):767–775
53. Ahluwalia R, Wang X, Rousseau A, Kumar R (2003) Fuel economy of hydrogen fuel cell vehicles. J Power Source 130:192–201
54. Rajashekara, K (2000) Propulsion System Strategies for Fuel Cell Vehicles, Fuel Cell Power for Transportation (Part C & D), pp 1–9. doi: 10.4271/2000-01-0369. ISSN 0148-7191
55. Rajashekara K (2003) Power conversion and control strategies for fuel cell vehicles. Ind Electron Soc 3:2865–2870
56. Santi E, Franzoni D, Monti A, Patterson D, Barry N (2002) A fuel cell based domestic uninterruptible power supply. Applied power electronic conference and exposition, pp 605–613
57. Choi DK, Lee BK, Choi SW, Won CY, Yoo DW (2005) A novel power conversion circuit for cost-effective battery-fuel cell hybrid systems. J Power Source 152(1):245–255
58. Drolia A, Jose P, Mohan N (2003) An approach to connect ultracapacitor to fuel cell powered electric vehicle and emulating fuel cell electrical characteristics using switched mode converter. Ind Electron Soc 1:897–901
59. Atwood P, Gurski S, Nelson DJ, Wipke KB (2001) Degree of hybridization modeling of a fuel cell hybrid electric sport utility vehicle. SAE Transactions. 110(3):93–100

60. Friedman D (1999) Maximizing direct-hydrogen PEM fuel cell vehicle efficiency-is hybridization necessary? Fuel Cell Power for Transportation. Society of Automotive Engineers, 400 Commonwealth Dr, Warrendale
61. Akella S, Sivashankar N, Gopalswamy S, Inc E, Plymouth M (2001) Model-based systems analysis of a hybrid fuel cell vehicle configuration. Am Control Conf. Proc, 3
62. Moore R, Hauer K, Ramaswamy S, Cunningham J (2006) Energy utilization an efficiency analysis for hydrogen fuel cell vehicles. J Power Source 159:1214–1230
63. Thounthong P, Raël S, Davat B (2006) Control strategy of fuel cell/supercapacitors hybrid power sources for electric vehicle. J Power Source 158(1):806–814
64. Paganelli G, Guezennec Y, Rizzoni G. (2002) Optimizing control strategy for hybrid fuel cell vehicle. pp 71–79
65. Rodatz P, Garcia O, Guzzella L, Büchi F, Bärschi M, Tsukada T, Dietrich P, Kötz R, Schreder G, Woukan A (2003) Performance and operational characteristics of hybrid vehicle powered by fuel cell and supercapacitors. Fuel cell power for transportation, pp 77–84
66. Kötz R, Müller S, Bärtschi M, Schnyder B, Dietrich P, Büchi FN, Tsukada A, Scherer GG, Rodatz P, Garcia O, et al (2001) Supercapacitors for peak-power demand in fuel-cell-driven cars. ECS Electro-Chemical Society, 52nd Meeting, San Francisco, pp 2–7
67. Caux S, Lachaize J, Fadel M, Shott P, Nicod L (2005) Modelling and control of a fuel cell system and storage elements in transport applications. J Process Control 15:481–491
68. Caux S, Lachaize J, Fadel M, Shott P, Nicod L (2005) Energy management of fuel cell system and supercaps elements. In Proceedings of the 16*th* IFAC World Congress, Prague
69. Jung J, Lee Y, Loo J, Kim H (2003) Power control strategy for fuel cell hybrid electric vehicles. Fuel cell for transportation, pp 201–205
70. Kim M, Peng H (2007) Power management and design optimization of fuel cell/battery hybrid vehicles. J Power Source 165:819–832
71. Schiffer J, Bohlen O, de Doncker RW, Sauer DU (2005) Optimized energy management for fuel cell-super cap hybrid electric vehicles VPP track 4: energy storage components/systems. Vehicle power and propulsion, 2005 IEEE Conference, pp 716–723
72. Meng Ni, Dennis Y.C. Leung, Michael K.H. Leung (2007) A review on reforming bio-ethanol for hydrogen production. International Journal of Hydrogen Energy, 32(15):3238–3247 International symposium on solar-hydrogen-fuel cells 2005
73. DOE Hydrogen Program. Hydrogen Production. Online, 2010. http://www.hydrogen.energy.go.
74. Department of Energy EEUU. Fuel cell technologies programa. Web EERE, 2010. http://www.eere.energy.gov/hydrogenandfuelcells.
75. Fuel Cell Today. 35 mpa high pressure hydrogen without a compressor. Web, 2004. http://www.fuelcelltoday.com.
76. Ovonic. Advances of solid hydrogen storage systems. Web, 2010. http://www.ovonic.com.
77. Fractal Carbon. Sodium borohydride fuelled hydrogen generator. Web, 2010. http://www.fractalcarbon.com/HGen-E20manual.pdf.
78. Cheng X, Shi Z, Glass N, Zhang L, Zhang J, Song D, Liu ZS, Wang H, Shen J (2007) A review of PEM hydrogen fuel cell contamination: impacts, mechanisms, and mitigation. J Power Source 165(2):739–756
79. Ho D Lightner V (2004) Future directions for doe fuel processing r&d: results of the on-board fuel processing go/no-go decision for doe's office on hydrogen, fuel cells and infrastructure technologies. In: Fuel cell seminar, p 37
80. Asprey SP et al (1999) Kinetic studies using temperature-scanning: the steam-reforming of methanol. Appl Catalys A: Gen 179(1–2):51–70
81. Amphlett J, Mann R, Peppley B, Roberge P, Rodrigues A (1996) A model predicting transient responses of proton exchange membrane fuel cells. J Power Source 61:183–188
82. Liguras D, Goundani K, Verykios X (2004) Production of hydrogen for fuel cells by catalytic partial oxidation of ethanol over structured ni catalysts. J Power Source 130(1–2):30–37

83. Evans R, Czernik S, Magrini-Bair K (2004) DOE hydrogen program. FY2004 Progress Report, 65.
84. Qi A, Peppley B, Karan K (2007) Integrated fuel processors for fuel cell application: a review. Fuel Process Technol 88(1):3–22
85. Sanchez OJ, Cardona CA (2005) Producción biotecnológica de alcohol carburante i: obtención a partir de diferentes materias primas. INCI 30(11):671–678
86. Prakash R, Henham A, Krishnan Bhat I (1998) Net energy and gross pollution from bioethanol production in India. Fuel 77(14):1629–1633
87. Galbe M, Zacchi G (2002) A review of the production of ethanol from softwood. Appl Microbiol Biotechnol 59(6):618–628
88. Haryanto A, Fernando S, Murali N, Adhikari S (2005) Current status of hydrogen production techniques by steam reforming of ethanol: a review. Energy Fuels 19(5):2098–2106
89. Vaidya PD, Rodrigues AE (2006) Insight into steam reforming of ethanol to produce hydrogen for fuel cells. Chem Eng J 117:39–49
90. Luyben ML, Luyben WL (1997) Essentials of process control. McGraw-Hill, New York, USA
91. Francesconi JA, Mussati MC, Mato RO, Aguirre PA (2007) Analysis of the energy efficiency of an integrated ethanol processor for PEM fuel cell systems. J Power Source 167(1):151–161
92. Rivera DE (2007) Una metodología para la identificación integrada con el diseño de controladores IMC-PID. Revista Iberoamericana de Automática e Informática Industrial 4(4):129–134
93. Molina G, Zumoffen D, Basualdo M (2009) A new systematic approach to find plantwide control structures. Comput Aided Chem Eng 27:1599–1604
94. Zumoffen D, Basualdo M (2009) Optimal sensor location for chemical process accounting the best control configuration. Comput Aided Chem Eng 27:1593–1598
95. Zumoffen D, Basualdo M, Ruiz J (2009) Optimal multivariable control structure design for chemical plants. AIChE annual meeting, Nashville
96. Zumoffen D, Basualdo M (2008) From large chemical plant data to fault diagnosis integrated to decentralized fault-tolerant control: pulp mill process application. Ind Eng Chem Res 47(4):1201–1220
97. Zumoffen D, Basualdo M (2008) Improvements in fault tolerance characteristics for large chemical plants part I: waste water treatment plant with decentralized control.. Ind Eng Chem Res 47(15):5464–5481
98. Zumoffen D, Basualdo M, Molina G (2008) Improvements in fault tolerance characteristics for large chemical plants part II: pulp mill process with model predictive control.. Ind Eng Chem Res 47(15):5482–5500
99. Kourti T, MacGregor J (1995) Process anlysis, monitoring and diagnosis, using multivariable projection methods. Chemom Intell Lab Sys 28:3–21
100. Martin EB, Morris AJ, Lane S (2002) Monitoring process manufacturing performance. IEEE Control Sys Mag 22:26–39
101. Lane S, Martin EB, Morris AJ, Gower P (2003) Application of exponentially weighted principal component analysis for the monitoring of a polymer film manufacturing process. Transac Inst Meas Control 1(25):17–35
102. Yue HH, Qin SJ (2001) Reconstruction-based fault identification using a combined index. Ind Eng Chem Res 40:4403–4414

Description of PEM Fuel Cells System

Diego Feroldi and Marta Basualdo

Abstract This chapter provides a description of polymer electrolyte membrane (PEM) fuel cell-based systems and different modeling approaches. First, it shows the structure of a single cell, the advantages and disadvantages of this type of fuel cell, the expressions of the generated voltage and the efficiency, and the generic structure of a generation system based on PEM fuel cell. Second, the chapter provides a review of the principal models presented in the literature to describe the behavior of the system. Different types of PEM fuel cell models are presented, focusing on dynamic models suitable for control purposes. Particularly, this chapter describes in detail the dynamic model used as a base to represent the system in the subsequent chapters of the book. Then, the described model is used to study the optimal operation of a fuel cell at different loads, showing the benefits of an optimal operation in terms of hydrogen reduction and greater peak power.

1 Introduction

The polymer electrolyte membrane fuel cell (PEMFC), also known as proton exchange membrane fuel cell, takes its name from the type of electrolyte: a polymeric membrane with high proton conductivity when the membrane is conveniently

D. Feroldi (✉) · M. S. Basualdo
CAPEG-CIFASIS-(CONICET-UNR-UPCAM),
27 de Febrero 210 bis, S2000EZP Rosario, Argentina
e-mail: feroldi@cifasis-conicet.gov.ar

M. S. Basualdo
UTN-FRRo, Zeballos 1341, S2000BQA Rosario, Argentina
e-mail: basualdo@cifasis-conicet.gov.ar

D. Feroldi
DCC-FCEIA-UNR, Pellegrini 250, S2000BTP Rosario, Argentina

M. S. Basualdo et al. (eds.), *PEM Fuel Cells with Bio-Ethanol Processor Systems*, Green Energy and Technology, DOI: 10.1007/978-1-84996-184-4_2,

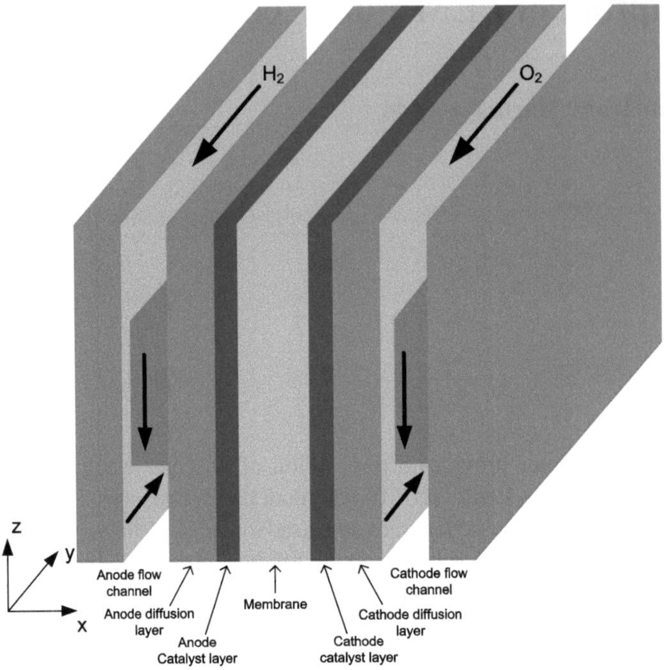

Fig. 1 Three-dimensional schematic diagram of a fuel cell

hydrated [1]. At the moment, the most common polymer used in these types of cells is the Nafion developed by *Du Pont* (USA), which is fabricated with chemically stabilized perfluorosulfonic acid copolymer [2].

1.1 Basic PEM Fuel Cell Structure

Basically, the physical structure of a PEMFC consists of seven components, according to Fig. 1 [3]: feeding channels, diffusion layer, and catalytic layer in the anode; membrane; catalytic layer, diffusion layer, and feeding channels in the cathode. The PEMFC combines in a very compact unit the electrodes and the electrolyte. This structure, well known as membrane electrode assembly (MEA), is not thicker than a few hundred microns. It is the heart of the fuel cell and is fed with hydrogen and oxygen,[1] generating electrical power with a power density of around $1\,\mathrm{W\,cm^{-2}}$ [5].

The polymeric solid electrolyte forms a thin electronic insulator and a barrier for gases between both electrodes, allowing fast proton transport and high current

[1] Usually, the fuel cell is fed with atmospheric air instead of pure oxygen. The oxygen mole fraction in atmospheric air is 0.21 [4].

density. The solid electrolyte has the advantage, as opposed to those of liquid type, that allows the FC to operate in any spatial position [1].

The electrodes consist of a catalytic layer of great superficial area on a substratum of coal, permeable to gases. Electrocatalyst materials are necessary to obtain a good operation, increasing the speed of the chemical reaction. In this way, the gases can react with a lower energy of activation, allowing the reaction to take place at a lower temperature [3]. The electrocatalyst used in PEMFC is platinum, which is one of the major drawbacks of this technology because of its high cost.

However, there are research advances of high temperature PEM fuel cells (HT-PEMFCs) in several fields because there are several reasons for operating at temperatures above 100°C [6]. First, the electrochemical kinetics for the reactions in cathode and anode are enhanced. Second, the water management issue can be simplified because there would be no liquid water. Third, the cooling system is simplified due to the increased temperature gradient between the fuel cell stack and the coolant. Fourth, the waste heat can be exploited using cogeneration. Fifth, the tolerance to CO is increased allowing the use of lower quality reformed hydrogen. Unfortunately, the area of HT-PEMFCs is incipient and still needs much research to be implemented in commercial applications.

1.2 Advantages and Disadvantages of PEM Fuel Cells

The main advantage of PEM fuel cells is their high efficiency compared with other energy conversion devices [7]. This allows the efficiency of a fuel cell vehicle using direct-hydrogen FC[2] to be twice that in a gasoline vehicle [8, 9]. Moreover, unlike the internal combustion engines where the efficiency is maximum with the highest loads, the FC efficiency is also high with partial loads. This is advantageous because in typical driving conditions, like urban and suburban scenarios, most of the time the vehicle is demanding a small fraction of the nominal FC power [10]. Thus, an FC vehicle will be working mostly at high efficiencies. At the same time, using direct-hydrogen FC, the local emissions problem in densely urban areas can be eliminated.

Another important advantage of PEMFC, in contrast to other types of fuel cells, is the low operation temperature (below 80°C) [11], allowing to reach the operation point quickly. In addition, the cost of the materials is smaller than for the high temperature fuel cells (except the catalyst, which is based on platinum) and their operation is safer. All these characteristics turn PEMFC particularly appropriate for applications in vehicles. Nevertheless, it is necessary to use better, and more economic, catalyst so that the reaction occurs at lower temperatures.

The main disadvantage of fuel cells is their high cost and the high production cost of hydrogen. Hydrogen is preferred because of the fast electrochemical reaction, and

---

[2] Direct-hydrogen FC refers to an FCS that is directly fed with hydrogen from a pressurized tank opposite to the case where the hydrogen is produced with an on-site reformer.

its high specific energy.[3] Nevertheless, as it was mentioned, hydrogen is not a primary fuel. Usually, it is produced from hydrocarbon reforming or water electrolysis [1, 13]. The use of electrolysis is advisable especially when some type of renewable energy is used, avoiding fossil fuel use. It is expected that the cost of fuel cells and hydrogen will diminish with the progress in technology. Thus, hydrogen has the possibilities of becoming an alternative to fossil fuels with the joint use of renewable energies.

1.3 Fuel Cell Voltage

The standard potential E^0 is a quantitative measurement of the maximum cell potential, i.e., the open circuit voltage. For a hydrogen–oxygen cell, in which there is a transfer of two electrons by each water molecule, $E^0 = 1.229\,\mathrm{V}$ if the produced water is in liquid state and $E^0 = 1.18\,\mathrm{V}$ if the produced water is in gaseous state [13]. These values correspond to normalized conditions: cell temperature (T_{fc}) equal to 298.5 K and partial pressures of oxygen (p_{O_2}) and hydrogen (p_{H_2}) equal to 1 atm. In the work of Amphlett et al. [14], the following expression of E^0 is given, depending on the temperature and the reactant partial pressures, which are used in several models of fuel cells e.g., [4, 15]:

$$E^0 = 1.229 - 8.5 \times 10^{-4}\,(T_{\mathrm{fc}} - 298.5) + 4.3085 \times 10^{-5} T_{\mathrm{fc}} \left[\ln\left(p_{\mathrm{H}_2}\right) + \frac{1}{2}\ln\left(p_{O_2}\right) \right]. \tag{1}$$

However, in practice the cell potential is significantly lower than the theoretical potential because there are some losses even when no external load is connected. Moreover, when a load is connected to the fuel cell, the voltage in the terminals decreases still more due to a number of factors, including polarization losses and interconnection losses between cells. The main voltage losses in a fuel cell are the following [1, 13, 16]:

Activation loss The *activation losses* v_{act} are caused by the slowness of the reaction that takes place on the surface of the electrodes. A proportion of the generated voltage is lost in maintaining the chemical reaction that transfers electrons from the negative electrode toward the positive electrode. This phenomenon is strongly nonlinear, and more important, it is at low current densities.

Fuel crossover and internal currents These *energy losses* result from the waste of fuel passing through the electroslyte and from electron conduction through the electrolyte. In PEMFC, the fuel losses and internal current are small and their effects are usually negligible.

---

[3] The Specific Energy of hydrogen at ambient pressure is 8890 W h kg^{-1}, meanwhile the corresponding one of petrol is 694 W h kg^{-1}. However, the necessary volume for its storage is greater. The energy density of petrol is 500 W h dm^{-3}, meanwhile for hydrogen (300 bar) is 55 W h dm^{-3} [12].

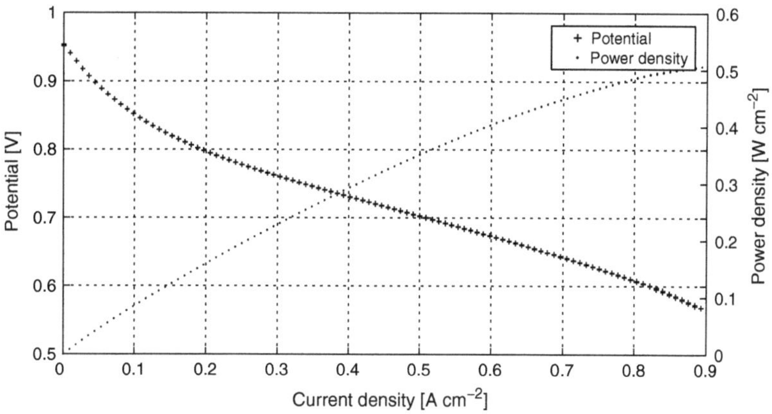

Fig. 2 Polarization curve showing the cell potential and the power density versus the cell current density at $p_{H_2} = p_{O_2} = 2.28$ atm and $T_{fc} = 353$ K

Ohmic loss The *ohmic losses* v_{ohm} are caused by the resistance to the transport of electrons through the electrodes and the different interconnections, and also to the passage of ions through the electrolyte. The behavior of v_{ohm} is approximately linear with the current density.

Concentration loss The *concentration losses* v_{conc} are caused by the diffusion of ions through the electrolyte which produces an increase in the concentration gradient, diminishing the speed of transport. The relation between the voltage of the cell and the current density is approximately linear upto a limit value, beyond which the losses grow quickly.

Therefore, the fuel cell voltage of a simple cell can be expressed as

$$v_{fc} = E^0 - v_{ohm} - v_{act} - v_{conc}. \tag{2}$$

A typical polarization curve showing the potential and power density as a function of the current density is shown in Fig. 2. This curve is obtained using the fuel cell model developed in [4]. In practice, a succession of cells are connected in series in order to provide the necessary voltage and power output, constituting a Fuel Cell Stack System.

1.4 Theoretical and Real Fuel Cell Efficiency

The efficiency of any energy conversion device is defined as the ratio between the useful energy output and the energy input. In a fuel cell, the useful energy output is the generated electrical energy and the energy input is the energy content in the mass of hydrogen supplied. The energy content of an energy carrier is called the Higher Heating Value (ΔH_{HHV}). The ΔH_{HHV} of hydrogen is 286.02 kJ mol^{-1} or 141.9 MJ kg^{-1}. This is the amount of heat that may be generated by a complete combustion of 1 mol or 1 kg of hydrogen, respectively. The ΔH_{HHV} of hydrogen is experimentally determined by reacting a stoichiometric mixture of hydrogen and oxygen in a steel container at 25°C. If hydrogen and oxygen are combined, water vapor emerges at high temperatures. Then, the container and its content are cooled down to the original 25°C and the ΔH_{HHV} is determined by measuring the heat released between the identical initial and final temperatures. On the contrary, if the cooling is stopped at 150°C, the reaction heat is only partially recovered (241.98 kJ mol^{-1} or 120.1 MJ kg^{-1}). This is known as the *lower heating value* (ΔH_{LHV}) of hydrogen [17].

Assuming that all the Gibbs free energy of hydrogen,[4] ΔG, can be converted into electrical energy, the maximum possible (theoretical) efficiency of a fuel cell is [16]

$$\eta_{\mathrm{HHV}} = -\Delta G / -\Delta H_{\mathrm{HHV}} = 237.34/286.02 = 83\%. \tag{3}$$

However, the ΔH_{LHV} is used very often to express the fuel cell efficiency to compare it with the internal combustion engine, whose efficiency has traditionally been expressed using the fuel ΔH_{LHV}. In this case, the maximum theoretical fuel cell efficiency results in a higher number:

$$\eta_{\mathrm{LHV}} = -\Delta G / -\Delta H_{\mathrm{LHV}} = 228.74/241.98 = 94.5\%. \tag{4}$$

If both ΔG and ΔH_{LHV} in (3) are divided by $2F$, where 2 is the number of electrons per molecule of H_2 and F is the Faraday number, the fuel cell efficiency may be expressed as a ratio of two potentials:

$$\eta_{\mathrm{HHV}} = \frac{-\Delta G}{-\Delta H_{\mathrm{HHV}}} = \frac{\frac{-\Delta G}{2F}}{\frac{-\Delta H_{\mathrm{HHV}}}{2F}} = \frac{1.23}{1.482} = 83\%, \tag{5}$$

where $\frac{-\Delta G}{2F} = 1.23\mathrm{V}$ is the theoretical cell potential, and $\frac{-\Delta H_{\mathrm{HHV}}}{2F} = 1.482\mathrm{V}$ is the potential corresponding to the ΔH_{LHV}, or thermoneutral potential.

In this section, we analyzed the theoretical fuel cell efficiency. However, as explained in the previous sections, in a real fuel cell the efficiency is quite lower and also depends on the fuel cell current. The fuel cell efficiency η_{FC} can also be defined as the ratio between the power produced and the power of hydrogen consumed [16]:

$$\eta_{\mathrm{HHV}} = \frac{P_{\mathrm{FC}}}{P_{H_2}} = \frac{\frac{V_{\mathrm{fc}} I_{\mathrm{fc}}}{-\Delta H_{\mathrm{HHV}} I_{\mathrm{fc}}}}{2F} = \frac{V_{\mathrm{fc}}}{1.482}, \tag{6}$$

$$\eta_{\mathrm{LHV}} = \frac{P_{\mathrm{FC}}}{P_{H_2}} = \frac{\frac{V_{\mathrm{fc}} I_{\mathrm{fc}}}{-\Delta H_{\mathrm{LHV}} I_{\mathrm{fc}}}}{2F} = \frac{V_{\mathrm{fc}}}{1.254}, \tag{7}$$

where V_{fc} is the generated voltage and I_{fc} is the fuel cell current. Thus, the FC efficiency is related to the actual voltage, which is related to the fuel cell current through the polarization curve.

---

[4] The Gibbs free energy is used to represent the available energy to do external work. The changes in Gibbs free energy ΔG are negative, which means that the energy is released from the reaction, and varies with both temperature and pressure [18].

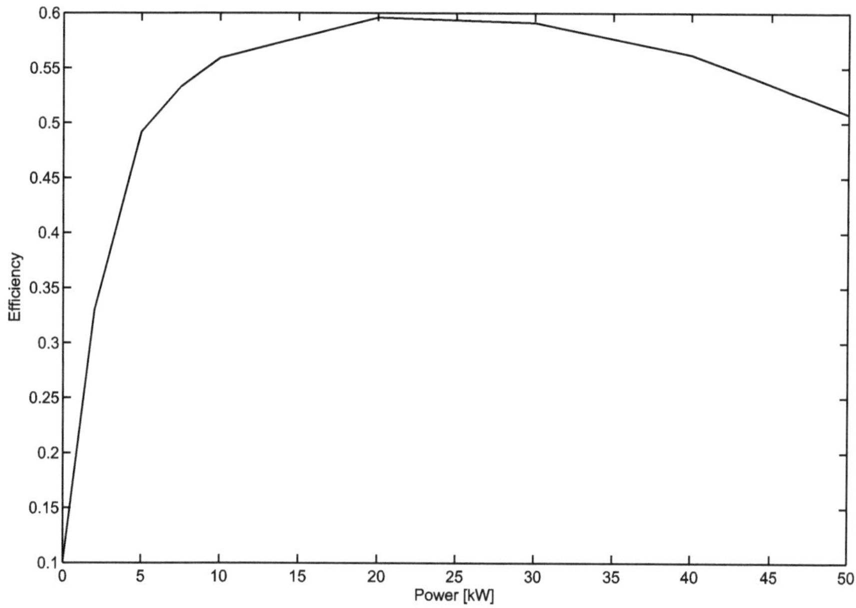

Fig. 3 Efficiency curve of a fuel cell system

Moreover, in a real system it is necessary to incorporate some auxiliary systems which consume a fraction of the generated power. As a result, the efficiency of the fuel cell system, η_{fcs}, is even lower than that expressed in Eq. 6:

$$\eta_{\text{fcs}} = \eta_{\text{HHV}} \frac{P_{\text{net}}}{P_{\text{fc}}} = \eta_{\text{HHV}} \frac{P_{\text{fc}} - P_{\text{aux}}}{P_{\text{fc}}} = \eta_{\text{HHV}} \left(1 - \frac{P_{\text{aux}}}{P_{\text{fc}}}\right), \tag{8}$$

where P_{net} is the net power output, P_{fc} is the fuel cell power, and P_{aux} is the power consumed by the auxiliary components, which include, in particular, the air compressor. The efficiency curve of a fuel cell system of 50 kW modelled in ADvanced VehIcle SimulatOR (ADVISOR)[5] is shown in Fig. 3.

1.5 Generic Structure of a Fuel Cell-Based Power Generation System

In order to be able to produce energy, it is necessary to integrate the fuel cell stack with other components to form a fuel cell-based power generation system. A generic scheme showing the interrelation between the main components of the power generation system is presented in Fig. 4. These components can be divided into the following subsystems [18]:

[5] ADVISOR is a toolbox developed by the national renewable energy laboratory with the aim of analyzing the performance and fuel economy of conventional, electric, and hybrid vehicles [19, 20].

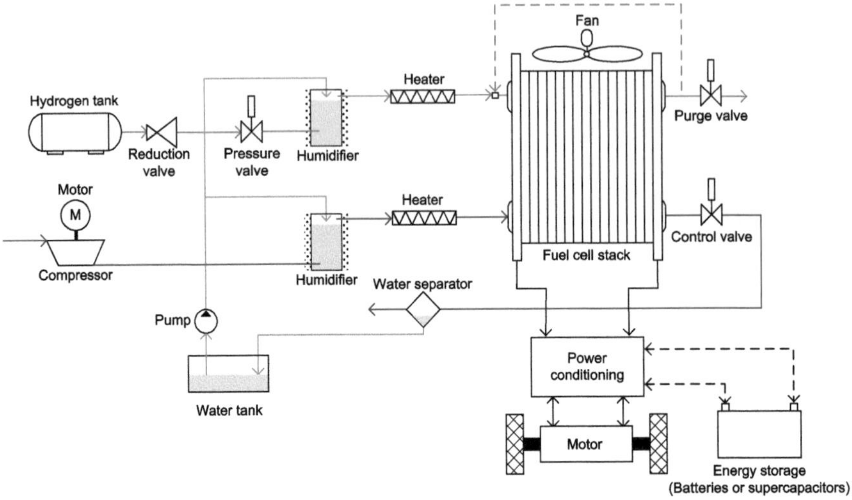

Fig. 4 General scheme of a FCS oriented to automotive applications

1.5.1 Reactant Flow Subsystem

The reactant flow subsystem consists of the hydrogen and air supply circuits. The objective is to supply the adequate reactant flow to ensure fast transient response and minimal auxiliary power consumption. The hydrogen supply circuit is generally composed of a pressurized tank with pure H_2 connected to the anode through a pressure-reduction valve and a pressure-controlled valve, meanwhile, the air supply circuit is generally composed of an air compressor which feeds the cathode with pressurized air from the atmosphere. The anode output is generally operated in *dead-ended* mode and a purge valve in the anode output is periodically opened to remove the water and accumulated nitrogen gas. In case the anode output is not closed it is possible to reinject the out-flowing hydrogen into the anode input. On the other hand, the cathode output is normally open through a fixed restriction. The cathode air supply will be studied in Chap. 3 where we propose to close the cathode output with a controlled valve.

1.5.2 Heat and Temperature Subsystem

The heat and temperature subsystem includes the fuel cell stack cooling system and the reactant heating system. The thermal management of the fuel cell is critical since the performance depends strongly on the temperature. The stack temperature control can be done using a fan or a water refrigeration subsystem.

1.5.3 Water Management Subsystem

The objective of the water management subsystem is to maintain an effective hydration of the polymer membrane and an adequate water balance, because the fuel cell performance is also strongly dependent on membrane hydration. Both the air and the hydrogen, are usually humidified before entering the fuel cell with humidifiers in both circuits. The water that leaves the cathode can be recovered in a water separator and reinjected in the humidifiers through a pump.

1.5.4 Power Conditioning Subsystem

Fuel cells generate an unregulated *DC* voltage which drops off when the current increases according to the polarization curve. In general, some power conditioning actions are necessary to supply the load properly. Such actions make necessary the use of *DC/DC* regulators and/or inverter regulators.

1.5.5 Power Management Subsystem

The power management subsystem controls the power drawn from the fuel cell stack. If no energy storage devices are used, the full load must be supplied by the fuel cell and no power management is necessary. However, if an energy storage system is included, such as batteries or supercapacitors, it is necessary to implement a power management between these two power sources. A review of fuel cell hybrid systems is described in the following section and a detailed study is done in Chap. 4.

1.6 Modeling of Polymer Electrolyte Membrane Fuel Cell Systems

Validated mathematical models provide powerful tools for the development and improvement of fuel cell-based systems. Mathematical models can be used to describe the fundamental phenomena that take place in the system to predict the behavior under different operating conditions and to design and optimize the control of the system.

The fuel cell system models describe quantitatively the physical and electrochemical phenomena that take place in to the cells. The models can be divided into two groups [3]: empirical models and mechanistic models. Most of the empirical models are focused on the prediction of the polarization curve, which is used to characterize the electrical operation of the FC, by means of empirical equations. The following empirical equation developed by Kim et al. [21] is used to calculate the voltage (E) at different current densities (J), fitting experimental data at several temperatures, pressures, and oxygen compositions in the cathode gas mixture:

$$E = E_0 - b \log J - RJ - m \exp(nJ), \tag{9}$$

where E_0 is the thermodynamic open-circuit voltage. The exponential term characterizes the mass-transport region of the polarization curve, the zone where the increase in slope of the pseudo-linear region and the subsequent rapid fall-off of the cell potential. The parameter n has more pronounced effects than the parameter m in this region. The terms E_0 and b yield the electrode kinetic parameters for oxygen reduction in the cell. R represents the resistance, predominantly ohmic, and, slightly, the charge-transfer resistance of the electro-oxidation of hydrogen.

Later, Squadrito et al. [22] used an empirical approach to account the mass transport limitation, modifying the model of Kim et al. by replacing the last term of (9) with $\left[mJ^n \ln\left(1 - J/J_{\text{lim}}\right)\right]$ where m and n are empirically determined constants and J_{lim} is the limiting current density obtained by fitting experimental data. The equation describes experimental data over the full range of current density taking into account possible mass transport limitations. The empirical equation was used to fit experimental data obtained from a single cell, showing good agreement between theoretical and experimental data.

One example of an empirical model widely used is the model developed by Amphlett et al. [14], which incorporates as much empirical properties as mechanistic to obtain a transient model able to predict the FC voltage as a function of the current, the FC temperature and the hydrogen and oxygen partial pressures for a 35-cell 5 kW PEM fuel cell. The model works well for the experimental system in study but can be generalized to any other system recalculating the values of the model.

On the other hand, the mechanistic model considers the fundamental phenomena in detail such as, heat and mass transport, forces, and electrochemical processes. Mechanistic modeling (single and multi-domain) has been utilized to study a wide range of phenomena including polarization effects (activation, ohmic and concentration overpotentials), water management, thermal management, CO kinetics, catalyst utilization and flow field geometry [23]. All these models are parametric in the sense that they account for the cell performance of various input parameters, typically temperature, pressure and humidity.

Many of the mechanistic models are one-dimensional where only the direction across the fuel cell is taken into account. Some of the first works dealing with this type of models are [24–26]. Then two-dimensional and three-dimensional models were developed. Some two-dimensional models describe the plane perpendicular through the flow channels [27, 28] while others simulate the fuel cell along the flow channels [29, 30]. Within the three-dimensional models can be mentioned the model developed by Nguyen et al. [31], which is a three-dimensional model of a PEM fuel cell with serpentine gas channels, and the model developed by Natarajan [32].

Some of the earlier works focused in the humidification problem, addressing the humidification requirements of the inlet gases to maintain the water balance [33]. Fuller and Newman [34] also addressed the water and thermal management, similar to Nguyen and White [29] who studied different forms of gas humidification and water management, analyzing the cell performance. The model of Wöhr et al. [35] showed that for fuel cell stacks water management becomes even more difficult and is strongly related to thermal management. The temperatures of the inner cells of the stack are higher than the outer cells resulting in membrane dehydration.

Another approach to fuel cell modeling is the use of equivalent circuits to represent the system behavior. In fact, one of the most attractive aspects of the Electrochemical Impedance Spectroscopy as a tool for investigating the electrical and electrochemical properties of materials and systems, is the direct connection that often exists between the behavior of a real system and that of an idealized model circuit consisting of discrete electrical components [36]. The procedure typically consist in the comparison or fitting the impedance data to an equivalent circuit, which is representative of the physical processes taking place in the system under investigation. There are analogies between the circuit elements and the electrochemical processes, so that the results of data fitting can be more easily converted into physical understanding.

Many authors have studied the modeling approach based on equivalent electrical circuits, consisting of an arrangement of different electrical components to have the same frequency response as that obtained by electrochemical impedance spectroscopy tests [37–40]. Some works present equivalent circuits using electrical elements, such as resistances, capacitances or inductance. However, other works use additional distributed elements that represent electrochemical or mass and ionic transport phenomena. For example, Warburg impedance represents the impedance of one-dimensional distributed diffusion of a species in an electrode. Another example is a constant phase element, used for describing a distributed charge accumulation on rough irregular electrode surfaces.

However, there are some problems in using analogies to describe electrochemical systems: one of them is that different equivalent circuits obtain the same frequency response, another problem is the overlapping frequency response of different phenomena and the dependence of the electrochemical phenomena on the operating conditions (temperature, current, pressure, etc.).

So far we have seen different types of models that study the cell behavior in both stationary and transient states. Nevertheless, there are few dynamic models suitable for control purposes. In the work of Pukrushpan et al. [4] a dynamic model for PEMFCS that is suitable for the control study has been developed. The model captures the transitory behavior of the air compressor, the gases filling dynamics (in the cathode and anode), and the effect of the membrane humidity. These variables affect the cell voltage and, therefore, the efficiency and the output power. The polarization curve in this model is a function of the hydrogen and oxygen partial pressures, the stack temperature and the membrane water content. This allows to evaluate the effect of variations of oxygen concentration and membrane humidity in the output voltage, which are necessary to make the control during transitory operation.

Another practical model oriented to control was developed by Del Real et al. [41] where the model parameters have been adjusted specifically for a 1.2 kW Ballard stack, which is considered a benchmark as it is widely used by research groups in the PEM fuel cell field. The model can predict both steady and transient states subject to variable loads (including flooding and anode purges), as well as the system start-up. The proposed model methodology is accurate since the simulated results show good agreement, compared to experimental data from the Ballard stack.

On the other hand, a dynamic model which incorporates the effects of charge double layer capacitance, the dynamics of flow and pressure in the anode and cathode channels and mass/heat transfer transient features in the fuel cell body has been presented [42]. This dynamic model can predict the transient response of cell voltage, temperature of the cell, hydrogen/oxygen out-flow rates and cathode and anode channel temperatures/pressures under sudden change in load current. The simulation results are analyzed and compared to benchmark results, reporting that a good agreement is found between tests and simulations. Similarly, a dynamic electrochemical model of a grid independent fuel cell power plant is presented in [43]. The model includes the methanol reformer, the PEM stack and the power conditioning unit. The model is used to predict the output voltage when subjected to rapid changes in a residential load connected to it, showing a high degree of accordance. More recently, a semi-empirical dynamic model for stack voltage, based on experimental investigation, was presented [44]. The proposed model can predict the transient response of stack voltage under step change in current with good agreement between tests and simulations.

The model developed by Pukrushpan et al. [4], which has been employed in several works for control purposes [45–51], is utilized in this book as a base to represent the behavior of a generic fuel cell system. The model is described in detail in Sects. 1 and 2.

1.7 Description of the Fuel Cell System Model

The model developed by Pukrushpan et al. [4] is used in several chapters of this book as a base to characterize the dynamic behavior and performance of PEM fuel cell systems. This model contains four main subsystems that interact with each other:

- FC voltage subsystem
- Membrane hydration subsystem
- Cathode flow subsystem
- Anode flow subsystem

The spatial variation of the parameters is not considered and, thus, they are treated as lumped parameters. On the other hand, the time constants of the electrochemical reactions are in the order of magnitude of 10^{-19} s despite another work [52] argues that this constants are significatively lower (10^{-9} s). In any case, all the literature agrees in the fastness of the electrochemical reactions, as remarked in [53] and [42]. Thus, for control purpose, these time constants can be assumed negligible compared to other constants much slower: temperature (10^{2} s) and dynamics of volume filling (10^{-1} s).

In this book, this base model has been modified to adapt it to the proposed control structures described in Sect. 2. In the original model there is only one control variable: the compressor motor voltage. In this book, it is proposed to add an extra variable, the throttle opening area in the cathode output (A_t), adding a control valve in the cathode output. The advantages of this new configuration are covered in detail in Sect. 2.3.

Usually, in model-based control, it is necessary to find simplified models, derived from the complete ones or from experimental data, to be used as inner models into the controller. Here, the step response is utilized to obtain a simplified model that is used as internal model to predict the future process response in the control strategy implemented in Sect. 2. The main advantage of this simplified model is that it is easily obtainable through experimental data with good agreement with respect to the original nonlinear system in the considered operating point.

1.8 Principal Equations in the Fuel Cell System Model

The model described in the previous section has the following governing equations where the mass of air in the supply manifold, the masses of oxygen, nitrogen and water in the cathode and the masses of hydrogen and water in the anode are defined using the principle of mass conservation [45]:

$$\frac{dm_{\mathrm{sm}}}{dt} = W_{\mathrm{cp}} - W_{\mathrm{sm,out}}, \tag{10}$$

$$\frac{dm_{\mathrm{O_2,ca}}}{dt} = W_{\mathrm{O_2,ca,in}} - W_{\mathrm{O_2,ca,out}} - W_{\mathrm{O_2,rct}}, \tag{11}$$

$$\frac{dm_{\mathrm{N_2,ca}}}{dt} = W_{\mathrm{N_2,ca,in}} - W_{\mathrm{N_2,ca,out}}, \tag{12}$$

$$\frac{dm_{w,\mathrm{ca}}}{dt} = W_{v,\mathrm{ca,in}} - W_{v,\mathrm{ca,out}} - W_{v,\mathrm{ca,gen}} + W_{v,m}, \tag{13}$$

$$\frac{dm_{\mathrm{H_2}}}{dt} = W_{\mathrm{H_2,an,in}} - W_{\mathrm{H_2,an,out}} - W_{\mathrm{H_2,rct}}, \tag{14}$$

$$\frac{dm_{w,\mathrm{an}}}{dt} = W_{v,\mathrm{an,in}} - W_{v,\mathrm{an,out}} - W_{v,m} - W_{l,\mathrm{an,out}}, \tag{15}$$

where

- W_{cp} is the compressor flow,
- $W_{\mathrm{sm,out}}$ is the outlet mass flow,
- $W_{\mathrm{O_2,ca,in}}$ is the mass flow rate of oxygen gas entering the cathode,
- $W_{\mathrm{O_2,ca,out}}$ is the mass flow rate of oxygen leaving the cathode,
- $W_{\mathrm{O_2,rct}}$ is the mass flow rate of oxygen reacted,
- $W_{\mathrm{N_2,ca,in}}$ is the mass flow rate of nitrogen gas entering the cathode,
- $W_{\mathrm{N_2,ca,out}}$ is the mass flow rate of nitrogen gas leaving the cathode,
- $W_{v,\mathrm{ca,in}}$ is the mass flow rate of vapor entering the cathode,
- $W_{v,\mathrm{ca,out}}$ is the mass flow rate of vapor leaving the cathode,
- $W_{v,\mathrm{ca,gen}}$ is the rate of vapor generated in the fuel cell reaction,

- $W_{v,\text{menbr}}$ is the mass flow rate of water across the fuel cell membrane,
- $W_{l,\text{ca,out}}$ is the mass flow rate of liquid water leaving the cathode,
- $W_{H_2,\text{an,in}}$ is the mass flow rate of hydrogen gas entering the anode,
- $W_{H_2,\text{an,out}}$ is the mass flow rate of hydrogen gas leaving the anode,
- $W_{H_2,\text{rct}}$ is the rate of hydrogen reacted,
- $W_{v,\text{an,in}}$ is the mass flow rate of vapor entering the anode,
- $W_{v,\text{an,out}}$ is the mass flow rate of vapor leaving the anode,
- $W_{v,m}$ is the mass flow rate of water transfer across the fuel cell membrane, and
- $W_{l,\text{an,out}}$ is the rate of liquid water leaving the anode.

The voltage of a fuel cell stack consisting of n fuel cells is given as

$$v_{\text{st}} = n \cdot v_{\text{fc}}, \tag{16}$$

where the voltage of a single fuel cell is defined as

$$v_{\text{fc}} = E - v_{\text{act}} - v_{\text{ohm}} - v_{\text{conc}} \tag{17}$$

with E being the open circuit voltage and v_{act}, v_{ohm} and v_{conc} being the activation, ohmic and concentration overpotentials, respectively. By fitting experimental data to the phenomenological model equations, the open circuit voltage and the three overpotentials are respectively defined as

$$\begin{aligned} E &= 1.229 - 0.85 \times 10^{-3}(T_{\text{fc}} - T_{\text{amb}}) \\ &\quad + 4.3085 \times 10^{-5} T_{\text{fc}} \times \left[\ln(1.01325 p_{H_2}) + \frac{1}{2} \ln(1.01325 p_{O_2}) \right]. \end{aligned} \tag{18}$$

The activation voltage is

$$v_{\text{act}} = v_0 + v_a(1 - e^{c_1 i}) \tag{19}$$

with

$$\begin{aligned} v_0 &= 0.279 - 8.5 \times 10^{-4}(T_{\text{fc}} - T_{\text{amb}}) + 4.308 \times 10^{-5} T_{\text{fc}} \\ &\quad \times \left[\ln\left(\frac{p_{\text{Ca}} - p_{\text{sat}}(T_{\text{fc}})}{1.01325} \right) + \frac{1}{2} \ln\left(\frac{0.1173(p_{\text{Ca}} - p_{\text{sat}}(T_{\text{fc}}))}{1.01325} \right) \right], \end{aligned} \tag{20}$$

$$\begin{aligned} v_a &= (-1.618 \times 10^{-5} T_{\text{fc}} + 1.618 \times 10^{-2}) \\ &\quad \times \left(\frac{p_{O_2}}{0.1173} + p_{\text{sat}}(T_{\text{fc}}) \right]^2 \\ &\quad + (1.8 \times 10^{-4} T_{\text{fc}} - 0.166) \left(\frac{p_{O_2}}{0.1173} + p_{\text{sat}}(T_{\text{fc}}) \right) \\ &\quad (-5.8 \times 10^{-4} T_{\text{fc}} + 0.5736), \end{aligned} \tag{21}$$

and

$$c_1 = 10. \tag{22}$$

The ohmic voltage is

$$v_{\text{ohm}} = i R_{\text{ohm}} \tag{23}$$

with the fuel cell electrical resistance

$$R_{\text{ohm}} = \frac{t_m}{\sigma_m}, \tag{24}$$

the membrane conductivity

$$\begin{aligned} &\sigma_m = (b_{11}\mu_m - b_{12}) \exp\left[b_2\left(\frac{1}{303} - \frac{1}{T_{\text{fc}}}\right)\right] \\ &b_{11} = 5.139 \times 10^{-3}, \qquad b_{12} = 3.26 \times 10^{-3}, \qquad b_{12} = 350 \end{aligned} \tag{25}$$

and

$$v_{\text{conc}} = i\left(c_2 \frac{i}{i_{\max}}\right)^{c_3} \tag{26}$$

with

$$c_2 = \begin{cases} (7.16 \times 10^{-4} T_{\text{fc}} - 0.622)\left(\frac{p_{O_2}}{0.1173} + p_{\text{sat}}(T_{\text{fc}})\right) \\ \quad +(-1.45 \times 10^{-3} T_{\text{fc}} + 1.68) \ \text{ for } \left(\frac{p_{O_2}}{0.1173} + p_{\text{sat}}(T_{\text{fc}})\right) < 2\,\text{atm} \\ (8.66 \times 10^{-5} T_{\text{fc}} - 0.068)\left(\frac{p_{O_2}}{0.1173} + p_{\text{sat}}(T_{\text{fc}})\right) \\ \quad +(-1.6 \times 10^{-4} T_{\text{fc}} + 0.54) \ \text{ for } \left(\frac{p_{O_2}}{0.1173} + p_{\text{sat}}(T_{\text{fc}})\right) \geq 2\,\text{atm} \end{cases} \tag{27}$$

and

$$i_{\max} = 2.2, \quad c_3 = 2. \tag{28}$$

The governing equations for the supply manifold pressure and the return manifold pressure are respectively defined using the energy conservation principle and the standard thermodynamics relationships as follows:

$$\frac{\mathrm{d}p_{\text{sm}}}{\mathrm{d}t} = \frac{\gamma R_a}{V_{\text{sm}}}(W_{\text{cp}} T_{\text{cp}} - W_{\text{sm,out}} T_{\text{sm}}), \tag{29}$$

$$\frac{\mathrm{d}p_{\text{rm}}}{\mathrm{d}t} = \frac{R_a T_{\text{rm}}}{V_{\text{rm}}}(W_{\text{ca,out}} - W_{\text{rm,out}}), \tag{30}$$

where V_{sm} is the supply manifold volume, V_{rm} is the return manifold volume, T_{sm} is the supply manifold air temperature, T_{rm} is the return manifold air temperature,

T_{cp} is the temperature of the air leaving the compressor, R_a is the air gas constant, and γ is the air specific heat ratio.

To express the governing equations in terms of the states, the closure relations Eqs. 31–38 are used. The supply manifold outlet air rate $W_{sm,out}$ is related to the supply manifold pressure p_{sm} and the cathode pressure p_{ca} via the linearized nozzle equation:

$$W_{sm,out} = k_{sm,out}(p_{sm} - p_{ca}), \tag{31}$$

where $k_{sm,out}$ is the supply manifold outlet orifice constant.

The inlet oxygen, nitrogen, and cathode vapour mass flow rates, $W_{O_2,in}$, $W_{N_2,in}$, and $W_{v,ca,in}$ are related to the cathode inlet air mass flowrate, the inlet air humidity and the mass fraction of oxygen and nitrogen in dry air using the ideal gas relations. The outlet oxygen, nitrogen and cathode vapour mass flow rates, $W_{O_2,out}$, $W_{N_2,out}$, and $W_{v,ca,out}$, are likewise related to the outlet air mass flowrate, the outlet air humidity and the mass fraction of the oxygen and nitrogen in dry air at the cathode outlet using the ideal gas relations. The reacted oxygen and hydrogen and generated water vapor (in the cathode) mass flow rates, $W_{O_2,rct}$, $W_{H_2,rct}$, and $W_{v,ca,gen}$, are related to the fuel cell current:

$$W_{O_2,rct} = M_{O_2}\frac{nI_{fc}}{4F}, \tag{32}$$

$$W_{H_2,rct} = M_{H_2}\frac{nI_{fc}}{2F}, \tag{33}$$

$$W_{v,ca,gen} = M_v\frac{nI_{fc}}{2F}, \tag{34}$$

where the constants 4 and 2 in the denominators denote the of electrons involved in the oxidation and the reduction half-reactions respectively, M_{O_2} is the molar mass of oxygen, M_{H_2} is the molar mass of hydrogen, M_v is the molar mass of vapor, and F is the Faraday constant.

The water mass flowrate through the membrane, $W_{v,m}$, is defined using the hydration model. The outlet hydrogen and water masses are assumed to be zero, that is, hydrogen is assumed to react completely in the anode, while water generated by the oxidation half-reaction is assumed to be transported via electro-osmosis through the membrane towards the cathode.

The governing equation for the rotational speed of the compressor is defined by the power conservation principle as

$$J_{cp}\frac{d\omega_{cp}}{dt} = \tau_{cm} - \tau_{cp}, \tag{35}$$

where J_{cp} is the combined inertia of the compressor and the motor (kg. m^2), ω_{cp} is the compressor speed (rad/sec), τ_{cm} is the compressor motor torque input (Nm)

calculated in Eq. 36, and τ_{cp} is the torque required to drive the compressor (Nm) calculated in Eq. 37.

The compressor motor torque τ_{cm} is related to the compressor motor voltage V_{cm} and the compressor motor rotational speed ω_{cp} by the static motor equation:

$$\tau_{cm} = \eta_{cm} \frac{k_t}{R_{cm}} \left(V_{cm} - k_v . \omega_{cp} \right), \tag{36}$$

where k_t, R_{cm}, and k_v are motor constants and η_{cm} is the motor mechanical efficiency. The steady state compressor torque τ_{cp} is related to the supply manifold pressure, the compressor motor rotational speed and the compressor air flowrate W_{cp} via the thermodynamic relations

$$\tau_{cp} = \frac{C_P}{\omega_{cp}} \frac{T_{atm}}{\eta_{cp}} \left[\left(\frac{p_{sm}}{p_{atm}} \right)^{(\gamma-1)/\gamma} - 1 \right] W_{cp}, \tag{37}$$

where C_p is the air specific heat, η_{cp} is the compressor efficiency, and T_{atm} and p_{atm} are the atmospheric temperature and pressure, respectively.

The air temperature in the compressor, T_{cp}, is defined using basic thermodynamic relations

$$T_{cp} = T_{atm} + \frac{T_{atm}}{\eta_{cp}} \left[\left(\frac{p_{sm}}{p_{atm}} \right)^{(\gamma-1)/\gamma} - 1 \right]. \tag{38}$$

The air temperature in the supply manifold, T_{sm}, is obtained from m_{sm}, p_{sm} and V_{sm} using the ideal gas law. The cathode outlet air flowrate $W_{ca,out}$ is related to the cathode pressure and return manifold pressure via a linearized nozzle equation analogous to Eq. 31. The return manifold outlet air flowrate $W_{rm,out}$ is defined using a non-linearized nozzle relation as discussed in Sect. 2.2, while the return manifold air temperature T_{rm} is considered to be constant and equal to the temperature of the fuel cell stack.

The flowrate of water through the membrane is controlled by two transport phenomena: electroosmotic drag of water molecules by the protons and back-diffusion from the cathode towards the anode. The transport phenomena is defined as

$$W_{v,m} = M_v A_{fc} n \left(n_d \frac{i}{F} - D_w \frac{(C_{v,ca} - C_{v,an})}{t_m} \right), \tag{39}$$

where the electroosmotic drag coefficient is given as

$$n_d = 0.0029 \mu_m^2 + 0.05 \mu_m - 3.4 \times 10^{-19} \tag{40}$$

and μ_m is the mean water content in the membrane. The water content is defined as

$$u_i = \begin{cases} 0.043 + 17.81 a_i - 39.85 a_i^2 + 36.0 a_i^3, & 0 < a_i \leq 1 \\ 14 + 1.4(a_i - 1), & 1 < a_i \leq 3 \\ & (i = m, \text{an}, \text{ca}) \end{cases} \tag{41}$$

where the water vapour activity is defined as

$$a_i = \frac{x_{v,i} p_i}{p_{\text{sat,i}}} = \frac{p_{v,i}}{p_{\text{sat,i}}} \quad (i = \text{an, ca}). \tag{42}$$

The average water vapour activity in the membrane is defined as

$$a_m = \frac{a_{\text{an}} + a_{\text{ca}}}{2} \tag{43}$$

and the water diffusion coefficient is given as

$$D_W = D_\lambda \exp\left[2416\left(\frac{1}{303} - \frac{1}{T_{\text{fc}}}\right)\right] \times 10^{-4} \tag{44}$$

with the preexponential term

$$D_\lambda = \begin{cases} 10^{-6}, & \mu_m < \quad 2 \\ 10^{-6}[1 + 2(\mu_m - 2)], & 2 \leq \mu_m \leq 3 \\ 10^{-6}[3 - 1.67(\mu_m - 3)], & 3 < \mu_m < 4.5 \\ 1.25 \times 10^{-6}, & \mu_m \geq \quad 4.5 \end{cases} \tag{45}$$

The water concentration at the membrane surfaces on anode and cathode sides in Eq. 39 is a function of the membrane water content

$$C_{v,i} = \frac{\rho_{m,\text{dry}}}{M_{m,\text{dry}}} \mu_i \quad (i = \text{an, ca}), \tag{46}$$

where $\rho_{m,\text{dry}}$ (kg/cm^3) is the membrane dry density and $M_{m,\text{dry}}$ (kg/mol) is the equivalent weight.

2 Auxiliary Equipment and System Modeling

While the fuel cell operates with oxygen as reactant in the cathode, it is more practical to use oxygen from air. Air is mainly composed of nitrogen (78.084%), oxygen (20.946%), and argon (0.9340%). The effect of using air instead of pure oxygen is a reduction of approximately 50 mV in the cell voltage [16]. Additionally, there is a reduction in the fuel cell efficiency because of the power consumption to pump the oxygen and almost four times that of nitrogen.

In a hydrogen-air fuel cell system the air is supplied by a fan, a blower or a compressor. In any case, there is an electric motor with a power consumption that implies a reduction in the efficiency of the fuel cell system. The net power output, P_{net}, that is actually available is the fuel cell power, P_{fc}, less the power consumed by the ancillary components, P_{aux}, which includes the compressor or the blower:

$$P_{\mathrm{net}} = P_{\mathrm{fc}} - P_{\mathrm{aux}}. \tag{47}$$

The principal consumption between the ancillary equipments is that corresponding to the electric motor that runs the compressor or the blower. Besides, it depends strongly on the operating conditions. Thus, it can be assumed that the consumption of the ancillary equipments is the consumption of this electric motor.

When the fuel cell operates at high pressures, the power output is higher. Thus, the most common scenario is a fuel cell where the cathode is supplied with air by a compressor. However, the power consumption of the compressor increases significantly with the pressure. Thus, when the compressor consumption is taken into account, there is a trade-off between pressure and efficiency.

The operation of the fuel cell system at high pressures increases the generated voltage as a result of the increase in the cathode oxygen partial pressure and anode hydrogen partial pressure (see Eq. 1). Especially, an increase in the cathode pressure produces an increase in the supply manifold pressure and thus, an increase in the pressure ratio across the compressor and in the compressor power consumption contributing to a reduction in the system efficiency. The power consumed by the air compressor is

$$P_{\mathrm{cp}} = \frac{C_p T_{\mathrm{atm}}}{\eta_{\mathrm{cp}}} \left[\left(\frac{p_{\mathrm{sm}}}{p_{\mathrm{atm}}}^{\frac{\gamma-1}{\gamma}} \right) - 1 \right] W_{\mathrm{cp}}, \tag{48}$$

where W_{cp} is the compressor air flow rate, P_{cp} is the compressor power, T_{atm} is the inlet air temperature in the compressor, η_{cp} is the compressor efficiency, p_{sm} is the supply manifold pressure, $C_p = 1{,}004\,\mathrm{J\,kg^{-1}\,K^{-1}}$ is the specific heat capacity of air, and $\gamma = 1.4$ is the ratio of the specific heat of air. Actually, the power consumed by the electric motor is higher because of the mechanical and electric inefficiencies:

$$P_{\mathrm{EM}} = \frac{P_{\mathrm{comp}}}{\eta_{\mathrm{mec}} \cdot \eta_{\mathrm{EM}}}, \tag{49}$$

where η_{mec} is the compressor mechanical efficiency and η_{EM} is the efficiency of the electric motor. In Chap. 3 will be seen how this power consumption affects the efficiency of the system. Besides, a solution from the point of view of control will be studied in detail.

3 Optimal Operation of the FCS

The FCS model is useful to study the optimal operation of an FCS, especially at low loads. An adequate operation produces important benefits, increasing the system efficiency in terms of hydrogen reduction and allowing a greater peak power. In a direct-hydrogen FCS with the cathode supplied with air through a compressor, the air supply subsystem has a crucial role in the improvement of the performance of the system [10]. In fact, there are two external variables that have greater impact on the polarization curve: the air pressure and the air stoichiometry.

The air pressure and the air stoichiometry control the oxygen partial pressure in the catalytic layer of the cathode, which determines the cathode polarization and, therefore, the efficiency. In [54], it is also stated the importance of the air pressure control to improve the FCS efficiency. The efficiency improvement for a given load is based on a trade-off between the increase of the air pressure and the air stoichiometry, and the increase of the parasitic compressor power.

In the work of Friedman and Moore [55], it is shown that an FCS can be optimized to obtain high peak power and high efficiency over a broad range of output powers. The key to obtain this objective is to vary the pressure and the air flow. Based on this result, it can be concluded that an FCS must be operated to the greater possible pressure and air stoichiometry. Nevertheless, if the energy necessary to compress the air is considered, the result is different: for a fixed air flow, the compressor power consumption increases significantly when the pressure is increased. This means that it is possible to find an optimal combination of pressure and air flow.

A similar conclusion is stated in [45]. In this work, it is assumed that the FCS net power P_{net} can be approximately defined as the difference between the power produced by the FCS, P_{fcs}, and the consumed compressor power P_{cm}. For each load current I_{st}, an air flow increment increases the cathode pressure and, therefore, increases the oxygen partial pressure, increasing the FCS voltage. This also leads to an increment in the oxygen excess ratio in the cathode, λ_{O_2}. The initial increase in the oxygen excess is translated into an FCS power increase and a net FCS power. Nevertheless, if a limit λ_{O_2} is exceeded, an excessive compressor power is produced so that the net power falls.

Following this analysis and using the model, it is possible to find the optimal relation among maximum net power, P_{max}^{net}, optimal oxygen ratio, $\lambda_{O_2}^{opt}$, optimal supply manifold air pressure, p_{SM}^{opt}, and optimal compressor motor voltage, v_{CM}^{opt}, with the stack current, I_{st} [45]. These relations are shown in Eqs. 50–53 and also in Fig. 5.

$$P_{max}^{net} = -2.99 \times 10^{-4} I_{st}^2 + 0.271 I_{st} - 0.871 \quad (50)$$

$$\lambda_{O_2}^{opt} = -2.7633 \times 10^{-6} I_{st}^2 - 15.931 \times 10^{-4} I_{st} + 2.7331 \quad (51)$$

$$p_{SM}^{opt} = 2.9996 I_{st}^2 + 736.4 I_{st} + 8.932 \times 10^4 \quad (52)$$

$$v_{CM}^{opt} = -1.36 \times 10^{-3} I_{st}^2 + 1.17 I_{st} + 14.3 \quad (53)$$

The Eq. 51 expresses that at each level of the stack current there is an optimal value of λ_{O_2} at which P_{net} has the maximum value. This optimal value in the oxygen excess ratio is achieved by applying a compressor voltage according to Eq. 53. The Eqs. 50–53 are obtained by solving the model equations previously shown under steady-state condition and the analysis corresponds to the typical fuel cell operating conditions (fuel cell temperature $T_{fc} = 353$ K and relative humidity $\phi_{Ca} = 1$).

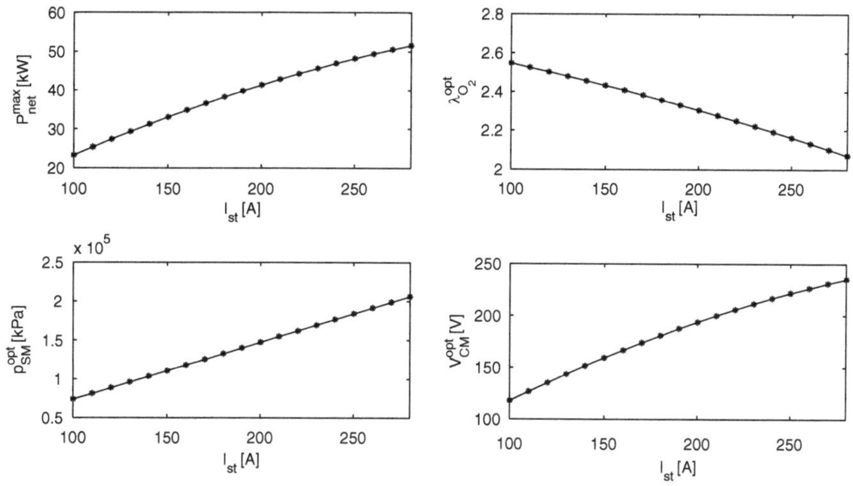

Fig. 5 Optimal relations of P_{max}^{net}, $\lambda_{O_2}^{opt}$, p_{SM}^{opt}, and v_{CM}^{opt} as a function of the stack current I_{st}

4 Conclusions

This chapter presents the main concepts about PEM fuel cell systems, showing the structure of these systems, the advantages and disadvantages, and, particularly, the expressions corresponding to the fuel cell voltage and efficiency. Also, different approaches for modeling are presented. Accordingly, a review of PEM fuel cell model has been done. Fuel cell models may be classified into one of three categories: analytic, semi-empirical or mechanistic (theoretical). Among these models, a model widely used in the literature for control purposes is described in detail. Using this model, the influence of the auxiliary equipment in the system efficiency is analyzed and the model is also used to study the optimal operation of a PEM fuel cell system, finding an optimal compressor voltage for each current load that maximizes the net output power. Similarly, this dynamic model is used as a base model in the other chapters of this book.

References

1. Laughton MA (2002) Fuel cells. Power Eng J 16:37–47
2. Tang H, Peikang S, Jiang SP, Wang F, Pan M (2007) A degradation study of Nafion proton exchange membrane of PEM fuel cells. J Power Sources 170(1):85–92
3. Yao K, Karan K, McAuley K, Oosthuizen P, Peppley B, Xie T (2004) A review of mathematical models for hydrogen and direct methanol polymer electrolyte membrane fuel cells. Fuel Cells 4(1–2):3–29
4. Pukrushpan JT, Peng AGH (2002) Modeling and control for PEM fuel cell stack system. American Control Conference, Proceedings of the 2002, 4:3117–3122

5. D'Arco S, Ianuzzi D, Pagano M, Tricoli P (2005) Design criteria optimizing the use of fuel cell source in electric power system. In: Proceedings of the 16th IFAC World Congress, Prague
6. Zhang J, Xie Z, Zhang J, Tang Y, Song C, Navessin T, Shi Z, Song D, Wang H, Wilkinson DP, Liu Z-S, Holdcroft S (2006) High temperature pem fuel cells. J Power Sources 160(2):872–891
7. Carrette L, Friedrich K, Stimming U (2001) Fuel cells-fundamentals and applications. Fuel Cells 1(1):5–39
8. Rajashekara K (2000) Propulsion system strategies for fuel cell vehicles. Fuel Cell Technol for Vehicles 1:179–187
9. Jeong K, Oh B (2002) Fuel economic and life-cycle cost analysis of a fuel cell hybrid vehicle. J Power Sources 105:58–65
10. Friedman D, Moore R (1998) PEM fuel cell system optimization. In: Proc Electrochem Soc 27:407–423
11. EG & G Technical Services, Inc. Science Applications International Corporation (2002) IN: U.S. Department of Energy, Fuel Cell Handbook, 6th edition, Morgantown, West Virginia, USA
12. Flipsen S (2006) Power sources compared: the ultimate truth? J Power Sources 162:927–934
13. Larminie J, Dicks A (2003) Fuel Cell Systems Explained, 2nd edn Wiley, New York, USA
14. Amphlett J, Mann R, Peppley B, Roberge P, Rodrigues A (1996) A model predicting transient responses of proton exchange membrane fuel cells. J Power Sources 61:183–188
15. Correa J, Farret F, Canha L, Simoes M (2004) An electrochemical-based fuel-cell model suitable for electrical engineering automation approach. Ind Electron, IEEE Trans on 51(5):1103–1112
16. Barbir F (2005) PEM fuel cells: theory and practice. Elsevier, Burlington, MA, USA
17. Bossel U (2003) Well-to-wheel studies, heating values, and the energy conservation principle. IN:Ulf Bossel (ed) European Fuel Cell Forum, October 2003, Oberrohrdorf, Switzerland 1:1–5
18. Pukrushpan J (2003) Modelling and control of fuel cell systems and fuel processors. PhD thesis, University of Michigan, Ann Arbor, Michigan, USA. http://www-personal.umich.edu/~annastef/FuelCellPdf/pukrushpan_thesis.pdf
19. Wipke K, Cuddy M, Burch S (1999) Advisor 2.1: a user-friendly advanced powertrain simulation using a combined backward/forward approach. IEEE Trans on Vehicular Technol 48:1751–1761
20. Markel T, Brooker A, Hendricks T, Johnson V, Kelly K, Kramer B, O'Keefe M, Sprik S, Wipke K (2002) Advisor: a system analysis tool for advanced vehicle modeling. J Power Sources 110:255–266
21. Kim J, Lee SM, Srinivasan S, Chamberlin CE (1995) Modeling of proton exchange membrane fuel cell performance with an empirical equation. J Electrochem Soc 142:2670
22. Squadrito G, Maggio G, Passalacqua E, Lufrano F, Patti A (1999) An empirical equation for polymer electrolyte fuel cell (PEFC) behaviour. J Appl Electrochem 29(12):1449–1455
23. Cheddie D, Munroe N (2005) Review and comparison of approaches to proton exchange membrane fuel cell modeling. J Power Sources 147(1–2):72–84
24. Bernardi DM, Verbrugge MW (1991) Mathematical model of a gas diffusion electrode bonded to a polymer electrolyte. AIChE J 37(8):1151–1163
25. Bernardi DM, Verbrugge MW (1992) A Mathematical model of the solid-polymer-electrolyte fuel cell. J Electrochem Soc 139:2477
26. Springer TE, Zawodzinski TA, Gottesfeld S (1991) Polymer electrolyte fuel cell model. J Electrochem Soc 138(8):2334–2342
27. He W, Yi JS, Van Nguyen T (2000) Two-phase flow model of the cathode of PEM fuel cells using interdigitated flow fields. AIChE J 46(10):2053–2064
28. Natarajan D, Van Nguyen T (2001) A two-dimensional, two-phase, multicomponent, transient model for the cathode of a proton exchange membrane fuel cell using conventional gas distributors. J Electrochem Soc 148:A1324
29. Nguyen TV, White RE (1993) A water and heat management model for proton-exchange-membrane fuel cells. J Electrochem Soc 140:2178

30. Wang ZH, Wang CY, Chen KS (2001) Two-phase flow and transport in the air cathode of proton exchange membrane fuel cells. J Power Sources 94(1):40–50
31. Nguyen PT, Berning T, Djilali N (2004) Computational model of a PEM fuel cell with serpentine gas flow channels. J Power Sources 130(1–2):149–157
32. Natarajan D, Van Nguyen T (2003) Three-dimensional effects of liquid water flooding in the cathode of a PEM fuel cell. J Power Sources 115(1):66–80
33. Bernardi DM (1990) Water-balance calculations for solid-polymer-electrolyte fuel cells. J Electrochem Soc 137:3344
34. Fuller TF, Newman J (1993) Water and thermal management in solid-polymer-electrolyte fuel cells. J Electrochem Soc 140:1218
35. Wohr M, Bolwin K, Schnurnberger W, Fischer M, Neubrand W, Eigenberger G (1998) Dynamic modelling and simulation of a polymer membrane fuel cell including mass transport limitation. Int J Hydrogen Energy 23(3):213–218
36. Barsoukov En, Macdonald JR (2005) Impedance spectroscopy: theory, experiment, and applications. Wiley, Hoboken, New Jersey, USA
37. Andreaus B, McEvoy AJ, Scherer GG (2002) Analysis of performance losses in polymer electrolyte fuel cells at high current densities by impedance spectroscopy. Electrochimica Acta 47(13–14):2223–2229
38. Ciureanu M, Mikhailenko SD, Kaliaguine S (2003) PEM fuel cells as membrane reactors: kinetic analysis by impedance spectroscopy. Catal Today 82(1–4):195–206
39. Macdonald DD (2006) Reflections on the history of electrochemical impedance spectroscopy. Electrochimica Acta 51(8–9):1376–1388
40. Primucci M, Ferrer L, Serra M, Riera J (2008) Characterisation of fuel cell state using Electrochemical Impedance Spectroscopy analysis. Simposium Ibérico de Hidrógeno, Pilas de Combustible y Baterías Avanzadas (HYCELTEC)
41. Alejandro J, Real D, Arce A, Bordons C (2007) Development and experimental validation of a pem fuel cell dynamic model. J Power Sources 173(1):310–324
42. Pathapati PR, Xue X, Tang J (2005) A new dynamic model for predicting transient phenomena in a pem fuel cell system. Renew Energy 30(1):1–22
43. El-Sharkh M, Rahman A, Alam M, Byrne P, Sakla A, Thomas T (2004) A dynamic model for a stand-alone PEM fuel cell power plant for residential applications. J Power Sources 138:199–204
44. An improved dynamic voltage model of pem fuel cell stack (2010) Int J Hydrogen Energy 35(20):11154–11160 Hyceltec 2009 Conference
45. Grujicic M, Chittajallu KM, Law EH, Pukrushpan JT (2004) Model-based control strategies in the dynamic interaction of air supply and fuel cell. Proc Inst Mech Eng, Part A: J Power Energy 218(7):487–499
46. Vahidi A, Peng A (2004) Model predictive control for starvation prevention in a hybrid fuel cell system. Am Control Conf, Proc 2004 1:834–839
47. Varigonda S, Pukrushpan JT, Stefanopoulou AG (2003) American Institute of Chemical Engineers. Challenges in fuel cell power plant control: the role of system level dynamic models. Am Inst Chem Eng 1:101–110
48. Gelfi S, Stefanopoulou AG, Pukrushpan JT, Huei Peng (2003) Dynamics of low-pressure and high-pressure fuel cell air supply systems. In: Am Control Conf, 2003. Proceedings of the 2003 3:2049–2054
49. Caux S, Lachaize J, Fadel M, Shott P, Nicod L (2005) Energy management of fuel cell system and supercaps elements. In: Proceedings of the 16th IFAC World Congress, Prague
50. Serra M, Aguado J, Ansade X, Riera J (2005) Controllability analysis of decentralized linear controllers for polymeric fuel cells. J Power Sources 151:93–102
51. Feroldi D, Serra M, Riera J (2007) Performance improvement of a pemfc system controlling the cathode outlet air flow. J Power Sources 169(1):205–212
52. Zenith F, Skogestad S (2007) Control of fuel cell power output. J Process Control 17(4):333–347

53. Ceraolo M, Miulli C, Pozio A (2003) Modelling static and dynamic behaviour of proton exchange membrane fuel cells on the basis of electro-chemical description. J Power Sources 113(1):131–144
54. Yang W, Bates B, Fletcher N, Pow R (1998) Control challenges and methodologies in fuel cell vehicles development. Fuel Cell Technol for Vehicles 1:249–256
55. Friedman D (1999) Maximizing direct-hydrogen PEM fuel cell vehicle efficiency-is hybridization necessary? SAE Int 1:265–272

Advanced Control Strategies for the Oxygen in the Cathode

Diego Feroldi, David Zumoffen and Marta Basualdo

Abstract This chapter presents two advanced control strategies based on model predictive control to control the oxygen level in the cathode of a proton exchange membrane (PEM) fuel cell system. The objectives are to achieve a better efficiency and to maintain the necessary level of the oxygen in the cathode to prevent short circuit and membrane damage. First, a methodology of control based on dynamic matrix control (DMC) is proposed. This strategy includes a stationary and dynamic study of the advantages of using a regulating valve for the cathode outlet flow in combination with the compressor motor voltage as manipulated variables in a PEM fuel cell system. The influence of this input variable is exploited by implementing a predictive control strategy based on DMC, using these manipulated variables. The objectives of this control strategy are to regulate both the fuel cell voltage and oxygen excess ratio in the cathode, and thus, to improve the system performance. Second, a methodology of control based on adaptive predictive control with robust filter (APCWRF) is proposed. The APCWRF is designed for controlling the compressor motor voltage. Because of the wide working range the algorithm is improved with three different zones supported by three nominal linear models.

D. Feroldi · D. Zumoffen · M. Basualdo (✉)
CAPEG-CIFASIS-CONICET-UNR-UPCAM,
27 de Febrero 210 bis,
S2000EZP Rosario, Argentina
e-mail: basualdo@cifasis-conicet.gov.ar

D. Zumoffen
e-mail: zumoffen@cifasis-conicet.gov.ar

D. Feroldi
DCC-FCEIA-UNR,
Pellegrini 250, S2000BTP Rosario, Argentina
e-mail: feroldi@cifasis-conicet.gov.ar

D. Zumoffen · M. Basualdo
FRRo-UTN,
Zeballos 1341, S2000BQA Rosario, Argentina

M. S. Basualdo et al. (eds.), *PEM Fuel Cells with Bio-Ethanol Processor Systems*, Green Energy and Technology, DOI: 10.1007/978-1-84996-184-4_3,

1 Introduction

Proton exchange membrane (PEM) fuel cell systems are efficient devices which allow the transformation of the chemical energy stored in hydrogen into electric energy. The behavior of the generated power in a fuel cell system depends on the hydrogen and air flows, the regulation of the pressure and the management of water and heat [1]. In order to obtain this transformation efficiently, all the subsystems must work properly. The air supply is one of these subsystems, which has a great influence in the system efficiency. In fact, one of the most important challenges in fuel cell control is to assure sufficient amount of oxygen in the cathode when current is abruptly drawn from the fuel cell stack.

Model-based predictive control (MPC) is the only advanced control strategy that has a significative impact on industrial process control. The main reason is that it is the only methodology that can handle both equipment and safety routinely.

The MPC application to industrial practice was supported primarily by the following characteristics:

- the idea behind the strategy is easy to understand,
- the basic formulation is directly extended to multivariable cases practically without modification,
- is more powerful than a PID control policy, still in a single loop without constraints and with the same tuning complexity,
- it considers the actuators limits.

MPC has been developed and used in industrial applications during the last twenty years. The improvements in both hardware (computing speed) and software (optimization algorithms) in the recent years make this control strategy feasible for processes with slow dynamics [2].

In this study, the air supply subsystem is composed by an air compressor, which supplies air to the cathode. On the other hand, the hydrogen supply subsystem relies on a hydrogen-pressurized tank and the hydrogen inlet flow rate is regulated by an independent control loop to maintain the working pressure in the anode close to the pressure in the cathode. The control of the fuel cell system without any energy storage is approached in this chapter with two methodologies based on MPC. First, a methodology of control based on Dynamic Matrix Control is proposed. Second, a methodology of control based on adaptive predictive control with robust filter is proposed.

2 Methodology Based on Dynamic Matrix Control

DMC [3] is a particular type of predictive control strategy developed in the late seventies by Cutler and Ramaker with great acceptance in the industry [3]. The process model employed in this formulation is the step response of the plant, while the disturbance is assumed to keep constant along the prediction horizon. This formulation uses the step response to determine the so-called *dynamic matrix* **G**, which is interpreted as a model of the process that takes into account only the h_p first samples until the response tends to a constant value, assuming therefore that the process is asymptotically stable. Thus, the predicted output can be expressed as

$$y = \mathbf{G}u + f, \tag{1}$$

where y is the time-vector of predicted outputs, $y = \left[y_1 \,\vdots\, y_2 \,\vdots\, \ldots \,\vdots\, y_p\right]^T$, with y_i representing the time-vector of the i-th output

$$y_i = \left[y_i(k+1|k), \ldots, y_i(k+h_p|k)\right]^T, \tag{2}$$

where $y_i(k+1|k)$ denotes the prediction of y_i at time $k+1$ done in time k along the prediction horizon h_p. The control vector $u = \left[u_1 \,\vdots\, u_2 \,\vdots\, \ldots \,\vdots\, u_m\right]^T$ is the sequence of future control actions where each vector u_i represents the h_m-dimensional vector of control movements, corresponding to the ith control variable, along the control horizon h_m

$$u_i = [u_j(k), u_j(k+1), \ldots, u_j(k+h_m-1)]^T, \tag{3}$$

and $f = \left[f_1 \,\vdots\, f_2 \,\vdots\, \ldots \,\vdots\, f_p\right]^T$ is the free response vector, which means that, the response that does not depend on future control movements, where

$$f_i = \left[f_i(k+1|k), \ldots, f_i(k+h_p|k)\right]^T. \tag{4}$$

A graphical description of a general predictive control strategy can be seen in Fig. 1.

The dynamic matrix **G** is constructed from the coefficients obtained from the step response with prediction horizon h_p and control horizon h_m:

$$\mathbf{G} = \begin{bmatrix} \mathbf{G}_{11} & \cdots & \mathbf{G}_{1m} \\ \vdots & \ddots & \vdots \\ \mathbf{G}_{p1} & \cdots & \mathbf{G}_{pm} \end{bmatrix}, \tag{5}$$

and each matrix $\mathbf{G}_{\mathrm{ij}}$ of dimension $h_p \times h_m$ contains the coefficients of the step response of the ith output corresponding to the jth input:

$$\mathbf{G}_{ij} = \begin{bmatrix} g_1 & 0 & \cdots & 0 \\ g_2 & g_1 & \cdots & 0 \\ \vdots & \vdots & \ddots & \vdots \\ g_{h_m} & g_{h_m-1} & \cdots & g_1 \\ \vdots & \vdots & \ddots & \vdots \\ g_{h_p} & g_{h_p-1} & \cdots & g_{h_p-h_m+1} \end{bmatrix}. \tag{6}$$

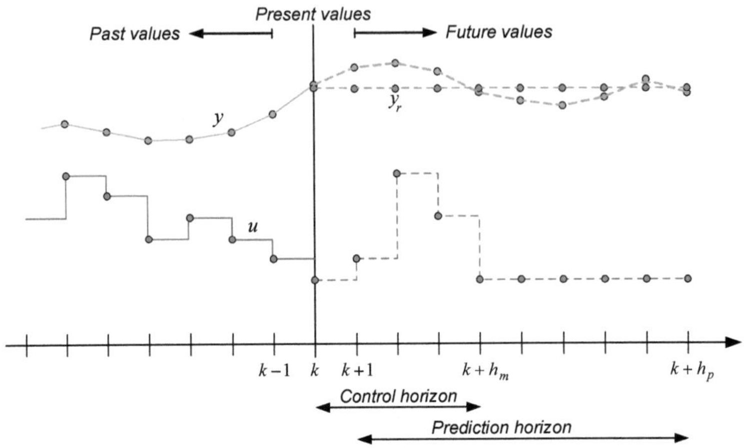

Fig. 1 Predictive control strategy

The objective of the DMC controller is to minimize the difference between the references y_{r_1} and y_{r_2}, and the predictions of the process outputs y_1 and y_2 over a horizon h_p in a least square sense with the possibility of including a penalty term on the control signal large movements:

$$\min_{u} \mathbf{J}(k). \tag{7}$$

The cost function $\mathbf{J}(k)$ is defined as

$$\begin{aligned}\mathbf{J}(k) = &\sum_{j=1}^{h_p} \left[y\left(k+j|k\right) - y_r\right]^T \mathbf{R} \left[y\left(k+j|k\right) - y_r\right] \\ &+ \sum_{j=0}^{h_m} \left[\Delta u\left(k+j|k\right)\right]^T Q \left[\Delta u\left(k+j|k\right)\right],\end{aligned} \tag{8}$$

where $\Delta u(k+j|k) \stackrel{\Delta}{=} u(k+j|k) - u(k-1+j|k)$, $\mathbf{R}$ is a matrix to compensate the different ranges of values of the process outputs, and $\mathbf{Q}$ is a matrix that allows to penalize the control effort.

If there are no restrictions in the manipulated variables, the minimization of the cost function $\mathbf{J}(k)$ can be realized making its derivative equal to zero, resulting

$$\Delta u = \left(\mathbf{G}^T\mathbf{R}\mathbf{G} + \mathbf{Q}\right)^{-1} \mathbf{G}^T\mathbf{R}e, \tag{9}$$

where**R** and **Q** are the following diagonal matrices:

$$\mathbf{R} = \text{diag}[\underbrace{r_1, \ldots, r_1}_{h_p}, \underbrace{r_2, \ldots, r_2}_{h_p}, \ldots, \underbrace{r_p, \ldots, r_p}_{h_p}], \tag{10}$$

$$\mathbf{Q} = \text{diag}[\underbrace{q_1, \ldots, q_1}_{h_m}, \underbrace{q_2, \ldots, q_2}_{h_m}, \ldots, \underbrace{q_m, \cdots, q_m}_{h_m}], \tag{11}$$

and e is the vector of future errors composed of the difference between each output y_i and the corresponding reference y_{r_i} along the prediction horizon:

$$e = [y_{r_1} - y_{p_1}(k), y_{r_1} - y_1(k+1|k), \ldots, y_{r_1} - y_1(k+h_p-1|k),$$
$$y_{r_2} - y_{p_2}(k), y_{r_2} - y_2(k+1|k), \ldots, y_{r_2} - y_2(k+h_p-1|k), \ldots]^T, \tag{12}$$

where y_{p_i} is the i-th plant output measured at present instant k. The measurement of the plant outputs provides a compensation mechanism to rectify the inevitable model errors and deal with non-measured disturbances, and thus, assures zero error at steady state.

In all the predictive control strategies, only the first element of each calculated control sequence u_j is applied to the plant and, in the next iteration, the sequence of control is calculated again using actualized information from the plant. The theoretical bases of the method are given in [3].

2.1 Control Objectives

The main objectives of fuel cell systems control found in the literature were exposed in Sect. 2.1 of Chap. 1 and, as mentioned, the control of the air supply is the more critical with regard to the efficiency improvement since the compressor consumption is the major parasitic power. The efficiency of FCS was defined in Eq. 8 of Chap. 2.

At a given fuel cell current I_{fcs}, if the compressor motor voltage is increased, the compressor flow rate augments the cathode pressure and, in turn, this also leads to a higher level the oxygen excess ratio λ_{O_2} in the cathode. However, an increment of λ_{O_2} above a certain limit generally requires an excessive increase in the compressor motor power consumption which causes a reduction in the net power since $P_{\text{net}} = P_{\text{fcs}} - P_{\text{cp}}$. Thus, it is concluded that at each level of the fuel cell current there is an optimal value of λ_{O_2} at which the net power is maximum [4]. Thus, the control of λ_{O_2} is an indirect way to improve the system efficiency.

On the other hand, since the efficiency is a function of the fuel cell voltage V_{fc}, any voltage reduction is translated into an efficiency degradation. Thus, it is necessary to regulate the fuel cell voltage. Therefore, the control objectives proposed are:

1. to regulate the oxygen excess ratio in the cathode, λ_{O_2}
2. to regulate the generated voltage in the fuel cell system, V_{fcs}

2.2 Control Variables

The only variable used to control the air supply in the cathode, from the literature review, is the compressor motor voltage [4–7]. The cathode outlet flow $W_{\mathrm{rm,out}}$ is a function of the upstream pressure and downstream pressures at the cathode nozzle. The downstream pressure is assumed to be fixed at p_{atm}. If the pressure difference is small, the flow rate can be approximated by the linearized expression

$$W_{\mathrm{rm,out}} = k\left(p_{\mathrm{rm}} - p_{\mathrm{atm}}\right). \tag{13}$$

On the contrary, a more complex expression is necessary to calculate the outlet mass flow as a function of the return manifold pressure p_{rm} as shown in Eq. 14 and Eq. 15 [5]:

$$W_{\mathrm{rm,out}} = \frac{C_{\mathrm{D,rm}} A_t\, p_{\mathrm{rm}}}{\sqrt{R T_{\mathrm{rm}}}} \left(\frac{p_{\mathrm{atm}}}{p_{\mathrm{rm}}}\right)^{\frac{1}{\gamma}} \left\{\frac{2\gamma}{\gamma-1}\left[1-\left(\frac{p_{\mathrm{atm}}}{p_{\mathrm{rm}}}\right)^{\frac{\gamma-1}{\gamma}}\right]\right\}^{\frac{1}{2}}$$
$$\text{for } \frac{p_{\mathrm{atm}}}{p_{\mathrm{rm}}} > \left(\frac{2}{\gamma+1}\right)^{\frac{\gamma}{\gamma-1}} \tag{14}$$

and

$$W_{\mathrm{rm,out}} = \frac{C_{\mathrm{D,rm}} A_t\, p_{\mathrm{rm}}}{\sqrt{R T_{\mathrm{rm}}}} \gamma^{\frac{1}{2}} \left(\frac{2}{\gamma+1}\right)^{\frac{\gamma+1}{2(\gamma-1)}} \quad \text{for } \frac{p_{\mathrm{atm}}}{p_{\mathrm{rm}}} \le \left(\frac{2}{\gamma+1}\right)^{\frac{\gamma}{\gamma-1}}, \tag{15}$$

where $C_{\mathrm{D,rm}}$ is the return manifold throttle discharge coefficient ($C_{\mathrm{D,rm}} = 0.0124$ in the used model), $R = 8.3145\,\mathrm{J\,mol^{-1}\,K^{-1}}$ is the universal gas constant, and $\gamma = 1.4$ is the ratio of the specific heat of air.

As mentioned, the compressor power consumption is the greater parasitic power and has a major effect on the system efficiency. Therefore, it is important to find efficient procedures to control the cathode air flow. We propose in this work to consider the throttle opening area A_t as a variable area. This is achieved adding a proportional valve at the cathode output. Thus, this new variable is a new control variable that regulates the outlet air flow in conjunction with the compressor motor voltage. Note that this control variable by itself is not capable to regulate the air flow since the compressor is always necessary to get pressure in the supply manifold and in the cathode. The advantages of this configuration are analyzed in Sect. 2.3. Therefore, the manipulated variables are $u_1 = V_{\mathrm{cm}}$ (compressor motor voltage) and $u_2 = A_t$ (outlet valve area). A schematic diagram of the system is shown in Fig. 2.

2.3 Performance Analysis Using an Extra Control Variable

In this section, the performance improvement achieved through the use of the extra control variable described before is analyzed. The analysis is done in two parts.

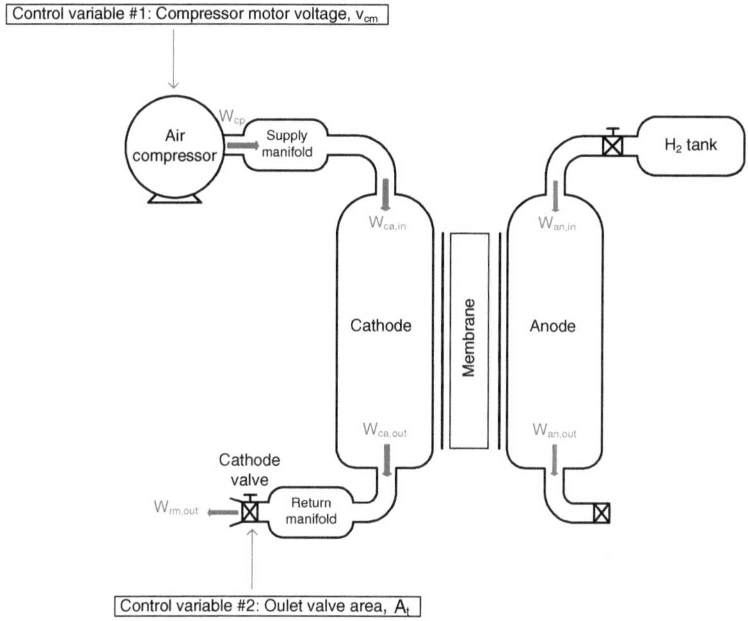

Fig. 2 PEMFC reactant flow scheme with the variables proposed for control purposes

First, a stationary analysis is done with the aim of showing the advantages in steady state using this new control variable. Second, a transient analysis is done in order to show the dynamic improvement.

2.3.1 Stationary Analysis

As mentioned before, the air supply subsystem has a strong influence the fuel cell performance. On the one hand, an insufficient air supply may cause oxygen starvation in the cathode, which causes voltage reduction and membrane life shortening. On the other hand, the compressor operation implies a power consumption that diminishes the system efficiency (see Eq. 8 of Chap. 2). The compressor motor voltage as a control input allows to regulate the oxygen partial pressure in the cathode. Augmenting the compressor voltage, the oxygen partial pressure increases. However, the compressor power consumption also increases.

A diminution of the area of the valve that closes the cathode air flow with a fixed V_{cm} contributes to increase the cathode pressure and, at the same time, contributes to modify the input air flow, the stoichiometry, and the oxygen concentration. The power consumption of the compressor has two opposite trends when the valve is partially closed: the trend to increase due to the pressure rise and the trend to decrease due to the flow reduction. When all these effects are taken into account, there is a positive balance in the total efficiency diminishing the valve area as we can observe analyzing

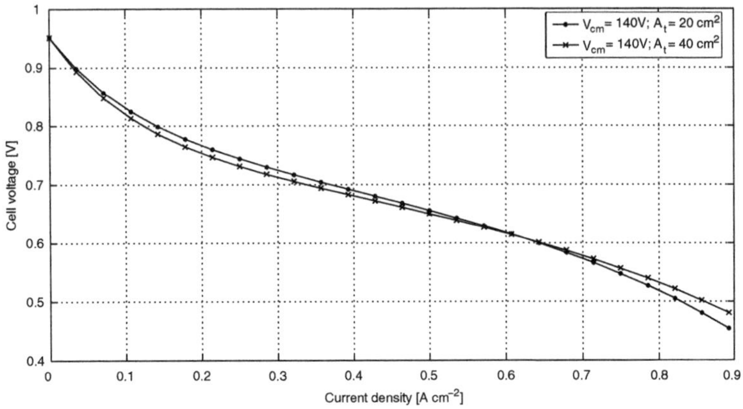

Fig. 3 Polarization curves for $V_{cm} = 140\,V$ and different valve areas

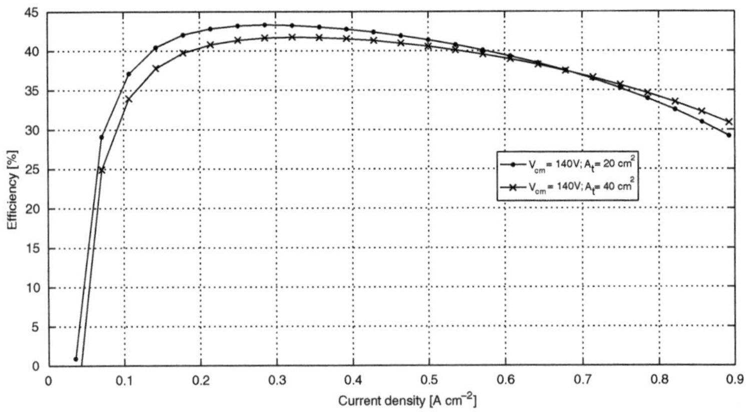

Fig. 4 Efficiency curves for $V_{cm} = 140\,V$ and different valve areas

the curves of polarization (see Fig. 3), efficiency (see Fig. 4), power compressor consumption (see Fig. 5) and oxygen partial pressure (see Fig. 6) plotted for two different valve areas and a certain compressor voltage ($V_{cm} = 140\,V$).

It is important to note that the increase in the efficiency is not for all current densities. When the current density is high, the flow and concentration reduction have a greater influence than the pressure increase and the result is a decrease in the oxygen partial pressure. In effect, the oxygen consumption is higher at higher current densities:

$$W_{O_2,rct} = \frac{M_{O_2} n_c A}{4F} i, \tag{16}$$

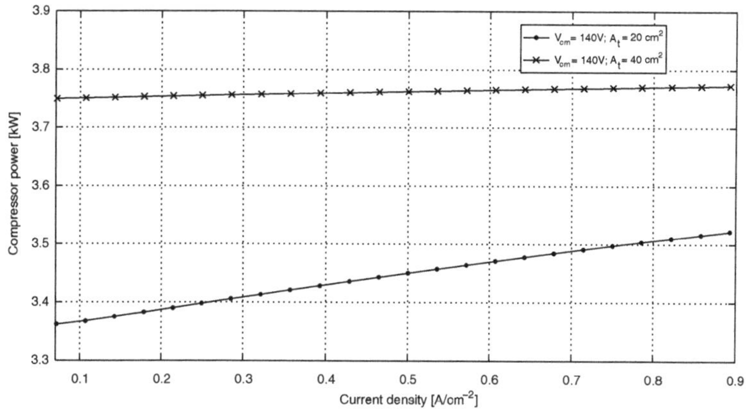

Fig. 5 Compressor power consumption curves for $V_{cm} = 140$ V and different valve areas

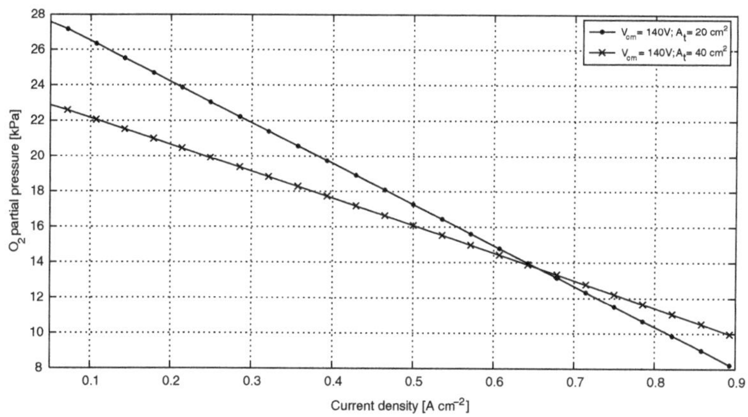

Fig. 6 Oxygen partial pressure curves for $V_{cm} = 140$ V and different valve areas

where $W_{O_2,\mathrm{rct}}$ is the rate of oxygen reacted, $M_{O_2} = 32 \times 10^{-3}\,\mathrm{kg\,mol^{-1}}$ is the molar mass of oxygen, n_c is the number of cells, A is the active cell area equal to $381\,\mathrm{cm}^2$, $F = 96485\,\mathrm{C\,mol^{-1}}$ is Faraday number, and i is the current density. Depending on the operating pressure, the increment in efficiency happens along different current density ranges. With low operating pressures (lower compressor voltage V_{cm}) the system efficiency increment due to the valve closure occurs only at low current densities, whereas with higher operating pressures (higher compressor voltage) the increment in system efficiency occurs along greater current density ranges. This can be seen in Figs. 7, 8, 9. For $V_{cm} = 100$ V the efficiency increment occurs only for current densities below $0.34\,\mathrm{A\,cm^{-2}}$ (see Fig. 7), for $V_{cm} = 140$ V the increment occurs for current densities below $0.67\,\mathrm{A\,cm^{-2}}$ (see Fig. 4), while for $V_{cm} = 180$ V the increment occurs for all the current densities analyzed (see Fig. 8).

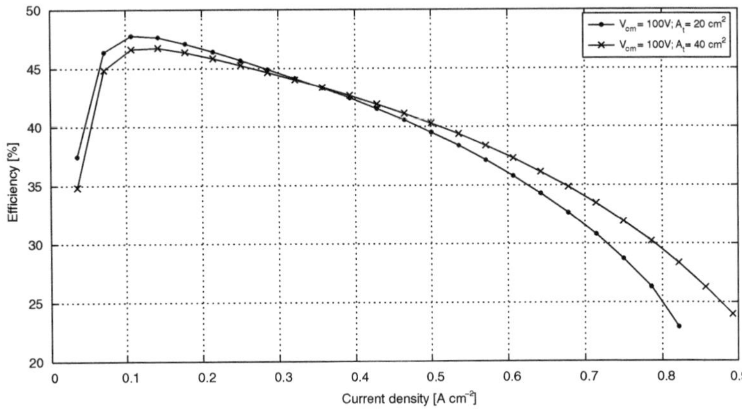

Fig. 7 Efficiency curves for $V_{cm} = 100\,\text{V}$ and different valve areas

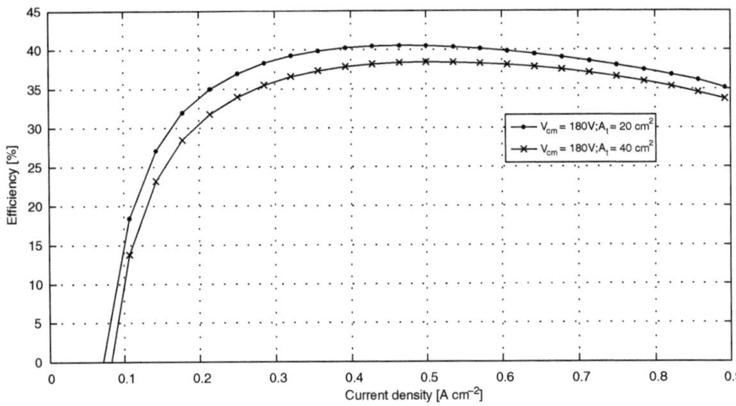

Fig. 8 Efficiency curves for $V_{cm} = 180\,\text{V}$ and different valve areas

As can be seen in Fig. 9 for $V_{cm} = 180\,\text{V}$, the oxygen partial pressure rise, as a result of diminishing the valve area, occurs at all current densities. In Fig. 10, it can be seen how the efficiency changes with A_t for values between 20 and 40 cm^2. As a conclusion of this stationary analysis, it can be stated that the cathode output valve area as well as the compressor motor voltage have to be adjusted in order to have high efficiency.

2.3.2 Transient Analysis

Besides the possible performance improvement observed in the stationary analysis, there is also an improvement in the transient behavior using the cathode air flow valve as a manipulated variable. A preliminary transient analysis is made employing a DMC

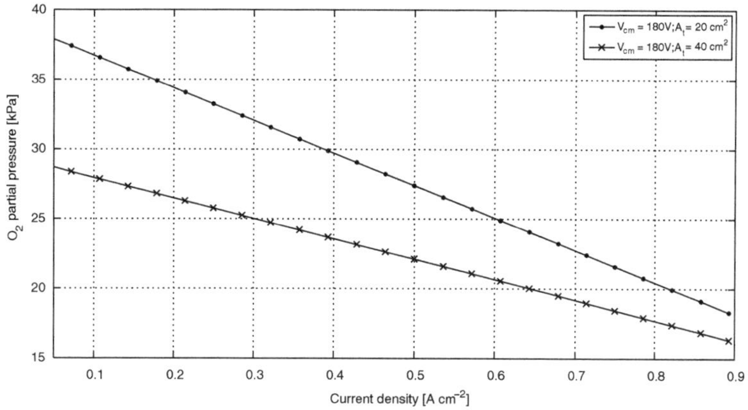

Fig. 9 Oxygen partial curves for $V_{cm} = 180$ V and different valve areas

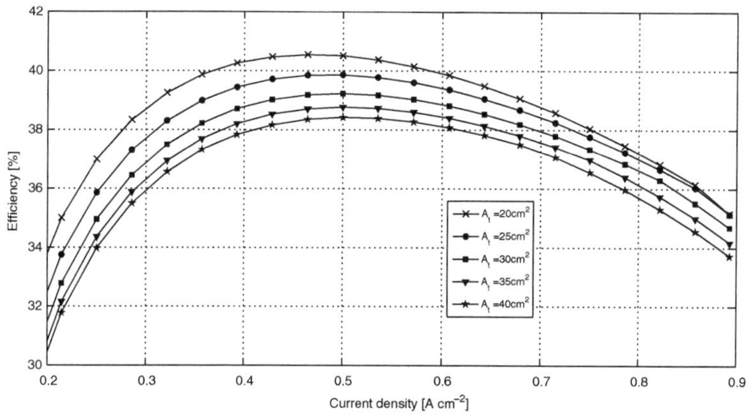

Fig. 10 Efficiency curves for $V_{cm} = 180$ V and different valve areas

strategy (the details of this control strategy are explained in Sect. 2). The controlled variable in this analysis is the stack voltage. A stack current disturbance from 200 to 210 A is applied to the fuel cell system in order to compare the disturbance rejection capability of the system. A comparison between the transient responses obtained using only the compressor voltage as a control variable and that obtained using the opening valve area in combination with the compressor voltage is shown in Fig. 11. The latter shows a better behavior with a reduction in the stack voltage time response from 0.6 to 0.4 s.

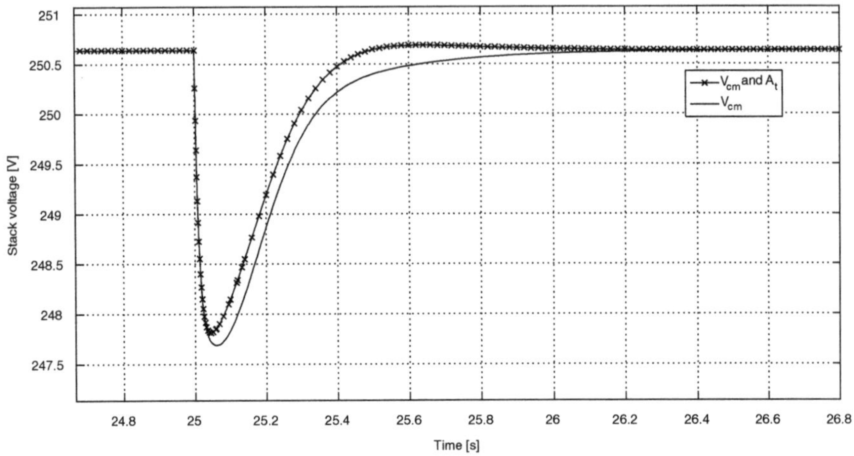

Fig. 11 Comparative between stack voltage responses

2.4 Simulation Results of DMC

The internal model utilized by the DMC controller has been obtained through the linearization of the nonlinear model described in Sect. 2.1 of Chap. 2 around the operating point corresponding to $u_1 = V_{\text{cm}} = 187.5\,\text{V}$, $u_2 = A_t = 20\,\text{cm}^2$, $w = I_{\text{st}} = 190\,\text{A}$, which gives a net power of 40 kW. The reference values for the outputs are $y_{r_1} = \lambda_{\text{O}_2} = 2.36$ and $y_{r_2} = V_{\text{fc}} = 250.6\,\text{V}$

The control horizon h_m, the prediction control h_p, and the matrices **R** and **Q** were adjusted to achieve an adequate dynamic behavior of the fuel cell system. The values of **R** and **Q** have a strong influence on the transient response obtained. This is especially true for the weight matrix **Q**, which reduces the control effort. The higher are the values in **Q**, the lower the control effort, but the response time is greater. Thus, the values chosen for the elements of the weight matrices are: $q_1 = 1$, $q_2 = 0.5$, $r_1 = 5$, $r_2 = 1$. The values of h_p and h_m are 100 and 15, respectively.

The simulation results of the controlled system with variations in the load current Fig. 12a are presented in Fig. 12b, c, d, e. Figure 12b, c show the controlled variables (λ_{O_2} and V_{fc},) while Fig. 12d, e show the manipulated variables (V_{cm} and A_t).

As can be observed in the figures, the implemented controller has a good disturbance rejection: a 17.5% of variation in the stack current at $t = 20\,\text{s}$ is demanded, $I_{\text{fc}} + \Delta I_{\text{fc}} = 170\,\text{A} + 30\,\text{A}$, the stack voltage response has a peak $\Delta V_{\text{fc}}/V_{\text{fc}} = -3.5\%$, which disappears in less than 0.5 s. At the same time, the oxygen excess ratio response presents a peak $\Delta\lambda_{\text{O}_2}/\lambda_{\text{O}_2} = -27.9\%$ that vanishes in 0.62 s. On the contrary, if only V_{cm} is controlled, λ_{O_2} has error in steady-state and the transitory response is inadequate, especially when the step in the stack current is negative (e.g., in $t = 10\,\text{s}$,) as can be seen in Fig. 13.

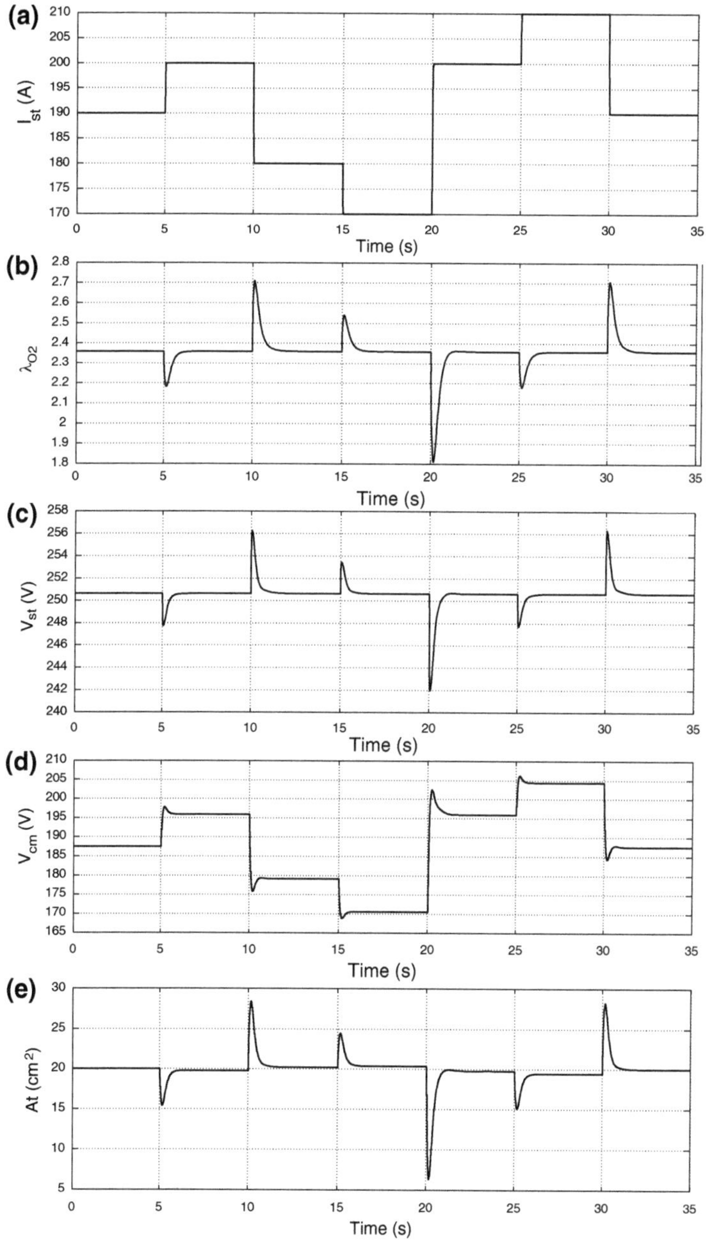

Fig. 12 Simulation results with DMC: disturbance profile (fuel cell stack current), controlled variables (oxygen excess ratio and fuel cell stack voltage), and manipulated variables (compressor motor voltage and cathode output valve area). **a** fuel cell stack current, **b** oxygen excess ratio, **c** fuel cell stack voltage, **d** compressor motor voltage, **e** cathode output valve area

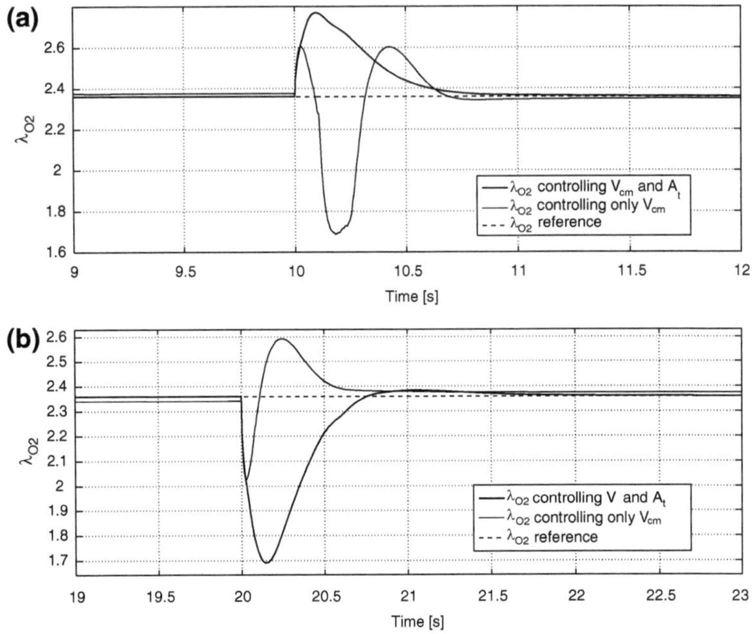

Fig. 13 Comparative between the λ_{O_2} response controlling V_{cm} and A_t respect to the case where only V_{cm} is controlled. **a** λ_{O_2} response when the step current is negative **b** λ_{O_2} response when the step current is positive

3 Methodology Based on Adaptive Predictive Control With Robust Filter

In this section, a methodology based on adaptive predictive control with robust filter is presented. First, a brief review of model predictive control is done. The Fig. 14 shows the basic idea in model predictive control for a single input–single output (SISO) case. Discrete time domain is assumed with $t = kT_s$, where k is a integer and T_s the sample time. Considering the current time, k, Fig. 14 displays the variables $y(k)$, $u(k)$ and $s(k)$ as the controlled variable (CV), manipulated variable (MV) and the set point evolutions respectively until the current time (known history). The future control actions/movements, $\hat{u}(k)$, required by the process can be computed by comparing the predicted outputs, $\hat{y}(k)$, with the set point (or reference $r(k)$) in the so-called prediction horizon $[H_w, H_p]$.

The reference trajectory, $r(k)$, is not the same as the set point trajectory, $s(k)$. In fact, the former starts in $y(k)$ (actual measured output) and defines an ideal trajectory through which the plant should return to the desired set point. Generally, $r(k)$ tends toward the set point exponentially from the current $y(k)$, with a time constant T_{ref}. Considering the current error in Eq. 17.

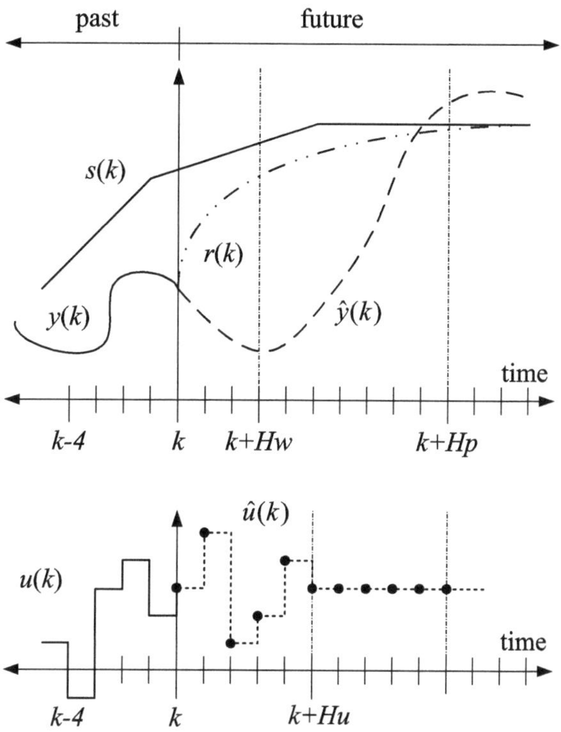

Fig. 14 MPC basic idea

$$\varepsilon(k) = s(k) - y(k), \tag{17}$$

then, considering i samples in the future,

$$\varepsilon(k+i) = e^{-iT_s/T_{\text{ref}}}\varepsilon(k) = \lambda^i \varepsilon(k), \tag{18}$$

where $\lambda = e^{-T_s/T_{\text{ref}}}$. The reference trajectory can be defined as shown in Eq. 19.

$$r(k+i) = s(k+i) - \varepsilon(k+i) = s(k+i) - e^{-iT_s/T_{\text{ref}}}\varepsilon(k) \tag{19}$$

Frequently, it is useful to consider some simple structures for future MV movements within the prediction horizon. In this case, a parameterization with a few control actions is proposed. Only H_u future movements are considered for $\hat{u}(k)$ and constant for the $H_p - H_u$ samples [2, 8]. H_u is called the control horizon. In the case represented in Fig. 14, $\hat{u}(k)$ changes the first five samples, $\hat{u}(k), \ldots, \hat{u}(k+4)$, and the remaining are held constant, $\hat{u}(k+5) = \hat{u}(k+6) = \ldots = \hat{u}(k+H_p)$.

The computing of the MV trajectory is based on the control objectives. Considering a linear model of the process, quadratic cost function, constraints given as linear inequalities and time invariant relationships, the overall control problem can be stated as shown in Eq. 20:

$$\mathbf{J}(k) = \sum_{i=H_w}^{H_p} \alpha_i^2 \left[\hat{y}(k+i) - r(k+i)\right]^2 + \sum_{i=0}^{H_u-1} \beta_i^2 \left[\hat{u}(k+i)\right]^2, \quad (20)$$

where α_i and β_i are the penalization coefficients for prediction error and signal control respectively for each future instant i [9–12]. Here, the magnitudes of $\hat{u}(k)$ are penalized rather than their change values, $\Delta u(k)$. In addition the prediction error, $\hat{y}(k+i) - r(k+i)$, is also penalized where, $\hat{y}(k+i)$ is the output prediction given by the linear model and $r(k+i)$ the reference trajectory, both evaluated within the prediction horizon. This horizon has a length of H_p samples but is not necessary to start the penalization immediately ($H_w = 1$). Considering a process with dead time or inverse response the beginning of the prediction horizon should be adjusted to $H_w > T_d$ to avoid excessive control action in this period, with T_d the delay for example. Note that, $H_u \leq H_p$ and $\Delta\hat{u}(k+i) = 0 \; \forall i > H_u$.

The closed-loop behavior of the process depends on several parameters: the beginning and end of the prediction horizon (H_w and H_p), the control horizon (H_u), the penalization coefficients (α_i and β_i), the reference trajectory ($r(k+i)$), and the linear model.

The functional cost in Eq. 21 can also consider the constraints tied to process operation ranges and/or limitations in the process devices. In this case, the following inequalities must be taken into account:

$$\begin{aligned} \mathbf{E}\left[\Delta\hat{u}(k), \ldots, \Delta\hat{u}(k+H_u-1), 1\right]^T &\leq [0, \ldots, 0]^T, \\ \mathbf{F}\left[\hat{u}(k), \ldots, \hat{u}(k+H_u-1), 1\right]^T &\leq [0, \ldots, 0]^T, \\ \mathbf{G}_r\left[\hat{y}(k+H_w), \ldots, \hat{y}(k+H_p), 1\right]^T &\leq [0, \ldots, 0]^T, \end{aligned} \quad (21)$$

where $\mathbf{E}$, $\mathbf{F}$ and $\mathbf{G}_r$ are matrices with suitable dimensions. This structure can be used to represent the potential limitations in both the range and rate of change in actuators and constraints on the process-controlled outputs.

3.1 Prediction with FIR Model

Considering a SISO process, the input–output behavior can be modeled by a finite impulse response (FIR) linear model as shown in Eq. 22,

$$y(k) = g(1)u(k-1) + g(2)u(k-2) + \ldots + g(N)u(k-N) = \sum_{i=1}^{N} g(i)u(k-i) \quad (22)$$

where $[g(1), g(2), \ldots, g(N)]$ are the FIR coefficients and N the model order. In this context, and considering the plant-model mismatch, $\eta(k)$, the prediction of the model representation in Eq. 23 can be expressed as

$$\hat{y}(k+i) = \sum_{j=1}^{N} g(j)u(k+i-j) + \eta(k+i)$$
$$= \underbrace{\sum_{j=1}^{i} g(j)u(k+i-j)}_{\text{future}} + \underbrace{\sum_{j=i+1}^{N} g(j)u(k+i-j)}_{\text{past}} + \eta(k+i), \quad (23)$$

where $u(k+i)$ is the control signal and $\hat{y}(k+i)$ the model prediction for a future instant i. The plant-model mismatch can be easily estimated by considering a constant behavior within the prediction horizon, $\eta(k+i) = \eta(k) = y(k) - \hat{y}(k)$ for $i = H_w, \ldots, H_p$. Expressing the predictions in Eq. 23 in a matrix way, the Eq. 24 can be obtained,

$$\hat{\mathbf{y}}(k) = \mathbf{T}_1\mathbf{G}_{fir}\mathbf{T}_2\hat{\mathbf{u}}(k) + \mathbf{T}_3\mathbf{S}_{fir}\mathbf{T}_4\psi(k) + \hat{\boldsymbol{\eta}}(k), \quad (24)$$

where $\psi(k)$ considers the past control actions and $\hat{\mathbf{u}}(k)$ the future control movements, with

$$\hat{\mathbf{y}}(k) = \left[\hat{y}(k+H_w), \ldots, \hat{y}(k+H_p)\right]^T, \quad (25)$$

$$\hat{\boldsymbol{\eta}}(k) = [1, \ldots, 1]^T \left[y(k) - \hat{y}(k)\right], \quad (26)$$

$$\hat{\mathbf{u}}(k) = \left[\hat{u}(k), \ldots, \hat{u}(k+H_u-1)\right]^T, \quad (27)$$

$$\psi(k) = [u(k-1), \ldots, u(k-N+H_w)]^T, \quad (28)$$

$$\mathbf{G}_{fir} = \begin{bmatrix} g(1) & 0 & \ldots & 0 \\ g(2) & \ddots & \ddots & \vdots \\ \vdots & \ddots & g(1) & 0 \\ g(N) & \ddots & g(2) & g(1) \\ 0 & \ddots & \vdots & g(1)+g(2) \\ \vdots & \ddots & g(N) & \vdots \\ 0 & \ldots & 0 & \sum_1^N g(i) \end{bmatrix}, \quad (29)$$

$$\mathbf{S}_{fir} = \begin{bmatrix} g(2) & g(3) & g(4) & \ldots & g(N) \\ g(3) & g(4) & \vdots & \cdot^{\cdot^{\cdot}} & 0 \\ g(4) & \vdots & g(N) & \cdot^{\cdot^{\cdot}} & 0 \\ \vdots & g(N) & \ddots & \cdot^{\cdot^{\cdot}} & \vdots \\ g(N) & 0 & 0 & \ldots & 0 \end{bmatrix}. \quad (30)$$

The matrices $\mathbf{T}_1$, $\mathbf{T}_2$, $\mathbf{T}_3$ and $\mathbf{T}_4$ contain zeros and ones and they allow to select specific parts of $\mathbf{G}_{fir}$ and $\mathbf{S}_{fir}$. The transformation matrices are functions of the prediction and control horizon length and the FIR model order N as shown in Eq. 31:

$$\begin{aligned} \mathbf{T}_1 &= \begin{bmatrix} \mathbf{0}\ \mathbf{I}\ \mathbf{0} \end{bmatrix}_{(H_p-H_w+1)\times(2N-1)}, & \mathbf{T}_2 &= \begin{bmatrix} \mathbf{0} \\ \mathbf{I} \end{bmatrix}_{N\times H_u}, \\ \mathbf{T}_3 &= \begin{bmatrix} \mathbf{I}\ \mathbf{0} \end{bmatrix}_{(H_p-H_w+1)\times(N-1)}, & \mathbf{T}_4 &= \begin{bmatrix} \mathbf{0} \\ \mathbf{I} \end{bmatrix}_{(N-1)\times(N-H_w)}, \end{aligned} \tag{31}$$

where $\mathbf{0}$ is a zero matrix and $\mathbf{I}$ the identity matrix, both with suitable dimensions.

In this context, the functional cost in Eq. 20 can be expressed in matrix form as

$$\mathbf{J}(k) = \mathbf{e}^T(k)\mathbf{A}^2\mathbf{e}(k) + \hat{\mathbf{u}}^T(k)\mathbf{B}^2\hat{\mathbf{u}}(k), \tag{32}$$

where $\mathbf{e}(k)$ is the prediction error vector in the horizon $[H_w, H_p]$ and

$$\mathbf{A} = \mathrm{diag}(\alpha_{H_w}, \ldots, \alpha_{H_p}), \quad \mathbf{B} = \mathrm{diag}(\beta_1, \ldots, \beta_{H_u}) \tag{33}$$

the penalization matrices. The expression of $\mathbf{e}(k)$ can be seen in Eq. 34:

$$\mathbf{e}(k) = \mathbf{y}_r(k) - \hat{\mathbf{y}}(k) = \mathbf{y}_r(k) - \mathbf{T}_1\mathbf{G}_{fir}\mathbf{T}_2\hat{\mathbf{u}}(k) - \mathbf{T}_3\mathbf{S}_{fir}\mathbf{T}_4\psi(k) - \hat{\eta}(k), \tag{34}$$

with $\mathbf{y}_r(k) = [y_r(k+H_w), \ldots, y_r(k+H_p)]^T$ the reference trajectory vector.

Replacing Eq. 34 in Eq. 32 the functional cost can be expressed as,

$$\begin{aligned} \mathbf{J}(k) = {} & \mathbf{C}^T(k)\mathbf{A}^2\mathbf{C}(k) - 2\hat{\mathbf{u}}^T(k)\mathbf{T}_2^T\mathbf{G}_{fir}^T\mathbf{T}_1^T\mathbf{A}^2\mathbf{C} + \hat{\mathbf{u}}^T(k) \\ & \times \left[\mathbf{T}_2^T\mathbf{G}_{fir}^T\mathbf{T}_1^T\mathbf{A}^2\mathbf{T}_1\mathbf{G}_{fir}\mathbf{T}_2 + \mathbf{B}^2\right]\hat{\mathbf{u}}(k) \end{aligned} \tag{35}$$

with $\mathbf{C}(k) = \mathbf{y}_r(k) - \hat{\eta}(k) - \mathbf{T}_3\mathbf{S}_{fir}\mathbf{T}_4\psi(k)$.

If constraints are not considered the optimal solution for $\hat{\mathbf{u}}(k)$ can be obtained by making $\partial\mathbf{J}(k)/\partial\hat{\mathbf{u}}(k) = 0$, and then

$$\begin{aligned} \hat{\mathbf{u}}(k) = {} & \left[\mathbf{T}_2^T\mathbf{G}_{fir}^T\mathbf{T}_1^T\mathbf{A}^2\mathbf{T}_1\mathbf{G}_{fir}\mathbf{T}_2 + \mathbf{B}^2\right]^{-1}\mathbf{T}_2^T\mathbf{G}_{fir}^T\mathbf{T}_1^T\mathbf{A}^2 \\ & \times \left[\mathbf{y}_r(k) - \hat{\eta}(k) - \mathbf{T}_3\mathbf{S}_{fir}\mathbf{T}_4\psi(k)\right]. \end{aligned} \tag{36}$$

Note that $\hat{\mathbf{u}}(k)$ is a vector with H_u components.

The predictive control philosophy (moving/sliding window) suggests that only the first component of $\hat{\mathbf{u}}(k)$ must be applied at k instant to the controlled process and so the complete procedure is repeated in the next sample time $k+1$. Thus, the current MV sample is

$$\hat{u}(k) = [1, 0, \ldots, 0]\hat{\mathbf{u}}(k) = \mathbf{R}\left[\mathbf{y}_r(k) - \eta(k)\right] - \mathbf{D}\psi(k) \tag{37}$$

with

$$\begin{bmatrix} r_1 \\ r_2 \\ \vdots \\ r_{H_p-H_w+1} \end{bmatrix}^T = [1, 0, \dots, 0] \left[\mathbf{T}_2^T \mathbf{G}_{fir}^T \mathbf{T}_1^T \mathbf{A}^2 \mathbf{T}_1 \mathbf{G}_{fir} \mathbf{T}_2 + \mathbf{B}^2 \right]^{-1} \times \left[\mathbf{T}_2^T \mathbf{G}_{fir}^T \mathbf{T}_1^T \mathbf{A}^2 \right] \quad (38)$$

$$\begin{bmatrix} d_1 \\ d_2 \\ \vdots \\ d_{N-H_w} \end{bmatrix}^T = [r_1, r_2, \dots, r_{H_p-H_w+1}] \mathbf{T}_3 \mathbf{S}_{fir} \mathbf{T}_4.$$

In this context the current control sample can be expressed using the Z transform as

$$\hat{u}(k) \left[1 + \sum_{i=1}^{N-H_w} d_i z^{-i} \right] = z^{H_w} y_r(k) \left[\sum_{i=1}^{H_p-H_w+1} r_i z^{i-1} \right] - \left[y(k) - \hat{y}(k) \right] \times \left[\sum_{i=1}^{H_p-H_w+1} r_i \right], \quad (39)$$

now called

$$D^*(z^{-1}) = 1 + \sum_{i=1}^{N-H_w} d_i z^{-i}, \quad R(z) = \sum_{i=1}^{H_p-H_w+1} r_i z^{i-1}, \quad R(1) = \sum_{i=1}^{H_p-H_w+1} r_i. \quad (40)$$

Then, by using Eq. 39 and Eq. 40 can be obtained by the control law shown in Eq. 41,

$$\hat{U}(z) = \frac{R(1)}{D^*(z^{-1})} \left[z^{H_w} \frac{R(z)}{R(1)} Y_r(z) - \left(Y(z) - \hat{Y}(z) \right) \right]. \quad (41)$$

Equation 41 requires static compensation to achieve zero tracking error at steady state. Thus the final control policy can be observed in Eq. 42,

$$\hat{U}(z) = \frac{D^*(1) K_g}{D^*(z^{-1})} \left[z^{H_w} \frac{R(z)}{R(1)} Y_r(z) - \left(Y(z) - \hat{Y}(z) \right) \right], \quad (42)$$

where $D^*(1) = 1 + \sum_{i=1}^{N-H_w} d_i$ and $K_g = 1/\sum_{i=1}^{N} g(i)$.

The control structure proposed by Eq. 42 has similarities to the classic internal model control (IMC) strategy and can be represented as shown in Fig. 15, where $F(z^{-1})$ is a low-pass filter to smooth the set point trajectory, $w(k)$, and to produce the reference trajectory $y_r(k)$.

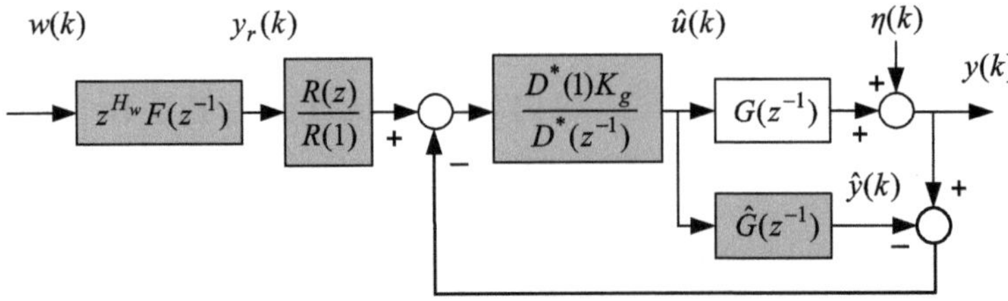

Fig. 15 MPC structure with FIR model

Several applications exist in the literature of this MPC strategy [8, 9, 11, 12]. A detailed review of this technique can be found in [13].

Remembering Eqs. 24–30, 38 and the control law in Eq. 42 it can be observed that the controller parameters ($R(z)$, $D^*(z^{-1})$, K_g) are given by the opportunely identified FIR model. This fact suggests a potential indirect adaptive control implementation by updating recursively the linear model. In the following sections this topic is addressed in detail.

3.2 *Adaptive Predictive Control*

The well-known indirect adaptive control (IAC) strategy is a classical methodology to obtain controllers with adaptive parameters. The philosophy behind the IAC is to parameterize the controller structure using a linear model of the process. Thus, any model adaptation indirectly produces a controller parameters update.

A particular strategy can be developed using the concepts of IAC and the MPC theory described in Sect. 3.1. Thus, another classical structure can be obtained called adaptive predictive control (APC)

Rewriting the FIR model expression in Eq. 22 as a linear regression

$$\hat{y}(k) = \psi(k)^T \hat{\theta}(k) + \eta(k), \tag{43}$$

where

$$\begin{aligned} \psi(k) &= [u(k-1), \ldots, u(k-N)]^T \\ \hat{\theta}(k) &= [\hat{g}(1,k), \ldots, \hat{g}(N,k)]^T \end{aligned} \tag{44}$$

are the regressor and model parameters vectors respectively, a recursive identification systems strategy can be applied to update the FIR model in each sample time. Details about classical recursive identification algorithms can be found in the appendix.

Supposing that the FIR model recursive adaptation is performed correctly, then Eq. 22 can be written as

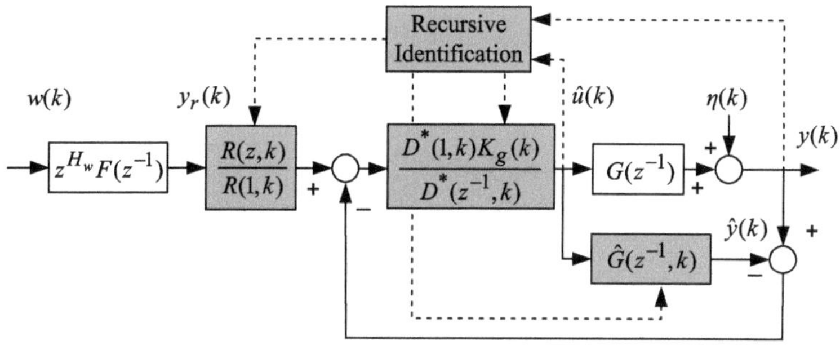

Fig. 16 APC structure with FIR model

$$\begin{aligned} \hat{y}(k) &= \hat{g}(1,k)u(k-1) + \hat{g}(2,k)u(k-2) + \ldots + \hat{g}(N,k)u(k-N) \\ &= \sum_{i=1}^{N} \hat{g}(i,k)u(k-i), \end{aligned} \tag{45}$$

where the sample time k shows explicitly the time dependence in the linear model (time variant).

In this context the control law in Eq. 42 becomes

$$\hat{U}(z) = \frac{D^*(1,k)K_g(k)}{D^*(z^{-1},k)} \left[z^{H_w} \frac{R(z,k)}{R(1,k)} Y_r(z) - \left(Y(z) - \hat{Y}(z) \right) \right]. \tag{46}$$

and eventually the blocks diagram displayed in the Fig. 15 is modified to include the recursive estimation algorithm as shown in Fig. 16 for the APC structure.

The APC (direct and indirect) topic has been addressed by several authors over many years. In Sastry and Bodson [14] stability, convergence and robustness are analyzed exhaustively. On the other hand, Jordan [8] and Kofahl [15] propose new strategies to supervise robustness and convergence problems in APC. In these works can be found both simulations and experimental results. For processes with strong nonlinearities also there exist different approaches [16, 17].

In the recent years, APC acquired a significant relevance due to its direct application in active fault-tolerant control systems. The APC adaptation properties allow to detect process abnormal events and provide a direct mechanism to update the controller automatically. These properties attracted renewed interest on these techniques. Several authors and applications use APC like active fault-tolerant control systems [12, 13, 18–21] to address problems such as faulty sensors and actuators and dynamic process modifications (Fig. 17).

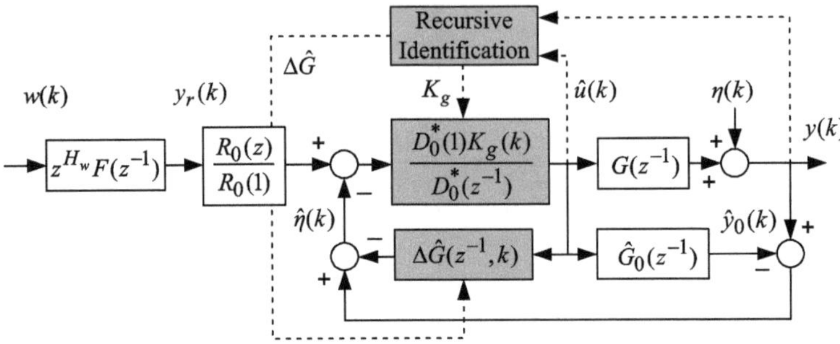

Fig. 17 APRFC structure with FIR model

3.3 Adaptive Predictive Robust Filter Control

A modification of the APC strategy is called adaptive predictive robust filter control (APRFC) [8, 9, 11]. The basic idea in this type of controller is to modify slightly the strategy shown in Sect. 3.2 by introducing an adaptive correction for the prediction given by the nominal FIR model $\hat{G}_0(z^{-1})$. In this methodology the controller parameters are not adaptive. Undoubtedly, this approach achieves a degree of robustness higher than APC, improving the closed-loop behavior even under sudden changes.

The FIR model can be considered as consisting of two effects, (1) the nominal stable part $\hat{G}_0(z^{-1})$ and (2) the adaptive stable one $\Delta\hat{G}(z^{-1},k)$ as shown in (47)

$$\hat{G}(z^{-1},k) = \Delta\hat{G}(z^{-1},k) + G_0(z^{-1}), \tag{47}$$

where

$$\begin{aligned} \Delta\hat{G}(z^{-1},k) &= \Delta\hat{g}(1,k)z^{-1} + \cdots + \Delta\hat{g}(N,k)z^{-N} \\ \hat{G}_0(z^{-1}) &= \hat{g}_0(1)z^{-1} + \cdots + \hat{g}_0(N)z^{-N} \end{aligned} \tag{48}$$

being $\hat{g}_0(i) = [h(i) - h(i-1)]/\Delta u(k)$, with $h(k)$ the process step response to $u(k)$ with magnitude $\Delta u(k)$. The nominal FIR model ($\hat{G}_0(z^{-1})$) generates, according to Eq. 42, a nominal stable predictive controller characterized by $D_0^*(z^{-1})$ and $R_0(z)$.

Rewriting the FIR model predictions as

$$\hat{y}(k) = \sum_{i=1}^{N} \Delta\hat{g}(i)u(k-i) + \sum_{i=1}^{N} \hat{g}_0(i)u(k-i) + \eta(k) \tag{49}$$

and by considering any recursive identification technique (see appendix) the feedback filter can be updated using the linear regression shown in Eq. 50:

$$\begin{aligned} \psi(k) &= [u(k-1), \ldots, u(k-N)]^T, \\ \Delta\hat{\theta}(k) &= [\Delta\hat{g}(1,k), \ldots, \Delta\hat{g}(N,k)]^T. \end{aligned} \tag{50}$$

Note that Eqs. 44 and 50 have the same regressor vector, so the APC and APRFC strategies can be coupled to work collaboratively.

Anyway, the APRFC methodology also requires static compensation to achieve zero static tracking error as shown in Eq. 51. This parameter is the only adaptive component in the controller structure.

$$K_g(k) = \frac{1}{\Delta\hat{G}(1,k) + \hat{G}_0(1)} \tag{51}$$

In [8–12] several APC and APRFC applications are presented based on FIR models both academics and experimental results.

3.4 Adaptive Predictive Control with Robust Filter

This section addressed the main concepts about the APC and APRFC interconnection and some issues related to improvements in recursive estimation and controller update algorithms.

First, a brief analysis of the inclusion of the nominal conditions in the model predictions is given. This allows to take into account the potential working point modifications in the APC algorithm implicitly.

Remembering the FIR model predictions in Eq. 22, but in this case using the nominal signal versions we have

$$\begin{aligned}\hat{y}(k) &= \sum_{j=1+d}^{N} \hat{g}(j)\,[u(k-j) - u_{00}] + y_{00} + \hat{\eta}(k)\\ &= \sum_{j=1+d}^{N} \hat{g}(j)u(k-j) + \left(y_{00} - \sum_{j=d+1}^{N} \hat{g}(j)u_{00}\right) + \hat{\eta}(k)\\ &= \sum_{j=1+d}^{N} \hat{g}(j)u(k-j) + \hat{p}_o + \hat{\eta}(k),\end{aligned} \tag{52}$$

where $\hat{g}(j)$ are the estimated FIR model coefficients, $u(k)$ is the real signal control applied to the process, $\hat{y}(k)$ the model prediction, $\hat{\eta}(k) = y(k) - \hat{y}(k)$ the plant-model mismatch, $\hat{p}_o = y_{00} - \sum_{j=d+1}^{N} \hat{g}(j)u_{00}$ is the operating point parameter, which takes into account the nominal working point $[y_{00}, u_{00}]$, and d is the offline estimation of the process dead time.

Equation 52 can be written as an augmented version of that opportunely given in Eq. 43 and eventually updated recursively:

$$\hat{y}(k) = \psi_{\text{au}}(k)^T \hat{\theta}_{\text{au}}(k) + \eta(k), \tag{53}$$

where

$$\begin{aligned}\psi_{\text{au}}(k) &= [u(k-1), \ldots, u(k-N), 1]^T, \text{ and} \\ \hat{\theta}_{\text{au}}(k) &= [\hat{g}(1,k), \ldots, \hat{g}(N,k), \hat{p}_o(k)]^T\end{aligned} \tag{54}$$

are the regressor and parameters vector in this case. Note the inclusion of the operating point parameter $\hat{p}_o$ in the parameters vector $\hat{\theta}_{\text{au}}(k)$.

Using the augmented version of the FIR predictions in Eq. 52 and the control policy defined in Eq. 37 it can be observed that [9–12]

$$\begin{aligned}\hat{u}(k) &= \mathbf{R}(k)\left[\mathbf{y}_r(k) - [1,\ldots,1]^T\left(y(k) - \psi_{\text{au}}^T(k)\theta_{\text{au}}(k)\right)\right] - \mathbf{D}(k)\psi(k) \\ &= \mathbf{R}(k)\left[\mathbf{y}_r(k) - [1,\ldots,1]^T\left(y(k) - \psi^T(k)\theta(k) - \hat{p}_o(k)\right)\right] - \mathbf{D}(k)\psi(k) \\ &= \mathbf{R}(k)\left[\mathbf{y}_r(k) + [1,\ldots,1]^T\hat{p}_o(k) - [1,\ldots,1]^T\left(y(k) - \psi^T(k)\theta(k)\right)\right] \\ &\quad - \mathbf{D}(k)\psi(k).\end{aligned} \tag{55}$$

The last expression in Eq. 55 allows to understand how the $\hat{p}_o(k)$ parameter introduces the information about the current process state. If there are no changes in the operating point the $\hat{p}_o(k)$ parameter is constant and gives the adequate compensation to the normalized (no augmented) FIR predictions $\psi^T(k)\theta(k)$. On other hand, if the operating point is modified either by disturbances or sensor faults (offset, drift, etc.) the $\hat{p}_o(k)$ parameter will take this into account recursively and give the suitable correction. This characteristic causes the APC algorithm to be an excellent choice to develop fault-tolerant control systems.

Second, the APC structure with FIR model has another important advantage, the process dead time (and/or inverse response) can be estimated online directly from the adaptive FIR coefficients by thresholding as shown in Eq. 56,

$$\hat{d}(k) = \max\left(i : \hat{g}(i,k) < \rho\right), \quad \text{with} \quad i = 1, \ldots, N, \tag{56}$$

where ρ is the threshold value.

This allows to update the beginning of the prediction horizon as $H_w(k) = \hat{d}(k)+1$ and to avoid excessive control action due to $\hat{d}(k)$. Thus, this strategy generates that the APC algorithm not only updates the FIR model and controller parameters but it also updates the order of the controller via $H_w(k) = \hat{d}(k) + 1$. Eventually $H_p(k) = \nu + H_w(k)$ can be updated to maintain the prediction horizon size, ν.

Another important issue is the time instant to perform the model and controller updating. It is well known that recursive identification algorithms suffer some problems when the regressor is not a sufficiently persistent excitation. In order to avoid unwanted updates when this persistence is lost it is necessary to develop some excitation monitoring index. In this book an error covariance matrix ($\mathbf{P}(k-1) = \mathbf{U}_*(k-1)\mathbf{D}_*(k-1)\mathbf{U}_*(k-1)^T$, see Appendix) based index is proposed as shown in Eq. 57 [8, 15]

$$z_n(k) = \frac{\lambda}{\lambda + \psi^T(k)\mathbf{P}(k-1)\psi(k)} \tag{57}$$

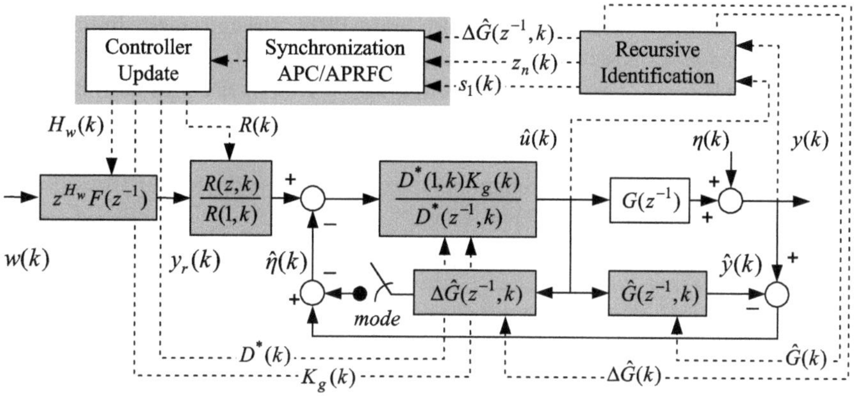

Fig. 18 APCWRF structure with FIR model

and its filtered version as displayed in Eq. 58:

$$s_1(k) = \begin{cases} 0.7s_1(k-1) + 0.3z_n^2(k), & \text{if } s_1(k-1) \le 0.8z_n(k) \\ 0.99s_1(k-1) + 0.01z_n^2(k), & \text{in other case.} \end{cases} \tag{58}$$

thus, Eqs. 57 and 58 conform the updating law stated in Eq. 59. This suggests that the FIR model will be updated if

$$z_n(k) < s_1(k) \tag{59}$$

is fulfilled, where $0 \le z_n(k) \le 1$. For a well-excited system $z_n(k)$ is near to zero and the opposite for system with poor excitation. In addition, the controller parameters are updated during N samples while Eq. 59 is fulfilled, with N the FIR model order.

Finally, the basic idea behind the APCWRF approach [8–12] is to maintain all the properties from both APC and APRFC methodologies in a single algorithm. In other words, coupling together the disturbances and sudden changes in robustness from APRFC and the tracking and convergence properties under normal operation from APC.

The APCWRF structure can be observed in Fig. 18. The recursive identification block, based on either classic RLS or RLS with UD-factorization, gives information about the model actualization $\Delta\hat{G}(z^{-1},k)$ and the persistent excitation indexes $z_n(k)$ and $s_1(k)$. This information is analyzed by the synchronization block which decides whether

- a model adaptation is necessary,
- APC or APWRFC structure is applied,
- the controller parameters must be updated.

The synchronization block is displayed in Fig. 19. The initialization step, $k = 0$, involves several activities as the offline FIR model identification of the process, parameters and matrices adjust for the APC and APRFC structures, etc. Initially, the binary variable *mode* is set to zero indicating that the APC structure is used. On the other hand, when $mode = 1$ it indicates that the APRFC is used. In each iteration the *mode* variable is analyzed, if $mode = 0$ then APC has been applied in the last iteration and the stability test on the controller polynomial, $D^*(z^{-1}, k)$, is allowed. In this context, it is necessary to remember the origin of this polynomial. Considering the control law in Eq. 42 and $\eta(k) = 0$ then

$$U(z) = \frac{D^*(1)K_g Y_w(z)}{D^*(z^{-1}) + \left[G(z^{-1}) - \hat{G}(z^{-1})\right] D^*(1)K_g}, \tag{60}$$

where $Y_w(z) = z^{H_w} F(z^{-1}) R(z) w(z)/R(1)$ is both the filtered and advanced versions of $w(z)$. Thus, the controller stability depends on the roots of the following polynomial [8, 9]:

$$D^*(z^{-1}) + \left[G(z^{-1}) - \hat{G}(z^{-1})\right] D^*(1)K_g. \tag{61}$$

Assuming that the model updates are performed correctly, thus the plant-model mismatch $[G(z^{-1}) - \hat{G}(z^{-1})] \to 0$ when $t \to \infty$. Thus the controller stability can be tied to $D^*(z^{-1})$ only. This is an approximation, in fact this polynomial can be stable but due to the plant-model differences the controller polynomial in Eq. 61 may be unstable. To avoid problems with the above approximation the condition that the model estimates are reliable,is also added. As seen from Fig. 19 if $D^*(z^{-1})$ is stable and a new test is added to guarantee the regressor excitation degree by $z_n(k) < s_1(k)$. If this is true, then the synchronization algorithm decides to stay in the APC strategy and both model and controller are updated. In the case that $D^*(z^{-1}, k)$ becomes unstable or $z_n(k) \geq s_1(k)$ the APC strategy is turned off and the APRFC is selected with $mode = 1$ indicating this situation. The APRFC remains active during N samples before returning to the APC policy and to update the controller matrices. This guarantees that the adaptation is made with stable data.

3.5 Application Results of APCWRF

In this section, a PEMFC system under an improved APCWRF is analyzed using a nonlinear dynamic model. Sudden changes in the stack current are associated with the abrupt changes in the power demanded by the electric motor of a vehicle, powered by the PEMFC. In this case two types of DC motors are considered. The APCWRF is designed for controlling the compressor motor voltage. Because of the wide working range the APC algorithm is improved accounting for three different zones supported by three nominal linear models. It is especially thought to achieve a better efficiency and to maintain the necessary level of the oxygen in the cathode to prevent short circuit and membrane damage. A DC/DC converter is connected to the electric motor.

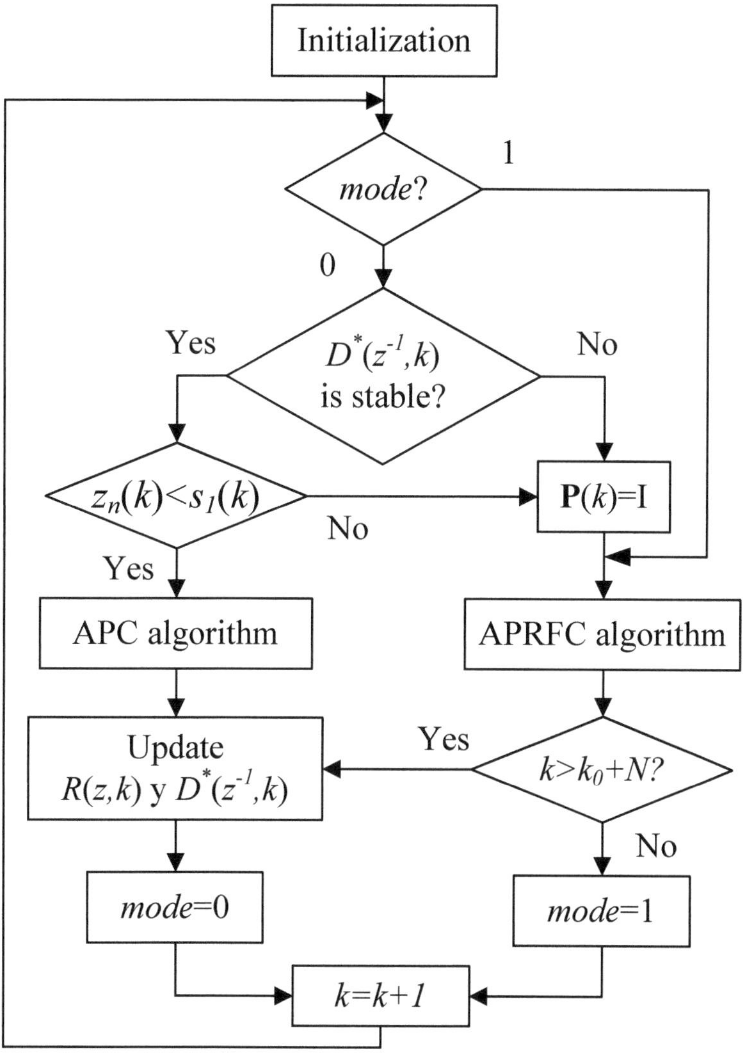

Fig. 19 APCWRF synchronization algorithm

In an FC system there are three main control systems to regulate: (a) the air/fuel supply; (b) the water supply; and (c) the heat management. The main assumptions are the following: a perfect air/fuel humidifier and perfect air and stack cooler. In addition, a fast proportional feedback fuel-flow controller, which ensures a zero pressure difference across the membrane, is assumed. Therefore, the control problem discussed in the present work focuses on the regulation of the air (i.e. oxygen) supply to the cathode. Remembering the optimal nonlinear mapping of the most important variables when the fuel cell operates under standard conditions [22–24], then from

Eq. 50 of Chap. 2 can be observed the optimal value of the compressor motor voltage v_{cm}^{opt} which maximizes the stack net power, P_{net}^{max}. It was found quite successfully to represent the variation of the optimal v_{cm}^{opt} with I_{st}. When the current demand from the load suddenly increases, the oxygen consumption in the cathode also increases leading the oxygen partial pressure to drops. The accompanying drop in the FC voltage may lead to a short circuit and/or membrane damage; this phenomenon is known as oxygen starvation. To prevent this, the air supply must be promptly increased to replenish the cathode with oxygen. Also, the fuel supplied to the anode must be quickly adjusted to ensure a minimal pressure difference across the FC membrane. A similar control of the FC system is required during a sudden drop in the stack current. Thus, the FC system control problem can be defined as

- manipulated variable (MV): compressor input voltage, v_{cm}.
- controlled variable (CV): compressor air mass flow rate, W_{cp}.
- disturbance (D): stack current, I_{st}.
- performance variables (PVs): power net difference, $z_1 = P_{net} - P_{net}^{max}$ and excess ratio difference, $z_2 = \lambda_{O_2} - \lambda_{O_2}^{opt}$

3.5.1 Disturbances to the PEMFC

The DC motors considered here present different excitation modes (field coil connections), and eventually different control structures as shown in Fig. 20. These DC motors type are briefly described as follows:

- **Permanent magnet DC (PMDC) motor:** In this case the poles are formed by a permanent magnet with constant magnetic flux. A cascade control structure is used for the speed control (master loop) that gives the current reference signal, $I_{a,ref}$, for the induced current controller (slave). In addition, this current controller defines the induced voltage feed, v_{ref}, for the DC/DC converter. The overall control structure is shown in Fig. 20a.
- **Excited separately DC (SEDC) motor:** In this case the field circuit (I_e, U_e) is excited independent of the armature circuit (I_a, U_a) and the speed control can manipulate both armature and excitation voltage. Basically, a combined strategy is proposed here allowing to operate the DC machine in both nominal and over-speed (field weakening) zones. In this control structure can be observed two control loops. By one hand, the armature loop, containing a current control of the internal armature and an external speed controller connected in cascade. On the other hand, the excitation loop, which also presents two hierarchical cascade controllers: the internal excitation current controller and the external electromotive force (emf) controller. This control structure can be observed in Fig. 20b.

All the controllers used in these subsystems are proportional/integral (PI) type with anti-windup. The currents and voltages in the armature and excitation circuits are limited considering their maximum allowed time of currents overshoot and 1.2 times the nominal value for voltages.

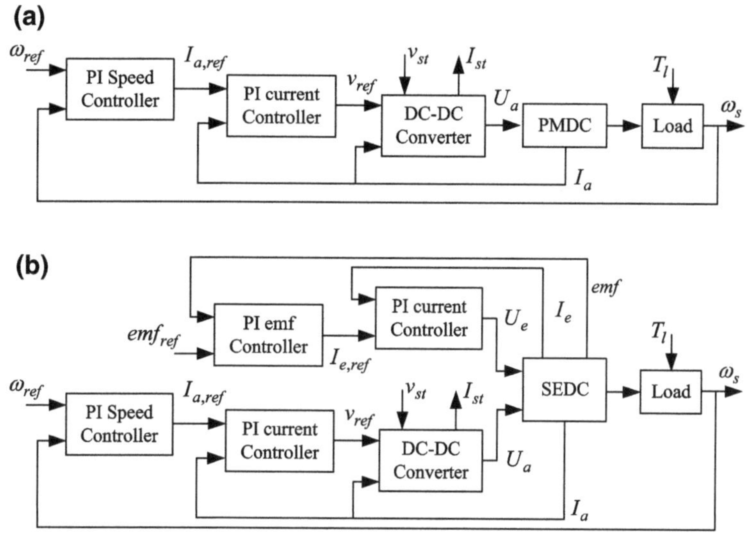

Fig. 20 Speed control strategies for DC motors **a** PMDC motor—armature control, **b** SEDC motor—armature/field weakening control

The only way to control the speed in PMDC motors is by means of the armature circuit (the field excitation is fixed). The zone in which the excitation is constant and the armature voltage, U_a, can be modified is generally called the *constant torque region*. The maximum value for U_a is given by its nominal magnitude, U_{aN}, and generates the nominal speed, ω_N. The maximum achievable speed in these conditions. On the other hand, the SEDC motors have an additional circuit with implications on the speed, the excitation circuit. It allows to achieve different degrees of the excitation in an independent way. Then, in SEDC motors the speed can achieve higher values than the nominal one by means of a suitable excitation field weakening procedure. This approach defines the so-called *constant power region*, due to both the armature voltage and current at the nominal magnitudes. The above stated regions are common in DC machines analysis by using the capability curves. Based on this, the speed control structure with both armature and excitation circuit manipulations present flexibility in the operational range.

The control structures shown in Fig. 20 are the load effect connected to PEMFC. This load is represented by both the stack voltage, v_{st}, and current, I_{st}. Thus, different range of operations for the DC motors represent different load scenarios from the PEMFC point of view.

In Fig. 21 the overall system with its control structure is summarized. Basically, it consists of the FC with its local controllers (feedforward and feedback) connected to the DC–DC converter, the DC motor as load disturbance and the speed controller constituting the traction control system (TCS). The load disturbance T_l [Nm], that affects the TCS, presents an arbitrary temporal profile as can be observed in Eq. 62. The DC motor reaches the desired speed without load ($t < 30$ s), avoiding an

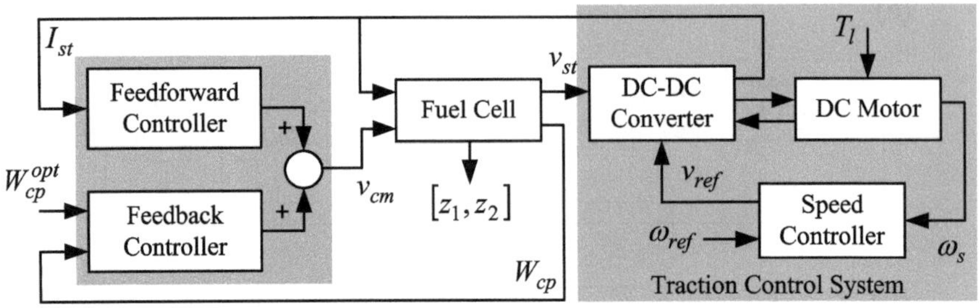

Fig. 21 Overall control structure of PEMFC and traction system

excessive fuel cell current transient requirement in this period.

$$T_l = \begin{cases} 0, & t < 30\,\mathrm{s} \\ 30, & 30\,\mathrm{s} \le t < 60\,\mathrm{s} \\ 75, & 60\,\mathrm{s} \le t < 90\,\mathrm{s} \\ 30, & 90\,\mathrm{s} \le t < 120\,\mathrm{s} \\ 60, & t \ge 120\,\mathrm{s} \end{cases} \tag{62}$$

The control structure proposed here performs two different actions. One of them considers the feedforward contribution given by an optimal static nonlinear mapping according to Eqs. 50–53 of Chap. 2, and the other takes into account the feedback contribution given by an APC by zones strategy. This last approach provides the suitable servo behavior for the optimal compressor air flow reference in Eq. 50 of Chap. 2 manipulating the compressor input voltage. Thus, the optimal excess ratio of oxygen and the maximum net power are guaranteed although the disturbances and plant-model mismatch occur.

The feedback controller block in Fig. 21 is implemented with an APCWRF approach as described in Sect. 3.4. In this particular case, this strategy has been modified by including a zone commutation taking into account the excursion range of the load current demanded I_{st} (disturbance). Thus, the blocks labeled as controller update, synchronization and recursive identification in Fig. 18 are augmented with a *zone* signal. In this context, three FIR models, obtained around three strategic operating points inside the corresponding ranges: zone 1: 100–150 A, zone 2: 150–200 A and zone 3: 200–250 A are proposed. The initial conditions for each zone were taking into account in the control algorithm so as to be switched together in the transition stages. This modification gave an interesting improvement on the dynamic behavior for the wide working range imposed by the electric motors considered here [12]. Thus, the commutation between zones involves selecting the corresponding model and controller parameterization, which is performed together with a set point modification in the same direction. This methodology allows a smooth dynamic behavior when sudden changes in the tracking and prediction error occur and prevents abrupt transitions in the manipulated variable. Hence, only slight transient effects during the commutation stage can be detected.

The stability aspects about the classical APCWRF approach has been discussed in [8, 9]. In this case, the stability performance of APCWRF by zones is guaranteed by a suitable initialization of the algorithm when a commutation must be performed. Each zone is characterized by a nominal stable FIR model together with a stable control structure parameterization. Thus, when a zone change takes place the actual control problem commutes to another one previously defined by their corresponding nominal FIR model and a stable controller. In other words, the overall problem can be seen as three asymptotic stable APCWRF control problems. Anyway, the switching control strategies are still an open and challenging problem for medium-and large-scale processes.

In this section several simulation results are included in order to demonstrate the potentiality of the control strategy described above. First, a brief comment on the controller tuning parameters is performed. For the case study given here, both control strategies (APC and APRFC) have the same parameters shown in Table 1. The prediction horizon $[H_w, H_p]$ was adopted as ten samples with $H_w = 1$ and $H_p = 11$ because it provides a good trade-off between transient response and prediction quality. Otherwise, for large values, the effect is mainly observed in the steady-state behavior. Furthermore, if the process presents dead time and/or inverse response, the prediction horizon must begin after the delay and/or the period with inverse response so as to spend less control energy. The end of the prediction horizon, H_p, has direct impact on the stability. Large values of H_p allow to have an extended vision of the future behavior so that corrective actions can be taken immediately. The main problem involved with these parameters selection is the computational cost associated with them. The control horizon, H_u, in this case was selected as two movements. In this case allowing fewer number of movements (small control horizon) means more quick and aggressive response. Another important parameter is the FIR model order, N, in this case it was adjusted to 30 samples. A practical rule of thumb [8] suggests no more than 40 samples, once the sampling time, T_s, is adopted. The predictive functional cost is also parameterized by the penalty coefficients α_i and β_i which weigh the prediction error and the control energy respectively. These coefficients can be represented in matrix form as $\mathbf{A} = \text{diag}(\alpha_{H_w}, \ldots, \alpha_{H_p})$ and $\mathbf{B} = \text{diag}(\beta_0, \ldots, \beta_{H_u-1})$. In this case these matrices are adjusted to an unitary diagonal matrix and a cero matrix respectively weighing only the prediction error. The tuning implications of these matrices are direct and opposite, large values of $\mathbf{A}$ generate aggressive responses and large values of $\mathbf{B}$ produce smooth manipulated movements. Finally, the main parameter in the recursive estimation is the forgetting factor, λ. This produces an exponential weighting of the data in the recursive algorithm. A rule of thumb [25] suggests that typical values are in the range [0.96, 0.995]; clearly this selection affects the convergence rate of the adaptive identification.

The results are grouped into two sets. Initially, the PMDC motor under the load disturbance profile of the Eq. 62 is presented in Figs. 22, 23, 24. The Figs. 25, 26, 27 summarize the same simulation condition but, in this case, connected to the SEDC motor. In both cases, the TCS guarantees the desired shaft speed of 10 rad/s under several load changes. In addition, graphical and numerical comparisons between

Table 1 APCWRF parameters setting

Variable	H_w [samp.]	H_p [samp.]	H_u [samp.]	N	α_r	**A**	**B**	λ	T_s [s]
Value	1	11	2	30	0.09	**I**	**0**	0.995	0.1

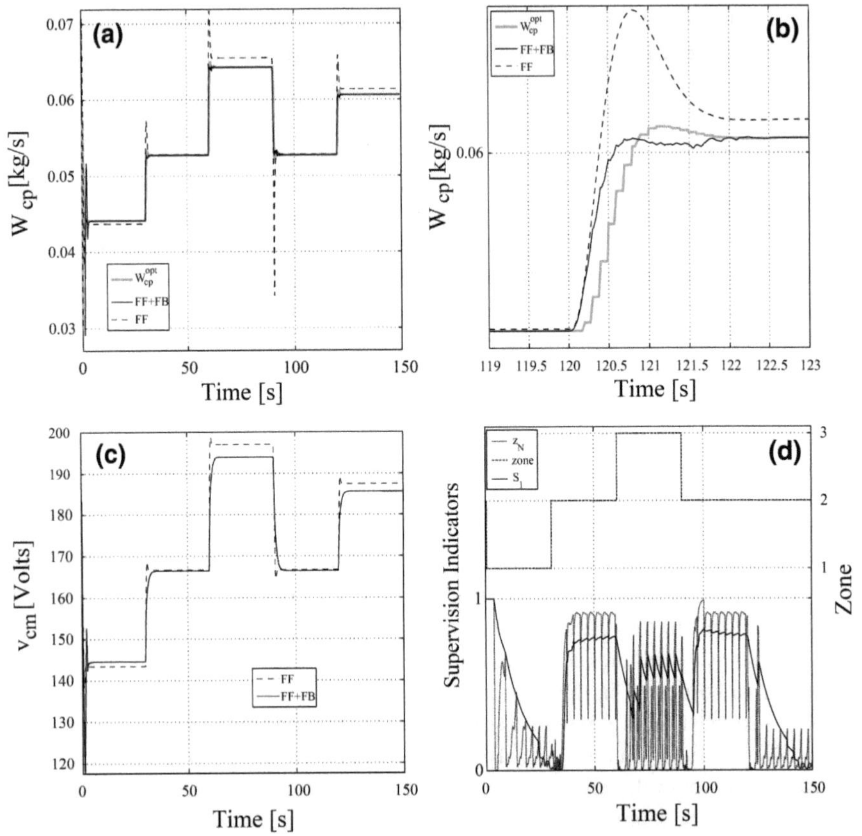

Fig. 22 Controller variables—PMDC motor **a** compressor air flow, **b** compressor air flow with zoom, **c** compressor motor voltage, **d** indicators and zone

the control strategy proposed here and the optimal static feedforward responses are included.

The Fig. 22 summarizes the principal variables involved in the fuel cell control strategy. First, in Fig. 22a,b can be observed as the optimal compressor air flow, W_{cp}^{opt}, set point, derived from the instantaneous, I_{st} and the measured air flow, W_{cp}, when both a static feedforward (FF) and a FF in combination with the APCWRF approach (FF+FB) are implemented. Clearly, the static FF presents offset problems and considerable overshoot in some responses. The FF+FB strategy proposed here displays a suitable behavior without offset and reduces the aggressive evolutions on,

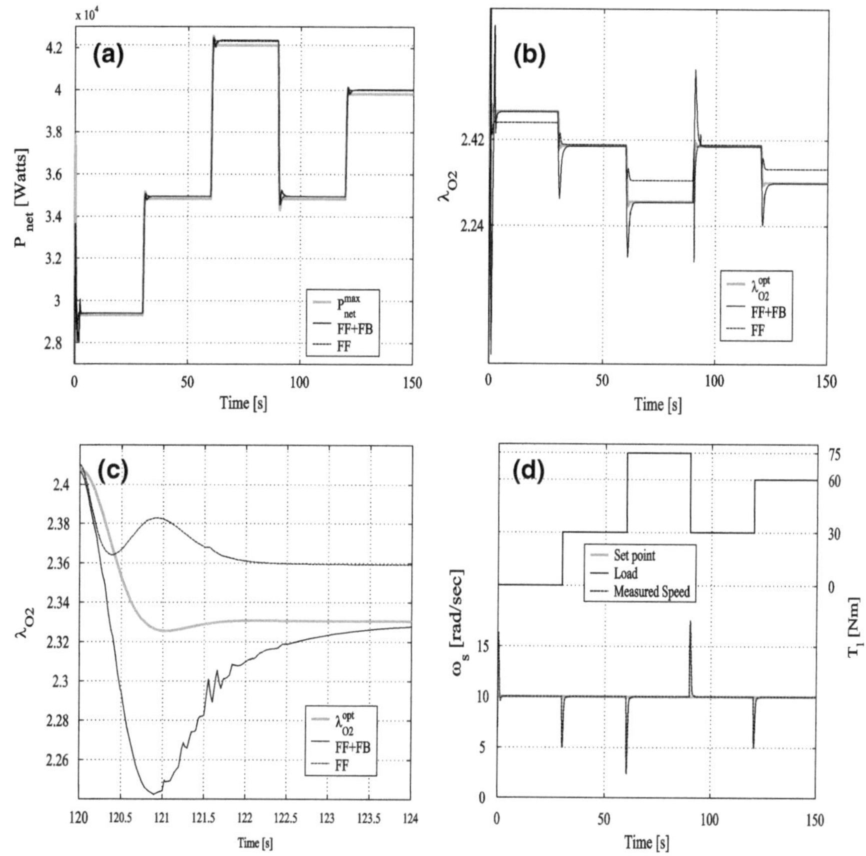

Fig. 23 Performance and load variables—PMDC motor **a** net power, **b** excess ratio, **c** excess ratio zoom, **d** shaft speed and torque

W_{cp}. It can be observed too in Fig. 22c where the manipulated variable, υ_{cm}, profile is presented for both control structures. The Fig. 22d shows the principal variables in the APCWRF approach, the supervision indicators of Eqs. 57 and 58, for deciding the use of the robust filter or not, and the operating zone evolution that drives to the proper FIR model selection.

On the other hand, in Fig. 23 can be seen both the fuel cell performance and TCS variables in the PMDC motor case. The fuel cell net power, P_{net}, in Fig. 23 shows good behavior in agreement with the optimal reference in Eq. 50 of Chap. 2. Although the net power is an indirect controlled variable (ICV), in this case both FF and FF+FB control strategies present a suitable performance. Another ICV is the excess ratio of the amount of oxygen, λ_{O_2}. The desired evolution of this variable is represented by the optimal one given in Eq. 51 of Chap. 2 Figure 23b, c (zoom) shows these behaviors when both the FF and FF+FB control strategies are implemented. Clearly, the FF approach presents important offset problems degrading its performance. Meanwhile,

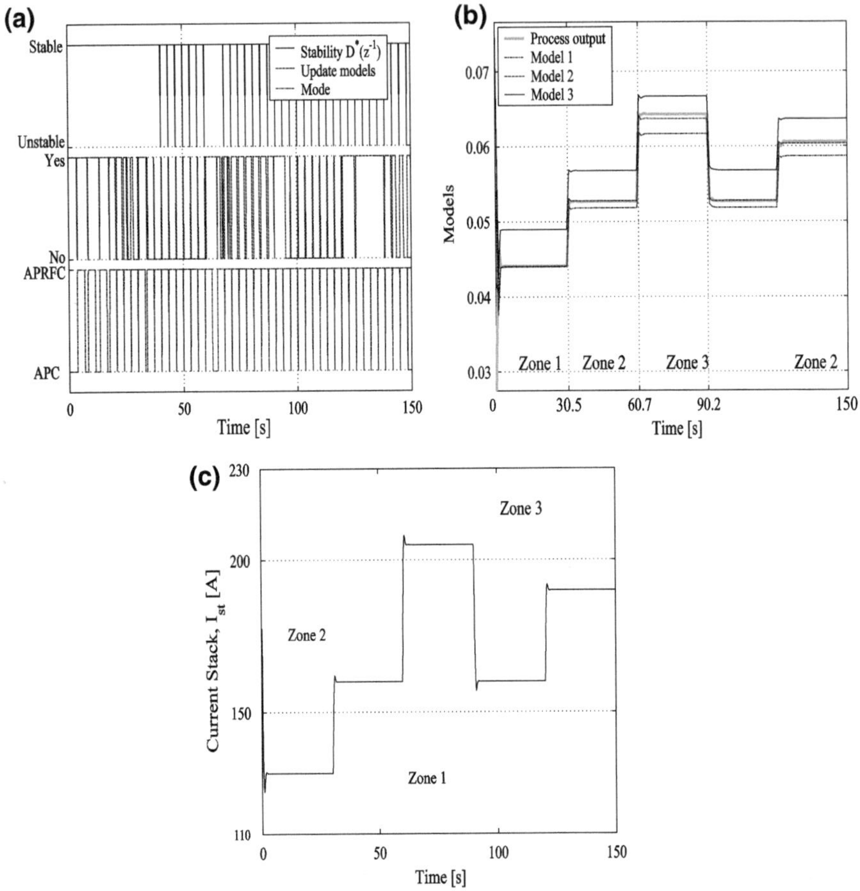

Fig. 24 Controller signals—PMDC motor **a** synchronization signals, **b** model predictions (compressor air flow), **c** stack current

the FF+FB strategy allows a suitable excess ratio evolution (without offset problems) guaranteeing the optimal one although they pass through different operating conditions. Figure 23d displays the principal variables in the TCS, the load torque changes and the controlled shaft speed under this condition.

On the other hand, Fig. 24 summarizes the different useful signals tied to the adaptive predictive controller. Figure 24 shows the synchronization signals used in the APCWRF approach (Fig. 19) along the simulation, like the stability check of $D^*(k^{-1})$, the FIR model update permission and the mode variable (the control approach used). Can be observed as both stability and permission signals define the control approach used, APC or APRF. Figure 24b displays the three nominal FIR predictions of the compressor air flow along the different working zones. Each model is the most representative of its zone. The stack current, I_{st}, excursions

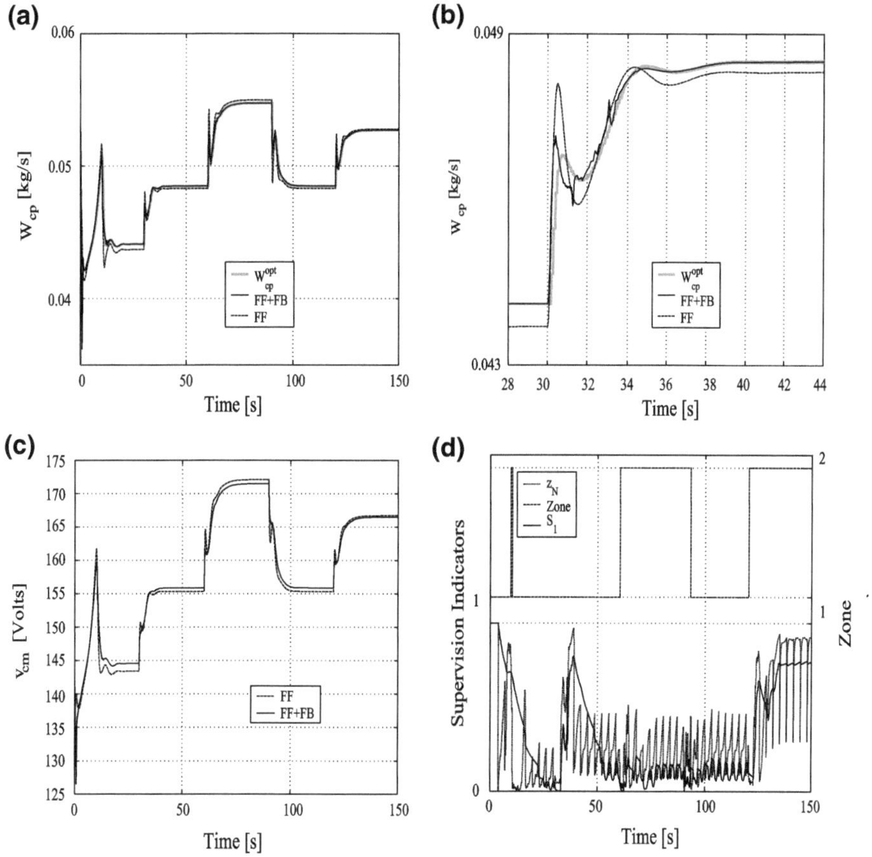

Fig. 25 Controller variables—SEDC motor **a** compressor air flow, **b** compressor air flow zoom, **c** compressor motor voltage, **d** indicators and zone

are shown in Fig. 24c. This variable defines when the zone changes occur for the controller. It is important to note that in the limit of two zones the nominal FIR model corresponding to each zone is equally valid, giving good predictions in both cases. Anyway, the zones definition were made assuming to cover a wide range of the stack current excursions. This may be the case when more demanding torque loads are applied to the motors or when the motor type is changed.

The second group of simulations considers the same operation conditions but in this case using an SEDC motor in the TCS. Figure 25 presents the principal variables involved in the fuel cell control strategy. First, in Fig. 25a and its zoom in Fig. 25b can be observed the optimal compressor air flow, W_{cp}^{opt} and the measured air flow, W_{cp}, when both FF and FF+FB control are implemented. The static FF presents offset problems with respect to the optimal reference trajectory. The FF+FB strategy proposed here demonstrates a suitable behavior without offset, handling well the plant-model mismatch and nonlinear problem by adaptation. This can also be

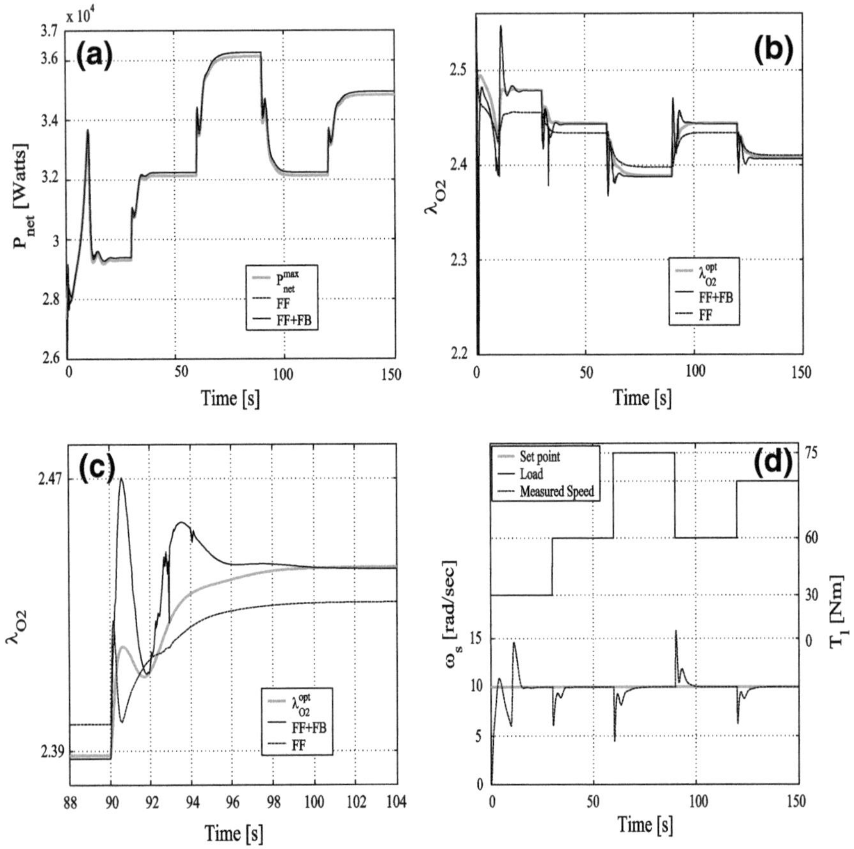

Fig. 26 Performance and load variables—SEDC motor **a** net power, **b** excess ratio, **c** excess ratio zoom, **d** shaft speed and torque

observed in Fig. 25c where the manipulated variable, v_{cm}, profile is presented for both control structures. Figure 25d shows the principal variables in the APCWRF approach, the supervision indicators of the Eqs. 57 and 58.

The last simulation presented in Fig. 26 displays both the fuel cell performance and TCS variables with SEDC motor. The fuel cell net power shown in Fig. 26a presents a good evolution in agreement with the given optimal reference with both control structures. The net power and the excess ratio of the amount of oxygen are ICVs and are strongly related to the way the compressor air flow is able to follow the optimal profile. In fact, the FF strategy presents considerable offset to control W_{cp} as can be observed in Fig. 25. Meanwhile, the suitable control of W_{cp}, made by the FF+FB approach, guarantees a correct evolution keeping the λ_{O_2} very close to the optimal trajectory as can be observed in Fig. 26b, c (zoom). Figure 26d displays the main variables in the TCS, the load torque changes and the controlled shaft speed under this condition.

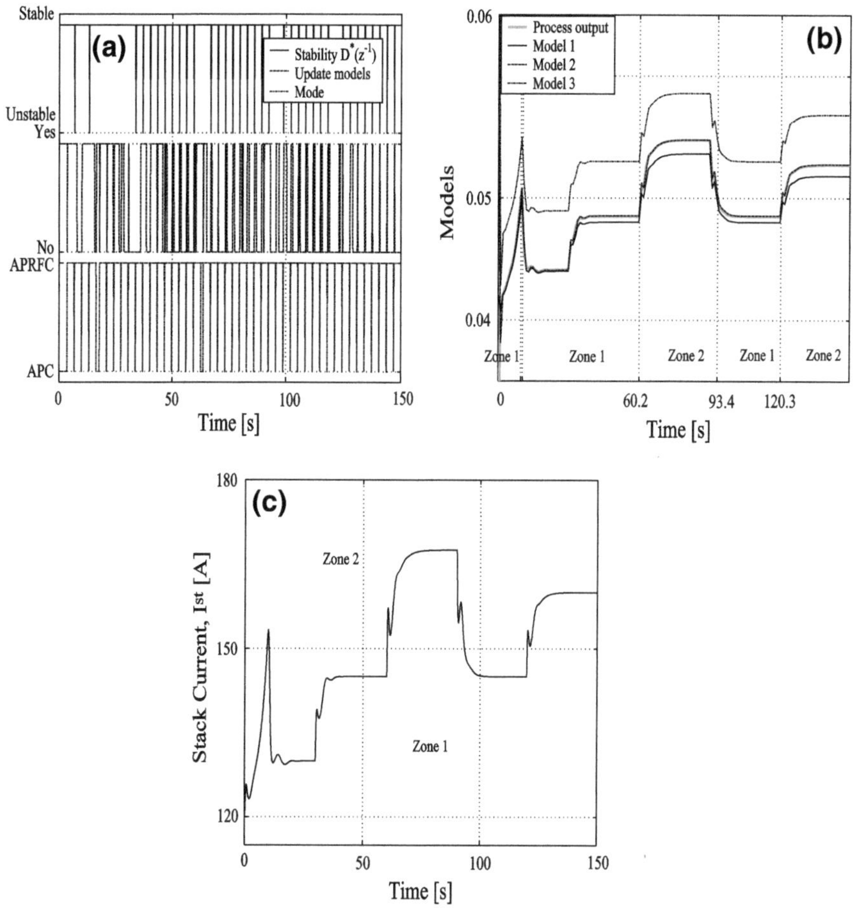

Fig. 27 Controller signals—SEDC motor **a** synchronization signals, **b** model predictions, **c** stack current

Finally, Fig. 27 summarizes the different useful signals tied to the controller. Figure 27a shows the synchronization signals used in the APCWRF approach (Fig. 19) along the simulation, such as the stability check of $D^*(k^{-1})$, the FIR model update permission and the mode variable (defining the used control approach). Figure 27b displays the three nominal model predictions of the compressor air flow along the different working zones. The stack current, I_{st}, excursions define when the zone changes happen to be processed by the controller, they are shown in Fig. 27c. In this case, the working points pass through zones 1 and 2.

The most important variables such as the compressor air flow, the net power and the excess ratio are evaluated using the integral absolute error (IAE) between their instantaneous evolutions and their corresponding optimal references. The results are shown in Table 2 for both control structures, FF and FF+FB and both kinds of motors.

Table 2 Performance Index - IAE

Motor type	PMDC		SEDC	
Control policy	FF	FF+FB	FF	FF+FB
Wcp [kg/s]	0.113	0.033	0.052	0.017
P_{net} [kW]	27.2	25.5	18.0	11.7
λ_{O_2}	3.408	1.335	1.70	1.005

Clearly, the approach proposed here presents an improvement in this performance index assuring the correct operation of the PEMFC.

4 Conclusion

High efficiency level, long durability and good transient behavior are fundamental issues for the success of fuel cell systems in the energy and automotive market. Thus, the adequate control of the system is a crucial topic to ensure a high performance. With this aim, two methodologies of control based on model predictive control were proposed and tested successfully.

In Sect. 2, a methodology based on dynamic matrix control (DMC) was presented. First, a steady-state analysis was done. With this analysis, it is shown that the system efficiency, in most of the current density ranges, can be improved by manipulating the cathode outlet air flow valve. Second, a dynamic analysis shows a transient response improvement with the additional manipulated variable.

Then, taking advantage of the results of the two previously mentioned analyses, the control strategy based on predictive control (DMC) was proposed, using the compressor motor voltage together with the cathode air flow valve area as manipulated variables. The controlled variables are the stack voltage and the oxygen excess ratio. To predict the future process response, the control strategy makes use of a process model easily obtainable through step response.

The simulation results show an appropriate dynamic response, which can still be improved with the inclusion of the disturbance model. Moreover, the control objectives have been accomplished with reduced control effort. This effort can be further reduced modifying the values of the matrix $\mathbf{Q}$. This is particularly important because of practical limitations in the manipulated variables.

In Sect. 3, a methodology based on adaptive predictive control with robust filter was presented. From the simulated results it is clear that the improved APCWRF is able to handle well the PEMFC requirements in a wide range of working zones produced by exigent load demands of different electric motors. The APCWRF presents a good performance and achieves the proposed objectives set for the complete system (PEMFC connected to the electric motors).

The combined action with the feedforward contribution allows the APCWRF to be able to present a well regulated behavior inspite of the complex disturbance scenarios.

In addition, the feedforward action is useful for setting the optimal compressor input voltage as a function of the demanded load. Another important advantage is that indirect controlled variables (ICVs) named as "performance variables" can be kept close to the optimal values according to the imposed load changes.

Acknowledgments The first author is grateful to Maria Serra and Jordi Riera from the Institut de Robòtica i Informàtica Industrial (IRI-UPC-CSIC) for their contribution in the first part of this chapter, a methodology based on dynamic matrix control for the oxygen in the cathode.

Appendix

Recursive Identification Algorithms

In this section a brief description about recursive system identification algorithms is presented.

Remembering the linear regression form in Eq. 43

$$y(k) = \psi(k)^T \theta \tag{63}$$

and supposing the parameters vector θ unknown but the input–output data were collected suitably from an identification experiment, so

$$Z^N = \{u(1), y(1), \ldots, u(N), y(N)\} \tag{64}$$

are known and N is the data base dimension. Then, the predictions in Eq. 63 can be used to fit the data base observations in a least squares sense by mean of cost criterion presented in Eq. 65

$$\min_{\theta} V_N\left(\hat{\theta}, Z^N\right), \tag{65}$$

where $\hat{\theta}$ is the estimated parameters vector and

$$V_N\left(\hat{\theta}, Z^N\right) = \frac{1}{N}\sum_{k=1}^{N}\left(y(k) - \hat{y}(k)\right)^2 = \frac{1}{N}\sum_{k=1}^{N}\left(y(k) - \psi(k)^T\hat{\theta}\right)^2. \tag{66}$$

In this case, the least square estimate can be represented as

$$\hat{\theta}_N = \arg\min_{\hat{\theta}} V_N\left(\hat{\theta}, Z^N\right), \tag{67}$$

where $\hat{\theta}_N$ indicates that N input–output samples were used.

Being V_N a quadratic function in $\hat{\theta}$ the optimal solution can be easily found by

$$0 = \frac{\mathrm{d}V_N\left(\hat{\theta}, Z^N\right)}{\mathrm{d}\hat{\theta}} = \frac{2}{N}\sum_{k=1}^{N}\psi(k)\left(y(k) - \psi(k)^T\hat{\theta}\right). \tag{68}$$

Reordering:

$$\sum_{k=1}^{N}\psi(k)y(k) = \sum_{k=1}^{N}\psi(k)\psi(k)^T\hat{\theta}, \tag{69}$$

or similarly:

$$\hat{\theta}_N = \left[\sum_{k=1}^{N}\psi(k)\psi(k)^T\right]^{-1}\sum_{k=1}^{N}\psi(k)y(k). \tag{70}$$

Recursive Least Squares

The least squares methodology presented in the previous section can be modified to update the parameters vector recursively as new input–output data are acquired from the process [8, 25].

Rewriting the covariance matrix in Eq. 70 as

$$P_N = \left[\sum_{k=1}^{N}\psi(k)\psi(k)^T\right]^{-1}. \tag{71}$$

Then,

$$P_N^{-1} = \sum_{k=1}^{N-1}\psi(k)\psi(k)^T + \psi(N)\psi(N)^T = P_{N-1}^{-1} + \psi(N)\psi(N)^T \tag{72}$$

and the estimated parameters vector

$$\begin{aligned}\hat{\theta}_N &= P_N\left[\sum_{k=1}^{N-1}\psi(k)y(k) + \psi(N)y(N)\right]\\ &= P_N\left[P_{N-1}^{-1}\hat{\theta}_{N-1} + \psi(N)y(N)\right]\\ &= \hat{\theta}_{N-1} + P_N\psi(N)\left[y(N) - \psi(N)^T\hat{\theta}_{N-1}\right].\end{aligned} \tag{73}$$

Thus, the following recursive algorithm can be obtained:

$$\begin{aligned}\hat{\theta}_N &= \hat{\theta}_{N-1} + K(N)\left[y(N) - \psi(N)^T\hat{\theta}_{N-1}\right]\\ K(N) &= P_N\psi(N)\\ \varepsilon(N) &= y(N) - \psi(N)^T\hat{\theta}_{N-1}.\end{aligned} \tag{74}$$

From Eq. 74, $\varepsilon(N)$ represents the output prediction error in $k = N$ computed by using the parameters vector estimated in the previous sample time $k = N - 1$. The first equation shows that the parameters vector estimate is computed by mean of the previous one and a correction factor that depends on the prediction error. The $K(N)$ term indicates how the previous estimate must be modified in the updating procedure.

The main problem with the recursive algorithm shown in Eq. 74 is the matrix inversion in P_N for each sample instant. This problem can be avoided by using the matrix inversion lemma, which defines that

$$[A + BCD]^{-1} = A^{-1} - A^{-1}B\left[DA^{-1}B + C^{-1}\right]^{-1} DA^{-1}. \tag{75}$$

Rewriting Eq. 71:

$$P_N = \left[P_{N-1}^{-1} + \psi(N)\psi(N)^T\right]^{-1} \tag{76}$$

and applying the previous lemma:

$$P_N = P_{N-1} - \frac{P_{N-1}\psi(N)\psi(N)^T P_{N-1}}{1 + \psi(N)^T P_{N-1}\psi(N)}, \tag{77}$$

that is a recursive expression for P_N. In this context, the recursive least squares algorithm is

$$\begin{aligned}
\hat{\theta}_N &= \hat{\theta}_{N-1} + K(N)\left[y(N) - \psi(N)^T\hat{\theta}_{N-1}\right] \\
K(N) &= P_N\psi(N) \\
\varepsilon(N) &= y(N) - \psi(N)^T\hat{\theta}_{N-1} \\
P_N &= P_{N-1} - \frac{P_{N-1}\psi(N)\psi(N)^T P_{N-1}}{1 + \psi(N)^T P_{N-1}\psi(N)}
\end{aligned} \tag{78}$$

An interesting variant of this algorithms is the called RLS with forgetting factor. In this strategy the functional cost stated in Eq. 66 is augmented by introducing a parameter λ as is shown in Eq. 79:

$$V_N\left(\hat{\theta}, Z^N\right) = \frac{1}{N}\sum_{k=1}^{N}\lambda^{N-k}\left(y(k) - \psi(k)^T\hat{\theta}\right)^2 \tag{79}$$

with $0 < \lambda \leq 1$. Thus, an exponential weighting is made on the prediction error giving more importance to the new data (λ^{N-k} near to 1) and less emphasis to the old ones (λ^{N-k} near to 0). Applying the same procedure as for the RLS case, the following algorithm can be obtained,

$$\begin{aligned}
\hat{\theta}_N &= \hat{\theta}_{N-1} + K(N)\left[y(N) - \psi(N)^T\hat{\theta}_{N-1}\right] \\
K(N) &= P_N\psi(N) \\
\varepsilon(N) &= y(N) - \psi(N)^T\hat{\theta}_{N-1} \\
P_N &= \frac{1}{\lambda}\left[P_{N-1} - \frac{P_{N-1}\psi(N)\psi(N)^T P_{N-1}}{\lambda + \psi(N)^T P_{N-1}\psi(N)}\right].
\end{aligned} \tag{80}$$

This approach is called RLS with forgetting factor. Note the only difference with the classical RLS in Eq. 78.

Recursive Least Squares with UD-Factorization

It is very helpful, from a numerical point of view, to represent the matrices by factorization. This avoids problems with large ill-conditioned matrices. A classical method is the UD-factorization. The basic idea is to represent the covariance matrix error in the RLS algorithms as is shown in Eq. 81:

$$\mathbf{P}(k) = \mathbf{U}_*(k)\mathbf{D}_*(k)\mathbf{U}_*(k), \tag{81}$$

where $\mathbf{U}_*(k)$ is a triangular superior matrix with unitary principal diagonal and y $\mathbf{D}_*(k)$ a diagonal matrix [8, 9, 25, 26].

The RLS with forgetting factor and UD-factorization is

A Initialization, $k = 0$

$$\mathbf{D}_*(0) = \delta\mathbf{I}, \text{ con}\,\delta = 10^2 - 10^4$$

B For $k > 0$

$$\begin{aligned}
\mathbf{f}(k) &= \mathbf{U}_*^T(k-1)\psi(k) \\
\mathbf{v}(k) &= \mathbf{D}_*(k-1)f(k) \\
\alpha_0(k) &= \lambda
\end{aligned}$$

1. **For** $j = 1, 2, \ldots, N$

$$\begin{aligned}
\alpha_j(k) &= \alpha_{j-1}(k) + f_j(k)v_j(k) \\
D_{jj}(k) &= \alpha_{j-1}(k)D_{jj}(k-1)/\alpha_j(k)\lambda \\
\tilde{\gamma}_j(k) &= v_j(k) \\
u_j(k) &= -f_j(k)/\alpha_{j-1}(k)
\end{aligned}$$

2. **For** $i = 1, \ldots, j-1$

$$\begin{aligned}
U_{ij}(k) &= U_{ij}(k-1) + \tilde{\gamma}_i(k)u_j(k) \\
\tilde{\gamma}_i(k) &\leftarrow \tilde{\gamma}_i(k) + U_{ij}(k-1)\tilde{\gamma}_j(k)
\end{aligned}$$

3. **Compute**

$$\gamma(k) = \frac{1}{\alpha_N(k)}\left[\tilde{\gamma}_1(k), \tilde{\gamma}_2(k), \ldots, \tilde{\gamma}_N(k)\right]^T$$

4. **Update**

$$\hat{\theta}(k) = \hat{\theta}(k-1) + \gamma(k)\left[y(k) - \psi^T(k)\hat{\theta}(k-1)\right]$$

5. **Store** $\mathbf{U}_*(k)$, $\mathbf{D}_*(k)$ and $\hat{\theta}(k)$

where $y(k)$ is the process output, $\psi(k)$ the regressor vector (inputs for FIR model) and λ the forgetting factor. With this algorithm the covariance matrix, $\mathbf{P}(k)$, is updated recursively by actualizing its factorization $\mathbf{U}_*(k)$ and $\mathbf{D}_*(k)$. Moreover, the parameters vector is estimated recursively also.

References

1. Yang W, Bates B, Fletcher N, Pow R (1998) Control challenges and methodologies in fuel cell vehicles development. Fuel Cell Technol Vehicles, vol 1. pp 249–256
2. Maciejowski JM (2002) Predictive control with constraints. Prentice Hall, Harlow, UK
3. Camacho E, Bordons C (1999) Model predictive control. Springer,
4. Grujicic M, Chittajallu KM, Law EH, Pukrushpan JT (2004) Model-based control strategies in the dynamic interaction of air supply and fuel cell. Proc Inst Mech Engs, Part A: J Power and Energy 218(7):487–499
5. Pukrushpan JT and Peng AGH (2002) Modeling and control for PEM fuel cell stack system. American Control Conference. Proceedings of the 2002, 4
6. Vahidi A and Peng A (2004) Model predictive control for starvation prevention in a hybrid fuel cell system. American Control Conference. Proceedings of the 2004, 1:834–839
7. Golbert J, Lewin D (2004) Model-based control of fuel cells: (1) regulatory control. J Power Sources 135:135–151
8. Jordán MA (1990) Digitale adaptive Regelung mit linearen nichtparametrischen Modellen. PhD thesis, Darmstad, Germany
9. Jordán M, Basualdo M, Zumoffen D (2006) An approach to improve the performance of adaptive predictive control systems: theory, simulations and experiments. Int J Control 79(10):1216–1236
10. Zumoffen D, Basualdo M, Jordán M and Ceccatto A (2006) Robust adaptive predictive fault-tolerant control linked with fault diagnosis system applied on a nonlinear chemical process. In: Proceedings of the 45th IEEE Conference on Decision and Control, San Diego, CA, USA, pp 3512–3517
11. Zumoffen D, Basualdo M, Jordán M, Ceccatto A (2007) Robust adaptive predictive fault-tolerant control integrated to a fault-detection system applied to a nonlinear chemical process. Ind Eng Chem Res 46(22):7152–7163
12. Zumoffen D and Basualdo M (2009) Advanced control for fuel cells connected to a DC/DC converter and an electric motor. Comp Chem Eng. 2010, 34(5), 643–655
13. Zumoffen D (2008) Desarrollo de Sistemas de Diagnóstico de Fallas Integrado al Diseño de Control Tolerante a Fallas en Procesos Químicos. PhD thesis, Universidad Nacional de Rosario, Argentina. http://www.cifasis-conicet.gov.ar/zumoffen/

14. Sastry S, Bodson M (1989) Adaptive control: stability, convergence and robustness. Prentice-Hall, Upper Saddle River, NJ, US
15. Kofahl R (1992) Robustness in parameter adaptive control (Cap XIII). In: Isermann R, Lachmann KH and Matko D (eds) Adaptive Control Systems. Prentice Hall, New York, pp 358–384.
16. Gao W, Ma GF, Zhou ML, Li YC, and Ying Li (2005) Parameter identification and adaptive predictive control of time-varying delay systems. In: Proceedings of the fourth international conference on machine learning and cybernetics, pp: 609–613, Guangzhou, August 18/21
17. Luo X, Zuo X and Du D (2005) Varying model based adaptive predictive control of highly nonlinear chemical process. International conference on control and automation (ICCA2005), Budapest, Hungary, June 27/29
18. Tao G, Joshi S, Ma X (2001) Adaptive state feedback and tracking control of systems with actuator failures. IEEE Trans Autom Control 46(1):78–95
19. Tao G, Chen S, Joshi SM (2002) An adaptive actuator failure compensation controller using output feedback. IEEE Trans Autom Control 47:506–511
20. Mahmoud MS, Boukas EK, Ismail A (2008) Robust adaptive control of uncertain discrete-time state-delay systems. Comp Math Apps 55:2887–2902
21. Zhang J, Liu G, Yu W, Ouyanga M (2008) Adaptive control of the airflow of a pem fuel cell system. J Power Sources 179:649–659
22. Pukrushpan J, Stefanopoulou A, Peng H (2004) Control of fuel cell power systems: principles, modeling, analysis and feedback design. Springer,
23. Grujicic M, Chittajallu K, Law E, Pukrushpan J (2004) Model-based control strategies in the dynamic interaction of air supply and fuel cell. Proco Inst Mech Engs. J Power Energy A:1–13
24. Grujicic M, Chittajallu K, Pukrushpan J (2004) Control of the transient behavior of polymer electrolyte membrane fuel cell systems. Proc Inst Mech Engs. Autom Eng D:1–12
25. Ljung L (1999) System Identification (Theory for the user). 2nd edn. Prentice Hall, Upper Saddle River, NJ, USA.
26. Zhang Y, Li X (1999) A fast UD factorization-based learning algorithm with applications to nonlinear system modeling and identification. IEEE Trans Neural Netw 10(4):930–938

Simulation and Control of PEMFCS Thermal Behavior

Abdelkrim Salah, Rachid Outbib and Jaafar Gaber

Abstract This chapter is dedicated to the thermal behavior of PEM fuel cells (PEMFC). The goal is twofold. First, we develop an alternative modeling approach; to partial differential equations, for analysis and simulation of the thermal behavior of PEMFC. The obtained linear model is, in order to improve the accuracy, of large size. Thus, in this approach the simulations are based on parallel computations. The parallel algorithm is implemented using three different systems that are UPC, MPI and PVM. Then, we analyze the performances of each system by considering the computation times and the speedup of each implementation. Second, we propose a strategy to control the thermal behavior. More precisely, we establish a bilinear state model, which describes the behavior of the thermal aspect. Afterwards, we prove that the proposed nonlinear system can be made locally asymptotically stable around a desired temperature value. The control law is obtained by feedback stabilization approach. To illustrate the proposed methodologies, simulation results are given.

1 Introduction

The chemical conversion to electrical energy, using PEMFC, is accompanied by the production of an important thermal energy. This energy results mainly from the loss of exothermic reaction within the electrodes and by internal irreversibility, i.e.

A. Salah · R. Outbib (✉)
LSIS, Aix-Marseille University, av. Escadrille Normandie Niemen, 13397 Marseilles, France
e-mail: rachid.outbib@lsis.org

A. Salah
e-mail: abdelkrim.salah@lsis.org

J. Gaber
SET-UTBM, 90010 Belfort cedex, France
e-mail: jafaar.gaber@utbm.fr

M. S. Basualdo et al. (eds.), *PEM Fuel Cells with Bio-Ethanol Processor Systems*, Green Energy and Technology, DOI: 10.1007/978-1-84996-184-4_4,

activation, ohmic and concentration polarization. Consequently, the control of the thermal aspect is compulsory. Hence, thermal management and minimizing temperature constraints add complexity and cost to the PEMFC system. Thus, in order to increase the efficiency of PEMFC, the purpose is to operate the PEMFC at its maximum allowed internal temperature. The rise of the temperature shifts the polarization curve upward and more power is produced than at a lower temperature. Consequently, no additional fuel or oxidant are required.

Generally speaking, for PEMFC, a controller is available for cooling in order to maintain the fuel cell at its operating temperature. The control input generates a mass flow rate signal and sets the speed of the pump which permits the circulation of the cooling fluid (liquid or gas). The waste heat is rejected with exhaust. Moreover, the small temperature difference between the power plant and surroundings makes heat rejection difficult. Hence, the operating range of the power of the fuel cell and its lifetime are greatly influenced by the operating temperature. In order to provide efficient stack operation and to ensure a durability of the fuel cell, the temperature must be maintained without large changes even when it is subject to disturbances and output power changes.

The control of thermal behavior have attracted the interest of several authors (see for instance [1–4]). In general, the goal is to stabilize the value of the temperature around the membrane. The proposed approaches are mainly based on models that are continuous-time or discret-time and are linear or nonlinear. The goal is to maintain accurately the optimal temperature within a proton membrane fuel cell stack by controlling the coolant flow rate. However, the thermal behavior of PEMFC requires more understanding and development of new strategies of control.

This chapter is dedicated to simulation of thermal behavior and its control. The goal is twofold. The first part deals with the problem of simulation of thermal behavior. In order to obtain accurate results, large scale simulations are needed. However, this simulations can take unacceptably large execution time. This is especially true when using traditional models which are governed by heat equations. Thus, a first main contribution in this work is the development of an alternative modeling approach, based on parallel computing, to analyze and simulate the thermal behavior of the heterogeneous system that is PEMFC. We show, in particular, that the thermal behavior of PEMFC lends well itself to parallelism and can therefore benefit from advances in parallel. Furthermore, we analyze the performances of three systems (UPC, PVM and MPI) by considering the computation times and the speedup of each implementation.

The second part treats the problem of the control of thermal behavior. First, we will consider some areas of the stack that are important and we associate to each area a point. Second, we will establish a new non linear model which describes the behavior of thermal aspect at each point. Notice that the established model is of bilinear form. This is no surprising since for others applications, in the literature, the thermal aspect is also described by this kind of model. Afterwards, we will propose a strategy, based on feedback stabilization, to control the thermal behavior around a desired value. To illustrate the performances of the proposed methodologies, simulation results are given for open-loop and closed-loop system, respectively.

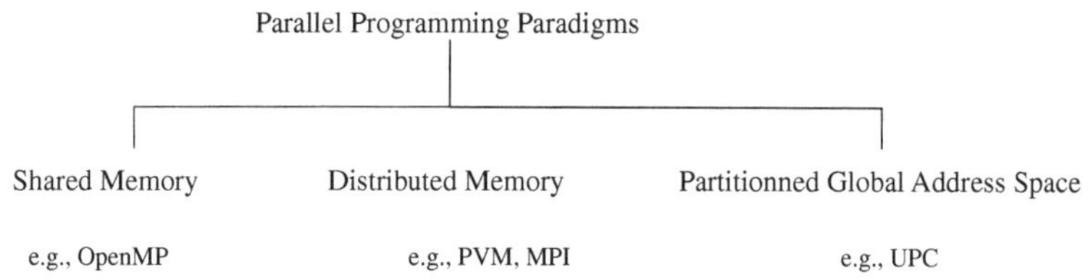

Fig. 1 Paradigm of parallel computing

2 Simulation of the Thermal Behavior

The aim of this section is to present methodologies for modeling and simulating the thermal behavior of a PEMFC [5, 6]. In the first subsection, the notion of parallel computers is introduced. The tools of thermal behavior modeling are given by the second subsection. The third subsection is constituted by descriptions of the parallel systems and by performances results on parallel implementations. The last subsection contains simulation results of thermal behavior of PEMFC.

2.1 Simulation Tools: Parallel Computers

The simulations of thermal behavior are, in general, obtained using models that are governed by heat equations and are based on computational tools that derive approximate solutions to partial differential equations. It requires computations that can take unacceptably large execution time. Such multi-model systems also require synchronization which results in overhead. Hence, this problematic requires new methodologies in order to obtain accurate results with an acceptable execution time. In this chapter we propose a new approach to model thermal behavior of PEMFC and we show how to use parallel calculation in order to optimize the execution time. The obtained data will be highly precise and accurate.

The goal of a parallel algorithm is to accelerate the execution of the calculations by means of several processes or several workstations. During the implementation, it is necessary to decompose the original problem in several tasks. These tasks can be dependent to each others and sharing the same data. Hence, the parallel programming is based on the model of communicating process. A process is in common memory, in distributed memory, or virtual shared memory (Fig. 1).

More precisely, in this parallel computers model, the calculation tasks are shared between several process having an independent access to the common memory. This kind of model requires multiprocessors and several ways of access to the memory. In the distributed memory model, several processes share the same memory space. In other words, the tasks are distributed between several processes which possesses reserved memory spaces and where every processor manages its process.

2.2 PEMFCS Thermal Behavior Modeling

The goal it to obtain a lumped parameter model which describes the behavior of thermal in representative areas of the PEMFC. The model will be obtained using the network nodal approach where each area is represented in the model by a scalar variable. Among the area of the PEMFC, the behavior of thermal aspect around the membrane will be considered for analysis. In fact it represents an important issue in the global control of PEMFC. Notice that the size (i.e. the number of variables) of the linear model, considered for simulation, is variable and will depend on the desired accurate.

2.2.1 Generalities on Thermal Behavior Modeling

In the literature, the thermal behavior problem, and independently of the fuel cell, is dealt by using tools and methodologies that are based on the classical heat equation applied to the considered process. Energy balance is used to quantify the energy used or produced by a system and is expressed using the following principle: the intake energy is the sum of the internal produced heat, the external work and the energy storage.

The obtained models are based on partial differential equations. For instance, consider the case of the thermal conduction. Let T be the temperature vector and k denotes the thermal conductivity. Let also ∇^2 designates the Laplacian. The quantity $-k\nabla^2 T$ represents the internal produced heat. Then the heat equation is as follows:

$$\rho \mathrm{cp} \frac{\partial T}{\partial t} = -k\nabla^2 T + P, \tag{1}$$

where P denotes the energy storage, ρ represents the density of the material and cp is the specific heat.

In the special case of the one-dimensional heat conduction, the equation describing the behavior of the variable T is given by:

$$\rho \mathrm{cp} \frac{\partial T}{\partial t} = -k \frac{\partial^2 T}{\partial x^2} Q, \tag{2}$$

where Q is the source term.

The finite difference method is one of the most widely used techniques to obtain an approximate solution to the partial differential equation describing the behavior of a physical system. In the rest part of this subsection, we use the discrete form of the formula above in order to obtain a discrete model which describes the behavior of temperature at a considered node.

For a node (i), we associate a source term denoted by Q_{i}. Using the central difference formula, the second derivative of heat equation is approximated, by the following Taylor series development:

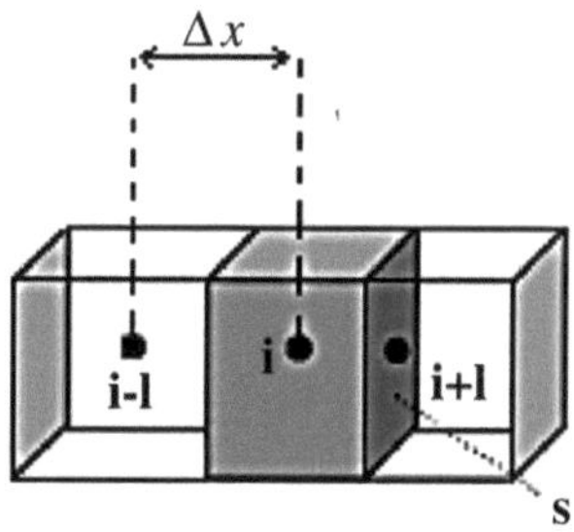

Fig. 2 Difference finite discretisation

$$\rho \mathrm{cp} \frac{\partial T_{\mathrm{i}}}{\partial t} = -k \frac{T_{\mathrm{i}+1} - 2T_{\mathrm{i}} + T_{\mathrm{i}-1}}{\Delta x^2} + Q_{\mathrm{i}} \tag{3}$$

or

$$\rho \mathrm{cp} \frac{\partial T_{\mathrm{i}}}{\partial t} = k \frac{(T_{\mathrm{i}} - T_{\mathrm{i}+1})}{\Delta x^2} + k \frac{(T_{\mathrm{i}} - T_{\mathrm{i}-1})}{\Delta x^2} + Q_{\mathrm{i}} \tag{4}$$

Let Δx be the distance between the two nodes $(i - 1)$ and (i). Let also s be the exchange surface between the two nodes then the volume v is defined by $v = S \cdot \Delta x$ (see Fig. 2). So, and using (4) we get

$$\rho \mathrm{cp} v \frac{\partial T_{\mathrm{i}}}{\partial t} = \frac{kS}{\Delta x}(T_{\mathrm{i}} - T_{\mathrm{i}+1}) + \frac{kS}{\Delta x}(T_{\mathrm{i}} - T_{\mathrm{i}-1}) + Q_{\mathrm{i}} v, \tag{5}$$

where $G = \frac{k.S}{\Delta x}$ denotes the conductance. The term $C_{\mathrm{i}} = \rho \mathrm{cp} v$ designates the thermal capacity and $P_{\mathrm{i}} = Q_{\mathrm{i}} \cdot v$ represents the source term of heat production.

2.2.2 Generalities on PEMFC

Fuel cells are composed by elementary electrochemical cells called membrane-electrode assembly (MEA), sandwiched between two bipolar plates (see Fig. 3). Each elementary cell comprises an electrolyte (i.e. membrane) and two electrodes (i.e. anode and cathode). Fuel oxidation process takes place at the anode

$$\text{Anode: } \mathrm{H_2} \rightarrow 2\mathrm{H^+} + 2e^-$$

while the reduction process takes place at the cathode

$$\text{Cathode: } \frac{1}{2}\mathrm{O_2} + 2\mathrm{H^+} + 2e^- \rightarrow \mathrm{H_2O}.$$

To obtain satisfactory tension, unit cells are stacked together in series to construct a fuel cell stack. Protons produced by the hydrogen oxidation move across the electrolyte, then combine with oxygen and the electrons coming from the external circuit to form water and heat

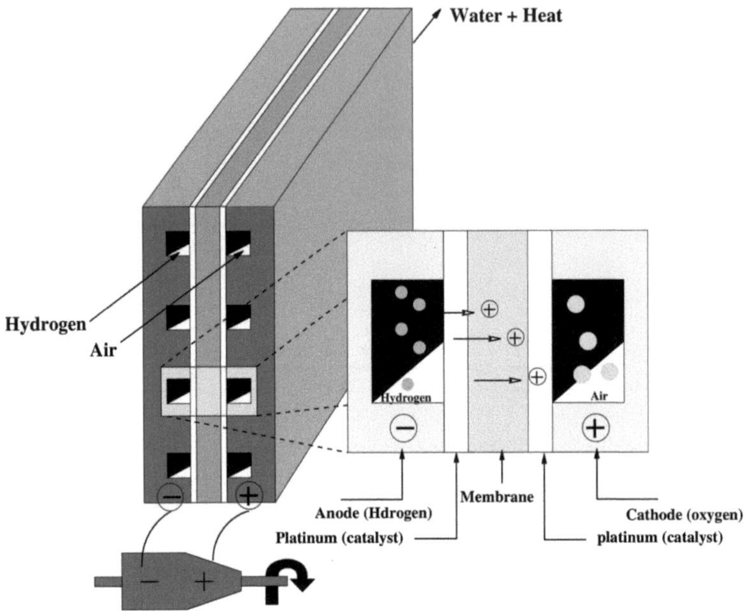

Fig. 3 PEM fuel cell operation

$$\frac{1}{2}O_2 + H_2 \rightarrow H_2O + \text{Heat}. \tag{6}$$

Bipolar plate that comprises channel gasses serves as both current collectors and separation barrier within electrochemical cell stack. To dissipate the heat produced by the fuel cell, separate cooling plates, through which circulate water, are used.

The thermal behavior of a fuel cell at the bipolar plates is governed by the conduction and convection phenomena. Distribution channel nodes (i.e. fluid volumes) exchange heat with the bipolar plate nodes by convection. By conductance convective movements, heat exchange occurs in channel nodes due to gases movement. The thermal phenomenon of conduction takes place between electrodes and membrane or plates. In the reactive zone of the MEA, the nodes receive the heat produced by the electrochemical reactions and the heat produced by water sorption at the membrane interfaces (see Fig. 4).

2.2.3 Network Nodal Approach

The goal of this part is to establish a model using the technical of network nodal modeling. The network nodal technical is similar to electrical–thermal analogy approach (see [7, 8]). This technical is based on local exchange heat, between nodes. To

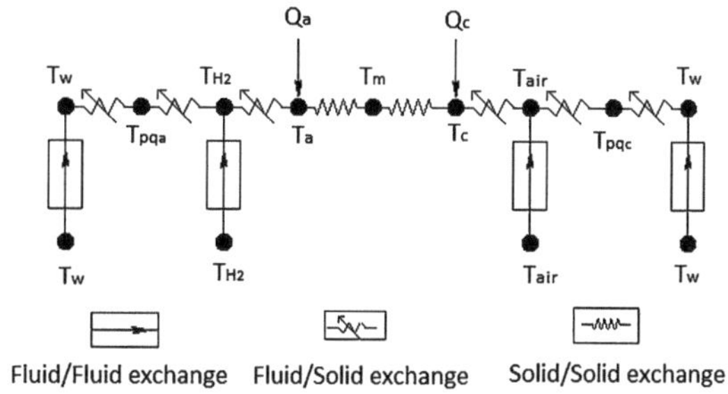

Fig. 4 Nodal network of a cell including gasses and water cooling

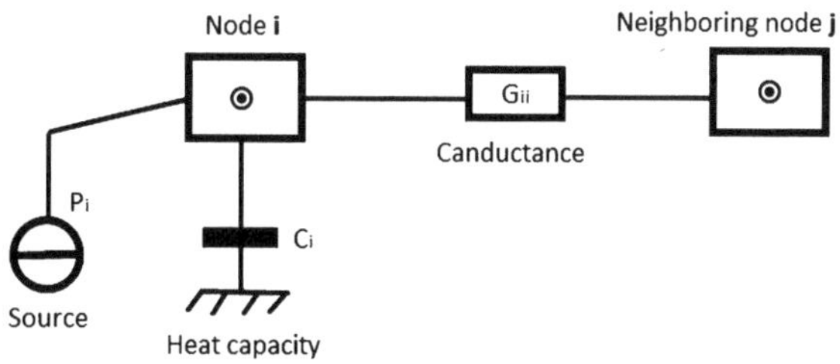

Fig. 5 Nodal principle

describe a thermal behavior of PEMFC using the nodal approach, we consider a sequence of nodes coupled with the reference node (see Fig. 5).

The behavior of the temperature at the node (i), taking into account the interconnection with the others nodes, can be described by the following formula:

$$C_{\mathrm{i}}\frac{\mathrm{d}T_{\mathrm{i}}}{\mathrm{d}t} = G_{\mathrm{ij}}(T_j - T_{\mathrm{i}}) + P_{\mathrm{i}}, \tag{7}$$

where G_{ij} describes the thermal exchange between the nodes (i) and (j) and P_{i} represents the source term.

The nodal network approach consists in dividing the physical system to be studied into elementary volumes (i.e. nodes) with isothermal temperature hypothesis. Connection between two given adjacent nodes is represented by a thermal conductance that depends on the nature of the thermal transfer. Two parameters are defined for each node (i). We use T_{i} to design the average temperature of the node and C_{i} to denote the heat-storage capacity. Note that the two quantities are related to thermo-physical properties of the material corresponding to the node. Recently, the nodal network approach coupled with FDM has been used to analyze the PEMFC thermal behavior (see for instance [7, 9]). More precisely, the thermal behavior of the fuel cell solid part has been analyzed with the nodal network approach, while the PDE was used to analyze the thermal behavior of gas channels (i.e. fluid part). The used methodology is based on application of FDM with sequential computations. In this work, the proposed nodal network approach do not require coupling with PDE model and the computations are done using parallel processing. In fact, both the solid part and the fluid part used to represent the channel gasses are divided into nodes with homogeneous temperature assumptions.

2.2.4 Mathematical Model

The temperature distribution in the nodal network approach is modeled as follows. A heat exchange between two adjacent nodes is given by a thermal conductance. Boundary conditions are described by specific conductance $G_{\mathrm{ij}}^{\mathrm{bound}}$. This conductance represents the combined heat transfer conduction–convection phenomenon. The reactive node of cell is represented by a source term Q_{i}. Using the thermal balance at each node we get the following linear system to describe the behavior of the temperature for the fuel cell:

$$C\frac{\mathrm{d}T}{\mathrm{d}t} = GT + B, \tag{8}$$

where T is the temperature vector, C is the diagonal matrix defined by $C_{\mathrm{ii}} = \rho_{\mathrm{i}} \mathrm{cp}_{\mathrm{i}} V_{\mathrm{i}}$ and $C_{\mathrm{ij}} = 0$ for $i \neq j$, ρ_{i} represents the density, cp_{i} is the specific heat of a fluid or a solid and Vi designates the elementary volume of node (i). B is a vector that represents both the source term and the boundary conditions which will defined hereafter. $G \in \mathscr{M}_{\mathrm{n}\times\mathrm{n}}$[1] is the conductance matrix defined by

$$G = G^{\mathrm{solid}} + G^{\mathrm{fluid}}. \tag{9}$$

The matrix G^{solid} designates the heat exchanges within the solid part of the cell. G^{fluid} is a matrix constituted by the conductances which represent the heat exchanges within the fluid part of the cell. Note that in order to ensure more data precision, the size of matrix G should be very large. Hence, the computation time needed to solve the linear system (8) becomes very huge. For that reason, parallelizing the simulation becomes a necessity to make it practical.

From now on, we use $G_{\mathrm{ij}}(i, j \in \{1, \ldots, n\})$, to denote the element of the matrix G at the ith line and jth column. G_{ij} represents the conductance value between a given node (i) and one of its neighbors (j). To define the value of each element G_{ij} of G we use the thermal balance principle and we distinguish two cases according the nature of the cell material.

A. Solid Part:

$$G_{\mathrm{ij}}^{\mathrm{solid}} = \begin{cases} -\sum_{\mathrm{j}}(G_{\mathrm{ij}}^{\mathrm{cond}} + G_{\mathrm{ij}}^{\mathrm{conv}} + G_{\mathrm{ij}}^{\mathrm{bound}}) & \text{if } i = j \\ \sum_{j}(G_{\mathrm{ij}}^{\mathrm{cond}} + G_{\mathrm{ij}}^{\mathrm{conv}}) & \text{if } i \neq j. \end{cases} \tag{10}$$

B. Fluid Part:

$$G_{\mathrm{ij}}^{\mathrm{fluid}} = \begin{cases} -G_{\mathrm{ij}}^{\mathrm{f}} & \text{if } i = j \\ G_{\mathrm{ij}}^{\mathrm{f}} - \sum_{\mathrm{j}} G_{\mathrm{ij}}^{\mathrm{conv}} & \text{if } i \neq j. \end{cases} \tag{11}$$

[1] $\mathscr{M}_{\mathrm{n}\times\mathrm{n}}$ denotes the set of the matrices with n lines and n columns where n is the size of the nodal network.

The matrix G^{cond} represents the heat exchanges by conduction (i.e. solid–solid exchange). G^{conv} is a matrix constituted by the convection–conductance (i.e. solid–fluid exchange). The matrix G^{f} is composed by conductance that models the mass motions. G^{bound} designates the boundary conductance. The vector B is expressed as follows

$$B = \begin{cases} \sum_{\text{j}} G_{\text{ij}}^{\text{bound}} T_{\text{i,j}}^{\text{amb}} + P_{\text{i}} & \text{for the solid part,} \\ G_{\text{bound}}^{\text{f}} T_{\text{amb}}^{\text{f}} & \text{for the fluid part,} \end{cases} \tag{12}$$

where T_{amb} represents the ambient temperature. G_{bound} is introduced to describe the heat exchange between the fuel cell and its environment and it is defined by

$$G^{\text{bound}} = \frac{2k_{\text{air}} \text{H}_{\text{air}} S_{\text{s,air}}}{2k_{\text{s}} + \text{H}_{\text{air}} d_{\text{s,air}}}, \tag{13}$$

where $d_{\text{s,air}}$ is the distance between the solid node and air. k_{s} and k_{air} represent respectively the thermal conductivity of the solid material and the H_{air} designates the heat exchange coefficient between fuel cell and the environment. $S_{\text{s,air}}$ is the transfer surface between a solid node and the environment. In the sequel, we define the matrix G_{cond}, G_{conv} and G_{f}. First, for G_{cond}, when the considered exchange is between two nodes constituted by the same material, the conduction is represented by a conductance expressed by (see [8])

$$G_{\text{ij}}^{\text{cond}} = \frac{k S_{\text{ij}}}{d_{\text{ij}}} \quad (i, j \in 1 \ldots 3), \tag{14}$$

where k is the thermal conductivity, S_{ij} denotes the heat transfer surface between two nodes and d_{ij} designates the distance between two nodes. In the case of two different materials, we consider the following equivalent conductance (see [7]):

$$G_{\text{ij}}^{\text{cond}} = \frac{2k}{k_{\text{i}} d_{\text{j}} + k_{\text{j}} d_{\text{i}}} S_{\text{ij}} \; (i, j \in 1 \ldots 3). \tag{15}$$

Second, concerning G^{conv}, the conductance–convection (i.e. fluid–solid) is given by the following equation (see [8]):

$$G_{\text{ij}}^{\text{conv}} = \text{H}_{\text{ij}} S_{\text{ij}} \; (i, j \in 1 \ldots 3), \tag{16}$$

where H_{ij} is the heat exchange coefficient between node (i) and node (j). Third, the conductance resulting from the mass motion of gasses and cooling water G^{f} (i.e. fluid–fluid) is defined by

$$G_{\text{ij}}^{\text{f}} = \dot{m}_{\text{node}} \text{cp}_{\text{f}} \qquad (i, j \in 1 \ldots 3), \tag{17}$$

where cp_{f} designates the specific heat of a fluid and $\dot{m}_{\text{node}}$ is the flow rate at each node. In the two following subsections, we define respectively the flow rate for the fluids and the source term. We distinguish three cases according the fluids (i.e., oxygen, hydrogen, and air).

C. Flow Rate Description: the goal of this subsection is to define the flow rate at each node. We first determine the flow rate for each fluid.

C.1. Water: notice that for PEMFC high-power, it is necessary to use liquid cooling in order to dissipate the heat within the fuel cell. The flow rate of the water is obtained by the following formula:

$$\dot{m}_{\mathrm{H_2O}} = \rho_{\mathrm{H_2O}} S v_{\mathrm{H_2O}}, \tag{18}$$

where $v_{\mathrm{H_2O}}$ denotes the speed of the water defined by the following equation:

$$v_{\mathrm{H_2O}} = \frac{R_{\mathrm{e}} \mu_{\mathrm{H_2O}}}{D_{\mathrm{h}}}, \tag{19}$$

where

$$R_{\mathrm{e}} = \left(\frac{Nu_{\mathrm{H_2O}}}{0.023 Pr^{0.4}} \right)^{\frac{1}{0.8}} \tag{20}$$

represents the number of Reynolds that is obtained in correlation with laminar flow within a channel (see for instance [10]). For the gas at anode and cathode, $\dot{m}_f$ is given by

$$\dot{m}_f = \frac{\dot{m}_f^{\mathrm{total}}}{n_{\mathrm{c}}} (f \in \{\mathrm{H_2}, \mathrm{air}\}), \tag{21}$$

where n_{c} is the number of flow channels in the bipolar plate, and the $\dot{m}_{\mathrm{f}}^{\mathrm{total}}$ represents the flow rate of fluids (gasses and water) in the inlet of a cell.

C.2. Hydrogen: the Hydrogen flow rate in the anode is controlled in an open-loop by the control speed given to the compressor according to the current delivered by the cell and a fixed ratio of stoichiometry. In general, it is computed from the following formula (see [11]):

$$\frac{M_{\mathrm{air}} 10^{-3}}{2F} ut_{\mathrm{H_2}} I. \tag{22}$$

C.3. Air: the air flow rate entering to the cathode is fixed by the consumption of the cell and by the flow rate of purging; it is also a function of the cell current density I as in the following formula (see [11]):

$$\frac{M_{\mathrm{air}} 10^{-3}}{4F} \frac{ut_{\mathrm{air}}}{x_{\mathrm{O_2}}} I, \tag{23}$$

where I is the current from the of fuel cell, F is Faraday's constant, ut_{air} and $ut_{\mathrm{H_2}}$ are respectively the air and hydrogen stoichiometric ratio and M_{air} is the molar mass of air.

Now we are in position to define the flow rate at node. Indeed, the gas flow at each node $\dot{m}_{\text{node}}$ is expressed by

$$\dot{m}_{\text{node}} = \dot{m}_f \frac{\dot{m}_{\text{node}}}{V_{\text{channel}}}, \tag{24}$$

where $\dot{m}_{\text{node}}$ and V_{channel} denote respectively the node's volume and channel volumes. Concerning the simulation, the expressions (14) and (29) are used for initialization. During the simulation, at each iteration, the expression of the flow rate changes in order to take into account the variation value of temperature. More precisely, using the perfect gas law, we get

$$\dot{m}_f = \frac{PM}{R} \frac{Sv_f}{T_f} \quad (f \in \{H_2, \text{air}\}), \tag{25}$$

where v_{f} represents the velocity of gas, which is defined in a similar way to v_{H_2O} (see (19)). R is the universal gas constant and P is the gas pressure.
D. Source Term Description: the heat produced within the cell is a result of electrodes over potentials, water evaporation and entropy of oxygen reduction (see [12]). The source terms are defined as follows.
D.1. Anode Reactive Node:

$$Q_{\text{a}} = L_{\text{v}} \alpha \frac{I}{2F} V_{\text{node}}, \tag{26}$$

where L_{v} denotes the latent heat of water evaporation.
D.2. Cathode Reactive Node:

$$Q_{\text{c}} = \left[LV_{H_2O} \left((1 + 2\alpha) + T\Delta S \right) \frac{1}{2F} + b_{O_2} \ln \left(\frac{I}{I^0_{O_2}} \right) \right] V_{\text{node}}, \tag{27}$$

where $T\Delta S$ represents the entropy change, $I^0_{O_2}$ designates the exchange current density, α denotes the water transport quantity coefficient across the membrane, and V_{node} is the isotherm volume.

2.3 Parallel Implementation

Programming paradigms such as message passing and shared memory, are developed to take full advantage of the high degree of parallelism in large-scale multiprocessors, PC clusters, and clusters of shared memory multiprocessors. Although these two programming paradigms represent the state of art today, message passing suffers from its high communication cost and difficulty to use, while the shared memory paradigm suffers from the problem associated with data locality. The distributed shared memory (DSM) model is receiving rising attention as it has been designed to

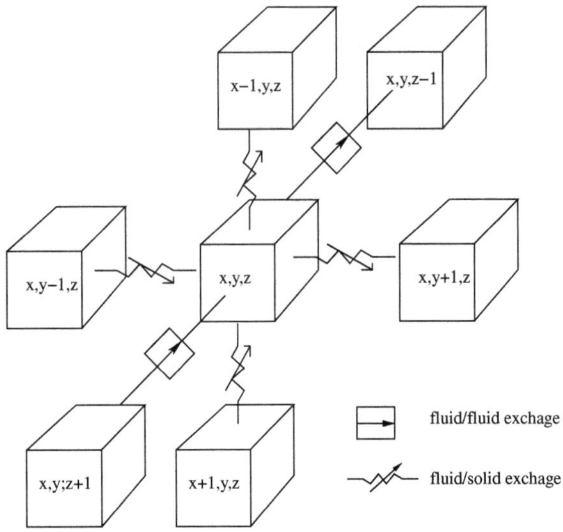

Fig. 6 Heat exchange between node (x, y, z) belonging to gas channel with its adjacent neighbors

leverage the ease of programming of the shared memory paradigm, and been able to solve the problem of data locality as in the message passing model [13, 14].

By parallelizing the nodal network, the performance can be greatly improved. Here, and after defining the computation model, we consider three parallel implementations: MPI, UPC and PVM.

2.3.1 The Computation Model

To obtain results that are highly precise and accurate, the physical system of PEMFC is divided into three-dimensional (3D) plans. We denote a 3D nodal network by *N (n, m, l)*, where *n* is the nodal network size along the horizontal axis, *m* is the size along the vertical axis, and *l* is the size along the third dimension (i.e. cell's representation in terms of depth). Each node *(x, y, z)* is surrounded by up to six neighbors; that are $(x, y-1, z)$; $(x, y+1, z)$; $(x-1, y, z)$; $(x+1, y, z)$, $(x, y, z-1)$ and $(x, y, z+1)$ (see Fig. 6).

The thermal model can be rewritten in the same form as in (8). Each node of GNM is assigned two values, *i* and *j*, which represent the indices of the neighboring nodes. The simulation algorithm performs a number of time steps. In each step *s*, for each node (*i*) in the nodal network the temperature is calculated and it is related to the values of its neighbor nodes according the following equation derived from (8):

$$\frac{T_i(s+\delta t) - T_i(s)}{\delta t} = \sum_j G_{ij} T_i(s) + B_i(s), \tag{28}$$

where G_{ij} are the entries of the matrix . Note that in order to obtain more data precision, the sparse matrix G becomes very large. Hence, the computation time needed to solve the linear system in (8) becomes very huge. Therefore, parallelizing the simulation becomes a necessity in order to make it practical.

2.3.2 Parallel Systems Under Consideration

A. Message Passing Interface: MPI is a portable message-passing library used to design parallel algorithms on both massively parallel nodes and node clusters [15]. In message passing, a problem can be logically divided among several processes (or tasks) and can all run in parallel on different nodes. The processes exchange data between each other by sending and receiving messages through the message passing library (e.g., MPI library). The most frequent type of parallel computing classification used is the single program, multiple data(SPMD) where all processors run the same program, while using their own data and cooperate with other by explicitly exchanging messages.

B. Unified Parallel C Parallel: UPC is a new language which is a parallel extension to ANSI C which was developed by a consortium of academia, industry, and government [16]. Thus, all the features of the C programming language are embodied in UPC. UPC is an explicit partitioned global address space (PGAS) language [13]. Each thread constituting the program is able to access a private and a shared space. The shared space is logically partitioned: threads have affinity to part of the space allowing it to exploit data locality. UPC has many features that make it easy to use for programmers. It uses a very simple syntax, an extension to C (e.g., C-like pointers usage over the global space), it is not library based which eliminates the use of many complex function calls and argument passing, application of consistent domain decomposition which facilitates greatly the conceptual effort [13, 14, 17].

C. Parallel Virtual Machine: parallel virtual machine(PVM) is a message passing system that enables a network of computers to be used as a single distributed memory parallel computer [15]. The PVM libraries have become widely available, making it much easier to write portable parallel programs. Such libraries allow to write complex parallel programs with a small set of functions for spawning and identifying thread, and for packing, sending, receiving and unpacking messages. This system has evolved in the past several years into a viable technology for distributed and parallel processing in a variety of disciplines. PVM is designed to link computing resources and provide users with a parallel platform for running their computer applications, irrespective of the number of different computers they use and where the computers are located [18].

2.3.3 Space Partition of the Fuel Cell

In order to process the nodal network in parallel, the fuel cell structure is partitioned into block strips across the processors constituting the cluster. Figure 7 schematically shows a PEM fuel cell divided into three sub-regions (i.e. blocks). The nodal network is divided into groups of complete planes that are nearly the same size (the

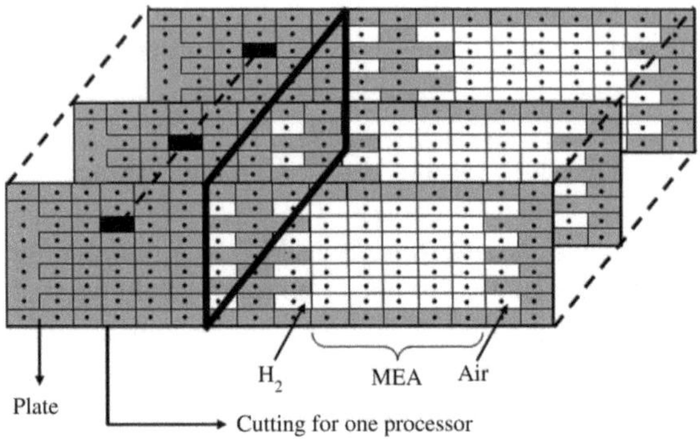

Fig. 7 Parallel implementation with block-striped partitioning of the nodal network

partition is uniformed). More precisely, the 3D representation of the fuel cell *N* (*n*, *m*, *l*) is partitioned among *p* processors so that every machine gets a block of size Ni (*n/p*, *m*, *l*). When parallelizing the nodal network, we should partition it in such a way such that the partitions can be assigned to different nodes. The heat exchanged between nodes on different nodes takes place by message passing or shared memory accesses (see Fig. 7). Each processor executes the following algorithm for an assigned node (*i*):

1. Set initial value for temperature: 353 K for solid part and 343 K for gas reactants.
2. Compute G_{ij} and B_i.
3. Exchange conductances with neighbor nodes.
4. Compute $\dot{m}_{\mathrm{f}}$ according to equation (25).
5. Compute T_{i} according to equation (28).
6. Exchange local temperature T_{i} with neighbor nodes.
7. If $T_{\mathrm{i}}(k+1) - T_{\mathrm{i}}(k) \geq 0.001$ go back to step 2.

2.3.4 Analysis of Parallel Systems Performances

In this section, we present the simulation results of the PEMFC thermal behavior obtained by running the parallel algorithm on a network of processing nodes. Figures 8–11 show the performance simulation results with different matrix sizes and with different platforms (PVM, MPI and UPC).

There are several ways to measure the performance of a parallel algorithm. The speedup is one of the best criteria used to measure the performance. Speedup *S* of a parallel algorithm is defined as the ratio of the sequential time T_{S}, to the parallel time T_{p} to run the algorithm, i.e, $S = T_{\mathrm{S}}/T_{\mathrm{p}}$. In order to achieve the optimum performance, we would like our parallel algorithm to run *p* times faster than the best sequential algorithm.

Notice that the goal here is to show that the simulations are improved by introducing of parallel computations. However, we would like to stress that the proposed

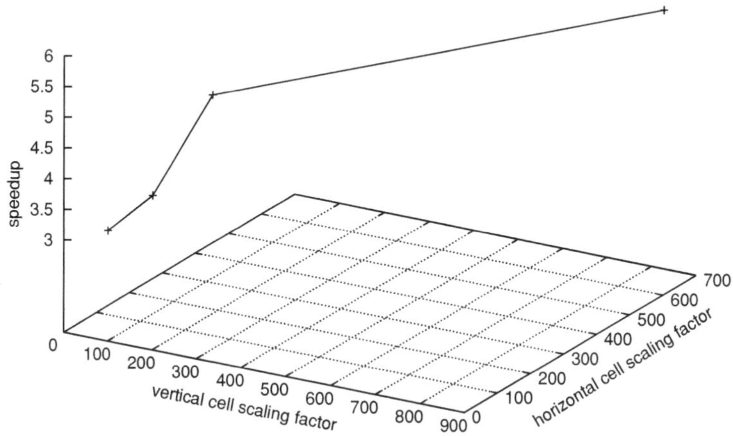

Fig. 8 Speedup according to the nodal network size with PVM on eight workstations connected by a 100 Mb ethernet LAN network

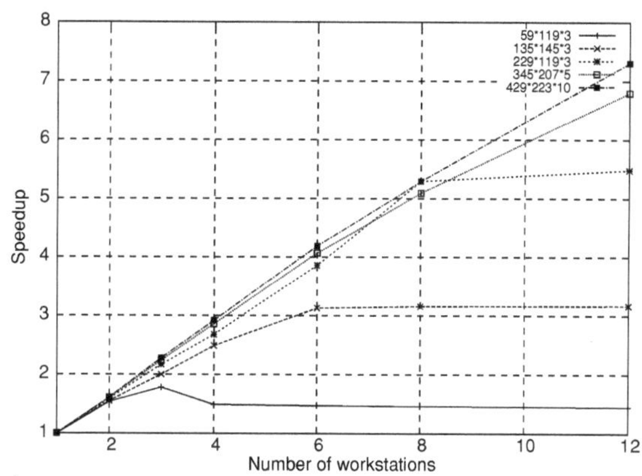

Fig. 9 Speedup on the cluster of nodes with MPI

results, hereafter, are not to be considered as a comparative analysis between the different systems that are UPC, MPI and PVM. In fact, here, the comparison has to be interpreted between the sequential computations and the parallel computations.

PVM Performances: the parallel algorithm has been implemented on a PVM system formed by a set of eight workstations interconnected by a 100 Mb ethernet LAN network. We remark that the speedup increases with the number of workstations involved as illustrated in Fig. 8.

In term of execution time, the parallel implementation of the nodal network based model results in best performance when the number of workstations involved together with the problem size is very large.

For instance, for a nodal network with size close to 10^6 nodes, the execution time is about 12 h with sequential calculation, while it can be reduced to 3 h with

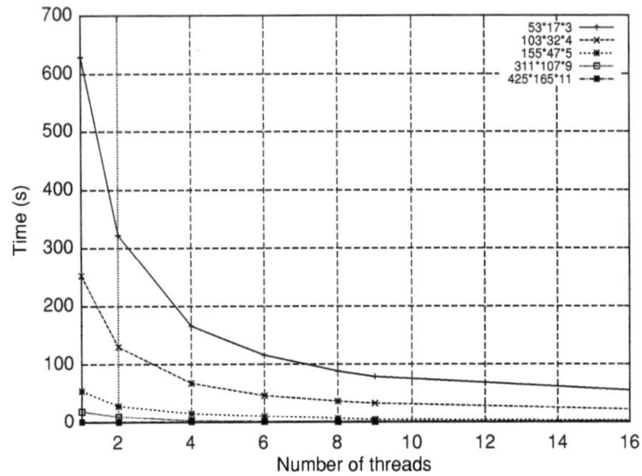

Fig. 10 Computation time on the myrinet cluster with UPC

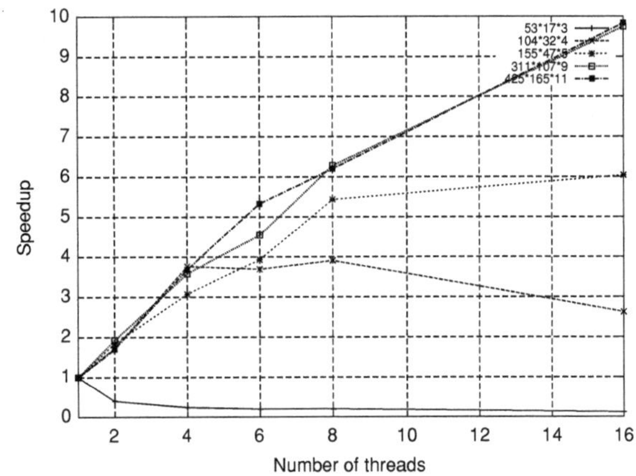

Fig. 11 Speedup on the myrinet cluster with UPC

8 workstations and to 7 min with 50 machines. The speedup is equal to 5.80 for 8 machines and can attain 25 for 50 machines.

MPI Performance: the experiments were also performed by using a much faster cluster of nine processing nodes, where each node contained two 1.8 GHz Xeon processors and 512 MB of RAM [19]. These nodes were interconnected through a high speed Myrinet network which provided a latency of 2.6 μs and a bandwidth of 2 GB s^{-1}.

Figure 9 shows the speedup obtained using MPI on different nodal network sizes expressed in kB. In terms of the execution time, the results show that the parallel implementation of the nodal network model achieved the best performance when the number of nodes involved together with the problem size was very large. However, when using smaller matrix sizes, the use of more nodes was useless, and could degrade the performance.

UPC performances: the experiments were executed using Berkeley UPC compiler on a cluster of eight nodes, where each node contains two 1.8 GHz Xeon processors and 512 MB RAM. The nodes are interconnected through a high speed Myrinet network which provides a 2.6 μs latency and a 2 GB s^{-1} bandwidth. To improve the performance of our parallel UPC programs, the code has been optimized by privatizing local accesses to the shared memory and by using aggregating communications [13].

Concerning UPC, Fig. 10 shows the execution time to run 20,000 iterations on different number of processors and different matrix sizes. The results show that for large matrices, the corresponding speedup (see Fig. 11) grows nearly linearly as the number of threads increase. For smaller matrices, the overhead time spent on communication increases faster and limits the performance gain.

2.4 Simulation Results

In this section, we present simulation results of the PEMFC thermal behavior obtained by nodal network approach implemented with the parallel library PVM. Results emphasize that temperature is not homogeneous throughout the stack and the cell also. This phenomenon has been also reported in [9, 20] wherein a PDE based simulation model has been used. The proposed nodal network simulation model, allow us both to observe with more precision the cell temperature evolution and identify the heat sensitive regions increase rapidly.

Figure 12 shows that the temperature of cell is affected by channel gas. In fact since thermal conductivity of bipolar plate is higher than cell thermal conductivity, and since the exchange surface between plate and gas is more important than the cell/gas exchange, temperature of the section crossed by the channel gasses is less hot than the section. This phenomenon affects strongly the membrane. Infrared microscopy showed that degradation is more important on the level of the gasses channels [12]. As illustrated by Fig. 12 area A' that interfaces the electrodes and the membrane designates the cell hot spots. Also, areas B (A, respectively) are more heated than areas C (A, respectively) Fig. 13.

3 Control of the Thermal Behavior

This section is devoted to control of thermal behavior of PEMFC. In the first subsection, generalities on control of the thermal behavior are given. The second subsection is dedicated to the modeling and a new model to describe the dynamic behavior of thermal behavior is established. In the third subsection, a result on control of thermal behavior, based on feedback stabilization, is proposed. The last subsection is constituted by simulation results.

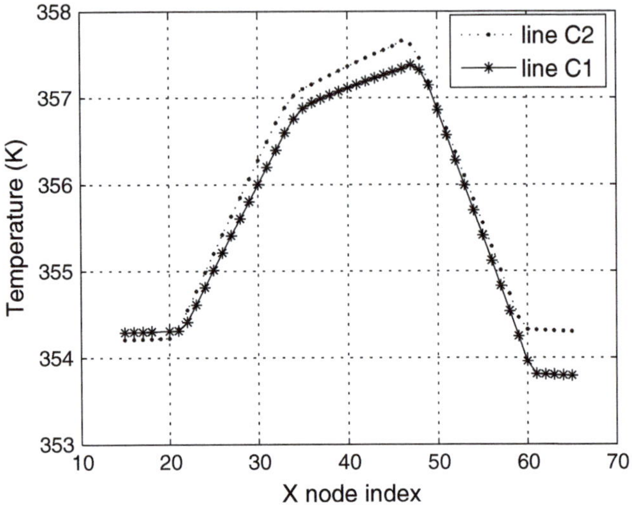

Fig. 12 Temperature fields of a stack-simulation based PVM

Fig. 13 3D temperature field in the first cell-simulation based PVM

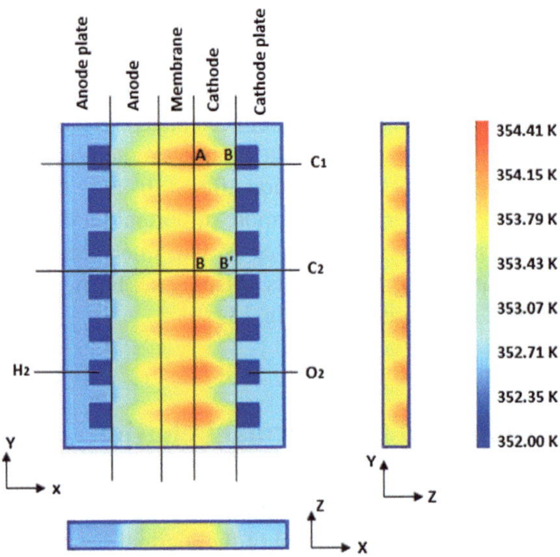

3.1 Generalities on Control of the Thermal Aspect

In the literature (see for instance [1]), the control of thermal behavior of PEM fuel cells is ensured by the use of fluid circulation process. Generally, a peddler fluid is used with a heat exchanger. The circulation is done until cooling is assured. A global thermal balance sheet of the stack allows its sizing. However, in the classi-

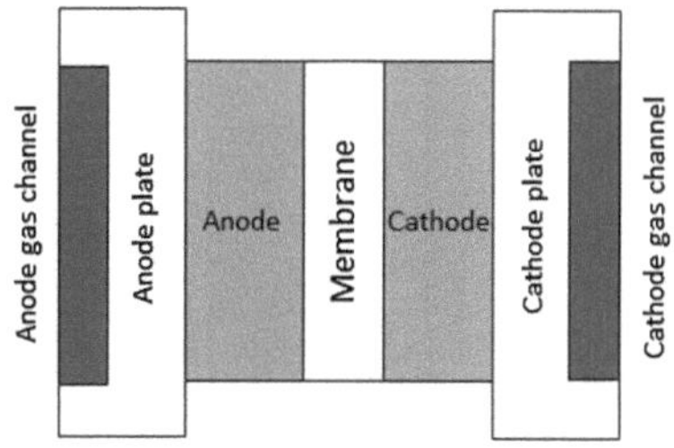

Fig. 14 Structure of the cells of a PEMFC

cal approach it is assumed that the temperature is uniform and its value is the same for each cell. In other word, the control of the temperature around specific values at each component of the fuel cells is not ensured. Here, a model is established to take into account the behavior of the temperature in each component of the fuel cells (membrane, plates, etc).

3.2 The Nonlinear Model Under Consideration

3.2.1 Assumptions Under Consideration

The main assumptions are as follows:

- Temperature is considered homogeneous in each block of cell (anode, cathode and membrane). Figure 14 depicts the structure of the cells of the stack for the PEMFC.
- All cells are at the same temperature during the functioning.
- Heat resulting by fluid movement within the cell is neglected.
- The heat losses in the manifolds are neglected.
- The inlet gas temperature is inferior to the operating temperature of the cell. Thus, there is no condensation of steam into water.
- The produced water at the cathode remains in the liquid state and hence the energy of water vaporization is zero.

3.2.2 Modeling Principle

The model is developed under the assumption that the PEM fuel cell can be decomposed into isothermal blocks called nodes. The considered model describes the behavior of the thermal at the following nodes (see Fig. 5.2):

- Nodes in the bipolar plate coolant channel (anode side and cathode side, respectively).
- Nodes in the gasses channels (anode side and cathode side, respectively).
- Nodes in the bipolar plates (anode side and cathode side, respectively).
- Nodes in the electrodes (anode side and cathode side, respectively).

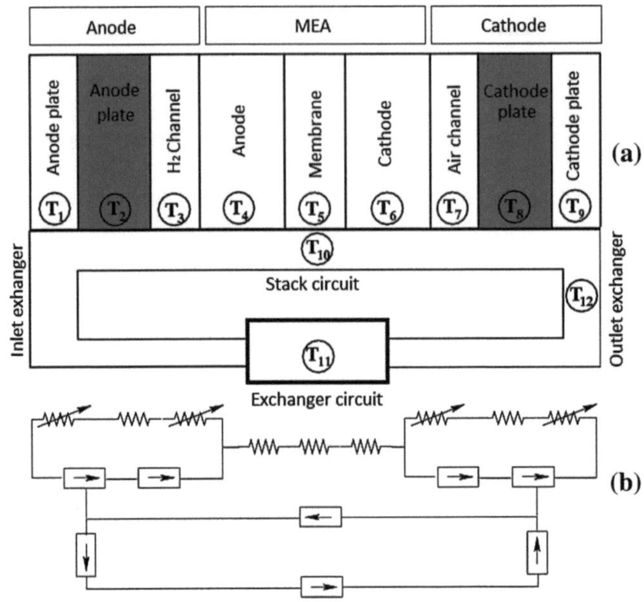

Fig. 15 Model under consideration

- Node in the membrane.
- Nodes at the inlet and at the outlet side of the fuel cell.
- Nodes at the entry and at the outlet side of the heat exchanger.

Notice that, and since the heterogeneous aspect of the fuel cells, in this approach we associate a node to each materiel part of a cell (see Fig. 4). Besides, and in order to take into account the heat exchange within the components of the fuel cell systems, a node is affected to every conduct of gasses (i.e. T_3 and T_7) and in the water conduct (i.e. T_1, T_9, T_{10} and T_{12}). The used heat exchanger which cools the stack and brings it back to the optimum temperature point is represented by a node T_{11} (see Fig. 5.2a).

All the nodes are connected by the conductances relating the thermal transfers and the temperature exchange is given by heat flow balance with the neighboring blocks. Indeed, temperature of each node i is given by the relationship between its two adjacent nodes $(i+1)$ and $(i-1)$ [8]. To describe the behavior of the temperature at a node, we have a similar formula that one given by (7):

$$C_{\mathrm{i}}\frac{\mathrm{d}T_{\mathrm{i}}}{\mathrm{d}t} = G_{\mathrm{i,i-1}}(T_{\mathrm{i}} - T_{\mathrm{i-1}}) + G_{\mathrm{i,i+1}}(T_{\mathrm{i}} - T_{\mathrm{i+1}}), \tag{29}$$

where C_{i} denotes the global heat capacity associated to node (i) and is determined by $C_{\mathrm{i}} = \rho \mathrm{cp} SL$, G the thermal conductivity between two nodes, S designates the heat exchange surface, L is the length between the nodes, ρ represents the density of the material, and cp is the specific heat Fig. 15.

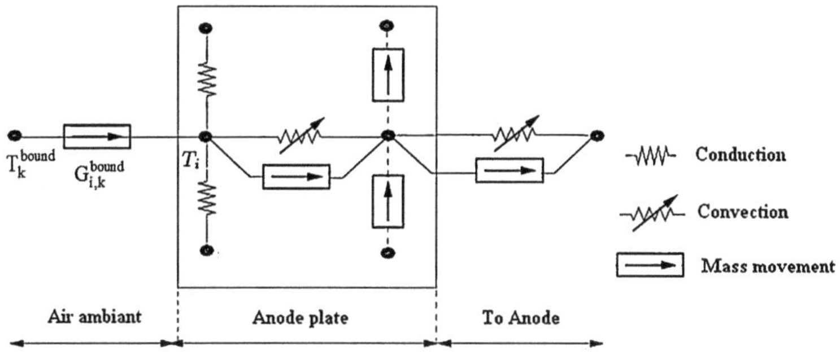

Fig. 16 Nodal network of the anode compartment

Notice that equation (29) can be completed, when it is necessary, by adjunct of a heat source or a sink P_i. In these cases, the boundary conditions are expressed Dirichlet's ones at the edge of the bipolar plates (i.e. $T = T_k^{bound}$). $G_{i,k}^{bound}$ represents the external heat exchange between the cell and the ambient temperature. k is the index of the node which stands for the boundary condition (see Fig. 16):

$$C_i \frac{dT_i}{dt} = \sum_j G_{i,j}(T_j - T_i) + \sum_k G_{i,k}^{bound}(T_k^{bound} - T_i) + P_i. \tag{30}$$

In the sequel, the goal is to control by feedback the thermal behavior of the cell. More precisely, the control consists in stabilizing the temperature around a desired value of the system composed by differential equations describing the behavior of the temperature at each node (see Fig. 2a): water anode (T_1), anode plate (T_2), channel hydrogen (T_3), anode (T_4), membrane (T_5), cathode (T_6), channel air (T_7), cathode plate (T_8), water cathode (T_9), stack water (T_{10}), heat (T_{11}), exchanger, (T_{12}) and exit exchanger (T_{12}).

Figure 5.2a depicts the heat transfers by conduction within the cell, as well as the convection transfers between gasses and plates of stack. The model takes into account the cooling exchanger and the energy resulting from the molecules of water and gas movement. T_{H_2}, T_{air} represent the hydrogen and air temperature, respectively. T_w designates the water cooling temperature. T_{pq_a}, T_{pq_c} are the anode and cathode plate temperatures. T_a, T_c, T_m denote temperatures of cell. Q_a, Q_c are the heats produced by the exothermic reactions at the electrodes. Figure 2b gives the electrical equivalent model of the thermal model.

In order to introduce the model describing the thermal behavior we distinguish four kinds of nodes.

A. Circuit of Liquid Circulation: the nodes related to the liquid circulation are $\{T_1, T_9, T_{10}, T_{11}\}$ and T_{12}. The behavior of the temperature depends on the energy of both movement of the water and the convective exchange between cooling water and the solid part of the cell. Hence, for instance concerning water anode (i.e. T_1), the dynamics is expressed by:

$$\underbrace{C_{\mathrm{a}}^{\mathrm{e}} \frac{\mathrm{d}T_1}{\mathrm{d}t}}_{\text{Accumulation}} = \underbrace{\dot{m}_{\mathrm{e,p}} \mathrm{cp}_{\mathrm{e}} (T_{10} - T_1)}_{\text{Movement energy}} + \underbrace{\mathrm{H_e} S_{\mathrm{pq,e}} (T_2 - T_{10})}_{\text{Convection}} . \tag{31}$$

B. Plates: the concerned nodes are T_2 and T_8. In this case, the exchanged energy, realized by conviction, is done between the node and those, representing, water and air ambient (or hydrogen). For instance, the behavior of the anode plate temperature (i.e. T_2) is described by:

$$\begin{aligned} \underbrace{C_{\mathrm{pq}} \frac{\mathrm{d}T_2}{\mathrm{d}t}}_{\text{Accumulation}} &= \underbrace{G_{2,\mathrm{H_2}} (T_{\mathrm{H_2}}^{\mathrm{in}} - T_2) + G_{3,2} (T_3 - T_2) + G_{2,\mathrm{e}} (T_1 - T_2)}_{\text{Convection}} \\ &+ \underbrace{G_{2,\mathrm{air}} (T_{\mathrm{amb}} - T_2)}_{\text{Boundary conditions}} . \end{aligned} \tag{32}$$

C. Membrane Electrode Assembly: the nodes related to the MEA are T_4, T_5 and T_6. For this kind of nodes, the energy is exchanged by thermal conduction. Besides, each node is endowed by a source term to express the local energy produced by the exothermic reactions at the electrodes. In the membrane, heat resulting by the proton transfer resistance. For instance, in the anode (i.e. for T_4) we obtain the following equation:

$$\underbrace{C_{\mathrm{a}} \frac{\mathrm{d}T_4}{\mathrm{d}t}}_{\text{Accumulation}} = \underbrace{G_{3,2} (T_2 - T_4) + G_{3,4} (T_5 - T_4)}_{\text{Conduction}} + \underbrace{\left(\alpha \frac{L_{\mathrm{v}}}{2F} \right) I}_{\text{Source}} . \tag{33}$$

D. Channels: in the channels, the nodes are T_3 and T_7. We apply a similar rule used in the first kind of nodes (i.e. liquid circulation) by adding the term of energy movement of steam of gasses humidification. We obtain for channel hydrogen (i.e. T_3) :

$$\begin{aligned} \underbrace{C_{\mathrm{H_2,a}} \frac{\mathrm{d}T_3}{\mathrm{d}t}}_{\text{Accumulation}} &= \underbrace{(T_{\mathrm{H_2}}^{\mathrm{in}} - T_3) \dot{m}_{\mathrm{H_2}} \mathrm{cp}_{\mathrm{H_2}}}_{\text{Gas movement energy}} + \underbrace{\dot{m}_{\mathrm{v}\,H_2} \mathrm{cp}_{\mathrm{H_2}} T_3}_{\text{Steam movement energy}} \\ &+ \underbrace{H_{\mathrm{H_2}} S_{pq,\mathrm{H_2}} (T_2 - T_{\mathrm{H_2}}^{\mathrm{in}})}_{\text{Convection}} . \end{aligned} \tag{34}$$

It should be noted that the T_{10} which designates the temperature at node in the exit of conduct water depends on the energy produced at the nodes corresponding to the temperatures T_1 and T_9, respectively. To take into account the effect of the convection, we assume that the temperature of the water in the circuit stack is the arithmetic average between the temperatures of the water in both the anode and the cathode. However, the temperature of the wall of the conduct will be assumed to be equal to the average between the temperature of the anode and cathode plate.

The temperature T_{11} at the exchanger, assumed to be isothermal fluid type, depends on the energy brought by the node represented by T_{10}. It is assumed that the temperature of the wall of the heat exchanger is identical to the temperature of the cooling water. The node, corresponding to T_{12}, represents the temperature of the outlet of water which depends on the contribution of energy of node represented by T_{11} and the exchange by convection with the ambient air.

3.2.3 The Considered Model

By applying the principle of thermal balance to all nodes, we deduce the following state model describing the behavior of the thermal aspect of PEMFC

$$\begin{cases} C_{\mathrm{a}}^{\mathrm{e}} \frac{\mathrm{d}T_1}{\mathrm{d}t} = \dot{m}_{\mathrm{e,p}} cp_{\mathrm{e}}(T_{10} - T_1) + \mathrm{H_e} S_{\mathrm{pq,e}}(T_2 - T_{10}) \\ C_{\mathrm{pq}} \frac{\mathrm{d}T_2}{\mathrm{d}t} = G_{2,\mathrm{H_2}}(T_{\mathrm{H_2}}^{\mathrm{in}} - T_2) + G_{3,2}(T_3 - T_2) + G_{2,\mathrm{e}}(T_1 - T_2) + G_{2,\mathrm{air}}(T_{\mathrm{amb}} - T_2) \\ C_{\mathrm{H_2,a}} \frac{\mathrm{d}T_3}{\mathrm{d}t} = (T_{\mathrm{H_2}}^{\mathrm{in}} - T_3)\dot{m}_{\mathrm{H_2}} \mathrm{cp}_{\mathrm{H_2}} + \dot{m}_{\mathrm{vH_2}} \mathrm{cp}_{\mathrm{H_2}} T_3 + \mathrm{H_{H_2}} S_{\mathrm{pq,H_2}}(T_2 - T_{\mathrm{H_2}}^{\mathrm{in}}) \\ C_{\mathrm{a}} \frac{\mathrm{d}T_4}{\mathrm{d}t} = G_{3,2}(T_2 - T_4) + G_{3,4}(T_5 - T_4) + \left(\alpha \frac{L_{\mathrm{v}}}{2F}\right) I \\ C_{\mathrm{m}} \frac{\mathrm{d}T_5}{\mathrm{d}t} = G_{3,4}(T_4 - T_5) + G_{4,5}(T_6 - T_5) + (\lambda_{\mathrm{m}}/\zeta_1 - \zeta_2 e^{(1/303 - 1/\mathrm{T_m})}) I \\ C_{\mathrm{c}} \frac{\mathrm{d}T_6}{\mathrm{d}t} = G_{4,5}(T_5 - T_6) + G_{5,6}(T_8 - T_6) + \left(\frac{L_{\mathrm{v}}(1+2\alpha) + T\Delta S}{2F}\right) I \\ C_{\mathrm{air,c}} \frac{\mathrm{d}T_7}{\mathrm{d}t} = (T_{\mathrm{air}}^{\mathrm{in}} - T_7)\dot{m}_{\mathrm{air,c}} \mathrm{cp}_{\mathrm{air}} + \dot{m}_{\mathrm{vair,c}} \mathrm{cp}_{\mathrm{air}} T_7 + \mathrm{H_{air,c}} S_{\mathrm{p,air}}(T_8 - T_{\mathrm{air}}^{\mathrm{in}}) \\ C_{\mathrm{p}} \frac{\mathrm{d}T_8}{\mathrm{d}t} = G_{6,\mathrm{air}}(T_7 - T_8) + G_{6,5}(T_6 - T_8) + G_{6,\mathrm{air}}(T_{\mathrm{amb}} - T_8) + G_{6,\mathrm{e}}(T_9 - T_8) \\ C_{\mathrm{e,c}} \frac{\mathrm{d}T_9}{\mathrm{d}t} = \dot{m}_{\mathrm{e,p}} \mathrm{cp_e}(T_{10} - T_9) + H_{\mathrm{e}} S_{\mathrm{pq,e}}(T_8 - T_{10}) \\ C_{\mathrm{e,p}} \frac{\mathrm{d}T_{10}}{\mathrm{d}t} = \dot{m}_{\mathrm{e,p}} \mathrm{cp_e} N(T_1 - T_{10}) + \dot{m}_{\mathrm{e,p}} \mathrm{cp_e} N(T_9 - T_{10}) - H_{\mathrm{e}} S_{\mathrm{sec}} \left[\left(\frac{T_1 + T_9}{2}\right) - \left(\frac{T_2 + T_8}{2}\right)\right] \\ C_{\mathrm{e,ec}} \frac{\mathrm{d}T_{11}}{\mathrm{d}t} = \dot{m}_{\mathrm{e,ec}} \mathrm{cp_e}(T_{10} - T_{11}) + \mathrm{H_e} S_{\mathrm{ec}}(T_{11} - T_{\mathrm{amb}}) \\ C_{\mathrm{e,sec}} \frac{\mathrm{d}T_{12}}{\mathrm{d}t} = \dot{m}_{\mathrm{e,ec}} \mathrm{cp_e}(T_{11} - T_{12}) + \mathrm{H_e} S_{\mathrm{t}}(T_{12} - T_{\mathrm{amb}}) \end{cases} \tag{35}$$

The system (35) can be rewritten in the bilinear compact form

$$\dot{z} = Az + \sum_{i=1}^{m} u_{\mathrm{i}}(B_{\mathrm{i}} z + b_{\mathrm{i}}), \tag{36}$$

where $z \in \mathscr{R}^{\mathrm{n}}$ denotes the state, $u = (u_1, \ldots, u_{\mathrm{m}})^{\mathrm{T}} \in \mathscr{R}^{\mathrm{m}}$ represents the control, A and B_{i} $(i = 1 \ldots m)$ are matrices with suitable dimensions, and b_{i} $(i = 1 \ldots m)$ are vectors in $\mathscr{R}^{\mathrm{n}}$.

Generally speaking, and regardless of fuel cells, feedback stabilization of (36) has attracted the interest of several authors (see for instance [21–26]). The interest is due to the fact that the behavior of a large number of process can be described using bilinear models. Indeed, it concerns process from thermal, nuclear, biology, immunological or chemical (see for instance [21, 27]). However, feedback stabilization of bilinear systems, despite their simple structure, remains an open question even for small dimensions, for instance, the 3D case or the homogeneous case; i.e.,

$$\dot{z} = Az + \sum_{i=1}^{m} u_i B_i z. \tag{37}$$

Results were proposed mainly when the free dynamics; i.e. $\dot{z} = Az$ is stable (see for instance [28–30]). Notice that the considered model for PEM fuel cells is no homogeneous; i.e. ($b_i \neq 0 pour i = 1 \ldots 3$), the state dimension is equal to twelve and the free dynamics is naturally unstable. Finally, there is no result, of the literature, that can be directly applied for feedback stabilization of thermal behavior using the considered model. Thus, in our approach, we focus on the class of bilinear systems that contains the model describing PEM fuel cells. More precisely, our analysis will depend on numerical values of parameters defining the models of PEM fuel cells and operating range.

3.2.4 A Model Transformation

The goal of this subsection is to rewrite the model in a suitable form. A simple reasoning shows the (35) can be rewritten in the compact form:

$$C\dot{T} = GT + (u_1 G_1 + u_2 G_2 + u_3 G_3)T + B_d \tag{38}$$

or

$$\dot{T} = C^{-1}GT + \sum_{i=1}^{3} u_i G_i T + C^{-1} B_d \tag{39}$$

with $T^2 = (T_1, T_2, T_3, T_4, T_5, T_6, T_7, T_8, T_9, T_{10}, T_{11}, T_{12})^T \in \mathscr{R}^{12}$. G and $G_i (i = 1 \ldots 3)$ are suitable matrices. The vector B_d is defined by

$$B_d = \begin{pmatrix} -H_e S_{p,e} T_{H_2}^{in} \\ G_{2,air} T_{amb} + G_{2,e} T_{H_2}^{in} \\ \left(\dot{m}_{H_2} cp_{H_2} - H_{H_2} S_{p,H_2}\right) T_e \\ \left(\alpha \frac{L_v}{2F}\right) I \\ \left(\lambda_m/\zeta_1 - \zeta_2 e^{(1/303 - 1/T_m)}\right) I \\ \left(\frac{L_v(1+2\alpha)+T\Delta S}{2F}\right) I \\ \left(\dot{m}_{air} cp_{air} - H_{air} S_{p,a}\right) T_{air,c}^{in} \\ G_{6,air} T_{amb} \\ -H_e S_{p,e} T_{H_2}^{in} \\ -H_{ext} S_{p,ec} T_{amb} \\ -H_{air} S_{tub} T_{amb} \end{pmatrix}. \tag{40}$$

[2] a^T designates the transpose vector of a.

It is possible to establish the relationship between steam flow made by moistening at the anode and cathode, respectively:

$$m_{\mathrm{vc}} = 2x_{\mathrm{O_2}} m_{\mathrm{va}} \frac{ut_{\mathrm{air}}}{ut_{\mathrm{H_2}}} \frac{P_{\mathrm{c}} - P_{\mathrm{sat}}}{P_{\mathrm{a}} - P_{\mathrm{sat}}}. \tag{41}$$

The water and steam flows are the control which will be used for feedback stabilization:

$$u_1 = \dot{m}_{\mathrm{e,pq}},\ u_2 = \dot{m}_{\mathrm{e,ec}},\ u_3 = \dot{m}_{\mathrm{va}}, \tag{42}$$

where the operating rang of u_{i} $(i = 1 \ldots 3)$ are defined by

$$\begin{cases} u_1 \in [u_1^{\mathrm{min}}, u_1^{\mathrm{max}}] \\ u_2 \in [u_2^{\mathrm{min}}, u_2^{\mathrm{max}}] \\ u_3 \in [u_3^{\mathrm{min}}, u_3^{\mathrm{max}}] \end{cases}. \tag{43}$$

Notice that for a numerical value of current density, the vector B_{d} is constant. Let A, B_i (= $i \ldots 3$) and D be the matrices defined by

$$\begin{cases} A = C^{-1}G \\ B_{\mathrm{i}} = C^{-1}G_{\mathrm{i}} \qquad (i = 1 \ldots 3). \\ D = C^{-1}B_{\mathrm{d}} \end{cases} \tag{44}$$

Then, (4.39) becomes

$$\dot{T} = AT + \sum_{i=1}^{3} u_{\mathrm{i}} B_{\mathrm{i}} T + D, \tag{45}$$

where u_{i} and ξ_{i} are defined by

$$\begin{cases} u_{\mathrm{i}} = u_{\mathrm{i}}^{\xi} + \xi_{\mathrm{i}} & (i = 1 \ldots 3) \\ \xi_{\mathrm{i}} = u_{\mathrm{i}}^{\mathrm{min}} & (i = 1 \ldots 3) \end{cases}. \tag{46}$$

Since (43), we have

$$u_{\mathrm{i}}^{\xi} \in [0, u_{\mathrm{i}}^{\mathrm{max}} - u_{\mathrm{i}}^{\mathrm{min}}]. \tag{47}$$

Replacing (46) in (45) yields

$$\dot{T} = AT + \sum_{i=1}^{3} (u_{\mathrm{i}}^{\xi} + \xi_{\mathrm{i}}) B_{\mathrm{i}} T + D \tag{48}$$

or

$$\dot{T} = (A + \xi_{\mathrm{i}} B_{\mathrm{i}})T + \sum_{i=1}^{3} u_{\mathrm{i}}^{\xi} B_{\mathrm{i}} T + D. \tag{49}$$

In the sequel, we use A_ξ to design the matrix defined by

$$A_\xi = A + \sum_{i=1}^{3} \xi_i B_i. \tag{50}$$

Then

$$\dot{T} = A_\xi T + \sum_{i=1}^{3} u_i^\xi B_i T + D. \tag{51}$$

Generally, the power issued from a fuel cells depends on its size. Hence, the power depends, but not only, on the surface of the membrane. Nowadays, for a surface of 1 cm^2 of membrane, the power issued by the fuel cells is about 1 W [31].

Based on this remark, for a sake of simplicity, we consider a simplifying assumption. More precisely, we assume that the power Q can be expressed as a linear function of the other physical quantities p_i of the fuel cells. Moreover, we suppose that:

$$p_i = \alpha_i Q, \tag{52}$$

where i denotes a geometric quantity of the fuel cells.

Thus, for a operating range of the power given by $[Q_{min}, Q_{max}]$ and for a parameter $p_i \in [p_{min}, p_{max}]$, we get the following expression:

$$\alpha_i = \frac{p_{max} - p_{min}}{Q_{max} - Q_{min}}. \tag{53}$$

Finally, and since (52) and (53), we obtain an estimation of the operating range of all the parameters using the equation:

$$p_i = p_{min} + \alpha_i (Q - Q_{min}). \tag{54}$$

We would like to stress that the considered assumption is taken into account in order to get a global result which will be valid for all the operating range of the power. Nevertheless, the proposed methodology still valid even if the parameters do not satisfy (53) and (54).

Under the considered assumption (see Eq. 54), the matrices depend on the value of the power of the fuel cells. Hence, the system (51) can be rewritten as

$$\dot{T} = A_\xi(Q)T + \sum_{i=1}^{3} u_i^\xi B_i(Q)T + D(Q). \tag{55}$$

Using a suitable change of coordinates the system (55) can be rewritten in the classical form given by (36). In order to ensure the stability of the system, the considered change of coordinates verifies $\mathscr{D}et^3(A_\xi(Q))$ is different from zero.

[3] For a matrix A, we use $\mathscr{D}et(\text{A})$ to denote the determinant.

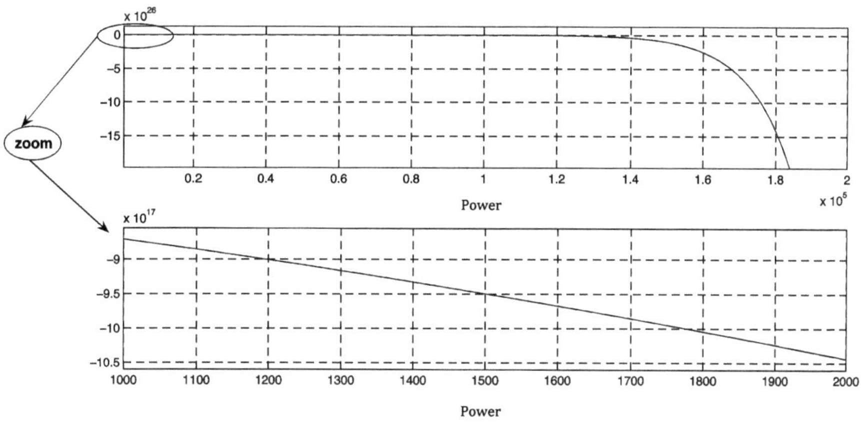

Fig. 17 Determinant of $A_\xi(Q)$ depending on Q

Figure 17 depicts $\mathscr{D}et(A_\xi)$ for the power Q in the operating range. Notice that the determinant do not vanish for any value the power Q. Hence, by continuity, and for $\sigma_{\rm i} \in \mathscr{R}(i = 1 \ldots 3)$ small enough, we obtain

$$\mathscr{D}et\left(A_\xi(Q) + \sum_{i=1}^{3} \sigma_{\rm i} B_{\rm i}(Q)\right) \neq 0. \tag{56}$$

We introduce a preliminary feedback $\sigma_{\rm i}$ in order to vary, and so to be able to chose, the operating point around which the feedback stabilization will be done. Let

$$\sigma_{\rm i} \in [0, u_{\rm i}^{\rm max} - u_{\rm i}^{\rm min}]. \tag{57}$$

We have

$$u_{\rm i}^{\sigma+\xi} \in [-\sigma_{\rm i}(u_{\rm i}^{\rm max} - u_{\rm i}^{\rm min}) - \sigma_{\rm i}] \quad \text{for} \quad i = 1...3, \tag{58}$$

where

$$u_{\rm i}^{\sigma+\xi} = u_{\rm i}^{\xi} - \sigma_{\rm i}. \tag{59}$$

Equation (51) becomes

$$\dot{T} = A_{\xi+\sigma}(Q)T + \sum_{i=1}^{3} u_{\rm i}^{\sigma+\xi} B_{\rm i}(Q)T + D(Q). \tag{60}$$

The system (60) can be rewritten in bilinear form (36). Indeed, let x be the new state variable defined by

$$x = T + (A_{\xi+\sigma}(Q))^{-1} D(Q). \tag{61}$$

Then,

$$
\begin{aligned}
\dot{x} &= \dot{T} \\
&= \left(A_{\xi}(Q) + \sum_{i=1}^{3} \sigma_{i} B_{i}(Q) \right) x \\
&\quad + \sum_{i=1}^{3} u_{i}^{\sigma+\xi} B_{i}(Q) \left[x - (A_{\xi+\sigma}(Q))^{-1} D(Q) \right] \\
&= A_{\xi+\sigma}(Q) x + \sum_{i=1}^{3} u_{i}^{\sigma+\xi} \left[B_{i}(Q) x - B_{i}(Q)(A_{\xi+\sigma}(Q))^{-1} D(Q) \right].
\end{aligned}
\tag{62}
$$

Finally, the system defined by (39) can be rewritten in the following bilinear form

$$
\dot{x} = A_{\xi+\sigma}(Q) x + \sum_{i=1}^{3} u_{i}^{\sigma+\xi} \left(B_{i}(Q) x + b_{i}^{\sigma}(Q) \right)
\tag{63}
$$

with

$$
b_{i}^{\sigma}(Q) = -B_{i}(Q)((A_{\sigma+\xi}(Q))^{-1} D(Q)).
\tag{64}
$$

Notice that the operating point, around witch the feedback stabilization will be realized, is defined by

$$
T^{o} = -((A_{\sigma+\xi}(Q))^{-1} D(Q)).
\tag{65}
$$

3.3 Feedback Stabilization

Let us consider the bilinear system defined by (36). We have the following result which concerns the local feedback stabilization around the origin.

Proposition 1 *If* $(A, [b_1 \dots b_m])$ *is controllable then the bilinear system defined by* (36) *can be render locally asymptotically stable around the origin.*

The proof is based on the fact that the controllability of the linear approximation, of a nonlinear system, implies its local feedback stabilization. It shall be omitted.

Since the result of the above proposition, to prove the local feedback stabilization of (36) it suffices to show that $(A_{\xi+\sigma}, [b_1 b_2 b_3])$ is controllable. Using straightforward computations, one can prove the following result:

Proposition 2 *For* $i = 1 \dots 12$ *and* $j = 1 \dots 3$, *we have* $A_{\xi+\sigma}^{i} b_{j} \in \mathscr{D}$ *with* $\mathscr{D} = Span$[4] $\{b_1, b_2, b_3, A_{\xi+\sigma}^{5} b_3, A_{\xi+\sigma}^{k} b_i, \quad k = 1 \dots 4, \quad i = 1, 3\}$.

Finally, $(A_{\xi+\sigma}, [b_1 b_2 b_3])$ is controllable if and only if the matrix

$$
\begin{aligned}
C = [&b_1, b_2, b_3, A_{\xi+\sigma} b_1, A_{\xi+\sigma}^{2} b_1, A_{\xi+\sigma}^{3} b_1, A_{\xi+\sigma}^{4} b_1, A_{\xi+\sigma} b_3, A_{\xi+\sigma}^{2} b_3, \\
&A_{\xi+\sigma}^{4} b_3, A_{\xi+\sigma}^{3} b_3, A_{\xi+\sigma}^{5} b_3]
\end{aligned}
$$

[4] We use *Span* to denote the space generated by vectors.

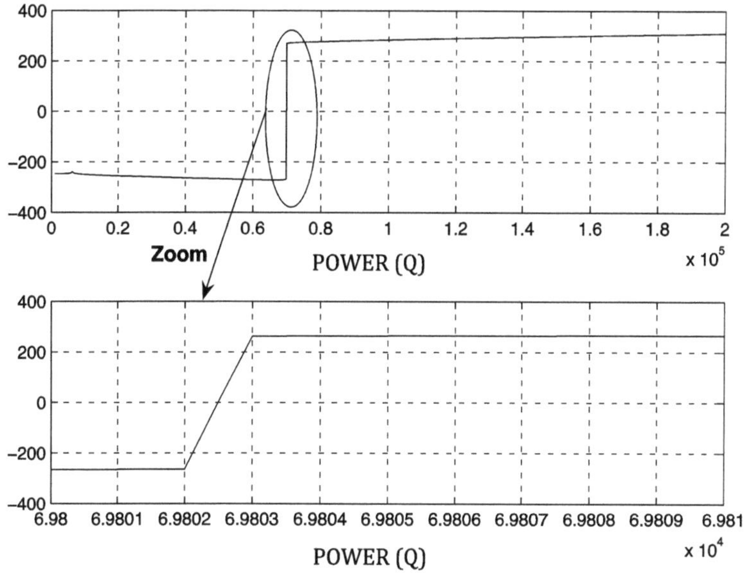

Fig. 18 Determinant of C depending on Q

has a full rank. Hence, we deduce the following result:

Proposition 3 *If* $\mathscr{D}et(C) \neq 0$ *then the system* (36) *can be render locally asymptotically stable at the origin.*

Figure 18 depicts the function $\mathscr{D}et(C(Q))$ depending on Q. We remark, and except for one value, the determinant do not vanish. Finally, we deduce that, in general, there exists a local feedback which made thermal behavior stable around a desired value (see Fig. 18).

3.4 Simulation Results

The first group of simulation results concern the temperature control for a PEMFC with a power equals to 20 W, a density of current equals to 0.3 A cm^{-2} and operating with an ambient temperature equals to 25°C. In order to avoid the heating problems, the conductivity is assumed to be high enough; of about 21 W m^{-1} K^{-1} (see [32, 33]). Notice that the values of the geometrical characteristics of the PEMFC are obtained from [34] and are summarized by Table 1.

The numerical simulation is obtained using the following gain matrix:

$$M = \begin{pmatrix} 0 & -0.05 & 0 & 29.96 & -59.82 & 29.96 & 0 & -0.05 & 0 & 0 & 0 & 0 \\ 0 & 0 & 0 & -0.05 & 0.1 & -0.05 & 0 & 0 & 0 & 0 & 0 & 0 \\ 0.03 & -278.33 & 0.1 & 0 & -3.13e^5 & 1.56e^5 & 0 & -281.9 & 0.03 & 0 & 0 & 0 \end{pmatrix} \quad (66)$$

and the control law is defined by $M \Delta T$.

Table 1 Simulation parameters

Parameter	Value	Unit	Parameter	Value	Unit
h_c	0.13	cm	h_m	16	cm
h_v	0.12	cm	T_{gaz}	70	°C
h_p	18	cm	T_{amb}	30	°C
L	18	–	I	1.5	$A\,cm^{-2}$
T	80	°C	T_e	30	°C
b_{O_2}	0.120	$V\,dec^{-1}$	$T\Delta S$	-48.7×10^3	$J\,kg^{-1}$
L_v	44×10^3	$J\,mol^{-1}$	F	96485.33	–
k_p	110	$W\,m^{-1}K$	k_m	0.34	$W\,m^{-1}K$
$cp_a = cp_c$	710	$J\,kg^{-1}K^{-1}$	cp_m	68.175	$J\,kg^{-1}K^{-1}$
H_{air}	76.17	$W\,m^{-2}K^{-1}$	H_{H_2}	73	$W\,m^{-2}K^{-1}$
H_e	10	$W\,m^{-2}K^{-1}$	N	200	–

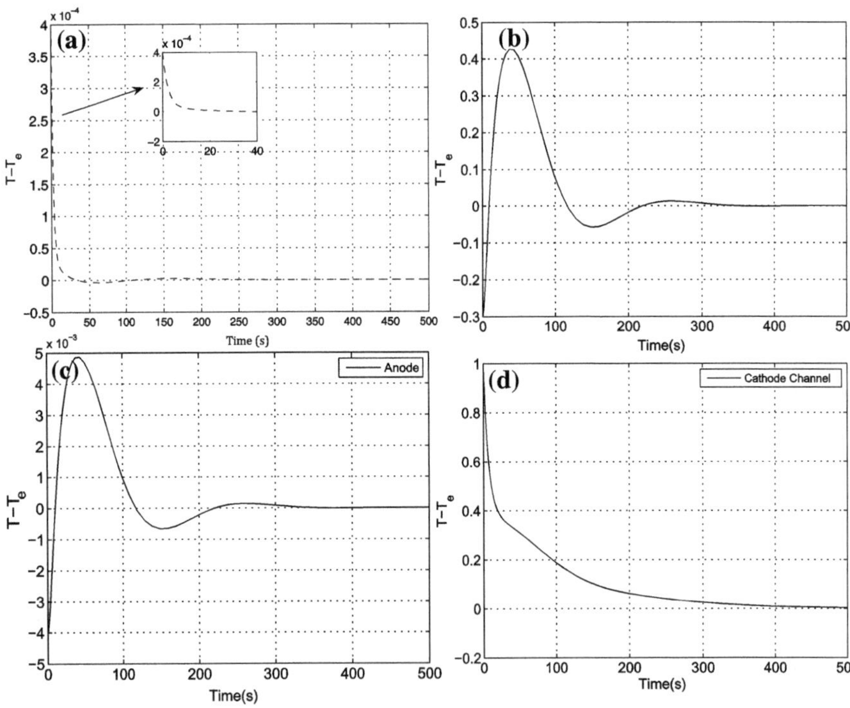

Fig. 19 Simulation results for an operating point where all the values are equal to 78°C

Figure 19 depicts the numerical results of the difference between the desired value and the temperature value at some nodes; i.e. membrane, water at the anode channel, anode and the cathode channel. This shows that the temperature value converges to the desired equilibrium point with good performances.

The second group of simulation results, depicted by Fig. 20, are dedicated to PEMFC with high power and concern the nodes that are the membrane, anode and anode plate.

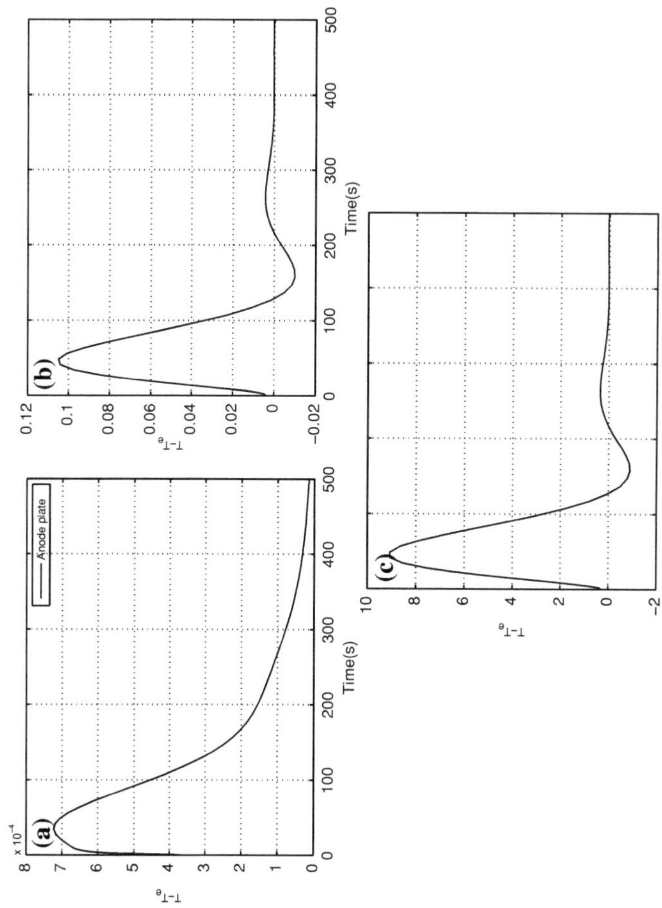

Fig. 20 Simulation results for an operating point where all the value are equal to 80°C

4 Conclusion

In this chapter, a new approach to describe the behavior of the PEMFC is presented without using the heat partial differential equation. The method of nodal network is adapted to thermal modeling of PEMFC in order to avoid the coupling problem between different codes which penalizing for the execution time. A strategy based on parallel computing is established for the simulation of thermal behavior of the PEMFC. The use of MPI and UPC with nodal network give good performances and the execution time is significantly improved. The second contribution concerns the control of thermal behavior. More precisely, we showed, first, that the thermal aspect of PEMFC can described by a bilinear model. Second, we proved that, for a desired value of the temperature, the system can made asymptotically stable using a linear feedback law.

References

1. Lachaise J (2004) Etudes des stratégies et des structures de commande pour le pilotage des systèmes energétique à pile à combustible destinés á la traction. PhD thesis, Institut National Polytechnique de Toulouse
2. Vega-Leal AP, Palomo FR, Barragán F, Garcia C, Brey JJ (2007) Design of control systems for portable PEM fuel cells. J Power Sources 169(1):194–197
3. Meyer RT, Yao B (2006) Control of a PEM fuel cell cooling system. In Proceedings of IMECE, volume 14151. Citeseer
4. Rakhtala SM, Fadian T, Shakeri M (2008) Temperature control system for PEM fuel cells. In: Fundamentals and Developments of Fuel Cells Conference FDFC. Nancy, France
5. Salah A, Gaber J, Outbib R, Serres O, El-Sayed H (2010) Modeling and simulation of pem fuel cell thermal behavior on parallel computers. Eng Convers IEEE Trans 25(3):768–777
6. Salah A (2008) Simulation et contrôle de l'aspect thermique d'une pile à combustible PEMFC. PhD thesis, Université de Technologie de Belfort-Montbéliard (UTBM)
7. Dumercy L (2004) Contribution à la Caractérisation Thermique et Fluidique d'une Pile à Combustible à Membrane échangeuse de Protons (PEMFC). PhD thesis, Université de Technologie de Belfort-Montbéliard (UTBM)
8. Saulnier J (1980) Modélisation thermique et ses applications aux transferts couplés at au contrôle acif. PhD thesis, Université de Poitier
9. Djilali N, Lu D (2002) Influence of heat transfer on gas and water transport in fuel cells. Int J Therm Sci 41(1):29–40
10. Baehr HD, Stephan K (2006) Heat and mass transfer. Springer, London
11. Larminie J, Dicks A (2003) Fuel cell systems explained. 2nd edn. Wiley, New York
12. Mayer J (2004) Etude des modes de dégradation de membranes polyimides sulfonés en pile à combustible H2/O2. PhD thesis, Université BLAISE PASCAL (U.F.R de Recherche Scientifique et Technique)
13. Cantonnet F, Yao Y, Zahran M, El-Ghazawi T (2004) Productivity analysis of the UPC language. In Parallel and Distributed Processing Symposium, 2004. Proceedings. 18th International 254. IEEE
14. Faghri A, Guo Z (2005) Challenges and opportunities of thermal management issues related to fuel cell technology and modeling. Int J Heat Mass Transf 48(19–20):3891–3920
15. Hwang K, Xu Z (1998) Scalable parallel computing: technology, architecture, programming. WCB/McGraw-Hill, New York
16. El-Ghazawi T, Carlson W (2005) UPC: distributed shared memory programming. Wiley-Interscience

17. Djilali N (2007) Computational modeling of polymer electrolyte membrane (PEM) fuel cells: challenges and opportunities. Energy 32(4):269–280
18. Al Geist AB, Dongarra J, Jiang W, Manchek R, Sunderam VS (1994) PVM: parallel virtual machine: a users' guide and tutorial for network parallel computing.
19. Salah A, Serres O, Gaber J, Outbib R, El-Sayed H (2008) Simulation of the Fuel Cell Thermal Behavior with Unified Parallel C. In: Signal Processing and Communications, 2007. ICSPC 2007. IEEE International Conference on, pages 149–152. IEEE
20. Ramousse J, Deseure J, Lottin O, Didierjean S, Maillet D (2005) Modeling of heat, mass and charge transfer in a PEMFC single cell. J Power Sources 145(2):416–427
21. Mohler RR (1973) Bilinear control processes: with applications to engineering, ecology, and medicine. Academic Press, XXX
22. Chen MS, Tsao ST (2002) Exponential stabilization of a class of unstable bilinear systems. Autom Control IEEE Trans 45(5):989–992
23. Bacciotti A, Ceragioli F (2006) Closed loop stabilization of planar bilinear switched systems. Int J Control 79(1):14–23
24. Gutman PO (2002) Stabilizing controllers for bilinear systems. Autom Control IEEE Trans 26(4):917–922
25. Longchamp R (2002) Stable feedback control of bilinear systems. Autom Control IEEE Trans 25(2):302–306
26. Wang YR (2002) Exponential Stabilization Of Bilinear Systems with Unstable Dynamics. Asian J Control 4:321–326
27. Espana M, Landau ID (1978) Reduced order bilinear models for distillation columns. Automatica 14(4):345–355
28. EP Ryan, NJ Buckingham (1983) On asymptotically stabilizing feedback control of bilinear systems. IEEE Trans Autom Control 28:863
29. Slemrod M (1978) Stabilization of Bilinear Control Systems with Applications to Nonconservatrice Problems in Elasticity. SIAM J Control 16:131–141
30. Jacobson DH (1977) Extensions of linear-quadratic control, optimization and matrix theory. Mathematics in science and engineering. Academic Press, XXX
31. Bonnet C, Didierjean S, Guillet N, Besse S, Colinart T, Carré P (2008) Design of an 80 kWe PEM fuel cell system: scale up effect investigation. J Power Sources 182(2):441–448
32. Wang L, Husar A, Zhou T, Liu H (2003) A parametric study of PEM fuel cell performances.. Int J Hydrogen Energy 28(11):1263–1272
33. Gurau V, Liu H, Kakaç S (1998) Two-dimensional model for proton exchange membrane fuel cells. AIChE J 44(11):2410–2422
34. Gerbaux L (1996) Modélisation d'une pile à combustible de type Hydrogène/air et validation expérimentale. PhD thesis, Institut National Polytechnique de Grenoble

Fault Diagnosis of PEM Fuel Cell

Andres Hernandez, Rachid Outbib and Daniel Hissel

Abstract This chapter deals with the problem of fault diagnosis for proton exchange membrane fuel cells (PEMFC). The goal is to present two strategies for fault diagnosis that are based, respectively, on the electrical equivalent technical and statistical approaches. The first strategy is carried out in two steps. First, an electrical equivalent model, which could be considered as a unifying approach to fuel cell systems, is established. Second, a technical approach to use the electrical model, for fuel cell system diagnosis, will be introduced. The second strategy is based on information from fuel cell conditions and operation modes. This makes the fault diagnosis procedure simpler and can be achieved by considering probability density functions of cell voltage. In this work, the main failure considered is flooding. To illustrate the performances of the two proposed approaches, experimental validations of the model and the diagnosis methodology are given.

1 Introduction

One of the main drawbacks of the proton exchange membrane fuel cell (PEMFC) is related to the stack lifetime and reliability when working at the strongly changing charge conditions that could be found in the applications. In order to overcome these drawbacks, an adequate fuel system control strategy is needed in order to avoid

R. Outbib (✉)
LSIS, Aix-Marseille University, av. Escadrille Normandie Niemen,
13397 Marseilles, France
e-mail: rachid.outbib@lsis.org

A. Hernandez
Escuela Colombiana de Ingeniería Julio, Garavito, Colombia
e-mail: andres.hernandez@escuelaing.edu.co

D. Hissel
FEMTO-FClab, rue Thierry MIEG 90010, Belfort cedex, France
e-mail: daniel.hissel@utbm.fr

M. S. Basualdo et al. (eds.), *PEM Fuel Cells with Bio-Ethanol Processor Systems*,
Green Energy and Technology, DOI: 10.1007/978-1-84996-184-4_5,

failure modes which could compromise the stack reliability, accelerate ageing or degradation processes or even compromise stack integrity. Thus, an adequate system diagnosis can provide valuable information either to optimize control strategies or to suggest preventive maintenance actions.

Generally speaking, the faults that may affect the fuel cell can be grouped into two classes: permanent faults and transient faults. Permanent faults are related to the damage of the fuel cell (membrane deterioration) and are therefore irreversible. After such dysfunction, the performances of the fuel cell will always be lower than those obtained in similar operating conditions before the occurrence of the fault. Transient failures are related to the fuel cell state and the control strategy used. These failures may be reversible. Hence, it is important to identify these failures as soon as possible. The corrective action is then based on the fuel cell control strategy and/or on the load operation. These failures can be grouped into two categories: those related to water (drying or flooding) and those related to contamination by gases (N_2, CO_2 or CO).

In the literature, several diagnosis methodologies for fuel cell stacks or electrochemical energy sources have already been proposed. Some of them are model based [1]. For instance, in Burford et al. [2], the authors propose to perform fuel cell stack temperature distribution measurements using nano-sized thermocouples embedded within the membrane electrode assembly. Other fuel cell diagnosis methodologies are based on gray or black-box models. These behavioral models can be obtained thanks to fuzzy logic [3], neural networks [4] or non-parametric identification using Markov parameters [5]. Then, derivation from the standard behavioral conditions is detected to evaluate the state-of-health of the fuel cell stack [4, 6]. Moreover, in Kozlowski et al. [1], the authors propose to fusion data coming from monitored and virtual sensors to evaluate this derivation. A last fuel cell stack diagnosis methodology is based on electrochemical impedance spectrometry [5]. This methodology is especially suitable for a better understanding of physico-chemical phenomena taking place in PEMFC and for the parameter determination of a PEMFC equivalent circuit description. It also permits to characterize, in a very efficient way, both the static and dynamical behaviors of fuel cell stacks.

This chapter deals with the problem of fault diagnosis for PEMFC. Two strategies of different kinds are presented. The first strategy is based on the use of the model. In fact, fuel cell systems are complex processes. Hence, several authors have chosen different approaches to model fuel cell systems. Electric equivalent models, state space models and bond graph representation models are the most representative approaches. In the first part of this work, the electric equivalent technical is considered and the strategy of fault diagnosis is carried out in two steps. First, an electrical equivalent model, which could be considered as a unifying approach to fuel cell systems, is established. Second, a technical approach to use the electrical model for fuel cell system diagnosis will be introduced, making special emphasis on fuel cell flooding detection. Generally, the use of the model allows a good understanding of the system.

The second strategy is motivated by some practical aspects. Indeed, the measured variables, in embedded systems, are relatively reduced due to technological instrumentation limitations and to the cost that can represent the sensors. As an immediate result, the available information is frequently insufficient to deduce failure origin. This strategy, which is not based on any model, gives additional information about fuel cell conditions and operation modes. This renders the fault diagnosis procedure simpler and can be achieved by considering the probability density functions of cell voltage. Note that this strategy is more suitable for real-time applications.

To illustrate the performances of the two approaches, experimental validations of the diagnosis strategies are also proposed. The chapter is organized as follows. The second section is dedicated to fault diagnosis using electrical equivalent models. The third section is devoted to a fault diagnosis strategy based on statistical analysis. The last section draws the conclusions.

2 MethodologyBased Electrical Equivalence Models

The aim of this section is two-fold. First, a strategy for modeling the fuel cell behavior is proposed. More precisely, it is based on electric–pneumatic analogy. Second, a methodology, using the electrical equivalent model, for faults diagnosis is given. The main results presented in this section were established in Hernandez et al. [7].

2.1 Modeling Approach

In classical physics, matter, charge and energy are preserved. Kirchoff's node and loop laws are representations of those fundamental principles applied to network analysis. Generalizing, these fundamental principles can be extrapolated to any network structure, including pneumatic and hydraulic networks [8–13]. Considering this fact, an electric equivalent model is built based on charge, matter and energy conservation laws. In this approach, each component of a hydraulic or pneumatic circuit is represented by its electrical equivalent in an electric circuit.

Comparing the physical behavior of pneumatic elements with those of electrical circuits, an analogy between these two isomorphs systems can be established to find the electrical equivalence of pneumatic systems.

Electrical current and fluid flows, share in their bases the same definition and thus the same mathematical expressions as illustrated by Eq. 1. This makes electrical current an ideal equivalent variable for either mass or molar flow:

$$\begin{aligned} &\text{Electrical environment:} \quad I = \frac{\mathrm{d}q}{\mathrm{d}t} \\ &\text{Pneumatic environment:} \quad Q_{m,\mathrm{net}} = \frac{\mathrm{d}n}{\mathrm{d}t} \end{aligned} \tag{1}$$

with

$$Q_{m,\mathrm{net}} = \sum Q_{m,\mathrm{in}} - \sum Q_{m,\mathrm{out}}, \tag{2}$$

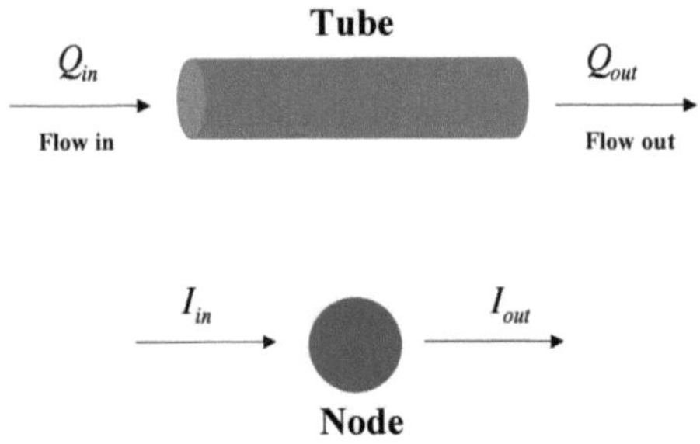

Fig. 1 Tube model

where n is the number of moles and Q_m denotes the molar flow in the membrane.

Using Bernoulli's continuity law, for incompressible liquid, non viscous and following a laminar flow the energy and mass flow are preserved when flowing through an ideal pipe. Under such considerations, the pneumatic conductor will behave as illustrated by Fig. 1.

In order to take into account the mechanical losses Poiseuille's law [14] is considered and an equivalence with Ohm's law can expressed by

$$\begin{aligned} \text{Pneumatic environment:} \quad Q_m &= \Delta P\left(\frac{\pi r^2}{8\mu l}\right), \\ \text{Electrical environment:} \quad I &= \frac{\Delta V}{R_f}, \end{aligned} \tag{3}$$

where r is the hydraulic radius and l designates the tube length.

Then, the expression of the pneumatic resistance is given by the following expression:

$$\begin{cases} R_f \equiv \frac{8\mu l}{\pi r^2} \\ \Delta P \equiv \Delta V \end{cases}. \tag{4}$$

From Eq. 3, one deduces that voltage can be considered as an electrical equivalent variable for pressure. It is also assumed that equivalent pneumatic resistance R_f is constant based on Poiseuille's law.

If a compressible fluid is taken into consideration, a variation of the fluid mass inside pipes can occur (i.e. $Q_{m,\text{net}} \neq 0$). In this case, the pneumatic conductor can no longer be modeled by a resistance and a variable storage device should therefore be added to the model.

Since it is ideal gas law, and for constant temperature and volume, one can deduce

$$\frac{\mathrm{d}P}{\mathrm{d}t} = \frac{RT}{v}\frac{\mathrm{d}n}{\mathrm{d}t}. \tag{5}$$

Considering the molar flow definition:

$$Q_m = \frac{\mathrm{d}n}{\mathrm{d}t}, \tag{6}$$

a change in pressure can be written as

$$\text{Pneumatic environment:} \quad \Delta P = \frac{RT}{v} \int_{t_0}^{t} Q_m(\tau)\,\mathrm{d}\tau. \tag{7}$$

This is an isomorfism to

$$\text{Electric environment:} \quad \Delta V = \frac{1}{C} \int_{t_0}^{t} I(\tau)\,\mathrm{d}\tau, \tag{8}$$

which represents a capacitor voltage in an RC circuit. A comparison between these equations leads to define voltage as the natural electrical equivalent for pressure. Additionally, it is possible to introduce C_f pneumatic capacitance defined by

$$C_f \equiv \frac{v}{RT} \tag{9}$$

complementary to the resistive properties described previously.

After superposing both the resistive and capacitive properties of pneumatic conductors in the same circuit, a resulting "T" quadripole model can be constructed as an electric equivalent for pneumatic conductors. This particular configuration (Fig. 2) will allow us to tune time constants for the proposed model, according to different flow profiles.

Other components of the fuel cell system (e.g. tanks, valves, etc.) can also be modeled by the same basic RC circuit under most operating conditions. While modeling tanks, a constant mean gas speed and a constant pressure can be considered inside the tank volume.

Experimentally, ΔP a pressure drop can be observed between the input and output quantities. Thus, and in order to take into account this phenomena, resistances (R_{in}, R_{out}) are introduced in the model.

Finally, in this model all components having a non-negligible volume are decoupled into three parts. The first part consists in modeling the volume, by a pneumatic capacitance C_f. The second part deals with the input nozzle, modeled by an input resistance R_{in}. The last part concerns the output nozzle represented by an output resistance R_{out}. All pressure drops are concentrated on component nozzle [15, 16].

2.2 Fuel Cell Model

The next step in the modeling procedure is to join the previous models, in order to obtain a representative electrical equivalent circuit for the fuel cell system. In this part, we will focus on the description of the stack model.

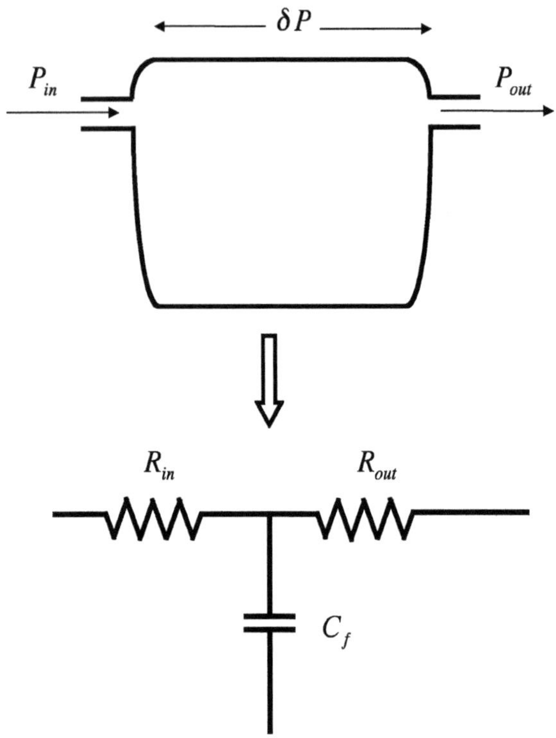

Fig. 2 Pneumatic component electric equivalent

2.2.1 Input Gas System

The system is composed mainly of pipes, regulation valves and a humidifying system [16]. All components are modeled by an RC circuit. The inter connectivity between components leads to a single equivalent RC circuit which integrates all capacitances and resistances involved in each component. Only the humidifying system deserves a closer look since adding steam $Q_{m,\text{vap}}$ into the dry gas flow $Q_{m,\text{in}}$ changes the molar flow. However, estimating steam partial pressure at stacks input is crucial for voltage prediction. The evaporation flow which is represented by $Q_{m,\text{vap}}$, depends on the dry air flow, temperature and relative humidity level desired. Water vaporization rate $Q_{m,\text{vap}}$ is modeled by a current source as shown in Fig. 3.

2.2.2 Fuel Cell Stack

Fuel cell gas circuits in the stack consist mainly of channels that can be modeled by an RC quadripole configuration. Note that RC circuit parameters depend on channels' geometry in the fuel cell stack. Molar flow is not preserved in the stack. At the cathode, steam balance $Q^{c}_{m,\text{vap}}$ is added to the air flow as a product of the electro

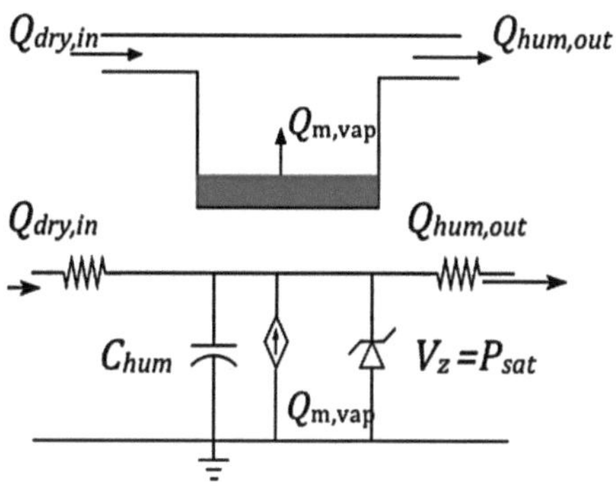

Fig. 3 Electric equivalent circuit for humidifier

chemical reaction given by

$$Q_{m,\mathrm{H_2O},r} = \frac{nI_{\mathrm{stack}}}{2F}, \tag{10}$$

electro osmotic drag defined by

$$Q_{m,\mathrm{H_2O},e} = D * nI, \tag{11}$$

and diffusion through the membrane

$$Q^{m}_{\mathrm{H_2O}} = Kd1\left(\left[\mathrm{H_2O}^{c}_{m}\right] - \left[\mathrm{H_2O}^{a}_{m}\right]\right) + Kd2\left(P^{c}_{\mathrm{H_2O}} - P^{a}_{\mathrm{H_2O}}\right), \tag{12}$$

where, and by considering ideal gas law,

$$P^{\alpha}_{\mathrm{H_2O}} = \frac{RT}{V}\left(n^{\alpha}_{\mathrm{H_2O_m}}\right), \quad \alpha \in \{a, c\}. \tag{13}$$

Now, using

$$\left[\mathrm{H_2O}^{\alpha}_{m}\right] = \frac{n^{\alpha}_{\mathrm{H_2O_m}}}{V}$$

we get

$$\left[\mathrm{H_2O}^{\alpha}_{m}\right] = \frac{P^{\alpha}_{\mathrm{H_2O}}}{RT}.$$

Finally, Eq. 12 can be rewritten as

$$Q^{m}_{\mathrm{H_2O}} = Kd\left(P^{c}_{\mathrm{H_2O}} - P^{a}_{\mathrm{H_2O}}\right) \tag{14}$$

with

$$Kd = Kd_2 + \frac{Kd_1}{RT}. \tag{15}$$

Nitrogen also diffuses according to pressure and concentration gradients, in a similar way as shown for water in Eqs. 12 and 14. However, a single RC circuit electric equivalent model is suitable only if gases are assumed to be mono component. Each specie is considered as a perfect and single gas and physical interactions between species are neglected. Under these considerations,the cathode and anode are modeled by three independent circuits.

Global pressure and flow at a given point "i" in each channel, are obtained by applying superposition properties to each individual gas circuit, according to gas composition:

$$\text{Cathode}\begin{cases} P^c(i) = \sum_k P_k^c(i) \\ Q_m^c(i) = \sum_k Q_{m,k}^c(i) \end{cases} \text{for } k \in \{N_2, O_2, H_2O\}, \tag{16}$$

$$\text{Anode}\begin{cases} P^a(i) = \sum_k P_k^a(i) \\ Q_m^a(i) = \sum_{m,k} Q_{m,k}^a(i) \end{cases} \text{for } k \in \{N_2, H_2, H_2O\}. \tag{17}$$

Note that the independent species partial pressure is usually smaller than the atmospheric pressure:

$$\text{Cathode}\begin{cases} P_{\text{atm}} \leq P_{N_2} + P_{O_2} + P_{H_2O} \\ P_{\text{atm}} \geq P_\alpha \quad \text{for } \alpha \in \{N_2, H_2, H_2O\} \end{cases}, \tag{18}$$

$$\text{Anode}\begin{cases} P_{\text{atm}} \leq P_{N_2} + P_{H_2} + P_{H_2O} \\ P_{\text{atm}} \geq P_\alpha \quad \text{for } \alpha \in \{N_2, H_2, H_2O\} \end{cases}. \tag{19}$$

The modeling principle is then to work with absolute pressures and to compensate each circuit output by a voltage source V_k^*, using the following expression:

$$V_k^{*j} = \frac{P_{k,\text{out}}^j}{P_{\text{out}}^j} P_{\text{atm}}. \tag{20}$$

Figures 4–6 depict the electrical model for oxygen, nitrogen and water vapour, respectively. Note that the hydrogen circuit is analogous to the oxygen circuit.

Under the hypothesis of homogeneity of gas distribution, the output flow is expressed by

$$Q_{m,k,\text{out}}^{a,c} = \frac{P_{k,\text{out}}}{P_{\text{out}}} Q_{m,\text{out}} \quad \text{for } k \in \{N_2, H_2, H_2O, O_2\}. \tag{21}$$

The effect of nitrogen diffusion, for "open mode" operation, is rather small. Hence, in several works (see for instance [11, 16, 17]), these phenomena are not considered

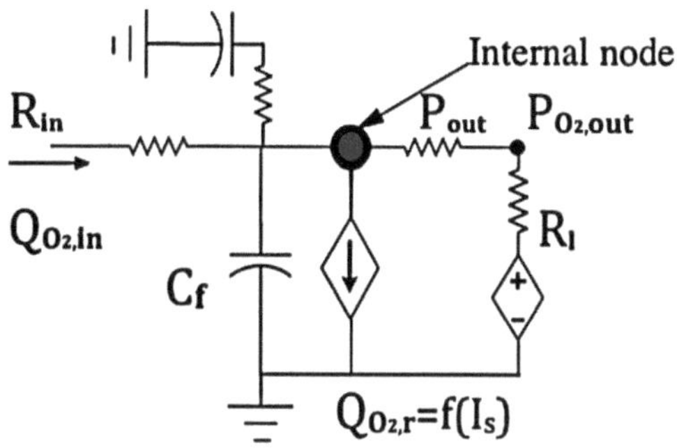

Fig. 4 Electric equivalent circuit for oxygen

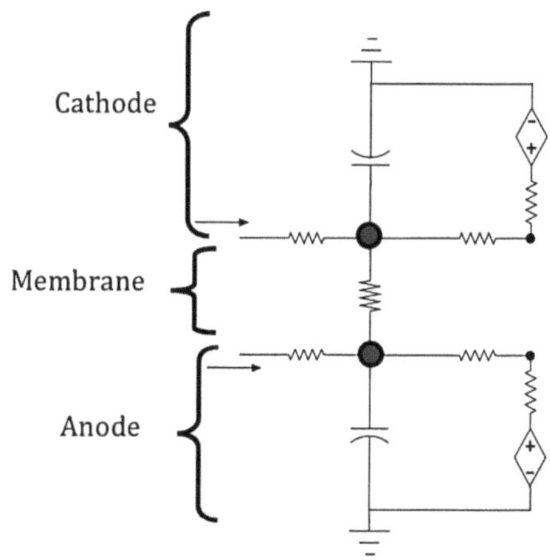

Fig. 5 Electric equivalent circuit for nitrogen

by assuming that they can be neglected. However, for "closed mode" operation it is important to take into account this phenomena and especially in the case of faults diagnosis.

In this approach, the phenomena of diffusion is represented, in the model, by resistances that can be constants or variables depending on the kind (i.e. local or global) of model considered . Note that a resistance joining both anode and cathode RC circuits, models nitrogen diffusion constant as depicted in Fig. 5.

To describe the behavior of water vapour it is important to take into account the phenomena of water production in cathode, electro osmotic drag (see Eqs. 10 and 11) and the saturation behavior that limits its partial pressure (see for instance [18]):

$$P^{\alpha}_{H_2O} \leq P_{sat} \text{ for } \alpha \in \{a, c\}. \tag{22}$$

Finally, if the fuel cell operates with internal vapor pressures close to saturation pressure P_{sat}, the existence of liquid water inside the stack should also be considered. In this case, condensation–evaporation phenomenon should also be included in the

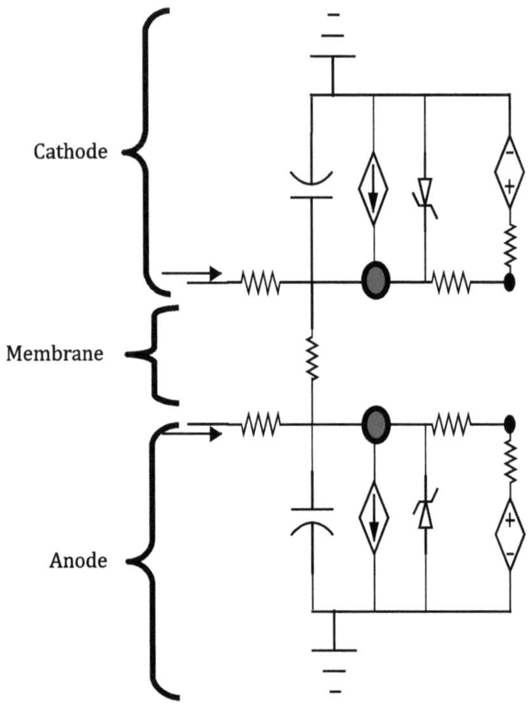

Fig. 6 Electric equivalent circuit for water vapour

model. For the sake of simplicity, a first model considering only the influence of the vapor phase water is proposed.

Both electro osmotic water and produced water flows $Q_{m,H_2O,e}$, $Q_{m,H_2O,r}$, defined by Eqs. 10 and 11, are dependent on stack's current. Hence, in the model, these quantities are depicted by current sources acting on the corresponding channels. As in the case of nitrogen, a resistance linking the cathode and anode circuit, models water diffusion. Water diffusion and electro osmotic drag coefficients are dependent on the membrane water content, which is usually estimated by the water activity on either channel. The resulting electrical equivalent model is presented in Fig. 6. When the water partial pressure reaches its saturation level [18], the system is no longer linear. In order to model vapour saturation pressure, a Zenner diode in parallel with the capacitance is introduced in the water equivalent circuit, to set a saturation voltage equivalent to vapor saturation pressure. In this way

$$V \leq V_{\text{Zenner}} \tag{23}$$

as

$$P_{H_2O} \leq P_{\text{sat}}. \tag{24}$$

In order to complete the model, a stack voltage model is built from the open circuit voltage given by Nernst equation:

$$E_{oc} = E_o - \left[\frac{RT}{nF} \ln \frac{P_{H_2O}}{P_{H_2}\sqrt{P_{O_2}}} \right], \tag{25}$$

with species partial pressures given by the equivalent voltages in the internal node of species circuits. Voltage losses under load conditions due to activation losses are described by Tafel's equation (see for instance [19]):

$$v_{act} = \left(\frac{RT}{\alpha nF} \right) \ln \left(\frac{i_o + I}{i_0} \right). \tag{26}$$

With regard to concentration and ohmic losses, we consider the equations

$$v_{conc} = \left(\frac{RT}{\alpha nF} \right) \ln \left(\frac{I_L - I}{I_L} \right) \tag{27}$$

and

$$V_{Ohm} = RI. \tag{28}$$

Remark 1 *Concerning the identification of parameters of the model and its experimental validation on a real process the reader is referred to* Hernandez et al. [7].

2.3 Diagnosis by Electrical Circuit Analysis

Electrical equivalent model representation presents characteristics that are suitable for fault diagnosis:

- A dynamical system, composed of differential equations, can be deduced from the electrical equivalent model. Hence, the obtained system can be easily linearized around the desired operating point. This can allow, for instance, the use observer for residual generation. This is an important fact, since measurements inside the stack are rather difficult to perform in real time, and the possibility of state reconstruction from the stack information available is highly desired.
- Failure modes can be associated with circuit parameters' variation. A parametric approach to diagnosis such as Kalman filtering is then feasible.
- existing software used for electrical network analysis, can easily be adapted.

2.3.1 Principal Failure Mode Description

Here, the three main types of failures considered are the following: flooding, drying and membrane deterioration. Both flooding and drying phenomena are transient conditions that can be reversed with an adequate control strategy, before a major failure occurs. Generally, membrane degradation is considered as a permanent, non reversible, failure.

A. Flooding

The accumulation of liquid water in diffusion layer pores and channel surfaces, thus limiting gas exchange between channels and reactive sites and between cathode and anode channels themselves, is called flooding. Flooding occurs mainly at the cathode side. It is the cumulative result of defective water management that can be summarized into three main causes:

- Insufficient water evacuation flow.
- Excess of input water vapour flow.
- Sudden temperature drop.

Seen from another point of view, active surface is reduced producing an increase in current density, increasing also ohmic losses and decreasing the fuel cell performance [8].

Water in its liquid state can partially limit gas flow through channels, increasing pressure losses between stacks input and output nozzles. Gas diffusion can also be perturbed by the presence of liquid water. In addition to the water influence on circuit resistances, stored liquid water also leads to decrease the available gas volume. Since stack storage properties are modeled by a capacitance, this capacitance is also affected by liquid water production.

B. Drying

In opposition to flooding, drying occurs when the water vapor balance is negative for a given time producing membrane dehydration. In a PEMFC stack, proton transfer is assured by the water molecules contained in the membrane. Membrane dehydration will increase the electrical resistivity of the stack and limit the maximal current density and electro-osmotic drag coefficient. Membrane width and thus gas diffusion constants will also be affected by dehydration.

C. Membrane Deterioration

Membrane degradation takes place with time. In the presence of this fault,the fuel cell performance is permanently compromised; it can be characterized by a system evolution towards new equilibrium points. As it can be easily inferred, only membrane parameters are affected in this case. Diffusion constants are significantly altered in most of the cases.The pressure gradient between cathode and anode channels drops, proportionally to failure magnitude.

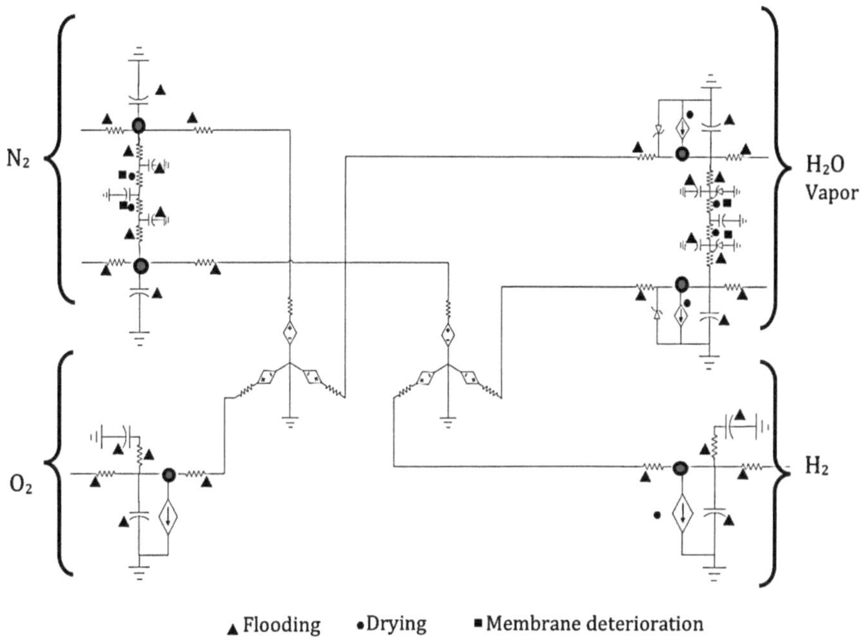

Fig. 7 Global model and fault related elements

2.3.2 Analysis of Parameters and Faults Diagnosis

The principle used for faults diagnosis is based on the analysis of the parameters. More precisely, in the proposed model, the presence of a fault can be represented by variations in circuit parameters Fig. 7. The diagnosis is then based on a residue generation between the identified parameters and the nominal values.

The flooding fault can be characterized by the change of parameters in the circuit as follows. First, increasing input and output resistances mainly in the part of the circuit representing the cathode side and membrane. Second, decreasing of circuit capacitance. In fact, this is due to the reduced effective volume.

The drying fault will increase the electrical resistivity and limits the current density and electro-osmotic drag coefficient. Hence, it is expected that this fault will be detected in the electrical circuit by a decreasing of the resistances representing the membrane and a change in the electro-osmotic current source are expected.

Under membrane deterioration, R_m is expected to drop, but no other components should be altered.

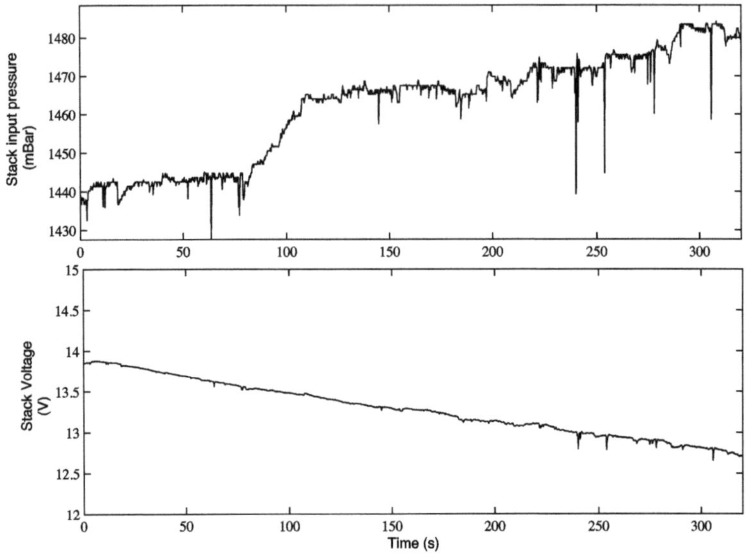

Fig. 8 Voltage degradation towards flooding fault [7]

2.3.3 Diagnosis Results

Failure tests proposed in Hernandez et al. [7] were conceived to carry the fuel cell stack into a flooding failure mode. In order to flood the stack, two different strategies were used: an increase in input air humidity level and a stack temperature drop.

The voltage degradation rate is constant as expected, since a direct relation between water content in the stack and voltage is supposed and given the experimental conditions, a constant water formation rate is expected.

The relation between stack voltage degradation and flow resistance increase is observed in Fig. 8.

From the flow point of view, an increase in cathode flow resistance and cathode stack pressure loss is observed in the cathode, whereas anode flow parameters remain relatively unchanged as expected. Experimental results illustrating the previous phenomena are shown in Figs. 9, 10.

Figure 11 shows that flooding induced by temperature, decreases or increases the input humidity level. A more continuous evolution of internal flow resistance is observed when reducing stack temperature, suggesting a more uniform water repartition in the stack.

Normally, pressure increase will produce an increase in voltage as described by the Nernst equation, however, experimental results in these tests show that a higher channel pressure is not necessarily related to a concentration increase on the reactional site.

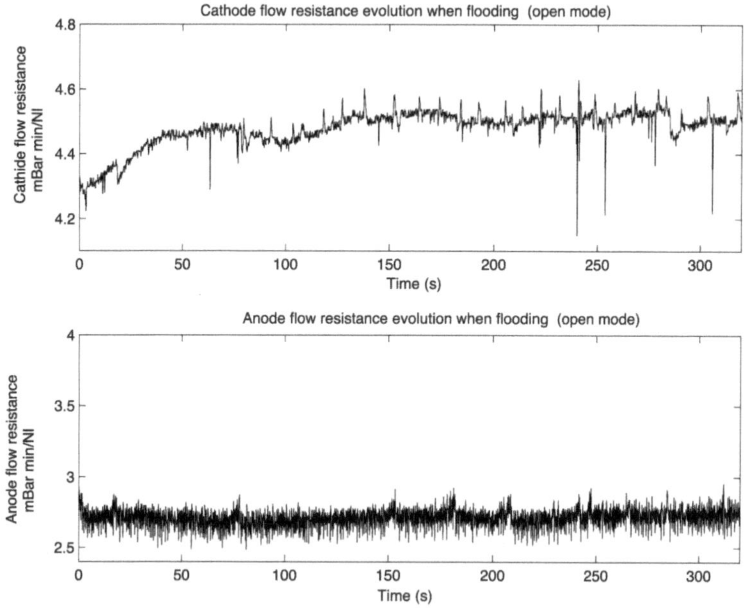

Fig. 9 Flooding due to an increase in humidity: evolution of flow resistance evolution [7]

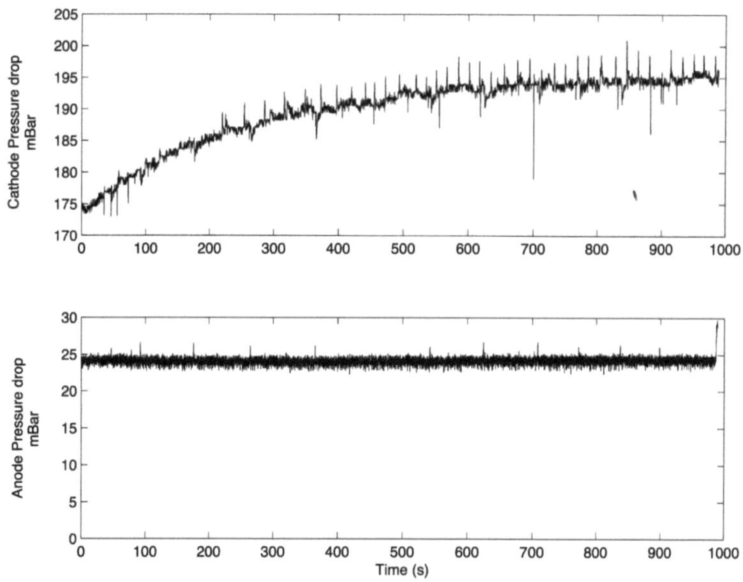

Fig. 10 Flooding due temperature decrease: evolution of pressure losses [7]

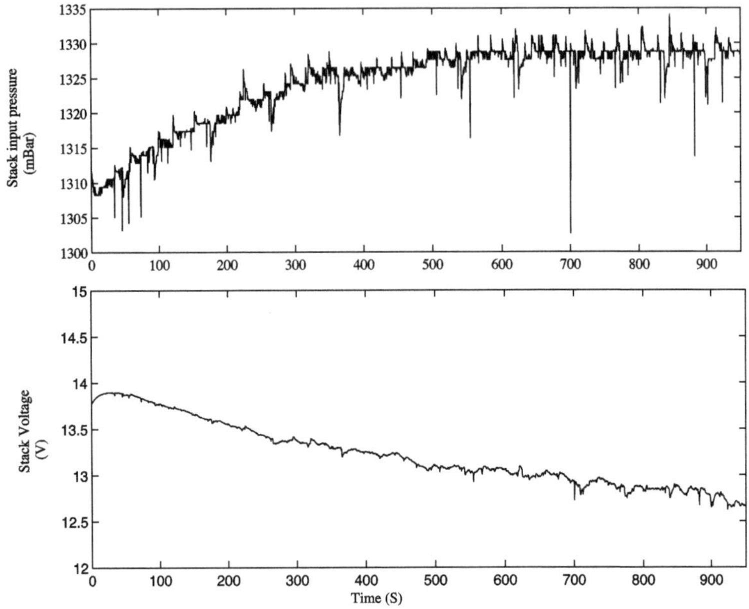

Fig. 11 Flooding due to stack temperature decrease: pressure losses evolution [7]

3 Statistical Approach

Due to the nature of the system, some important constraints must be respected when developing a diagnosis strategy. First, the computation time must be small since stack's electrical dynamics are fast, compared to system ancillary components, and thus adequate control strategies have to be taken rapidly in order to avoid a system performance degradation. Second, diagnosis should perform well in a wide operation range, including system normal degradation due to ageing. Finally, the diagnosis strategy should minimize the hardware costs if it is intended to serve mass production markets. In order to accomplish this last task, sensor choices should try to minimize their number and complexity. The diagnosis strategy suggested in this section takes into account the constraints exposed previously, including a stochastic approach to estimate fuel cell operation/failure mode, using the information provided by an individual cell voltage probability density function.

3.1 A Static Model

The more simpler model for a PEMFC is established using the relationship between the current and the voltage and is represented by the polarization curve. The classical form, depicted by Fig. 12, is defined by

Fig. 12 Polarization curve for a PEMFC

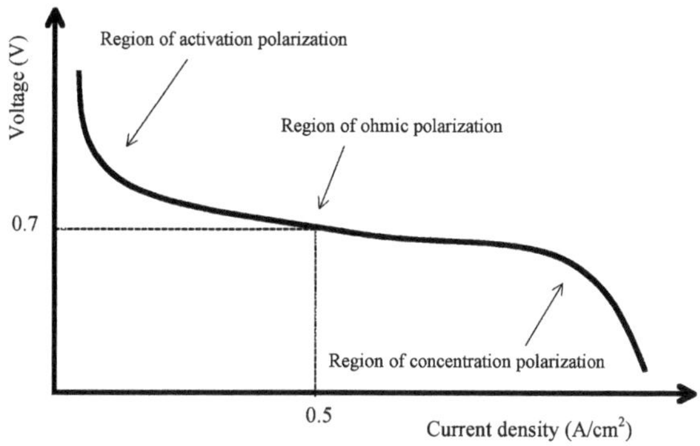

$$V_s = E_0 - \frac{RT}{2F}\ln\left(\frac{P_{H_2O}}{P_{H_2}\times P_{O_2}^{1/2}}\right) - \frac{RT}{2F}\ln\left(\frac{i_0+I}{i_0}\right) + \frac{RT}{2F}\ln\left(\frac{I_l-I}{I_l}\right) - IR, \quad (29)$$

where E_0, i_0, I_l, P_x (where $x \in \{H_2, O_2, H_2O\}$) and R are load voltage, the current exchange, the maximum current, the pressure expressed in atmospheres and the ohmic resistance, respectively. These settings are system specific and only load voltage is measured directly.

Since the model is static, its use assumes that the operating conditions of the system and dependents change only very slightly for a time long enough.[1] Thus, a fault (or dysfunction) can be detected from this model only after a long period of time resulting in delays between the fault and its detection.

A perfect stabilization of the operating conditions of the PEMFC system is rarely achieved due to the complexity inherent in the system. Variations of parameters, more or less important, such as temperature and humidity are still present. It therefore seems inappropriate to consider a single polarization curve to evaluate even a static operation of the system. A possible solution to take into account this phenomenon is to consider the notion of statistical dispersion due to disturbances on the system. The initial polarization curve is then substituted by a zone of polarization (see Fig. 13) that integrates the system perturbations and the bias introduced by sensors:

$$V(I) = \underbrace{\bar{V}(I)}_{\text{Average polarization curve}} + \underbrace{\theta_1(o, \sigma_1)}_{\text{Sensor bias}} + \underbrace{\theta_2(o, \sigma_2)}_{\text{System bias}}. \quad (30)$$

It is possible to define the area of polarization using the set of parametric functions of probability density f_{V_I}[2] of the random variable V for a current I with values in an interval characterized by the parameters $\mu = E(V_I)$, mathematical expectation

[1] Typically on the order of the minute for PEMFC.

[2] In the sequel we use f_x to denote the density of probability of the random variable X.

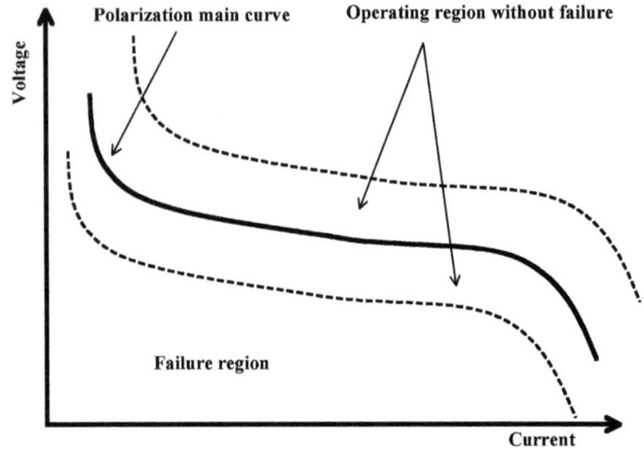

Fig. 13 Definition of areas of polarization and faults

centered on the theoretical polarization (see Fig. 12) and σ a dispersion (strongly linked to the physical process under consideration) that depends on system features and sensors.

The concept of a fault occurrence requires a preliminary formalization. To do this, we adopt two different approaches. In the first approach, the presence of a default is assumed if

$$p \notin [v_1, v_2], \tag{31}$$

where p denotes the operating point and $[v_1, v_2]$ is the interval obtained from trust parameter α defined by resolving the following equations (see Fig. 14):

$$\int_{-\infty}^{v1} f_{V_I}(\tau)\,\mathrm{d}\tau = \frac{\alpha}{2} \quad \text{and} \quad \int_{-\infty}^{v2} f_{V_I}(\tau)\,\mathrm{d}\tau = 1 - \frac{\alpha}{2}. \tag{32}$$

A second approach consists in assuming that the occurrence of a fault when

$$P(v) \leq P_{\min}, \tag{33}$$

where $P_{\min}$ is the value of a minimal probability. $P(v)$ is the probability defined compared to the interval $[v, v + \Delta v]$, where Δv designates a nonnegative real number depending on the sensors. This probability can be expressed by

$$P(v) = \int_{-\infty}^{v+\Delta v} f_{V_I}(\tau)\,\mathrm{d}\tau - \int_{-\infty}^{v} f_{V_I}(\tau)\,\mathrm{d}\tau. \tag{34}$$

It should be noted that in the literature similar approaches have already been proposed. For example, the use of logic and the concept of fuzzy membership degree to quantify distance from a point in the experimental polarization curve theory

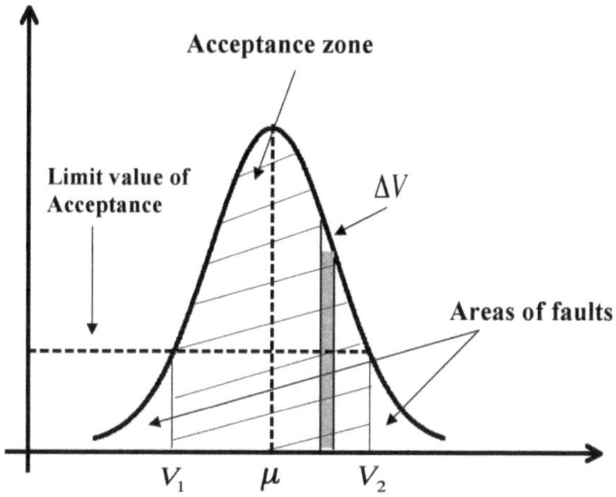

Fig. 14 Determination of nominal and fault zones

(cf. [3]). In general, methods based on the polarization curve can detect the probable presence of a dysfunction, but show significant limitations for diagnosis. Indeed, it is a necessary part to reach a quasi-steady state before identifying any malfunction. Moreover, information given by this method can be insufficient to establish the cause.

3.2 Diagnosis Using the Distribution of Cell Voltages

3.2.1 General Principle of the Approach

Generally, in the literature, for the proposed models (see for instance [19–21]), the voltage of fuel cell stack is first obtained by calculating the voltage of a unit cell. Second, it is generalized to all cells. This consists in using models that are of a global type. Note that these models are valid only under the assumption that the cell voltage in the fuel cell is homogeneous. Then the possible variations between the voltages cells are not representative of the operating mode of the stack. This approach can give satisfactory results for modeling the operating conditions of PEMFC where the nominal voltage difference cells remains low. Nonetheless, it does not take into account the fact that the cells behavior may be different, and mainly, in a degraded mode of operation. Thus, a significant amount of information with easy access (requiring only the sensors voltage) is, in fact, neglected.

Moreover, the use of so-called global models allows only a partial explanation of the complex phenomena related to the heart of fuel cells. In particular, those related to inaccuracies in the assembly, irregularities of materials, the non-uniformity of implementation of diffusion layers, the membrane or other parts of the stack. The impact of these phenomena can partly be found reflected in the behavior of individual cells of the PEMFC. Thus, the statistical approach given here takes into account and analyzes, unlike approaches based on global models, the information available for each individual cell, to explain and classify adequately patterns of dysfunction stack for different causes.

By considering Eqs. 29 and 30, it is possible to decompose the total voltage of the fuel cells in the following manner:

$$\bar{V}_p = \sum_1^n \bar{V}_i \quad \text{and} \quad \sigma_{V_p} = \sum_1^n \sigma_{V_p}. \tag{35}$$

This consists in considering the voltage of the n basic cells as being random variables. Further, we can note that from the statistical analysis of the cell tensions Eq. 35, it is possible to increase the information available for diagnosis.

It is important to emphasize that the fact of considering, the overall PEMFC system, as a stochastic system, can be explained naturally by the uncertainty of many variables and parameters in the models of the cell.

3.2.2 Mathematical Model

Generally, the flow in some sections of the channels is of turbulent nature [9]. Therefore, the pressure gradient is not homogeneous, causing local variations in pressure. These variations produce a pressure distribution that is specific to each mode of operation of the cell and can be observed from the distribution of the cell voltages.

By observing the voltages of each cell and their temporal evolution[3] (see Fig. 15) it seems clear that the operating conditions of each cell may be very different. Moreover, their behavior does not seem to follow specific laws; the effects of disturbances propagate through the system and are partially reflected through the measurement of the voltage of each elementary cell. In this approach, the image disturbances on the distribution voltage function for each individual cell is analyzed. The goal is to characterize some states and operating modes of the stack system.

By taking a sufficiently fine mesh over the scale to the considered phenomena, the number of samples of pressure is large enough to use the central limit theorem[4] and to claim that the samples of pressure are random variables according to a normal distribution[5]:

$$P \hookrightarrow \mathcal{N}\left(\bar{P}, \sigma_P\right). \tag{36}$$

Since the pressure variable is supposed to be a random variable, then the logarithm of pressure function can also be assumed to follow a random variable and thus from

[3] Experimental results are obtained with a stack (composed of 20 cells) of 500 W (composed of 20 cells).

[4] From 30 samples of a population, a distribution function can be tested in order to decide if eventually it can be assimilated to a normal distribution.

[5] In the sequel, we use $X \hookrightarrow \mathcal{N}(m, \sigma)$ to denote the random variable X which follows a normal distribution with mean m and standard deviation σ.

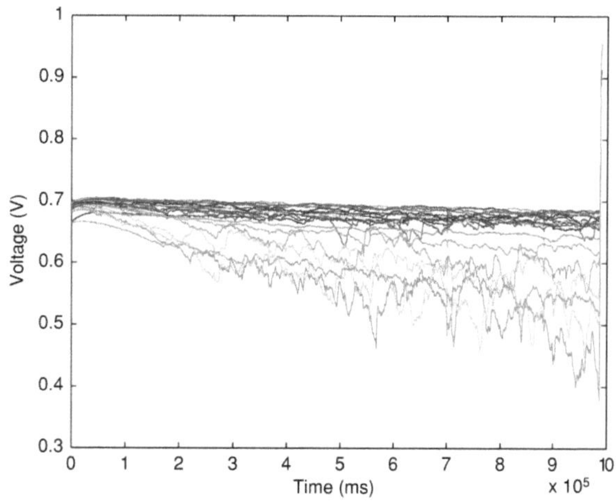

Fig. 15 Voltage behavior of cells in case of flooding

the equation Eq. 36 we get

$$\ln P \hookrightarrow \mathcal{N}\left(\ln \bar{P}, \sigma_{\ln P}\right). \tag{37}$$

When working in open circuit mode, the link between pressure distribution and voltage distribution is expressed through the Nernst equation [17]:

$$V_0 = E_0 + A \ln \left(\frac{\frac{P_{H_2O}}{P_{sat}}}{\frac{P_{H_2}}{P_{standard}} \cdot \left(\frac{P_{O_2}}{P_{standard}}\right)^{1/2}} \right). \tag{38}$$

The fact that the cell voltage follows a normal distribution is proved by using the hypothesis test of Kolmogorov–Smirnov (see e.g. [18]). In the case treated here the function of distribution of an empirical sample is compared to the theoretical distribution of normal function and the decision criterion is based on

$$\text{Max}\ (|F_n(x) - F(x)|), \tag{39}$$

where

$$F_n(x) = \frac{1}{n}\sum_{1}^{n} n_i \quad \text{with} \quad n_i = \begin{cases} 1 & \text{if } y_i \leq x \\ 0 & \text{if } y_i > x \end{cases}. \tag{40}$$

y_i designates the ith element of the sample and $F(x)$ denotes the repartition function of the normal law $\mathcal{N}(\sigma, \mu)$ where σ and μ are defined in an empirical manner and using a sample.

Finally, by assuming the following hypothesis :

- the gas composition is homogeneous for a given volume (i.e. the proportion of species is constant),
- the electrode potential is constant,

the expression of the voltage, when working in open circuit mode, can be transformed in the following form:

$$\begin{aligned} V_o &= E_o + A\left(\ln(K_1 \cdot P_{\text{air}}) - \ln(K_2 \cdot P_{\text{H}_2}) - \frac{1}{2}\ln(K_3^2 \cdot P_{\text{air}})\right) \\ &= E_o + A\left(\ln P_{\text{air}} - \ln P_{\text{H}_2} - \frac{1}{2}\ln P_{\text{air}} + \ln(K_1 K_2 K_3)\right) \\ &= E_o + A\left(\frac{1}{2}\ln P_{\text{air}} - \ln P_{\text{H}_2} + \ln(K_1 K_2 K_3)\right), \end{aligned} \tag{41}$$

where $K_i (i = 1, 2 \text{ or } 3)$ correspond to partial fractions of the water vapor, hydrogen and oxygen, respectively. Using Eqs. 36 and 37, the expression of the voltage, when working in the open loop mode, can be rewritten in the following form:

$$V_0 \hookrightarrow \mathcal{N}\left(\bar{V}_0, \sigma_{V_0}\right), \tag{42}$$

with

$$\bar{V}_0 = E_0 + A\left(\frac{1}{2}\ln\left(\frac{\bar{P}_{\text{air}}}{P_{\text{standard}}}\right) - \ln\left(\frac{\bar{P}_{\text{H}_2}}{P_{\text{standard}}}\right) + \ln(K_1 K_2 K_3)\right) \tag{43}$$

and

$$\sigma_{V_0} = \sigma_{\ln P_{\text{air}}} + \sigma_{\ln P_{\text{H}_2}}. \tag{44}$$

3.2.3 Methodology Used

A. Local Disturbances of Pressures

The system, even during normal operation, is subjected to local disturbances involving some dispersion of individual cell voltages (see for instance Figs. 16, 17 where the test results are realized on an open circuit mode). This dispersion, amplified by high flows, can be explained in part by local pressure disturbances caused by turbulence in the flow (see Fig. 18). However, it should be noted that if the system, even in a normal operation, is subjected to random perturbations (see for example the test results presented by Fig. 18 the voltage dispersion of cells generated cannot be considered as the unique indicator of the presence of a dysfunction or failure on the system stack.[6] Hence, the characterization of the distribution of cell voltages during normal operation is a prerequisite for the use of a statistical approach for diagnosis.

---

[6] Test results obtained with a PEMFC of 500 W.

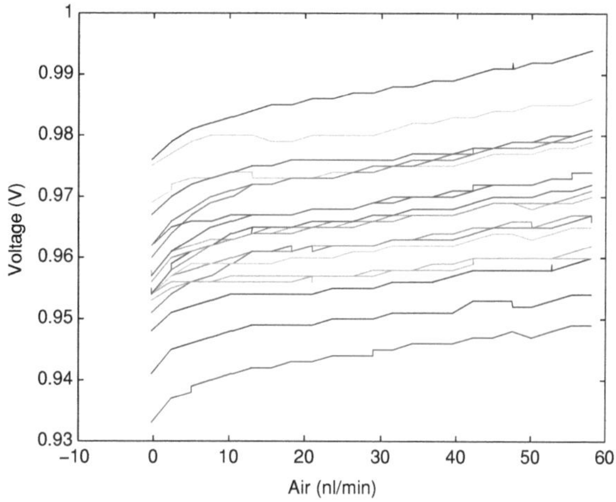

Fig. 16 Voltage dispersion of the cells: variable air flow (see footnote 6)

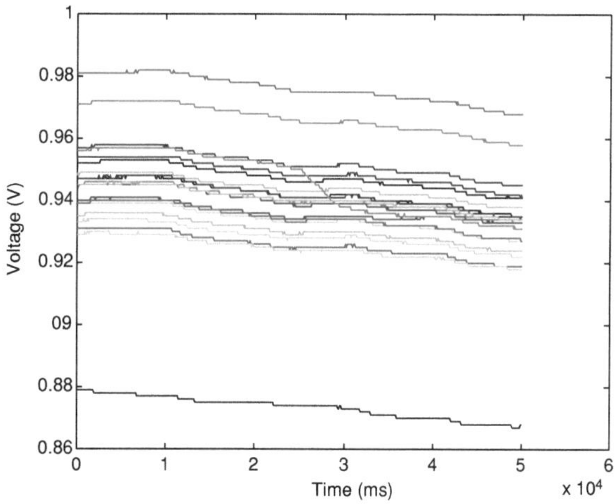

Fig. 17 Voltage dispersion of the cells: depending on time

B. Impact of Water in the Liquid State

Concerning PEMFC, it is well known that the presence of water in liquid phase inside the fuel cell is the main cause of dysfunction and faults (see for instance [22]). The condensation of water occurs when the partial pressure of water vapor reaches its saturation value (i.e. $P_{H_2O} \geq P_{sat}$). The impact of the presence of water in the liquid phase on the fuel cell is important. In fact, it causes an immediate voltage drop induced by the decreasing of the active surface for the exchange of gases. This causes an increase in the current density and then an increase in the temperature gradient which can damage the membrane. Also, the water vapor is present in the anode and cathode. Hence, its partial pressure, which is affected by disturbances, produces a no homogeneous distribution of pressures in the channels. Moreover, the saturation pressure is sensitive to temperature variations (see for instance [18]):

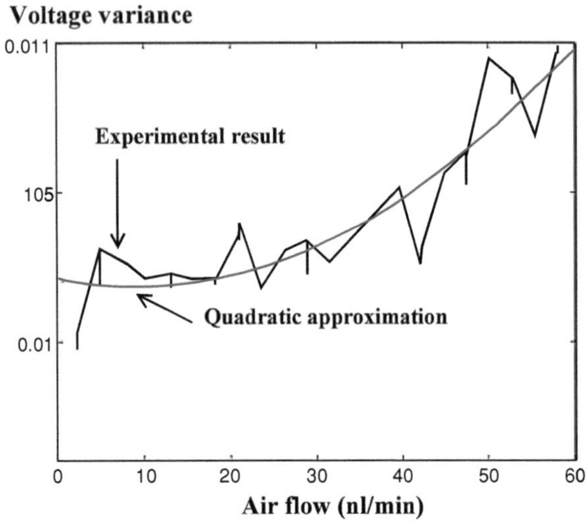

Fig. 18 Voltages variance of cells related to the air flow

Table 1 Numerical values for the computation of P_{sat} ([18])

$R = 22105649.25$	$D = 0.12558 \times 10^{-3}$
$A = -27405.526$	$E = -0.48502 \times 10^{-7}$
$B = 97.5413$	$F = 4.34903$
$C = 0.146244$	$G = 0.3938 \times 10^{-2}$

$$\ln\left(\frac{P_{sat}}{R}\right) = \frac{A + BT + CT^2 + DT^3 + ET^4}{FT - GT^2}, \tag{45}$$

with the parameters listed in Table 1.

Combining these two factors favors the formation of liquid water in some areas when some others are not affected. These phenomena were observed experimentally by several authors (see for instance [9, 23, 24]). Moreover, these works show that the formation of water is produced in a random way and is not homogeneous. Thus, the water condensation can be considered as a probabilistic event [24], imposing a water distribution in the fuel cell:

$$V^l_{H_2O} \hookrightarrow \mathcal{N}\left(\bar{V}^l_{H_2O}, \sigma_{V^l_{H_2O}}\right). \tag{46}$$

Note that the flooding phenomena in some zones of the fuel cell concern, in general, one or two cells when the others are only slightly affected. These phenomena are accompanied by an important voltage drop of the flooded cells and then an increase in the current density depending on the liquid water produced (see for instance [8]):

$$J = I_l\left(A\left(1 - K_d V^l_{H_2O}\right)\right)^{-1}, \tag{47}$$

with

J : Current density (A/cm^2),
A : Active surface (cm^2),
K_d : Relationship between volume and surface (cm^{-1}).

The following equation:

$$V_s = E_0 - \frac{RT}{2F} \ln\left(\frac{P_{H_2O}}{P_{H_2} \times P_{O_2}^{1/2}}\right) - \frac{RT}{2F} \ln\left(\frac{j_0 + J}{j_0}\right) + \frac{RT}{2F} \ln\left(\frac{J_l - J}{J_l}\right) - Jr \tag{48}$$

gives the relationship between the current density and the voltage of a cell. This expresses the relationship between voltage and the water distribution in the fuel cell. The probabilistic aspect of the cells voltage is then established.

C. Dispersion of Voltages of Cells

By considering the expressions Eqs. 29, 42, 46 and 47, the voltage of the fuel cell can be expressed by the equation Eq. 30. The obtained dispersion of voltages of cells is decomposed in the following way:

$$\sigma_{V_p} = \sigma_{cap} + \sigma_{\Sigma}, \tag{49}$$

where

- σ_{cap} concerns the variations introduced by the disturbances and sensors bias. The systemic disturbances, which are present for all operative conditions, correspond to inherent errors of the process and hence are assumed to have a constant amplitude.
- σ_{Σ} corresponds to bias and disturbances related to the PEMFC system. It can be decomposed into three parts:

$$\sigma_{\Sigma} = \sigma_p + \sigma_{H_2O} + \sigma_c, \tag{50}$$

with:

- σ_p as the dispersion related to pressure disturbances that are nonuniform in the gas channels. Note that the differences in pressures are mainly the result of the variations of gas flows and transitional modifications of channels' geometry causing liquid water accumulation.
- σ_{H_2O} denotes the dispersion related to a water distribution that is nonuniform inside the fuel cell. The water distribution inside the fuel cell is the result of random conditions good for condensation and for mechanisms which allow the movement of liquid water inside the channels. These mechanisms are strongly coupled to gas flows and depend on the operating conditions of the PEMFC system.
- σ_c correspond to a dispersion produced by the differences related to the physical states of the cells when constructing or assembly or as a result of irregular wear. It should be noted that these phenomena depend on the considered system and, generally, their time behaviors are very slow.

Fig. 19 Tree of variance decomposition

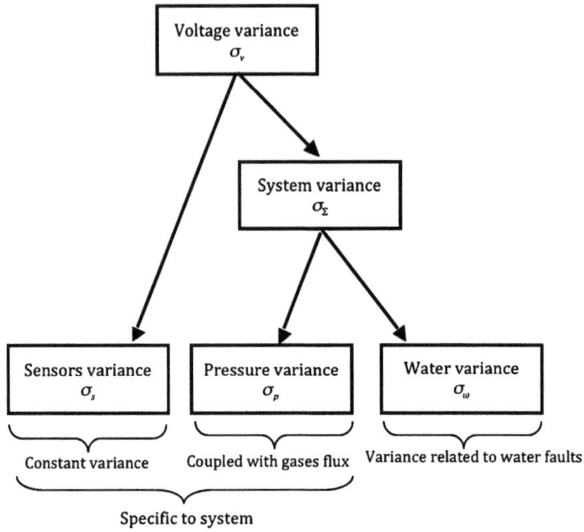

The description of voltage variance of the cells is depicted by Fig. 19.

To realize the diagnosis for PEMFC using a statistical approach based on the voltage distribution of cells, mainly when it concerns faults related to water, it is necessary to identify previously the voltage variance for cells (see for instance [25]).

The bias introduced by the sensors and the physical differences between the cells are assimilated to a white noise, then

$$\theta_{\text{cap}} \hookrightarrow \mathcal{N}\left(0, \sigma_{\text{cap}}\right). \tag{51}$$

The distribution variance of pressure is unknown, but the variance of the voltage distribution σ_{V_p} (see Eq. 42) is measurable experimentally. However, still coupled to sensors ones, for instance:

$$\sigma_{V_p} = \sigma_{\text{cap}} + \sigma_{\text{H}_2\text{O}} + \sigma_c. \tag{52}$$

For steady state, with flux and current equal to zero, the pressure inside the channels and the water distribution are homogeneous. Then, the associated variance can be neglected under the boundary conditions:

$$\lim_{D,I\to 0} \sigma_p = 0 \quad \text{and} \quad \lim_{D,I\to 0} \sigma_{\text{H}_2\text{O}} = 0, \tag{53}$$

where D designates the flux and I the current.

Finally, σ_{V_p} is associated to variances of sensors and the physical differences between the cells using the relationship defined by the following expression:

$$\lim_{D,I\to 0} \sigma_{V_p} = \sigma_{\text{cap}} + \sigma_c. \tag{54}$$

These inherent variations of the system are supposed to be real constants.

3.2.4 Operating Zone Definition

A. Principle

It is possible to build a polarization zone (around the theoretical polarization curve) which corresponds to the normal "operating" of the system. The polarization curve is hence transformed into a polarization zone (see Fig. 13) containing the disturbances phenomena which is inherent to the system. Thus, the value of cell voltage can be expressed by a probabilistic function Eq. 30 and not a unique value. The trust degree α is then used as the criteria for determining whether an operating point is outside the zone of "normal operating", which indicates the presence of fault or dysfunction as it is established by Eqs. 31 and 33. In the diagnosis strategy, we define n operating zones:

$$\begin{array}{lll} \text{Normal operating zone} & Z_0 & \text{defined as from } F_0\{\sigma, i_s, V\}, \\ \text{Fault zone number 1} & Z_1 & \text{defined as from } F_1\{\sigma, i_s, V\}, \\ \text{Fault zone number 2} & Z_2 & \text{defined as from } F_2\{\sigma, i_s, V\}, \\ \vdots & \vdots & \\ \text{Fault zone number n} & Z_n & \text{defined as from } F_n\{\sigma, i_s, V\}, \end{array} \tag{55}$$

where F_i $(i = 1 \dots n)$ are suitable functions that defined the zones Z_i $(i = 0 \dots n)$. Note that the zones Z_i $(i = 0 \dots n)$ can be defined using available knowledge about the fuel cell system obtained in an experimental way:

$$U = \{i_s, \sigma, V\}\ \forall i_s, \sigma, V\ \in \Re^+, \tag{56}$$

which can be defined in three dimensions as from i_s, σ and V. The parameters which define the operating mode in the presence of fault are chosen in an empiric way using experimentation (see Fig. 20) or by respecting safety norms or physical indictions from the constructor of the fuel cell.
In the deterministic approach, a fault can be identified in the following way:

$$p \in a_i \quad \text{if} \quad P - A_i \leq S_{a_i}(P - A_i), \tag{57}$$

where p is an operating point possessing a direction vector P, a_i correspond to i^{th} aggregate with the mass center determined by the direction vector A_i characterizing a fault, and $S_{a_i}(R)$ corresponding to a vector function which is the border surface to ith aggregate (see Fig. 20).

However, the random aspect of the voltage suggests that a statistical approach can be an adopted technique to diagnose for PEMFC. Indeed, using this approach, a probability to belong to an aggregate is defined depending on the density of the aggregate and the distance (meaning of Euclidian norm) to aggregate center:

$$P(p \in Z_i) = F(\delta_i(r_i)), \tag{58}$$

where $P_{p,i}$ designates the probability to belong to the aggregate i, $\delta_i(r_i)$ denotes the density of ith aggregate to r_i and F is the function which defines the probability to

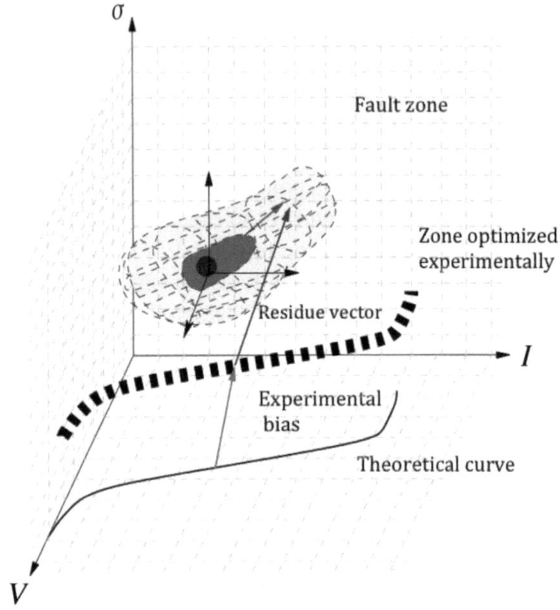

Fig. 20 Probability definition of a zone

belonging. However, it is possible to not be able to associate a point to a specific aggregate. The set of these points depicts the "transition zone":

$$Z_t = \{P/P\,(pinZ_i) < \epsilon \text{ for } i = 0\ldots n\}\,. \tag{59}$$

If greater precision is needed for the diagnosis strategy, then it is necessary to increment the dimension of the space in order to obtain a large dimension for the residue vector. This will allow to find the causing factor of the fault in a precise way. More parameters, as spatial distribution of cells' voltage and the trajectory of residue in experimental space (see Fig. 20 and Eq. 60), designate the possible alternatives for incrementing the space of decision.

B. Application

From the space definition of $\Re^3$ (see Eq. 56), it is possible to build a residue vector using the parameter (σ, V) of the distribution voltages of cells.

A first delimitation of zones by identifying simply the dangerous zones where a fast damage can occur because of the operating conditions (see Fig. 21):

- Minimal voltage for fuel cell $V \geq V_{\text{min}}$;
- Maximal current density $j \leq j_{\text{max}}$ (i.e., $i \leq i_{\text{max}}$);
- Maximal power $V \leq \frac{P_{\text{max}}}{I_{\text{max}}}$;
- Maximum dispersion of voltage of cells $\sigma \leq \sigma_{\text{max}}$.

However, such a delimitation of operating zones of systems do not take into account the system specificities and do not allow the identification of fault causes.

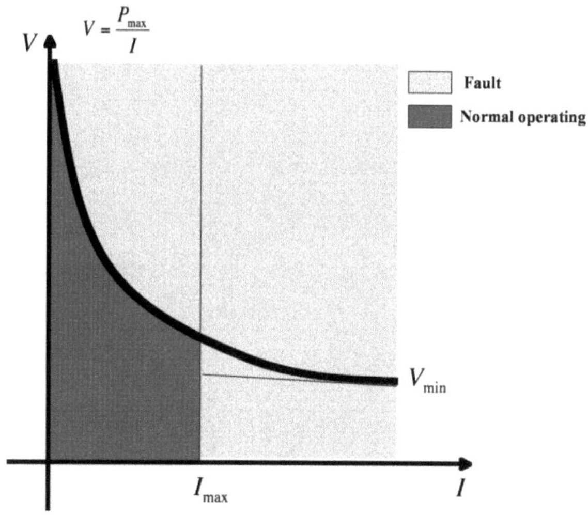

Fig. 21 Delimitation of the prohibited zone

Hence, it should be delimited more precisely the areas using experimental data and may be to create, in the considered space, specific aggregates for each kind of fault taking into account the specific characteristics of the system. Thus, the characteristics of voltages distribution of cells are first identified experimentally in normal operating conditions, Second, similar experimentation can be done in the presence of fault. Finally, the residue vector can be built in the following way:

$$\underbrace{\sigma_r\ V_r}_{\text{Residue } V,\sigma} = \underbrace{\sigma_o\,(T,\phi,I)\ \bar{V}_o\,(T,\phi,I)}_{\bar{V},\sigma \text{ Nominal conditions}} - \underbrace{\sigma_v\ \bar{V}}_{\text{Measured values } \bar{V},\sigma} . \tag{60}$$

The partial flooding of one or more areas of fuel cells leads to the loss of homogeneity of conditions and an increase in the variance. An important increase in the variance implicates the presence of a fault, here the flooding.

3.3 Experimental Results

In order to illustrate the proposed strategy of diagnosis, experimental tests were realized using different initial conditions in different zones: operating modes without dysfunction and operating modes with partial flooding. To realize the flooding the strategies adopted were:

- increasing the hygrometry of air entrance with constant temperature and constant load;
- decreasing the stack temperature with constant air hygrometry of entrance and constant load;
- increasing the load with constant air hygrometry of entrance and a constant stack temperature.

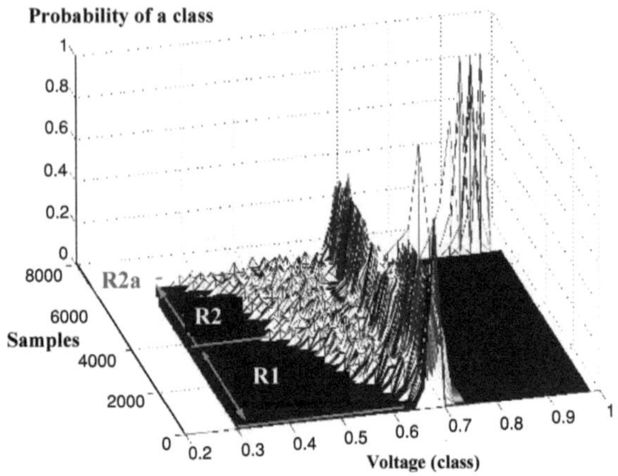

Fig. 22 Probability distribution function of class: Evolution to fault with a constant load

The experimental results show that the different operating modes are characterized by voltages distributions of cell voltages that are specific at each mode. For a normal operating mode, the variance is moderate and can be associated to the basic variance of the system (see Fig. 18 and Fig. 19).[7] The flooding inside the fuel cell is characterized by an important increase in variance and an average voltage drop of cells as depicted in Fig. 22. More precisely, three areas are to be distinguished. The zone R1 where the voltage is normal with a reduced dispersion. R2 is a zone characterized by an important voltage drop with an important dispersion. The third zone R2a where the voltage drop is critical with an extreme dispersion: the operating of the fuel cell must be stopped.

In the extreme cases, the behavior of voltage distribution loses its symmetry and with the test of Kolmogorov–Smirnov Eq. 39, the hypothesis of the normality can be rejected. This approach can be used for an early detection that the system is moving to fault zone.[8]

However, even if the fault produced by flooding is in general characterized by the parameters σ and $\bar{V}$, the cause of the dysfunction, namely the increasing of hygrometry in entrance or the increasing of the electrical load or the decreasing of the temperature of the stack is still difficult to determine. Nevertheless, if we observe the spatial distribution of the cell voltages, it is possible to detect significative differences between the flooding produced by a temperature drop (see Fig. 23) and a flooding caused by an increasing of hygrometry in entrance (Fig. 24). The localization of the areas which promote formation of the liquid water allows to define the spatial distribution of voltage in the fuel cell. Then, it is possible to identify the fault cause by establishing a relationship between the surface geometry of voltage distribution and the phenomena which produce the fault. In the first case (i.e. the situation where the hygrometry in entrance is higher), we observe an important decrease for the last

[7] Experimental test results on a fuel cell with 500 W

[8] Experimental test results on a fuel cell of 500 W

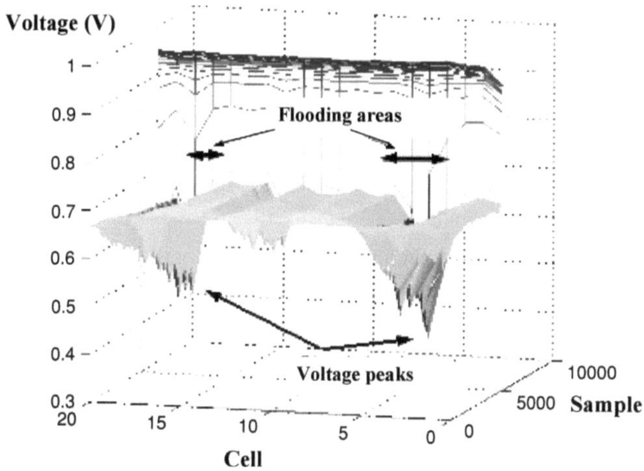

Fig. 23 Spatial distribution of cells voltages caused by a drop of temperature (see footnote 8): Temporal evolution. Two zones of flooding formed that are almost symmetrical

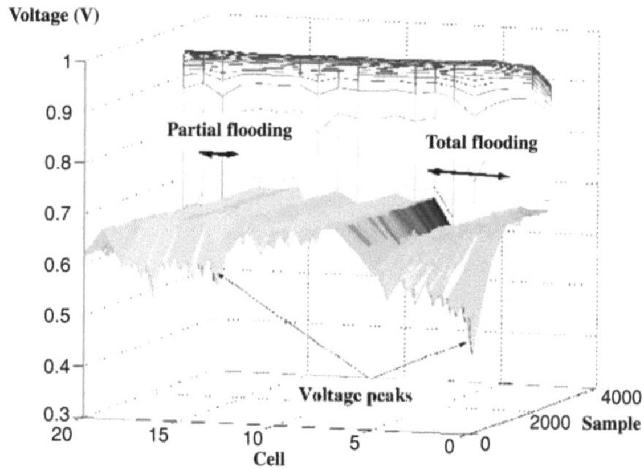

Fig. 24 Spatial distribution of cells voltages caused by an increasing of hygrometry of entrance: Temporal evolution. A partial zone of flooding and a total zone of flooding are formed with an asymmetric configuration

cells (i.e. the cells near the exit) thus, a less important decrease for the cells near the entrance of the gases. This can be explained by a more important concentration of steam in the last cells and by an accumulation of water near the exit, produced by the mass transport done by the gas fluxes, especially at the cathode.

In the second case (i.e. the temperature drop), a more important voltage drop occurs in the first cells. Thus, the air in the entrance of the fuel cell strongly charges steam with a steam pressure close to saturation pressure. When the air enters the cathode, it saturates and condenses quickly in the first cells. Then, in the intermediate cells the air is relatively dry which can explain the higher value and homogeneity of the voltage. Later, the air is charged and the pressure of steam increases and promotes a second condensation, less important, in the last cells. Parallel to the condensation phenomena, the water is transported and accumulated in the last cells generating a decrease in the voltage.

4 Conclusions

In this chapter, two strategies dedicated to faults diagnosis were presented. The first strategy based on the electrical equivalent model of a PEMFC which showed to be efficient in gas dynamic and voltage prediction has been developed. A diagnosis methodology based on the electrical circuit parameter identification has also been introduced. The proposed model can be improved to introduce saturated conditions to include liquid phase in the diffusion layer and to look at the liquid water effect on circuit parameters. The results in this field are expected to give important information about fuel cell fault diagnosis.

The second strategy is developed using a statistical approach and without using any model. Failure detection by stochastic parameters have shown to be a promising approach to fuel cell diagnosis. Different operation modes are clearly identified and experimentally tested. The proposed approach is an easy way to implement online a diagnosis strategy due to the simplicity of data calculations and the reduced cost of data acquisition system needed. Towards a diagnosis oriented to choose control strategies, a more complex but powerful alternative is to identify the residual vector path and speed.

References

1. Kozlowski JD, Byington CS, Garga AK, Watson MJ, Hay TA (2002) Model-based predictive diagnostics for electrochemical energy sources. In: Aerospace conference, 2001, IEEE Proceedings, vol 6. IEEE, pp 3149–3164
2. Burford D, Davis T, Mench MM (2004) Real-time electrolyte temperature measurement in an operating polymer electrolyte fuel cell. In: Adv Mater Fuel Cells Batteries Symp
3. Hissel D, Péra MC, Kauffmann JM (2004) Diagnosis of automotive fuel cell power generators. J Power Sources 128(2):239–246
4. Nitsche C, Schroedl S, Weiss W (2004) Onboard diagnostics concept for fuel cell vehicles using adaptive modelling. In:Intelligent Vehicles Symposium, 2004 IEEE. IEEE, pp 127–132
5. Brunetto C, Tina G, Squadrito G, Moschetto A (2004) PEMFC diagnostics and modelling by electrochemical impedance spectroscopy. In: Electrotechnical conference, 2004. MELECON 2004. Proceedings of the 12th IEEE Mediterranean, vol 3. IEEE, pp 1045–1050
6. Tsujioku Y, Iwase M, Hatakeyama S (2005) Analysis and modeling of a direct methanol fuel cell for failure diagnosis. In: Industrial electronics society, 2004. IECON 2004. 30th Annual Conference of IEEE, vol 3. IEEE, pp 2837–2842
7. Hernandez A, Hissel D, Outbib R (2010) Modeling and fault diagnosis of a polymer electrolyte fuel cell Using electrical equivalent analysis. . Energy Convers, IEEE Trans on 25(1):148–160
8. Hernandez A, Hissel D, Outbib R (2006) Non linear state space modelling of a PEMFC. Fuel Cells 6(1):38–46

9. Pekula N, Heller K, Chuang PA, Turhan A, Mench MM, Brenizer JS, Ünlü K (2005) Study of water distribution and transport in a polymer electrolyte fuel cell using neutron imaging. Nucl Instrum Methods Phys Res Section A: Accel Spectrom Detect Assoc Equip 542(1–3):134–141
10. Hernandez E, Diong B (2005) A small-signal equivalent circuit model for PEM fuel cells. In: Applied power electronics conference and exposition, 2005. APEC 2005. Twentieth Annual IEEE,vol 1. IEEE, pp 121–126
11. Famouri P, Gemmen RS (2004) Electrochemical circuit model of a PEM fuel cell. In: Power engineering society general meeting, 2003, IEEE, vol 3. IEEE
12. Yu D, Yuvarajan S (2005) A novel circuit model for PEM fuel cells. In: Applied power electronics conference and exposition, 2004. APEC'04. Nineteenth Annual IEEE, vol 1. IEEE, pp 362–366
13. Jung A (2002) A mathematical model of the hydrodynamical processes in the brain. In: Und N(ed) Nichtgleichgewicht in kondensierter Materie, Workshop Report II
14. Kerr R (2005) Fundamental fluidmechanics. In: Lectures for es30a/d, University of Warwick
15. Pukrushpan JT, Peng H, Stefanopoulou AG (2002) Simulation and analysis of transient fuel cell system performance based on a dynamic reactant flow model. In: Proceedings of 2002 ASME, Nov International mechanical engineering congress & exposition, pp 17–22
16. Pukrushpan J, Peng H, Stefanopoulou A (2004) Control-oriented modelling and analysis for automotive fuel cell systems. J Dyn Syst 126:14–25
17. Nguyen TV, White RE (1993) A water and heat management model for proton-exchange-membrane fuel cells. J Electrochem Soc 140:2178
18. ASAE. Psychometric Data SAE D271.2. American Society of Agricultural Standard, 1999
19. Yerramalla S, Davari A, Feliachi A, Biswas T (2003) Modeling and simulation of the dynamic behavior of a polymer electrolyte membrane fuel cell. J Power Sources 124(1):104–113
20. Golbert J, Lewin D (2004) Model-based control of fuel cells: (1) regulatory control. J Power Sources 135:135–151
21. Maggio G, Recupero V, Pino L (2001) Modeling polymer electrolyte fuel cells: an innovative approach. J Power Sources 101(2):275–286
22. Natarajan D, Van Nguyen T (2003) Three-dimensional effects of liquid water flooding in the cathode of a PEM fuel cell. J Power Sources 115(1):66–80
23. Hernandez A, Diagnostic d'une pile à combustible de type PEFC. PhD thesis, Université de Technologie de Belfort-Montbéliard (UTBM)
24. Fang G, Ward CA (1999) Examination of the statistical rate theory expression for liquid evaporation rates. Phys Rev E 59(1):441–453
25. Correa JM, Farret FA, Popov VB, Parizzi JB (2004) Influence of the modeling parameters on the simulation accuracy of proton exchange membrane fuel cells. In: Power tech conference proceedings, 2003 IEEE Bologna, vol 2. IEEE, p 8

Fault Diagnosis and Fault Tolerant Control of PEM Fuel Cell Systems

Diego Feroldi

Abstract The energy generation systems based on fuel cells are complex since they involve thermal, fluidic and electrochemical phenomena. Moreover, they need a set of auxiliary elements such as valves, compressor, sensors, regulators, etc. to make the fuel cell work at the pre-established optimal operating point. For these reasons, they are vulnerable to faults that can cause the stop or the permanent damage of the fuel cell. Therefore, it is useful to use systematic techniques, like the recent methods of fault-tolerant control (FTC), to guarantee the safe operation of the fuel cell systems. The Fault Diagnosis of PEM fuel cell systems using a model-based methodology is addressed in Sect. 2.2, whereas the FTC based on Model Predictive Control is addressed in Sect. 2.3). This chapter is complementary to Chap. 3, showing that the proposed control architecture with two actuators adds FTC capabilities to the fuel cell systems.

1 Introduction

The energy generation systems based on fuel cells are complex since they involve thermal, fluidic and electrochemical phenomena. Moreover, they need a set of auxiliary elements such as valves, compressor, sensors, regulators, etc. to make the fuel cell work at the pre-established optimal operating point. For these reasons, they are vulnerable to faults that can cause the stop or the permanent damage of the fuel cell. For example, the faults that have to be detected and diagnosed on fuel cell vehicle are very numerous [1], from major damages on the stack itself, which imply the replacement of the stack, to temporary variations of the system output values that can be easily corrected through control system. Therefore, it is useful to use

D. Feroldi (✉)
CAPEG-CIFASIS-(CONICET-UNR-UPCAM) and DCC-FCEIA-UNR,
27 de Febrero 210 bis, S2000EZP Rosario, Argentina
e-mail: feroldi@cifasis-conicet.gov.ar

M. S. Basualdo et al. (eds.), *PEM Fuel Cells with Bio-Ethanol Processor Systems*, Green Energy and Technology, DOI: 10.1007/978-1-84996-184-4_6,

systematic techniques of fault diagnosis and fault-tolerant control to guarantee the safe operation of the fuel cell systems [2–4].

The overall concept of fault diagnosis consists in the following three essential tasks [5]:

- Fault detection: consisting in the detection of the occurrence of faults in the system, which lead to undesired or intolerable behavior of the whole system.
- Fault isolation: localization or classification of different faults.
- Fault analysis or identification: determination of the type, magnitude and cause of the fault.

The model-based fault diagnosis techniques has developed notoriously from its origin, demonstrating a great efficiency in detecting faults in a great number of industrial processes applications. For example, model-based fault diagnosis systems are fully integrated into vehicle control systems, robots, transport systems, power systems, manufacturing processes, process control systems, etc.

Model-based fault diagnosis techniques are common in the explicit use of a process model, based on which algorithms are implemented for processing data that are on-line collected and recorded during the system operation. The major difference between the model-based fault diagnosis schemes are the adopted process model and, particularly, the applied algorithms, existing an intimate relationship between the model-based fault diagnosis technique and the modern control theory [5].

2 Model-based Fault Diagnosis in PEM Fuel Cell Systems

The first task to achieve active tolerant control consists of the inclusion of a fault diagnosis system operating in real-time. The diagnosis system should not only allow the fault detection and isolation but also the fault magnitude estimation. A central problem with the application of model-based fault diagnosis technique can be expressed as filtering/extracting the needed information about the faults of interests from the residual signals. To this end, two different strategies have been developed [5]:

- Designing the residual generator to achieve a decoupling of the fault of interests from the other faults, unknown disturbances and model uncertainties
- Extracting the information about the fault of interests from the residual signals by means of post-processing of the residuals. This procedure is called residual evaluation.

In this section, a model based fault diagnosis is proposed as a way of faults diagnosis in fuel cell systems. The model-based fault diagnosis is based on the on-line comparison between the real behavior of the monitored system by means of sensors and a dynamic model of the system. In case that a significant discrepancy (residual) is detected between the model and the measurements obtained by the sensors, the existence of a fault is assumed.

If a set of measurements is available, it is possible to generate a set of residuals (indicators) that present a different sensitivity to the set of possible faults. Then, analyzing in real-time the evolution of the residuals, it is possible, in some cases, to isolate the fault, and even in some cases, it is also possible to determine its magnitude. The innovation of the fault diagnosis methodology proposed is based on the use of a residual fault sensitivity analysis that allows to isolate faults that otherwise would not be separable.

2.1 Foundations of the Proposed Fault Diagnosis Methodology

The methodology of fault diagnosis that is proposed and applied to a fuel cell system is mainly based on the classic theory of model-based diagnosis described for example in [3], [6], [7] and [8]. Model based diagnosis can be divided in two subtasks: fault detection and fault isolation. The principle of model-based fault detection is to check the consistency of the observed behavior while fault isolation tries to isolate the component that is in fault.

The consistency check is based on computing residuals $r(k)$. The residuals are obtained from measured input signals $u(k)$ and outputs $y(k)$ and the analytical relationships obtained by system modeling:

$$r(k) = \Psi(y(k), u(k)), \tag{1}$$

where Ψ is the residuals generator function that depends on the type of detection strategy used (parity equation [6] or observer [9]). At each time step, k, the residual is compared with a threshold value, which is zero in ideal case or almost zero in real case. The threshold value is typically determined using statistical methods that take into account the effect of noise and model uncertainty [2]. When a residual is bigger than the threshold, it is determined that there is a fault in the system; otherwise, it is considered that the system is working properly. In practice, because of input and output noise, and modeling errors, robust residuals generators must be used. The robustness of a fault detection system means that it is only sensitive to faults, even in the presence of model-reality differences and noise [9].

Robustness can be achieved at residual generation phase (active) or at evaluation phase (passive). Most of the passive robust residual evaluation methods are based on an adaptive threshold changing in time according to the plant input signal and taking into account model uncertainty even in the time domain [10]. In this work, a passive method in time domain has been proposed for robust fault detection, where the detection threshold has been obtained using statistical techniques. Robust residual evaluation allows obtaining a set of *fault signatures* $\Phi(k) = [\phi_1(k), \phi_2(k), \ldots, \phi_{n_\Phi}(k)]$, where each indicator of fault is obtained as follows:

$$\phi_i(k) = \begin{cases} 0 & \text{if } |r_i(k)| \le \tau_i \\ 1 & \text{if } |r_i(k)| > \tau_i \end{cases}, \tag{2}$$

and τ_i is the threshold associated to the residual $r_i(k)$.

Fault isolation consists in identifying the faults affecting the system and it is carried out on the basis of fault signatures Φ that is generated by the detection module and their relation with all the considered faults $f(k) = \{f_1(k), f_2(k), \ldots, f_{n_f}(k)\}$. The method most often applied is a relation defined on the Cartesian product of the sets of faults FSM $\subset \phi \times f$, where FSM is the *theoretical signatures matrix* [6]. One element of that matrix FSM_{ij} is equal to one if the residual $r_i(k)$ is affected by the fault $f_j(k)$ and, in this case, the value of the fault indicator $\phi_i(k)$ must be equal to one when the fault appears in the monitored system. Otherwise, the element FSM_{ij} is zero.

The isolation approach previously presented use a set of binary detection tests to compose the observed fault signature. When applying this methodology to dynamic systems, the use of binary codification of the residual produces a loss of information, since they may exhibit symptoms with different dynamics [11]. This can be the origin of false isolation decisions, especially because some detection tests have a transient behavior in response to the faults, for example in dynamic slow/delayed systems. Also, in complex systems, some faults could present the same theoretical binary fault signature not allowing fault isolation. In both cases, it is possible to use other additional information associated with the relationship between the residuals and faults, as the sign, sensitivity, order or time activation, to improve the isolation results [11].

Here, it is proposed the use of information provided by the fault residual sensitivity in the design of the diagnosis system to increase the fault isolability. According to [6], the sensitivity of the residual to a fault is given by

$$S_f = \frac{\partial r}{\partial f}, \tag{3}$$

which is a transfer function that describes the effect on the residual r of a given fault f. Sensitivity provides a quantitative information of the effect of the fault on the residual and a qualitative information in their sense of variation (sign). The use of this information at the stage of diagnosis will allow to separate faults that even presenting the same theoretical binary fault signature, present, qualitatively or quantitatively, different sensitivities.

In order to perform diagnosis, the algorithm uses the theoretical signatures matrix FSMsensit with the residual sensitivity in the rows, the faults in columns, and each value of this matrix is noticed as $S_{r_i f_j}$. Although sensitivity depends on time in case of a dynamic system, here the steady-state value after a fault occurrence is considered as it is also suggested in Gertler and Inc [6]. The theoretical value of $S_{r_i f_j}$ describes how easily a fault will cause a violation of the threshold of the i-th residual and can be computed analytically using Eqs. 1, 3 or by simulation.

Besides, to perform real time diagnosis, the observed sensitivity $S^{\text{obs}}_{r_i f_j}$ should be computed using the value of the residuals at instant k, $r_i(k)$, and the magnitude of

Table 1 Observed fault signature matrix using relative sensitivity with respect to r_1

	f_1	f_2	$\cdots$	f_n
r_2/r_1	$S^{\text{rel, obs}}_{r_2 r_1 f_1}$	$S^{\text{rel, obs}}_{r_2 r_1 f_2}$	$\cdots$	$S^{\text{rel, obs}}_{r_2 r_1 f_n}$
r_3/r_1	$S^{\text{rel, obs}}_{r_3 r_1 f_1}$	$S^{\text{rel, obs}}_{r_3 r_1 f_2}$	$\cdots$	$S^{\text{rel, obs}}_{r_3 r_1 f_n}$
$\vdots$	$\vdots$	$\vdots$	$\ddots$	$\vdots$
r_m/r_1	$S^{\text{rel, obs}}_{r_m r_1 f_1}$	$S^{\text{rel, obs}}_{r_m r_1 f_2}$	$\cdots$	$S^{\text{rel, obs}}_{r_m r_1 f_n}$

the fault at the same instant, $f(k)$:

$$S^{\text{obs}}_{r_i f_j} = r_i(k)/f(k). \tag{4}$$

2.2 Fault Diagnosis Methodology based on Relative Sensitivity

From Eq. 4, it can be seen that using FSMsensit in real time requires the knowledge of the fault magnitude or making an estimation of it. In order to overcome this drawback, the diagnosis is designed using the new concept of relative sensitivity rather than the absolute sensitivity given in Eq. 3. The observed *relative fault sensitivity* is defined as

$$S^{\text{rel, obs}}_{r_i r_1 f_j} = \frac{S_{r_i f_j}}{S_{r_1 f_j}} = \frac{r_i(k)/f_j(k)}{r_1(k)/f_j(k)} = \frac{r_i(k)}{r_1(k)}, \tag{5}$$

which corresponds to the ratio of the i-th residue at instant k, $r_i(k)$, with another one, for example $r_1(k)$. Then, the relative sensitivity is not dependent on to the magnitude of that unknown fault. In the case of a set of n fault, a relative fault sensitivity matrix FSMsensit-rel as the one shown in Table 1 should be used.

The diagnostic algorithm is used to obtain real-time observed relative sensitivities using Eq. 5, as a ratio of residuals, which provides a vector of relative sensitivities. The generated vector is compared with the vectors of theoretical faults stored into the relative sensitivity matrix FSMsensit-rel. The theoretical fault signature vector with a minimum distance with respect to the fault observed vector is postulated as the possible fault:

$$\min\left\{d_{f_1}(k), \ldots, d_{f_n}(k)\right\}, \tag{6}$$

where the distance is calculated using the Euclidean distance between vectors

$$d_{f_i}(k) = \sqrt{\left(S^{\text{rel, obs}}_{r_2 r_1 f_i} - S^{\text{rel, teo}}_{r_2 r_1 f_i}\right)^2 + \cdots + \left(S^{\text{rel, obs}}_{r_m r_1 f_i} - S^{\text{rel, teo}}_{r_m r_1 f_i}\right)^2}. \tag{7}$$

Table 2 Description of the fault scenarios

Fault	Description
f_1	Increase of the compressor motor friction.
f_2	Overheating of the compressor motor.
f_3	The fluid resistance increases due to water blocking the channels or flooding in the diffusion layer.
f_4	Air leak in the air supply manifold.
f_5	Increase of the voltage value below which the compressor motor does not turn.
f_6	Increase of the stack temperature due to a failure in the temperature controller.

2.3 Application to a PEM Fuel Cell System

To show the validity of the proposed model based fault diagnosis approach when applied to a PEM fuel cell system (PEMFC), the model developed by Pukrushpan et al. [12] (described in Sect. 2.1) has been modified in order to include a set of typical faults. It was developed a fault simulator, with which it is possible to reproduce any of the faults presented in Table 2 at a given time.

The implementation of the fault diagnosis system is shown in Fig. 1 and, in the simulation set-up, the fuel cell is the modified fuel cell model. The set of available measurements is compared with its predicted value using a non-faulty fuel cell model. The differences between the predicted and measured values generate a set of residuals that are sent to the fault diagnosis system. When a fault appears, the residuals that are sensitive to this fault take a value different from zero. When some of the residual values cross the detection threshold, the fault diagnosis starts reasoning with all the violated residuals. The reasoning (described in Sect. 2.2) is based on computing the minimum distance between the observed fault signature and the theoretical one. The fault that approaches the most to one of the fault signatures is the one indicated as a possible fault.

The fault f_1 is simulated with an increment Δk_v in the compressor constant k_v and, similarly, the fault f_2 is simulated with an increment ΔR_{cm} in the compressor motor resistance R_{cm}. Both faults result in a change in the compressor torque τ_{cm}:

$$\tau_{cm} = \frac{\eta_{cm}\, k_t}{(R_{cm} + \Delta R_{cm})} \left(v_{cm} - (k_v + \Delta k_v)\omega_{cm}\right), \tag{8}$$

where η_{cm} is the motor mechanical efficiency, k_t is a motor constant, and ω_{cm} is the compressor speed.

The fault f_3 is simulated with an increment $\Delta k_{ca,\,out}$ in the orifice constant of the cathode output, $k_{ca,\,out}$, which produces a change in the outlet air flow in the cathode, $W_{ca,\,out}$:

$$W_{ca,\,out} = \left(k_{ca,\,out} + \Delta k_{ca,\,out}\right) \left(p_{ca} - p_{rm}\right), \tag{9}$$

where p_{ca} is the cathode pressure and p_{rm} is the return manifold pressure.

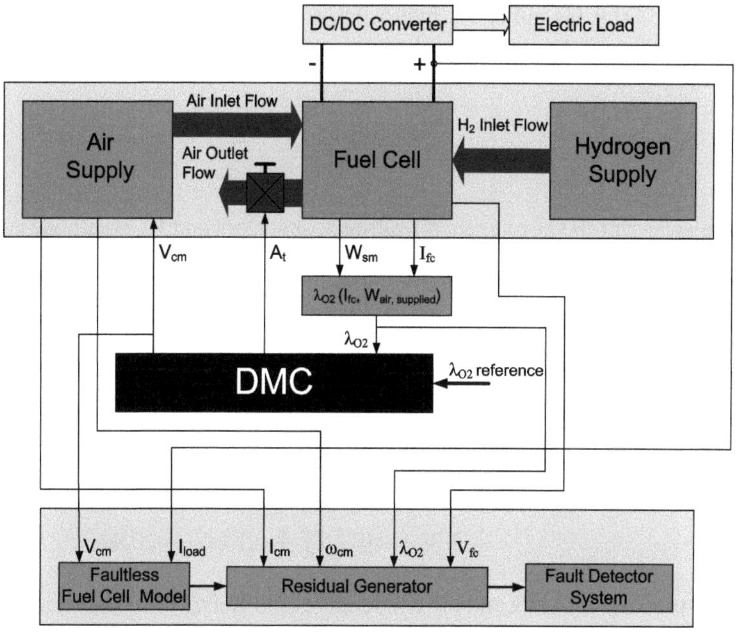

Fig. 1 Implementation of the fault diagnosis system

The fault f_4 is simulated with a diminution in the supply manifold outlet flow constant, which is translated into a change in the supply manifold outlet air flow $W_{\mathrm{sm,\,out}}$:

$$W_{\mathrm{sm,\,out}} = \left(k_{\mathrm{sm,\,out}} + \Delta k_{\mathrm{sm,\,out}}\right), \left(p_{\mathrm{sm}} - p_{\mathrm{ca}}\right), \tag{10}$$

where p_{sm} is the supply manifold pressure.

The fault f_5 is simulated with an increment in the voltage value below which the compressor motor does not turn, $V_{\mathrm{cm,\,low}}$, that the controller supplies to the compressor motor, a boundary that also influences the compressor torque in Eq. 8. The fault f_6 is simulated with an increment ΔT_{fc} in the stack temperature T_{fc}, which has an impact on the open circuit voltage of the stack, the partial pressure of gases, the relative humidity, and the water diffusion coefficient in the membrane. Therefore, the open circuit voltage E results

$$\begin{aligned} E = {} & 1.229 - 0.85 \times 10^{-3}\left(T_{\mathrm{fc}} + \Delta T_{\mathrm{fc}} - 298.15\right) \\ & + 4.3085 \times 10^{-5}\left(T_{\mathrm{fc}} + \Delta T_{\mathrm{fc}}\right)\left[\ln(p_{\mathrm{H_2}}) + 0.5\ln(p_{\mathrm{O_2}})\right], \end{aligned} \tag{11}$$

where $p_{\mathrm{H_2}}$ and $p_{\mathrm{O_2}}$ are the partial pressure of hydrogen and oxygen, respectively.

The partial pressure of gases p_i in the anode is

$$p_{i,\,\mathrm{an}} = \frac{\mathrm{M_{i,\,an}\,R_i(T_{fc} + \Delta T_{fc})}}{V_{\mathrm{an}}}, \tag{12}$$

where the subscript i is either H_2-hydrogen or v-vapor, M is the molar mass, R is the gas constant, and V_{an} is the anode volume. The partial pressure of gases in the cathode is

$$p_{i,\,ca} = \frac{M_{i,\,ca}\, R_i (T_{fc} + \Delta T_{fc})}{V_{ca}}, \tag{13}$$

where the subscript i is O_2-oxygen, N_2-nitrogen or v-vapor and V_{ca} is cathode volume. The relative humidity ϕ_j is

$$\phi_j = \frac{p_{vj}}{p_{sat}(T_{fc} + \Delta T_{fc})}, \tag{14}$$

where the subscript j is either an-anode or ca-cathode and the saturation pressure of vapor is calculated using the following expression:

$$\begin{aligned} log_{10}(p_{sat}) = & -1.69 \times 10^{-10}\, T_{fc}^4 + 3.85 \times 10^{-7}\, T_{fc}^3 \\ & - 3.39 \times 10^{-4}\, T_{fc}^2 + 0.14\, T_{fc} - 20.92, \end{aligned} \tag{15}$$

where the pressure is in kPa and the temperature is in Kelvin. The stack temperature also affects the water diffusion coefficient in the membrane:

$$D_w = D_\lambda \exp\left(2416 \left(\frac{1}{303} - \frac{1}{T_{fc} + \Delta T_{fc}}\right)\right), \tag{16}$$

where D_λ is a constant which depends on the water content in the membrane.

2.3.1 Residual Generation and Fault Sensitivity Analysis

The variables that are considered as measured and consequently can be used for residual generation are the oxygen excess ratio λ_{O_2}, the compressor motor rotational speed ω_{cm}, the compressor motor current I_{cm}, and the fuel cell system voltage V_{fc}. Using these variables and the non linear model presented in [12], four residuals can be derived:

$$\begin{aligned} r_1 &= \lambda_{O_2} - \hat{\lambda}_{O_2}, \\ r_2 &= V_{fc} - \hat{V}_{fc}, \\ r_3 &= I_{cm} - \hat{I}_{cm}, \\ r_4 &= \omega_{cm} - \hat{\omega}_{cm}. \end{aligned} \tag{17}$$

Using the PEMFC fault simulator previously described, it has been determined if the faults defined in Table 2 affect or not each of the previous residuals. From these results, the theoretical binary fault signature matrix presented in Table 3 can be derived. It can be noticed that all the considered faults affect all the residuals. Thus, the faults are not diagnosable. Even taking into account the sense (sign) in which the

Table 3 Theoretical fault signature matrix FSM using binary and sign information

	f_1	f_2	f_3	f_4	f_5	f_6
r_1	(−)1	(−)1	(−)1	(−)1	1	(−)1
r_2	(−)1	(−)1	(−)1	(−)1	1	1
r_3	(−)1	(−)1	(−)1	(−)1	1	(−)1
r_4	(−)1	(−)1	1	1	1	1

Table 4 Theoretical fault signature matrix FSMsensit-rel

	f_1	f_2	f_3	f_4	f_5	f_6
r_2/r_1	1	0.824	0.118	0.643	0.036	−0.221
r_3/r_1	0.854	1	0.197	0.206	0.039	0.151
r_4/r_1	1	0.937	−0.128	−0.134	0.168	−0.098

fault affect the residual not all the faults are diagnosable: f_1 can not be distinguished from f_2 and, similarly, f_3 from f_4.

Alternatively, using the relative fault sensitivity Eq. 5, the fault signature table FSMsensit-rel can be calculated. The values of this matrix are shown in Table 4.[1] It can be noticed that in this case all the considered faults can be isolated since the following condition is satisfied:

$$S^{\text{rel, teo}}_{r_i r_1 f_j} \neq S^{\text{rel, teo}}_{r_i r_1 f_k} \quad \text{for all } j \neq k. \tag{18}$$

2.4 Results

In order to evaluate the model-based fault diagnosis methodology proposed, the fault scenarios and fault simulator presented in Sect. 2.3 are used. As discussed in previous sections, the fault detection is based on checking at every time the difference (residual) between the signal monitored by a sensor and its estimation using the detection model 17.

The simulation results for the fault scenario f_1 are shown to illustrate how the methodology works: Figure 2 shows the temporal evolution of the residuals and the detection threshold for each of them, and the time that the diagnosis system takes for detecting and isolating the fault, respectevely. The fault is introduced into the system at time 50 s and some time after, all the fault signals cross their detection threshold (dash line) activating the four indicators of fault Eq. 2. The detection subsystem stores the fault at the time that one of the thresholds is violated by any of the residuals and, as soon as it is detected that a fault is present, the isolation process begins. The isolation process is based on evaluating the distance of the observed relative fault sensitivity vector to the theoretical fault sensitivity vector.

Figure 3 presents the Euclidean distance between the observer and the theoretical sensitivity fault signatures for each fault Eq. 7. It can be noticed that since fault f_1

[1] This table has been derived from the PEMFC model in the operating point described in Sect. 2.1.

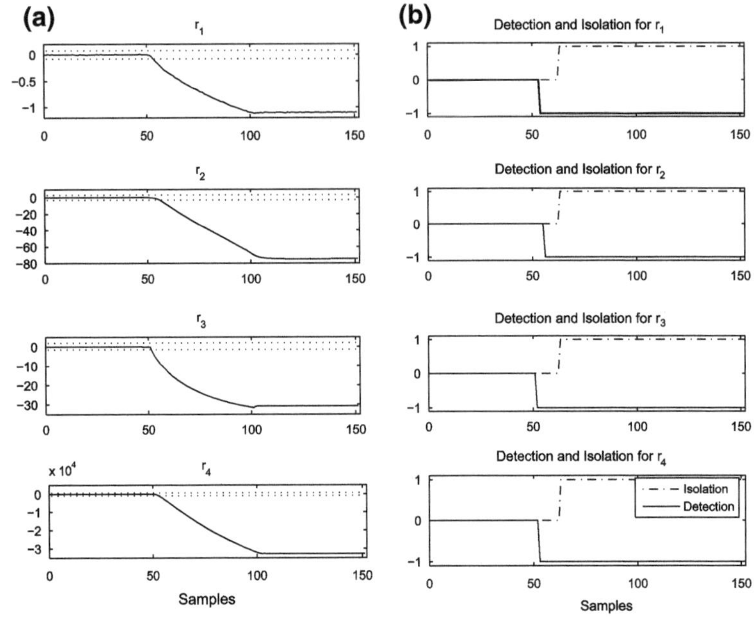

Fig. 2 Simulation results corresponding to fault scenario. f_1 **a** Time evolution of the residuals. **b** Time evolution of fault detection and isolation indicators

has a similar fault signature than f_2 (see also Table 4) at the beginning of the fault isolation process f_2 is the fault proposed as the possible fault, since presents a smaller distance than f_1. However, from time instant 82 s, f_1 can be isolated. It is seen that the proposed methodology, after some isolation time delay, allows isolate the true fault.

Similarly, it is possible to detect and isolate the other five fault scenarios using the proposed methodology. However, the time that the methodology takes to isolate is different for each fault scenario. It is remarkable that the process of isolation needs considerably more processing time with the fault scenarios f_1 and f_2 because of the similarity between their fault signatures.

3 Fault-Tolerant Model Predictive Control of PEM Fuel Cells Systems

A fuel cell system is not composed of the fuel cell stack alone but it integrates many other components. Several devices are included in the system and, thus, there are many control loops schemes, depending on the devices that must be controlled. The lower control level takes care of the main control loops inside the fuel cell, which are basically fuel/air feeding, humidity, pressure and temperature. The upper control level is in charge of the whole system, integrating the electrical conditioning, energy storage and possibly a fuel reformer [13].

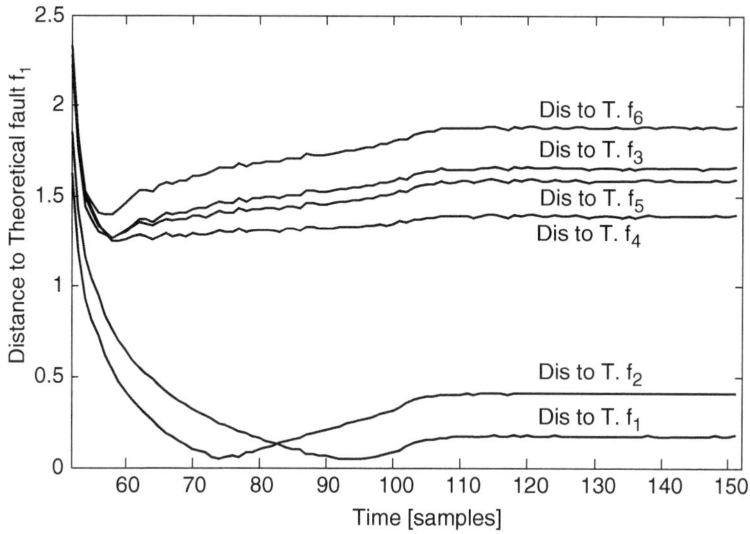

Fig. 3 Results of the fault diagnosis methodology proposed for fault scenario f_1

Many control strategies have been proposed in the literature, ranging from feed-forward control [14], LQR [15, 16], neural networks [17, 18] or model predictive control(MPC) [19, 20]. This section is focused on the control of the fuel cell using MPC to fulfil the oxygen supply. Two controller architectures will be compared. The first one uses only one actuator (the compressor) and the air feeding subsystem is controlled manipulating the compressor voltage. The second one uses two actuators (the compressor and the outlet air valve area) so additionally to the compressor voltage, the controller has an extra control variable based on the opening of the outlet air valve at the cathode [20]. The advantages and disadvantages of using this extra actuator are in study within the PEMFC community.

As it was mentioned in Chap. 3, one of the main control objective is to avoid a lack of oxygen in the cathode (a phenomenon that is known as "oxygen starvation") by maintaining the oxygen excess ratio close to a pre-established set-point, in spite of the disturbance introduced by the current load. However, because of the complexity of the fuel cell system, it is prone to suffer faults during the operation [21]. Therefore, some fault-tolerant capabilities should be added to the control system in order to maintain the fuel cell system under control even in the presence of faults.

This section explores the possibility of making use of the known inherent fault-tolerant capabilities of MPC when applied to PEMFC. In particular, it is shown that the controller architecture based on manipulating two actuators (the compressor and the outlet air valve) that was proposed in Sect. 2.2 offers better fault-tolerant capabilities and improves the system performance.

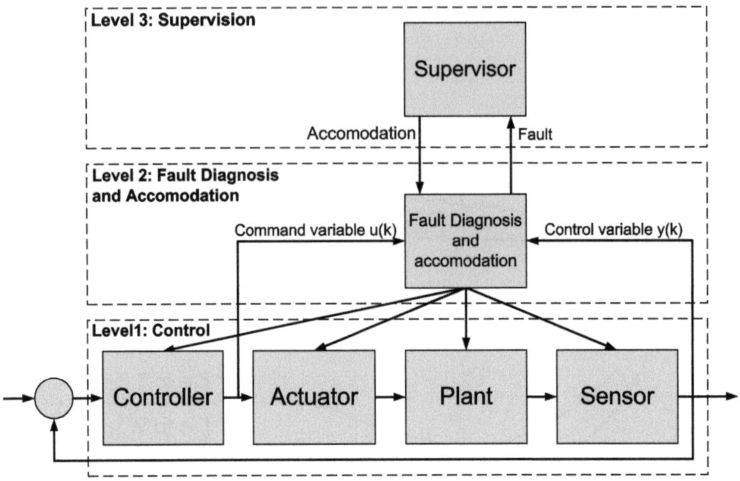

Fig. 4 Fault-tolerant architecture

3.1 Fault-Tolerant Model Predictive Control

Fault-tolerant MPC is an incipient research area in the automatic control field [2]. One way of achieving fault-tolerance is to employ a fault detection and isolation (FDI) scheme on-line. This system will generate a discrete event signal to a supervisor system when a fault is detected and isolated. The supervisor, in turn, will activate some accommodation action in response, which can be predetermined for each fault or obtained from real-time analysis and optimization. Because of the discrete-event nature of fault occurrence and the reconfiguration/accommodation, a FTC system is a hybrid system by nature. For design purposes of these systems, the hybrid nature has been traditionally neglected in order to facilitate a simple design, reliable implementation, and systematic testing. The whole FTC scheme can be expressed using the three-level architecture for FTC systems proposed by Blanke et al. [2] (see Fig. 4):

- Level 1: *Control Loop*. This level comprises a traditional control loop with sensors and actuator interfaces, signal conditioning and filtering, and the controller.
- Level 2: *Fault Diagnosis and Accommodation*. The second level comprises a given amount of detectors, usually one per each fault effect to be detected, and actuators that implement the desired reconfiguration or other corrective actions given by the autonomous supervisor.
- Level 3: *Supervision*. The supervisor is a discrete-event dynamical system which comprises state-event logic to describe the logical state of the controlled object. The supervisor functionality includes an interface with FDI detectors and generates remedial actions to accommodate a fault.

3.2 Inclusion of Fault Tolerance in Model Predictive Control

Fault tolerance can be embedded in MPC relatively easily [22]. This can be done in several ways:

1. Changing the constraints of the MPC in order to represent certain kinds of faults, being especially easy to adapt the algorithms for faults in actuators, assuming that the fault has been located and their effects have been estimated using a FDI module.
2. Modifying the internal plant model used by the MPC in order to reflect the fault influence over the plant using the information provided by the FDI module.
3. Relaxing the initial control objectives in order to reflect the system limitations under fault conditions.

3.3 Comparing two Model Predictive Control Architectures for Oxygen Control

In Chap. 3, it was explained that a common practice in the PEMFC control is to control the hydrogen supply using the anode inlet valve in such a way that the anode pressure tracks the cathode pressure whereas the control of the air supply is achieved controlling the oxygen excess ratio.

In this section, two architectures to control the oxygen excess ratio are compared: the common architecture in the literature with one actuator (the compressor motor voltage) and a proposed architecture with two actuators: the compressor voltage and the opening area in the cathode outlet valve. The two architectures are shown in Fig. 5. In Chap. 3, it was shown that the use of these two actuators offers better control performance and efficiency improvement in the case of regulation of the stack voltage and the oxygen excess ratio at the same time. Here, the two control structures are compared in terms of fault-tolerant capability. The objective is to regulate the oxygen excess ratio around the reference value $\lambda_{O_2,\text{ref}} = 2.3$, i.e., $\lim_{k\to\infty} |\lambda_{O_2}(k) - \lambda_{O_2,\text{ref}}| = 0$.

The control architectures are implemented using MATLAB MPC Toolbox and the fuel cell system linear model used to implement the MPC is derived through a linearization of the nonlinear model proposed in Pukrushpan et al. [12] at the selected operating point: $P_{\text{net}} = 40\,\text{kW}$, $\lambda_{O_2} = 2.3$, $V_{\text{fc}} = 235\,\text{V}$, $I_{\text{fc}} = 191\,\text{A}$, $V_{\text{cm}} = 164\,\text{V}$, and $A_t = 20\,\text{cm}^2$.

An open-loop optimal control problem is solved at each sampling time where the future control action is calculated over a finite horizon h_m, i.e.,

$$u = \begin{bmatrix} u_1(k),\ u_1(k+1), \ldots,\ u_1(k+h_m) \\ u_2(k),\ u_2(k+1), \ldots,\ u_2(k+h_m) \end{bmatrix}, \tag{19}$$

minimizing a cost function $J(k)$ over a finite prediction horizon h_p :

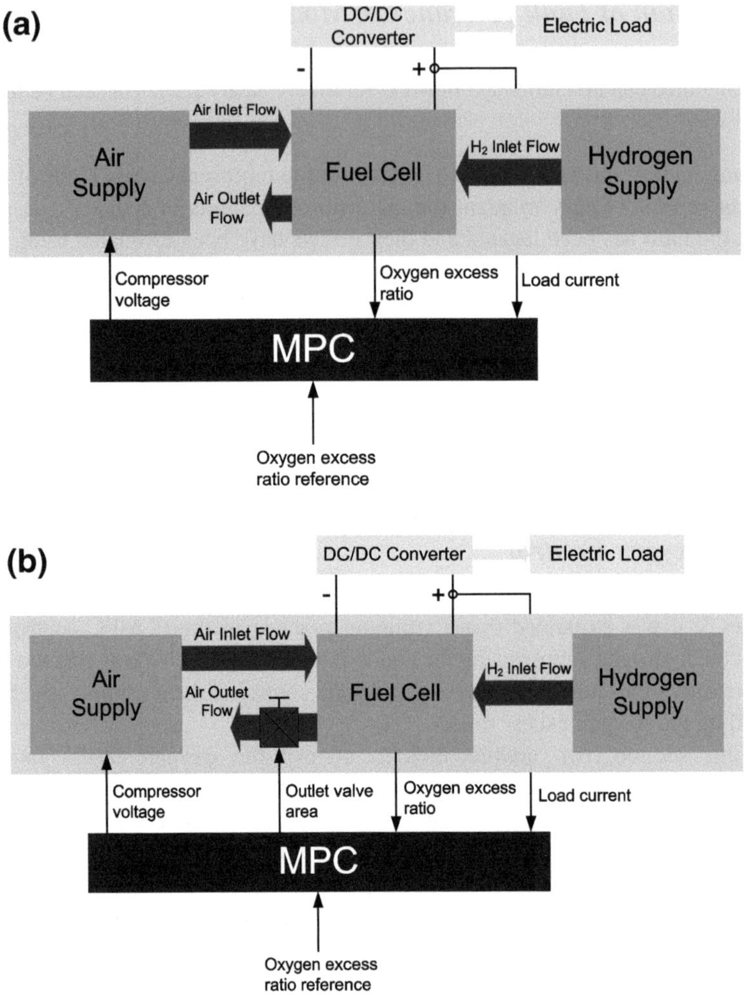

Fig. 5 One actuator versus two actuator MPC architecture for fuel cell control. **a** One actuator control architecture. **b** Two actuator control architecture

$$\min_{\Delta u} J(k), \tag{20}$$

subject to

$$\begin{aligned} x(k+1) &= A\,x(k) + B\,u(k), \\ y(k) &= C\,x(k), \\ u(k) &\in \left[\underline{u},\, \overline{u}\right], \text{ and} \\ y(k) &\in \left[\underline{y},\, \overline{y}\right] \end{aligned} \tag{21}$$

for $k = 0, 1, 2, \ldots$.

The cost function is defined as follows:

$$J(k) = \sum_{j=1}^{h_p} \lambda \left[y(k+j|k) - y_r)\right]^2 + \sum_{j=0}^{h_m} \Delta u(k+j|k)^T Q \Delta u(k+j|k), \qquad (22)$$

where $\Delta u(k+j|k) \triangleq u(k+j|k) - u(k-1+j|k)$, λ is a coefficient to weight the error, and Q is a diagonal matrix of dimension 2×2 to weight the control increments.

Only the first value of the calculated control sequence $u(k)$ is applied and the computation is repeated at the next sampling time starting from the new state and over a shifted horizon, leading to a moving horizon policy. The solution relies on a linear dynamic model, respecting all input and output constraints and optimizing a quadratic performance index. The MPC weights λ and Q are tuned to achieve the desired control goal, that is to maintain the oxygen excess ratio, penalizing the control effort. The parameter λ has been selected to have a value of 5 while the diagonal elements of Q have been selected to have a value of 0.1 for a good performance control, after some trial and error experimentation.

The air compressor voltage is considered as a constrained input due to physical limits: the maximum compressor voltage cannot exceed 230 V and the voltage value is never negative. The oxygen excess ratio is represented as a constrained output: the operating range is between 1.5 and 3 in order to avoid starvation and to obtain near optimal operation [23]. Note that this output constraint is implemented as a soft constraint in the MPC toolbox in order to prevent the infeasible solution.

Simulating different scenarios, it is found that the compressor voltage variations from the steady-state value to obtain the same control performance regarding the control of the oxygen excess ratio are smaller in the two actuator case than in the one actuator case. These smaller compressor voltage variations are compensated by the control of the outlet valve area: Fig. 6b shows the comparison of the oxygen excess ratio and the compressor voltage in case of using one actuator and two actuator control architectures when a series of step changes in the stack current are applied (Fig. 6a). This current is considered a measured disturbance for MPC controller. From this figure it can be clearly seen that for the same quality of regulation of the oxygen excess ratio, providing a maintained set point of 2.3, the two actuator architecture requires smaller voltage variations in the compressor.

Besides, two other basic variables of the fuel cell performance are also shown in Fig. 6a, c: the stack voltage and the oxygen partial pressure in the cathode. Even though these variables are not controlled in this analysis, their behavior must be checked in order to avoid high degradation of the efficiency or dangerous responses.

The simulation results show different trade-offs: the control architecture with one actuator maintains the stack voltage close to the nominal value and drives the oxygen partial pressure to higher variations while the control structure with two actuators maintains the oxygen partial pressure closer to the nominal value and drives the stack voltage to higher variations.

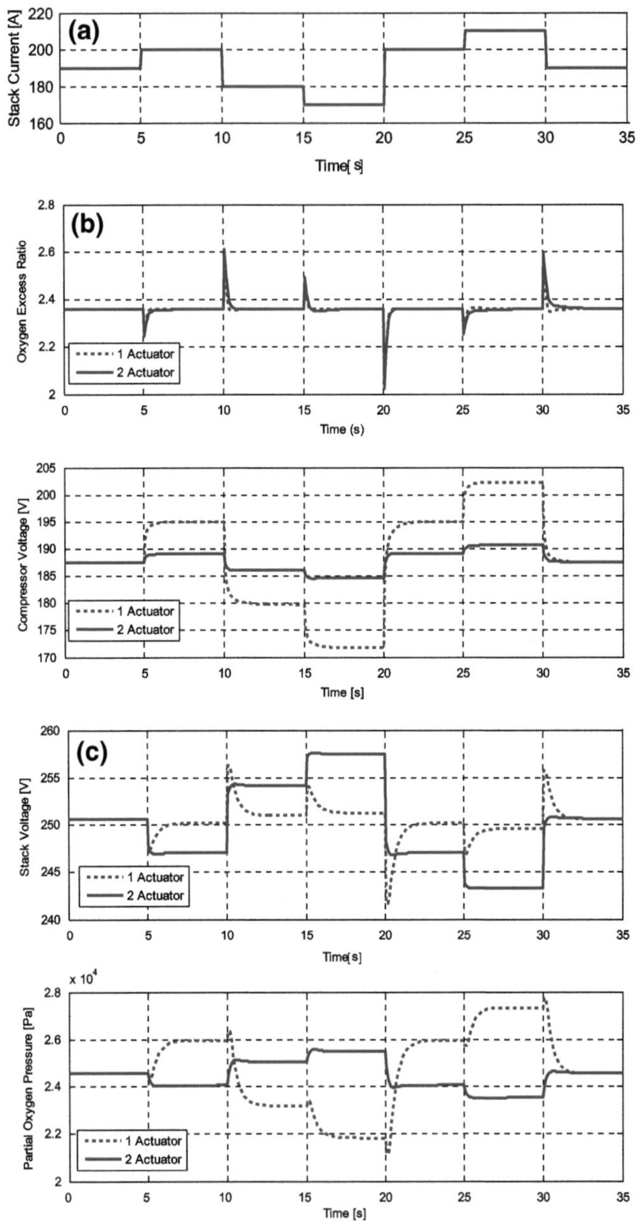

Fig. 6 Comparison of control results using both architectures. **a** Step changes in the stack current. **b** Evolution of the oxygen excess ratio and the compressor voltage. **c** Evolution of the stack voltage and the partial pressure of oxygen

3.4 Fault-Tolerance of the two Considered MPC Control Architectures

In this section, the advisability of implementing an active fault-tolerant scheme using the two actuators MPC control architecture is shown. As explained in previous sections, the MPC formulation allows to easily include active FTC capabilities in the control law. In this section, the case of actuator faults in the PEM fuel cell system is addressed. In a preliminary manner, the compressor faults are modelled through a reduction of the compressor voltage range. According to Sect. 3.3, the FDI module should provide the MPC controller the new limits of the compressor voltage range, once the fault has been detected, isolated and estimated. The active FTC control architecture is showed in Fig. 7, where the actuator operating range limits are estimated by the FDI module.

The FDI module (drawn in dashed line) is not implemented in this section and it is assumed that it is available and working perfectly. In order to take into account changes in the compressor voltage range due to the fault (actuator limits), the linear model for MPC design is modified by including the actuator limits as new states that will be estimated by the FDI module and can be written as

$$\begin{bmatrix} x_1(k+1) \\ \vdots \\ x_n(k+1) \\ x_{n+2}(k+1) \\ x_{n+2}(k+1) \end{bmatrix} = \begin{bmatrix} A & 0 \\ 0 & I \end{bmatrix} \begin{bmatrix} x_1(k) \\ \vdots \\ x_n(k) \\ x_{n+1}(k) \\ x_{n+2}(k) \end{bmatrix} + \begin{bmatrix} B \\ 0 \end{bmatrix} \begin{bmatrix} u_1(k) \\ u_2(k) \end{bmatrix} \tag{23}$$

$$\left[\begin{array}{c} y_1(k) \\ \vdots \\ y_m(k) \\ \hline y_{m+1}(k) \\ y_{m+2}(k) \end{array}\right] = \left[\begin{array}{c|c} C & 0 \\ \hline 0 & I \end{array}\right] \left[\begin{array}{c} x_1(k) \\ \vdots \\ x_n(k) \\ \hline x_{n+1}(k) \\ x_{n+2}(k) \end{array}\right] + \left[\begin{array}{c} D \\ \hline D' \end{array}\right] \begin{bmatrix} u_1(k) \\ u_2(k) \end{bmatrix}, \tag{24}$$

where x_i, $i = 1, \ldots, n$ and y_i, $i = 1, \ldots, m$ are the original states and outputs, respectively, and A, B, C, and D are the original matrices of the linearized fuel cell system model described in Sect. 2.1. Vector D' is defined as $D' = [\,-1\ -1\,]^T$ and I is the identity matrix of suitable dimensions.

In Eq. 23, x_{n+1} corresponds to the upper limit while x_{n+2} corresponds to the lower limit, and $x_{n+1}(k+1) = x_{n+1}(k)$ and $x_{n+2}(k+1) = x_{n+2}(k)$ means that the upper limit and the lower limit remains constant during the prediction horizon, respectively. Besides, additional output constraints have been added to the MPC controller: $y_{m+1}(k) \geq 0$ and $y_{m+2}(k) \leq 0\, \forall k$ to ensure that the computed control variable $u(k)$ will be into the range estimated by FDI module. Thus,

$$y_{m+1}(k) = x_{n+1}(k) - u(k) \geq 0 \Rightarrow u(k) \leq x_{n+1}(k), \tag{25}$$

$$y_{m+2}(k) = x_{n+2}(k) - u(k) \leq 0 \Rightarrow u(k) \geq x_{n+2}(k). \tag{26}$$

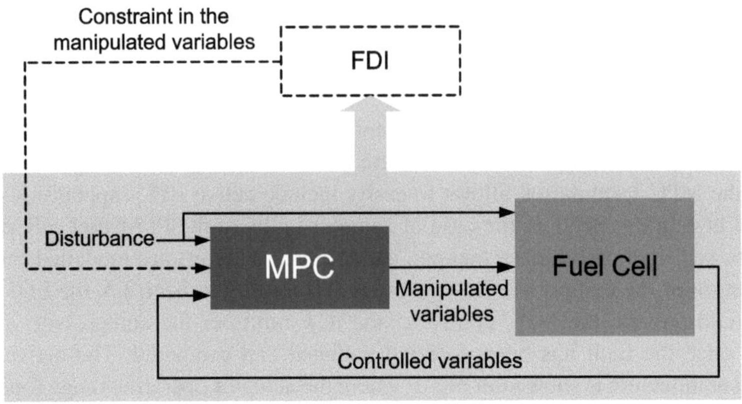

Fig. 7 Fault-tolerant MPC scheme for air compressor faults

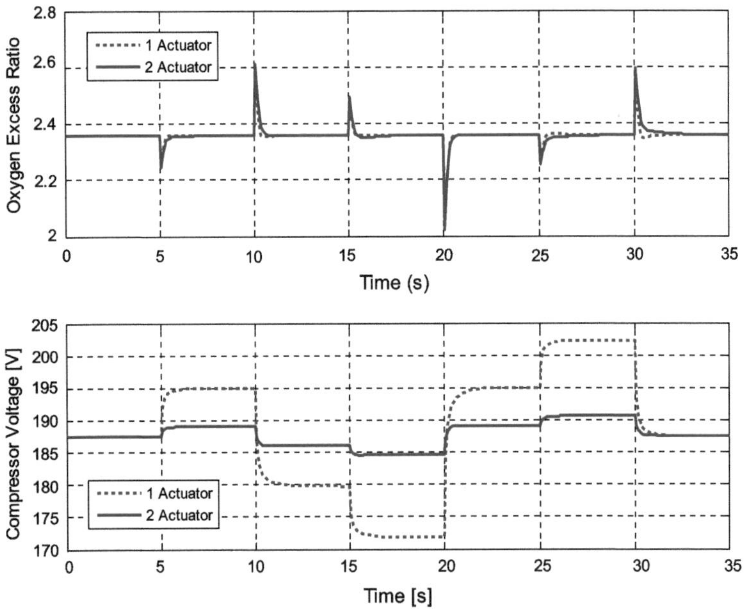

Fig. 8 Fault-tolerant MPC results in case of an actuator fault that limits the operating range to 0–75% of the original range

3.5 Results

Figures 8, 9, 10 show the simulation results of FTC scheme for several fault actuator scenarios considering the two control structures presented in Sect. 3.3. The current applied to the stack is the same than in the non-faulty scenario presented in Fig. 6a. The control action is shown in Fig. 8 when the actuator (air compressor) fault causes a

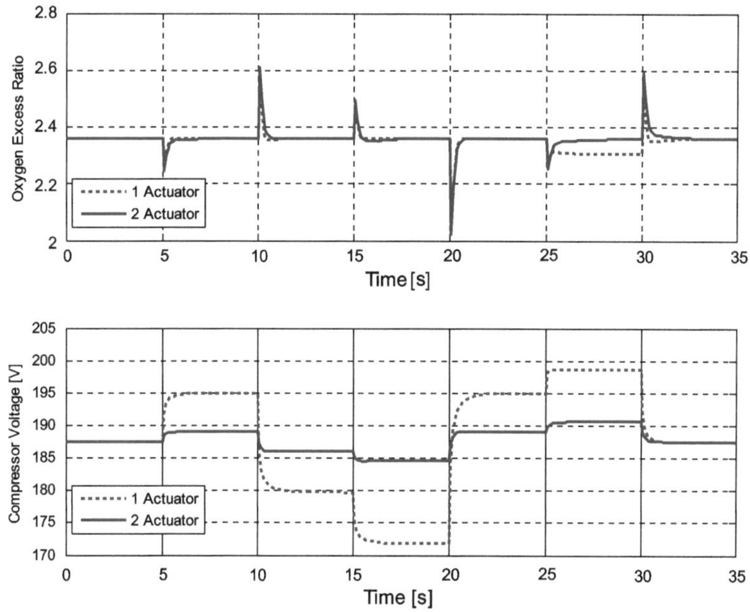

Fig. 9 Fault-tolerant MPC results in case of an actuator fault that limits the operating range to 0–50% of the original range

range reduction such that the upper limit of the range is reduced to 75% of the original one. In this case, there is no control degradation since the compressor voltage does not reach the upper range limit in any of the two considered controller architectures. Figure 9 shows the case corresponding to a reduction of the upper limit of 50% with respect to the original one. Now, the control degradation is visible in case of the controller architecture that uses only one actuator (the compressor) when the values of stack current are high. In the case of the two actuator architecture, since the voltage excursion of the controller does not reach the upper limit of the operating range, the controlled variable is not affected. Finally, the upper limit range is reduced to be 25% of the original one in Fig. 10. In this case, the control goal is highly affected in the case of using the one actuator architecture while in the case of the two actuators architecture the control results are still unaffected.

The results presented in the previous section show that the control architecture that uses two actuators offers a better fault tolerance against compressor faults. This is because to achieve the same control results, the compressor voltage changes are smaller. The inclusion of the cathode outlet valve opening adds additional degree of freedom to the system that can be exploited in case of non severe faults in the compressor.

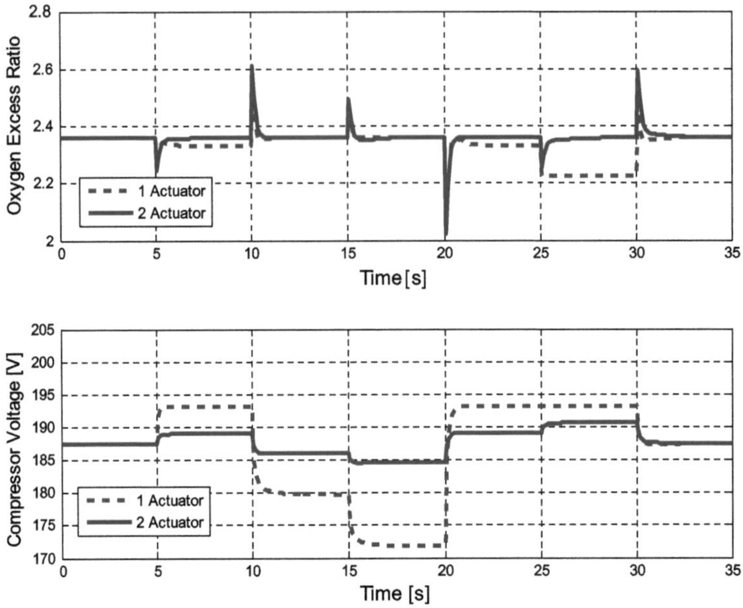

Fig. 10 Fault-tolerant MPC results in case of an actuator fault that limits the operating range to 0–25% of the original range

4 Conclusions

In the first section of this chapter, a new model-based fault diagnosis methodology based on the relative fault sensitivity has been presented and tested. An advantage of this new methodology is that it does not require the knowledge of the fault magnitude to provide a diagnostic. Furthermore, it allows to isolate faults although all the considered faults affect all the residuals, whenever the sensitivities are different.

To prove this methodology, a PEMFC simulator based on a model presented in the literature has been developed. The simulator was modified to include the proposed set of fault scenarios. This modified simulator allows imposing a determined fault scenario, within the considered set of faults, and analyzing its behavior. All the simulated faults have been tested with the new diagnosis methodology, which has diagnosed correctly the simulated faults in contrast with other well known methodologies using binary signature matrix of analytical residuals and faults, which do not permit to isolate the complete set of faults.

In the second section, fault-tolerance of MPC control for fuel cell systems has been addressed. MPC is a suitable control methodology to control fuel cell systems because of their multivariable and complex behavior. At the same time, MPC is one of the control methodologies that can introduce more easily fault-tolerance. However, the problem of including actuator fault-tolerance in the control loop of FCS had not been already addressed in the literature for those systems.

Here, a new control structure that not only uses the compressor voltage as a control variable but also the cathode outlet valve area has been proposed. It is shown that using this new control structure allows improving the control performance and at the same time allows introducing fault-tolerance against compressor faults. This result reinforces the idea of inclusion of the cathode output valve, which is a debatable aspect in the design of PEMFC. Finally, the proposed approach has been assessed on a simulation test bench based on a known PEMFC model.

Acknowledgments The author wish to thank to Joseba Quevedo, Viçenc Puig, Teresa Escobet and Salvador de Lira from the Sistemes Avançats de Control (SAC) group of the Universitat Politècnica de Catalunya and to Maria Serra and Jordi Riera from the Institut de Robòtica i Informàtica Industrial (IRI-UPC-CSIC).

References

1. Hissel D, Péra MC, Kauffmann JM (2004) Diagnosis of automotive fuel cell power generators. J Power Sour 128(2):239–246
2. Blanke M, Kinnaert M, Lunze J, Staroswiecki M (2006) Diagnosis and fault-tolerant control, 2nd edn. Springer, Heidelberg, Germany
3. Puig V, Quevedo J, Escobet T, Morcego B, Ocampo C (2004) Control tolerante a fallos (parte I): fundamentos y diagnostico de fallos. Revista Iberoamericana de Automatica e Informatica Industrial 1(1):15–31
4. Puig V, Quevedo J, Escobet T, Morcego B, Ocampo C (2004) Control tolerante a fallos (parte II): mecanismos de tolerancia y sistema supervisor. Revista Iberoamericana de Automatica e Informatica Industrial 1(2):5–21
5. Ding SX (2008) Model-based fault diagnosis techniques: design schemes, algorithms, and tools. Springer, Heidelberg, Germany
6. Gertler JJ, Inc NL (1998) Fault detection and diagnosis in engineering systems. Marcel Dekker, New York
7. Staroswiecki M, Comtet-Varga G (2001) Analytical redundancy relations for fault detection and isolation in algebraic dynamic systems. Automatica 37:687–699
8. Isermann R (2005) Model-based fault-detection and diagnosis-status and applications. Ann Rev Control 29(1):71–85
9. Chen J, Patton RJ (1999) Robust model-based fault diagnosis for dynamic systems. Kluwer Academics Publishers, Dordrecht, The Netherlands
10. Puig V, Quevedo J, Escobet T, Nejjari F, de las Heras S (2008) Passive robust fault detection of dynamic processes using interval models. Control Syst Technol IEEE Trans 16(5):1083–1089
11. Puig V, Quevedo J, Escobet T, Meseguer J (2006) Toward a better integration of passive robust interval-based fdi algorithms. In: In IFAC Safeprocess06, China
12. Pukrushpan J, Peng H, Stefanopoulou A (2004) Control-oriented modelling and analysis for automotive fuel cell systems. J Dyn Syst 126:14–25
13. Puig V, Feroldi D, Serra M, Quevedo J, Riera J (2008) Fault-tolerant MPC control of PEM fuel cells. In 17th World Congress IFAC, Seoul, Korea, p 11112–11117
14. Pukrushpan J, Stefanopoulou A, Peng H (2004) Control of fuel cell breathing: initial results on the oxygen starvation problem. IEEE Control Syst Mag 24:30–46
15. Pukrushpan JT, Peng H, Stefanopoulou AG (2002) Simulation and analysis of transient fuel cell system performance based on a dynamic reactant flow model. Proceedings of the ASME International Mechanical Engineering Congress & Exposition
16. Rodatz P, Paganelli G, Sciarretta A, Guzzella L (2004) Optimal power management of an experimental fuel cell/supercapacitors-powered hybrid vehicle. Control Eng Pract 13:41–53

17. Almeida PEM, Simoes MG (2005) Neural optimal control of PEM fuel cells with parametric CMAC networks. Ind Appl IEEE Trans 41(1):237–245
18. El-Sharkh M, Rahman A, Alam M, Byrne P, Sakla A, Thomas T (2004) A dynamic model for a stand-alone PEM fuel cell power plant for residential applications. J Power Sour 138:199–204
19. Bordons C, Arce A, del Real AJ (2006) Constrained predictive control strategies for PEM fuel cells. In: American Control Conference, 2006, p 6
20. Feroldi D, Serra M, Riera J (2007) Performance improvement of a pemfc system controlling the cathode outlet air flow. J Power Sour 169(1):205–212
21. Fouquet N, Doulet C, Nouillant C, Dauphin-Tanguy G, Ould-Bouamama B (2006) Model based PEM fuel cell state-of-health monitoring via ac impedance measurements. J Power Sour 159(2):905–913
22. Maciejowski JM (2002) Predictive control: with constraints. Prentice Hall, Harlow, UK
23. Pukrushpan JT, Stefanopoulou AG, Peng H (2004) Control of fuel cell power systems: principles, modeling, analysis and feedback design. Springer, Surrey, UK

Fuel Cell Hybrid Systems

Diego Feroldi

Abstract The control of fuel cell systems was studied in Chaps. 3, 6 analyzing the system composed by the fuel cell stack and its auxiliary subsystems (e.g., compressor, valves, etc.), with the following objectives: achieve high efficiency, reduce the hydrogen consumption, improve the dynamic behavior and guarantee its safe operation. We continue in this chapter and Chap. 8 with the study of fuel cell-based systems approaching the fuel cell hybrid systems with some energy storage. Hybridization has important advantages in fuel cell hybrid vehicles (FCHV), a fuel cell application that is central in this book. Therefore, the process of designing a hybrid system, or methodology of design, is addressed in this chapter. We concentrate our attention on FCHVs because this application is particularly attractive, although some general aspects studied in this chapter also apply to other applications such as stand-alone residential PEM fuel cell power systems.

1 Introduction

In a fuel cell hybrid system (FCHS), an energy storage system (ESS), e.g., a battery or a SC bank, is added to the fuel cell system in order to increase the efficiency and performance of the power generation system. The main concepts about hybridization were explained in Sect. 3 of Chap. 1, showing the advantages of this configuration. In particular, hybridization has important advantages in fuel cell hybrid vehicles (FCHV), a particular type among FCHS and a fuel cell application that is central in this book.

Therefore, the process of designing a hybrid system, addressed as a methodology of design oriented to automotive applications, is covered in this chapter. We concentrate our attention on FCHV because this application is particularly attractive,

D. Feroldi (✉)
CAPEG-CIFASIS-(CONICET-UNR-UPCAM) and DCC-FCEIA-UNR,
27 de Febrero 210 bis, S2000EZP Rosario, Argentine
e-mail: feroldi@cifasis-conicet.gov.ar

M. S. Basualdo et al. (eds.), *PEM Fuel Cells with Bio-Ethanol Processor Systems*, Green Energy and Technology, DOI: 10.1007/978-1-84996-184-4_7,

although some general aspects studied in this chapter also apply to other applications such as stand-alone residential PEMFC power systems.

2 Advantages of Hybridization in Fuel Cell Hybrid Vehicles

The hybridization in fuel cell-based systems should produce an increase in the efficiency over a large operating range. Particularly, three different mechanisms allow this efficiency improvement in FCHVs, which are described below:

Regenerative braking: In automotive applications, it is possible to recover energy from regenerative braking, allowing a considerable improvement in the hydrogen economy. The recovered energy from regenerative braking may represent an important fraction of the economy improvement. However, the amount of energy that it is possible to recover is strongly dependent on the driving profile. In urban driving cycles, in which there are abundant brakings, the percentage of regenerable energy is much greater than in non-urban cycles where this percentage is quite small.

Energy management strategy: The efficiency of an FCS is maximum at a fraction of the rated power. Therefore, it is convenient to operate the fuel cell in its maximum efficiency zone. Thus, taking advantage of an auxiliary power source and an adequate energy strategy, it is possible to maintain the fuel cell operating at its maximum efficiency region most of the time and reduce the hydrogen consumption.

Fuel cell downsizing: In a pure fuel cell-based power generation system, the FCS must be sized to meet the maximum load. When the load is strongly variable (e.g., in automotive applications or stand-alone residential applications), the FCS is working, in average, well below its rated power. One important advantage of hybridization is that it allows a diminution of the rated power of the fuel cell stack in these types of applications.

Therefore, it is possible to meet the demand with a smaller FCS in combination with an energy storage system. In addition to this advantage, downsizing the FCS has an impact on the hydrogen economy in three ways. In fact, downsizing leads to the following advantages: (1) improvements in the hydrogen economy when the peak efficiency of the FC stack is shifted to better match with the power requirements from a given driving cycle, (2) a smaller compressor, reducing the parasitic losses, and (3) a diminution in the fuel cell system mass, reducing the total mass.

Although the main motivation for introducing hybridization in fuel cell systems is to improve the hydrogen economy, the benefits are not limited to that. In fact, hybridization is useful to solve two important problems in fuel cell control, which are described below.

2.1 Improvement of the Transient Response

The dynamics of FCS are relatively slow, mainly because of the dynamics of the air compressor and the manifold-filling dynamics [1]. In this context, an additional energy flow from an energy storage device with high-specific power[1] is useful to compensate this drawback. In the literature, this issue is not completely resolved. In [2], it is concluded that it is feasible to use an FCS in a load-following mode without any storage device. In that work, the conclusions are based on results obtained from experimental data taken from a scaled-down FCS and a particular standard driving cycle. It is remarked that it is possible to take out short bursts of power from the stack, due to the *"charge double layer"*[2] CDL capacitance. However, other publications, for example [4–6], conclude that the use of energy storage devices is beneficial from both the system dynamics and the hydrogen economy.

2.2 *Prevention of "Oxygen Starvation"*

Oxygen starvation is a complicated phenomenon that occurs when the oxygen partial pressure falls below a critical level at any point in the cathode. It can be produced by an abrupt increase in the current demand (that causes a rapid increase in the oxygen consumption due to the cathode electrochemical reaction) and an insufficient air supply. This phenomenon can be observed through a rapid decrease in the cell voltage although in severe cases it can produce a short circuit and a hot spot on the surface of the membrane cell [7]. The phenomenon is not well described in the literature and more information from electrochemical studies is needed to be able to quantify the critical level of oxygen concentration below which permanent damage is produced or, perhaps more probably, the combination of level and time. Nevertheless, hybridization can help in avoiding oxygen starvation by means of the additional energy supply from the energy storage elements. With these elements, the peaks of actual current demand to the fuel cell in a hybrid system are lower than in the pure fuel cell case where no auxiliary power source is present.

3 Electrical Structure of Fuel Cell Hybrid System

The electrical structure of a fuel cell-based vehicle involves, essentially, a fuel cell stack with its auxiliary systems and the load, which is generally an alternating current (AC) electrical motor. The FCS by itself is an electrical power source, whose direct current (DC) output voltage drops with the current according to the polarization curve. Thus, it is necessary to incorporate power converters to convert the voltage from DC to AC with the appropriate frequency and voltage level.

---

[1] The specific power is defined as power per unit of mass (W kg^{-1}).

[2] The *"charge double layer"* (CDL) is an electrode phenomenon: when two different materials are in contact there is a build-up of charge on the surfaces or a charge transfer from one to the other. The effect is that if the current suddenly changes, the voltage shows an immediate change due to the internal resistance, but moves fairly slowly due to this accumulated charge. One way of modeling this is by using an equivalent circuit, with the CDL represented by an electrical capacitor [3].

The electrical structure for an FCHV also includes an energy storage system (e.g., a battery bank, a SC bank or a combination of battery and SCs), with the possibility of stocking energy recovered from braking and energy from the fuel cell. This feature requires that the converter that connects the ESS is a bidirectional converter to allow the charge and discharge of the ESS.

In this section, we approach the design of the electrical structure for an FCHV focusing on the determination of the electrical topology Sect. 3.1 and the selection of the energy storage system Sect. 3.2. The detailed design of the power converters and their control is beyond the scope of this work.

3.1 Topology of the Electrical Structure

The topology of an electric system is given by the interconnections of the system components. The selection of the most adequate topology is discussed in [8–11], where different structures and their appropriate controls are analyzed, also the advantages and disadvantages of each case are shown. Thus, several topologies are considered in the literature to connect the ESS and the FCS to the load, depending on the following issues:

- Characteristics of the load (DC or AC voltage, single-phase or three-phase and range of voltage).
- Possibility of energy recovering from the load (e.g., regenerative braking).
- Range of voltage in the ESS.
- Output voltage of the FCS.

In this work, we analyzed the better option for an FCHV and a number of options, represented by the decision tree shown in Fig. 1, were considered before choosing the topology of the electrical system, which is shown in Fig. 2.

The main decision was if the vehicle has regenerative braking. This electrical topology is used as the frame of reference for the rest of this chapter and for the study of the energy management strategies for FCHVs in Sect. 1 of Chap. 8.

In this topology, the FCS is connected to the DC bus through a step-up power converter (*Boost* converter) because the DC voltage bus is normally of high voltage, whereas the ESS is connected to the DC through a bi-directional power converter (*Buck-Boost* converter) allowing the charge and discharge of the ESS. With regard to the load, which is an AC induction motor, it is fed through a DC–AC inverter.

The power converter that connects the ESS to the DC bus is fundamental to implement the energy management strategy in the hybrid system: this converter acts as a "*switch*" that allows to regulate the energy flow between the ESS and the DC bus. In the same way, the converter that connects the FCS to the DC bus allows the regulation of the power flow from the FCS and, besides, has to cope with the variations in the FCS output voltage since the FCS does not act as an ideal voltage source.

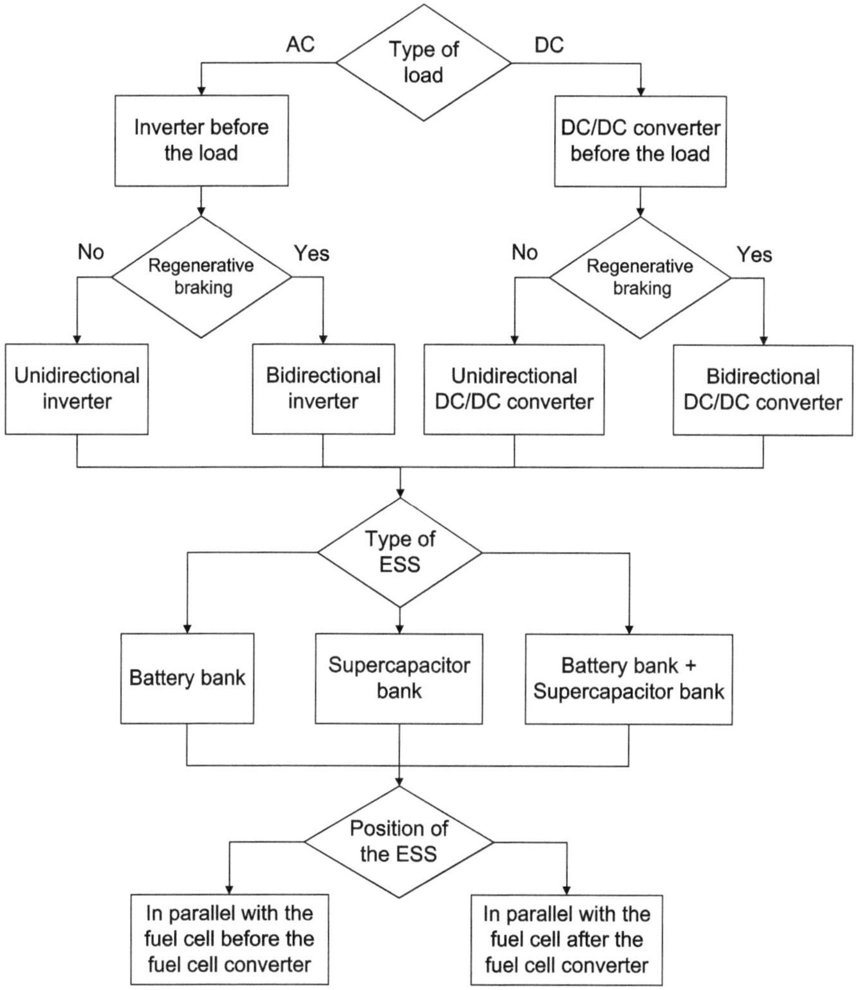

Fig. 1 Decision tree for the selection of the electrical structure for the FCVH

3.2 Energy Storage System: Batteries Versus Supercapacitors

The energy storage system (ESS) in the fuel cell hybrid system can be implemented either by a high-specific energy device like batteries or by a high-specific power device like Supercapacitors (SCs). There is also the possibility of a combined storage bank composed of batteries and capacitors [12]. The SCs present significantly better specific energy than conventional capacitors and much better specific power than conventional batteries. The capacitance of SCs may vary from a few Farads to several thousand Farads per cell [13].

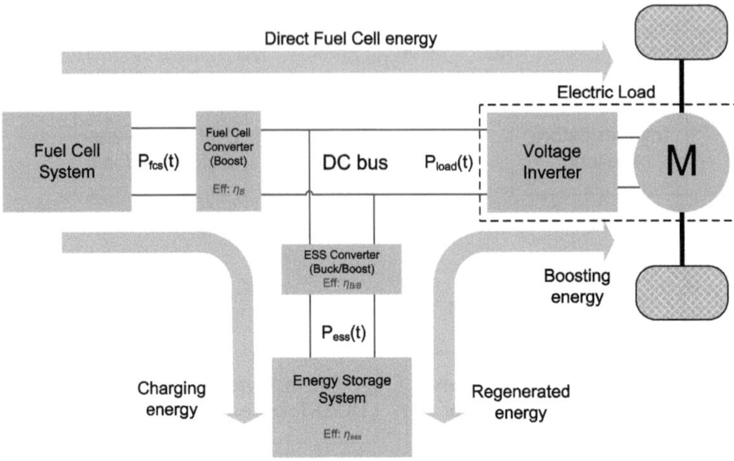

Fig. 2 Diagram showing the electrical topology and the energy and power flows in a FCHV

Before dealing with this analysis, two meaningful parameters concerning the ESS, are introduced: the hybridization degree (HD) and the power/energy storage system ratio (P/E).

3.2.1 Hybridization Degree

The *Hybridization degree* is defined by a relationship between two installed powers: the *maximum FCS power* ($P_{\text{fcs,max}}$) and the *maximum ESS power* ($P_{\text{ess,max}}$). We use the following expression:

$$\text{HD} = \frac{P_{\text{ess,max}}}{P_{\text{fcs,max}} + P_{\text{ess,max}}} \times 100\,[\%]. \tag{1}$$

With this definition a 100% HD indicates a vehicle without FC (only storage bank), and 0% HD indicates a pure fuel cell vehicle (no storage bank).

3.2.2 Power/Energy Storage System Ratio (P/E)

The power/energy storage system ratio is defined as the relationship between the *specific power (P)* and the *specific energy (E)* of the storage system [14]:

$$\frac{P}{E} = \frac{P[\text{W kg}^{-1}]}{E[\text{Wh kg}^{-1}]} \quad [\text{W Wh}^{-1}]. \tag{2}$$

Some relevant examples are collected from the literature [14] and reported in Table 1. Actually, for some SCs on the market, such as the model BCAP0350 from

Table 1 Characteristics of some relevant energy storage systems examples

	P/E [W W h^{-1}]	Specific power [W kg^{-1}]	Specific energy [W h kg^{-1}]
High power lead-acid batteries	10	300	30
Ni–Mh batteries	10	4–500	50–70
Supercapacitors	>100	up to 1,000	up to 5

Maxwell Technologies [15], the specific power is 3,900 W kg^{-1} and a recent study with asymetric double-layer SC reports advances in specific energy, approaching 40 W h kg^{-1} [16].

We choose SCs in this work because they are more attractive for applications with low energy and high power demand, particularly at low temperatures, although cost may remain an obstacle. SCs can be deeply charged and discharged at high rates for 500,000–1,000,000 cycles with a relatively small change in characteristics (10–20% of degradation in capacitance and resistance) and with high charge/discharge efficiency [13]. However, batteries and SCs do not compete but are complementary in the development of Hybrid Electric Vehicles (HEVs) because of the high-specific energy of the former and the high-specific power of the latter, notwithstanding the significantly longer cycle-life of SCs (more than two orders of magnitude [17]).

Different types of HEV require energy storage systems characterized by different specific power and energy, *P/E* ratio and cycle-life. In the work of Gao [18], two Fuel Cell Hybrid Vehicles (FCHVs) are studied, comparing a fuel cell-battery hybrid powertrain and a fuel cell-SC hybrid powertrain. The conclusion is that the second option is better since SCs can more effectively assist fuel cells to meet transient power demand. Since SCs have high specific power, they can supply great quantities of instantaneous power, allowing to boost the fuel cell in power-demanding situations, such as fast accelerations or climbing. Although the study of Gao is done with a fuel cell-battery hybrid powertrain composed of lead-acid batteries instead of other types of batteries with higher specific power (e.g., Li–ion or Ni–MH), the general conclusion still is valid because of the exceptional specific power of SCs. In [19], analyzing the requirements of batteries for city hybrid buses, it is remarked that to get a really significant fuel consumption improvement it is essential to be able to recover a good percentage of the available energy during braking and, for that end, devices with high power recharge capability are required.

With regard to the lifetime, it is concluded that actual batteries are not good for automotive applications. However, some modifications in the chemistry of conventional lead-acid batteries are being analyzed in order to operate in high-rate partial-state-of-charge mode with acceptable performance [20]. There are also ultrabatteries in development which are hybrid energy-storage devices combining an asymmetric supercapacitor[3] and a lead-acid battery in one unit taking advantage of the best from both technologies without the need for extra electronic controls [21].

[3] The term *asymmetric supercapacitors* stands for SCs with different types of electrodes.

Table 2 Vehicle specifications in the case of study

Specification	Symbol	Value	Unit
Vehicle total mass	m_{T}	1,380	kg
Vehicle mass[a]	m_{veh}	882	kg
Frontal area	A_{f}	2	m^2
Drag coefficient	C_{d}	0.335	–
Coefficient of rolling friction	f_{r}	0.009	–
Air density	ρ_{a}	1.2	kg m^{-3}
Gravity	g	9.8	m s^{-2}

[a] Vehicle mass without taking into account the FCS mass and the ESS mass

4 Modeling of Fuel Cell Hybrid Systems

In order to study FCHS it is necessary to rely on accurate and practical models to describe the system behavior. As it was mentioned, FCHV are one of the main applications of hybrid systems based on PEM fuel cell and, thus, this section is devoted to FCHS modeling oriented to automotive applications.

ADVISOR is a system analysis tool for vehicle modeling created by the National Renewable Energy Laboratory (USA) in the MATLAB/Simulink environment [22, 23]. It provides a flexible and robust set of models, data and script text files, which are used to quantify the fuel economy, the performance and the emissions of vehicles that use alternative technologies including fuel cells, batteries, SCs, electric motors and internal combustion engines in hybrid configurations.

In this platform, the vehicle is modeled in detail taking into account the several components that compose the vehicle drivetrain (i.e., vehicle, wheel and axle, final drive, gearbox, clutch, fuel converter, etc.). For example, Eq. 3 determines the mechanical power necessary to meet the desired speed v, with road slope α and acceleration $\dot{v}(t)$:

$$P_{\mathrm{mec}}(t) = \left(m_{\mathrm{T}} g f_{\mathrm{r}} + \frac{1}{2}\rho_{\mathrm{a}} C_d A_{\mathrm{f}} v(t)^2 + m_{\mathrm{T}} g \sin(\alpha(t)) + m_{\mathrm{T}} \dot{v}(t) \right) v(t). \quad (3)$$

This equation performs the force balance among the rolling resistance force, the aerodynamic drag force, the force of gravity that must be overcome to climb a grade and the force required to accelerate, respectively. The meaning of the constants in Eq. 3 are explained in Table 2. Actually, the necessary power supplied by the fuel cell is greater than that determined in Eq. 3 because of the power losses in the final drive, the gearbox and the motor. These losses depend on the efficiencies of the components and ADVISOR utilizes efficiency maps to compute the power losses according to the present speed.

In this work, we consider the performance of an FCHV based on a small car and the entire system being modeled in ADVISOR according to the principal parameters listed in Table 2. This model is used as a case of study in Sects. 5 and 6 to exemplify the methodologies proposed.

4.1 Fuel Cell System model

The FCS consists of the fuel cell stack, which is a serially layered pack of fuel cells, and the auxiliary system, a set of devices necessary for the FCS operation that include a compressor, cooling/heating devices and a water management system. Among the auxiliary components, the compressor is the one with the largest power consumption (up to 93.5% of the total auxiliary power consumption [24]). Therefore, the FCS is a complex and nonlinear system whose dynamical model, like the one developed by Pukrushpan et al. [25], is too sophisticated to be used in an energy management strategy. Even a much simpler model, like the one developed by Correa et al. [26], which does not contemplate the filling dynamics of cathode and anode and the compressor dynamics, is unsuitable for that purpose. Moreover, the standard driving cycles in ADVISOR are defined every 1 s and the transient response of the FCS generated power is shorter than 1 s. Thus, a static model as the one employed in the ADVISOR platform is adequate to represent the FCS when the objective is to implement adequate management strategies.

The most relevant characteristic of the FCS, with a major impact on its performance, is the hydrogen consumption: the relation between the delivered power and the hydrogen consumption (see Fig. 3). This consumption map constitutes a static characteristic of the FCS and it does not take into account any others. It is assumed that the FCS is controlled by its own controller, a control algorithm that deals with the FCS control variables to maintain the system operating properly with the load changes. This issue has been covered in Chap. 3.

Given that in the hybridization degree analysis it is necessary to account with models representing FCS of different sizes, it is needed to scale the original model to represent FCSs with different rated powers. In the work of Kim et al. [24], it is concluded that it is possible to linearly scale the efficiency map of the FCS. This conclusion is reinforced by the work of Ahluwalia et al. [27], where four FCSs with different rated powers in study are also linearly scaled, and is also the way that ADVISOR toolbox works.

This consideration of a linear relationship between different scaled FCS is possible because of the followings reasons. First, the rated power of the FCS depends on the number of cells and the active area of each cell of the fuel cell stack. The polarization curve is practically unaffected by the cell number in contrast to the active area of the cells, which affects the design of the reactant flow channels, influencing the humidity and thermal characteristics of the stack and, consequently, changing the polarization curve [24]. Thus, it is possible to account for fuel cell stacks of different rated powers and polarization curve linearly scalable by changing the cell number. Secondly, it is also possible to linearly scale the efficiency map of the compressor, which is, as already said, the auxiliary component with major consumption.

A set of hydrogen consumption maps and a set of efficiency maps are plotted in Fig. 3 for different FCS rated powers, using data from ADVISOR. The efficiency maps are derived from the consumption maps:

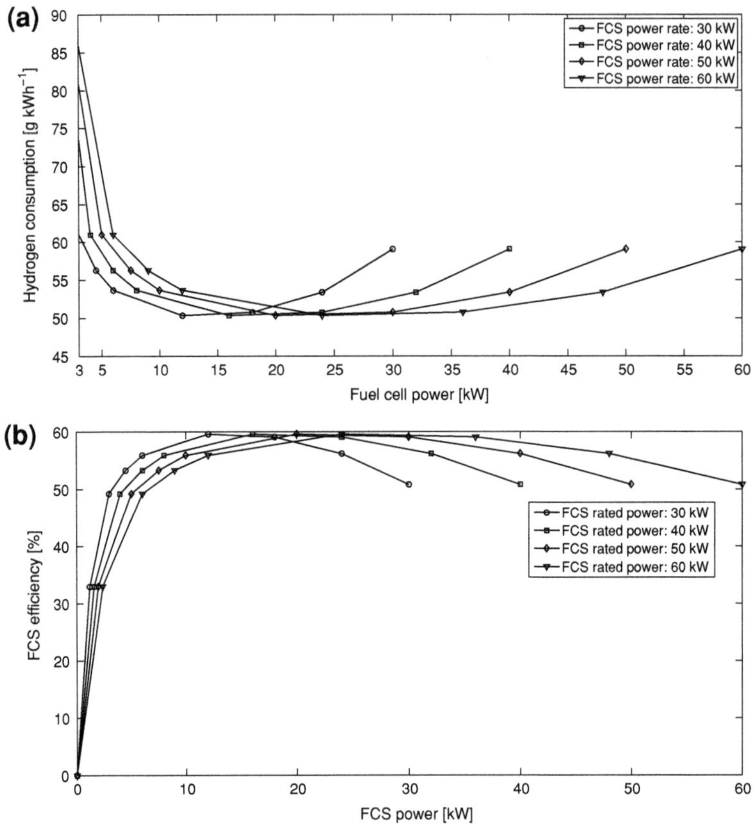

Fig. 3 Hydrogen consumption and efficiency maps for different rated powers. **(a)** Hydrogen consumption maps **(b)** Efficiency maps

$$\eta_{\text{fcs}}[\%] = \frac{360 \times 10^3}{\text{Cons}_{\text{H}_2} \times LHV_{\text{H}_2}}, \tag{4}$$

where the hydrogen consumption $Cons_{\text{H}_2}$ is expressed in g kW h^{-1} and LHV_{H2} is $120\,\text{kJ g}^{-1}$. In the four efficiency maps, the efficiency at rated power is 50% and the maximum efficiency 60% is at 25% of the rated power.

4.2 Energy Storage System model

The energy storage system, whether based on batteries or SCs, can be modeled with two subsystems. One subsystem takes into account the system efficiency, both during charging and discharging, and the other subsystem computes the actual energy level. The energy stored in the storage system, $E_{\text{ess}}(t)$, results in integrating $P_{\text{ess}}(t)$:

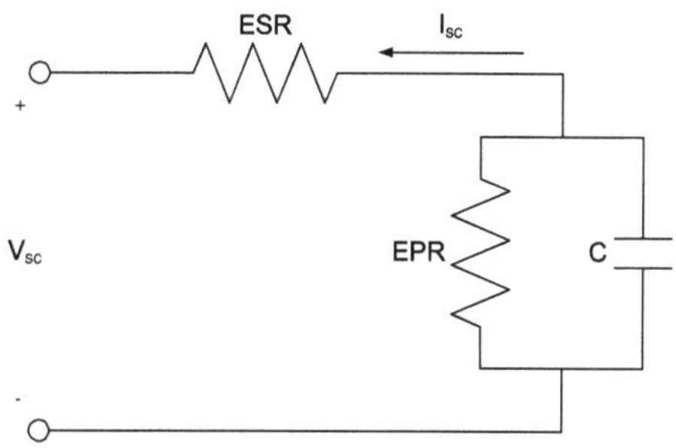

Fig. 4 Equivalent model for the supercapacitor unit

$$E_{\text{ess}}(t) = E_{\text{ess}}(0) + \int_0^{t_c} P_{\text{ess}}(t)dt. \tag{5}$$

It is useful to define a parameter that indicates the relative amount of energy in the ESS. The State of Energy, SoE(t), is defined as

$$\text{SoE}(t) = \frac{E_{\text{ess}}(t)}{E_{\text{cap}}} \times 100 \quad [\%], \tag{6}$$

where E_{cap} is the maximum energy that the storage system is capable of storing. In discrete time the storage system energy, at the time instant k, is

$$E_{\text{ess}}(k) = E_{\text{ess}}(0) + \sum_{k=1}^{N_c} P_{\text{ess}}(k)\Delta t, \tag{7}$$

where $N_c = t_c/\Delta t$, t_c is the duration of the driving cycle, and Δt is the sampling time.The SoE(k) results as

$$\text{SoE(k)} = \frac{E_{\text{ess}}(k)}{E_{\text{cap}}} \times 100 \quad [\%]. \tag{8}$$

Figure 4 shows an electrical equivalent model for an ESS composed of SCs. Actually, the equivalent circuit is more complex, like that used in [28], but for the purpose of designing and analyzing the performance of the system, the model in Fig. 4 is a good approximation. In fact, this model is similar to that employed in [29] for these purposes.

The ESS efficiency, in this case a SC bank, is related to the equivalent series resistance (ESR). For a pure capacitor, having zero ESR, the efficiency of charge or discharge would be 100%. However, in a real capacitor, having nonzero ESR, the irreversible dissipation of power is ESR $\times I_{\text{sc}}^2$, giving efficiencies lower than 100%. The equivalent parallel resistance (EPR) represents the self-discharge losses and only impacts the long-term energy storage performance of the SC and is extremely high [30]. Thus, the time constant of the period of charge/discharge can be expressed by

ESR × C. Since ESR is as low as 0.019 Ω and C is 58 F [31], it is possible to charge and discharge the SC in a very short time. The energy stored in the SC is directly proportional to the capacitance and the squared voltage:

$$E_{SC}(t) = \frac{1}{2} C V_{sc}^2(t). \tag{9}$$

Besides, the power that the SC is capable to deliver or store is inversely proportional to the internal resistance and directly proportional to the squared voltage:

$$P_{SC}(t) = \frac{k_{SC} V_{sc}^2(t)}{ESR}, \tag{10}$$

where k_{SC} is a constant depending on the particular device. Thus, it is possible to express the power as a function of the energy in the SC, instead of $V_{sc}^2(t)$:

$$P_{SC}(t) = \frac{2k E_{SC}(t)}{ESR \cdot C}, \tag{11}$$

an expression that is more convenient, because the proportional relationship between power and energy. In this way, we work with the state of energy, SoE(t), instead of the state of charge, SoC(t).[4]

The ESS can be conformed using several equal SCs and connecting them in series and parallel. The total capacitance and total resistance of the bank are calculated as

$$R_{\text{ess,total}} = \#s \frac{ESR}{\#p}, \tag{12}$$

$$C_{\text{ess,total}} = \#p \frac{C}{\#s}, \tag{13}$$

where $\#s$ is the number of SCs connected in series and $\#p$ is the number of series string connected in parallel (see Fig. 5).

The SC bank voltage is

$$V_{SC,\text{total}} = \#s \cdot V_{SC,\max} \tag{14}$$

and the total system energy results

$$E_{SC,\text{total}} = \#_{SC} \cdot E_{SC}, \tag{15}$$

where $\#_{SC} = \#p \cdot \#s$ is the total number of SCs connected in the bank.

[4] The state of charge, SoC(t), is defined as the relation between the actual charge voltage and the maximum charge voltage in a capacitor, whereas the SoE(t) is defined as the relation between the actual energy and the maximum charge energy.

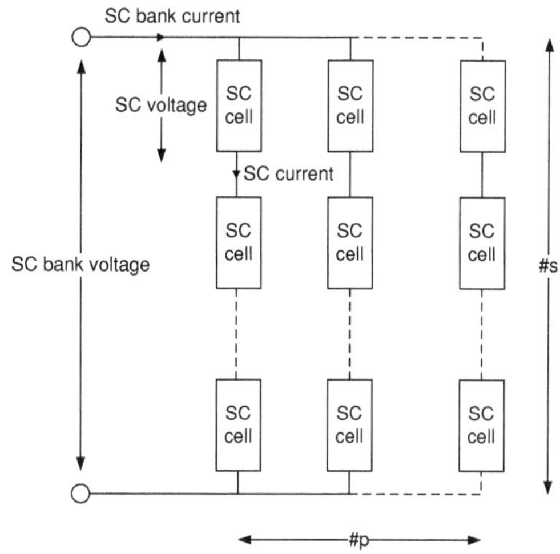

Fig. 5 Distribution of supercapacitors in a SC bank

4.3 Characteristic Load Demand

In order to evaluate the performance of a given FCHV, the standard drive cycles are widely utilized in the literature. These standardized speed profiles, representing urban and highway scenarios, were originally stated for measuring pollutant emissions and gasoline economy of engines [32]. Four standard driving cycles are plotted in Figs. 6, 7: the *new European driving cycle* (*NEDC*), the *urban dynamometer driving schedule* (*UDDS*),[5] the *federal test procedure* (*FTP*), and the *highway fuel economy cycle* (*HWFET*).

In each figure, two magnitudes are plotted: the speed profile and the demanded power to fulfill the profile. The calculation of the power demand is done exploiting the model developed in ADVISOR according to the parameter of the small vehicle described in Table 2 (vehicle total mass of 1,380 kg).

The NEDC cycle includes four intervals representing urban driving conditions, repeated without interruption, followed by one interval representing extra urban driving conditions with a more aggressive, high speed, profile. The UDDS cycle and the FTP represent city driving conditions, whereas the HWET cycle represents highway driving conditions. In Table 3, the key statistics about the driving cycles are shown concerning speed, acceleration and deceleration.

The power histogram corresponding to the four driving cycles are plotted in Fig. 8. It is observed that most of the time the FCS is operating in an unfavorable zone. This fact reaffirms the idea of implementing an adequate energy management strategy to overcome this drawback. This issue is studied in Sect. 1 of Chap. 8.

During a driving cycle, the power required by the vehicle varies strongly from very low powers to relatively high powers. High power requirements take place during

[5] It is also known as *federal urban driving schedule* (*FUDS*)

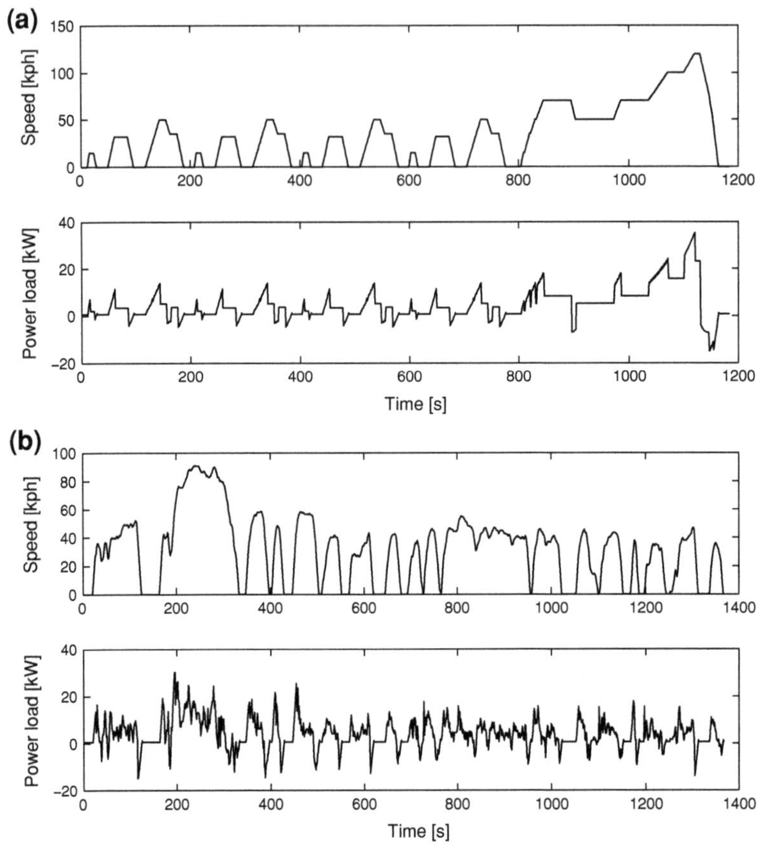

Fig. 6 Speed and power profiles for the NEDC and the UDDS standard driving cycles for a vehicle of total mass 1,380 kg. **a** new European driving cycle (NEDC) **b** urban dynamometer driving schedule (UDDS)

Table 3 Key statistics of driving cycles

	NEDC	UDDS	FTP	HWFET
Distance [km]	10.93	11.99	17.77	16.51
Maximum speed [km h^{-1}]	120	91.25	91.25	96.4
Average speed [km h^{-1}]	33.21	31.51	25.82	77.58
Maximum acceleration [m s^{-2}]	1.06	1.48	1.48	1.43
Maximum deceleration [m s^{-2}]	−1.39	−1.48	−1.48	−1.48
Average acceleration [m s^{-2}]	0.54	0.50	0.51	0.19
Average deceleration [m s^{-2}]	−0.79	−0.58	−0.58	−0.22

a relatively short fraction of time. If there is no energy storage, the FCS must meet the highest peak power and, therefore, the FCS is oversized most of the time. In addition, the efficiency of an FCS is strongly degraded at low powers (see Fig. 9).

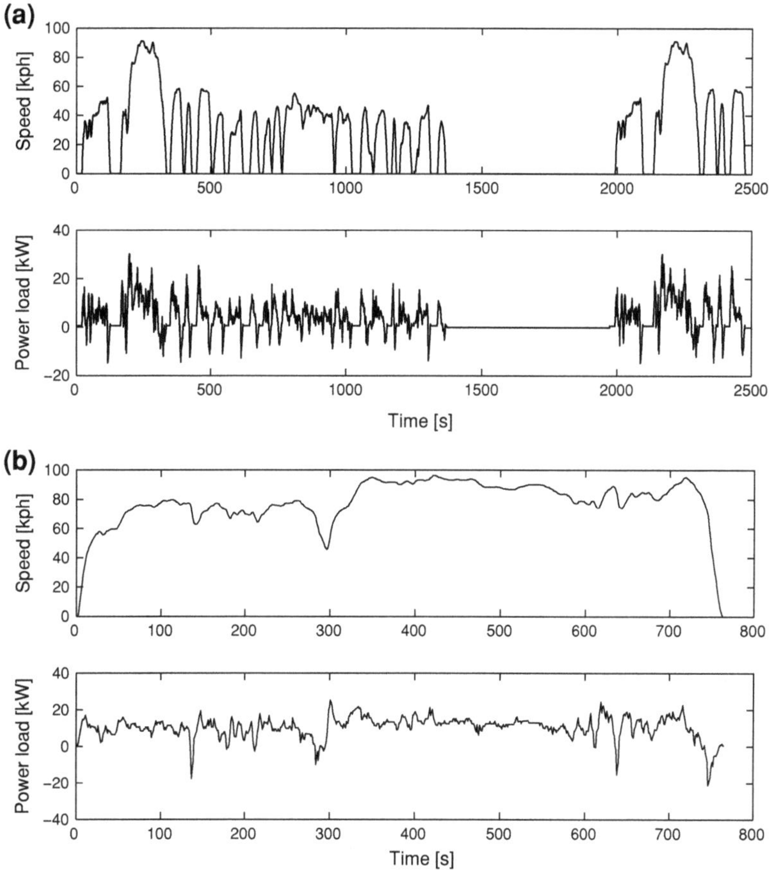

Fig. 7 Speed and power profiles for the FTP and the HWET standard driving cycles for a vehicle of total mass 1,380 kg. **a** federal test procedure (FTP) **b** highway fuel economy cycle (HWFET)

Thus, if no hybridization is present, the FCS has to work in large periods of time at a low efficiency zone. On the contrary, with an additional power source and a suitable energy management strategy it is possible to avoid these unfavorable operating zones. Thus, the hydrogen economy can be improved through hybridization.

Analyzing a typical efficiency curve for a 50-kW FCS according to ADVISOR model (see Fig. 9), it is concluded that the energy management strategy has to avoid the FCS operation at powers smaller than 5 kW because the efficiency is strongly degraded below 50%. With this model (Fig. 9) of FCS the maximum efficiency is obtained operating at 20 kW. In this case, the FCS efficiency is approximately 60%. Within the range of 5–50 kW, the efficiency is high enough, close to the maximum value. Therefore, a good strategy would be one that allows the FCHS to work the longest possible periods of time close to the point of maximum efficiency. This strategy would depend on the ESS state-of-energy. For example, if the load power is greater than the maximum efficiency power and there is no sufficient stored energy to

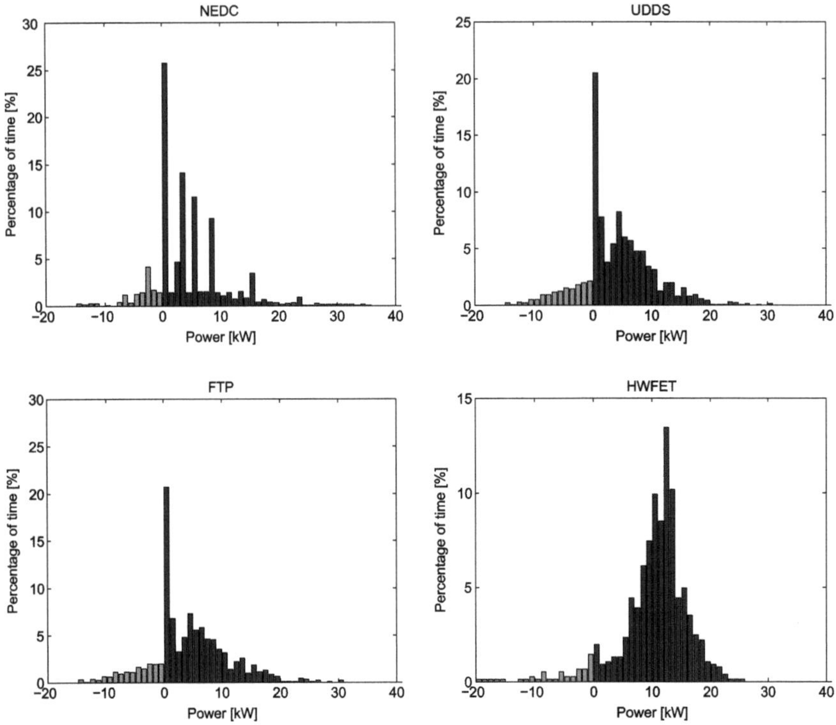

Fig. 8 Power histogram for the vehicle in study of total mass1,380 kg running in different standard cycles

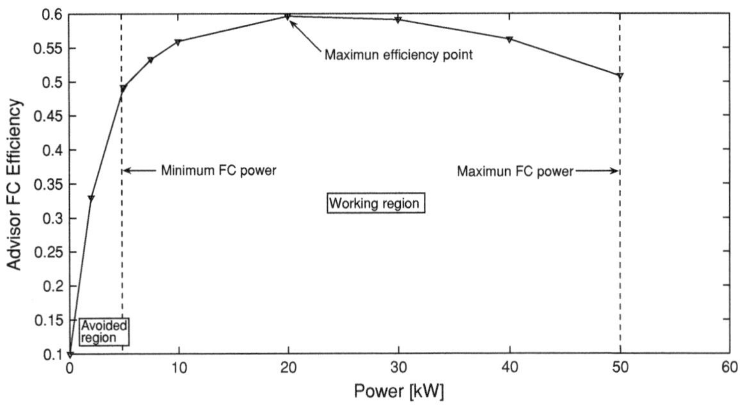

Fig. 9 Efficiency curve for a 50-kW FCS from ADVISOR model

meet the power balance, then, the strategy has to move the operating point toward the zone of higher power although the FCS efficiency is worse. The opposite situation takes place if the load power is lower than the maximum efficiency power and there is no sufficient storage capacity in the ESS.

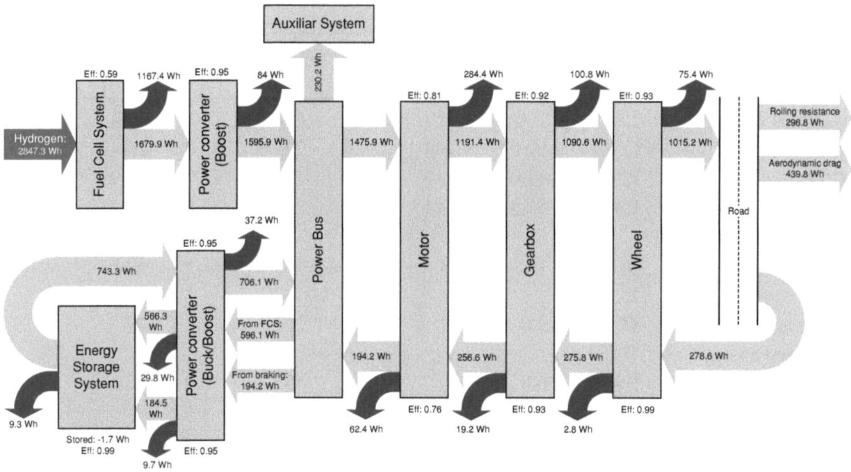

Fig. 10 Energy flow in the FCHV running on NEDC

The question is how to determine the most convenient operating point at every time. It is necessary to formulate a functional cost that quantifies the increase of the hydrogen consumption produced by the displacement from the maximum efficiency point. But in addition, it is necessary to relate this functional to the state of energy in the ESS, not only considering whether at the present time the state of energy is suitable to meet the power balance, but also considering the more foresighted point of view of the future power requirements. This is difficult to achieve because the driving cycle is not known a priori. This point is studied in Sect. 1 of Chap. 8.

5 Analysis of the Energy Flows in the Fuel Cell Hybrid System

In Sect. 3 of Chap. 1, it was stated that in an FCHV there are four principal energy flows between the fuel cell system, the energy storage system and the electric load. Entering into more detail, the energy flow from the hydrogen tank towards the wheels through the power bus, the electric motor and the gearbox can be seen in Fig. 10 for the case of study with the main parameters listed in Table 2, running on the NEDC. The number in the arrows corresponds to the total energy transferred between components or the total losses in the components along the load profile in study. Each component has its own losses, except the power bus which is considered ideal. Thus, the initial amount of energy from the hydrogen tank is degraded. Besides, the vehicle employs energy to overcome the rolling resistance and the aerodynamics drag force. In the same figure, it is also represented the wheel-to-ESS flow, the energy flow from regenerative braking to the ESS, which also diminishes due to the losses in the components.

Table 4 Energy flow summary in a 1,380 kg FCHV

	NEDC		UDDS		FTP		HWFET	
	[Wh]	[%][a]	[Wh]	[%][a]	[Wh]	[%][a]	[Wh]	[%][a]
Hydrogen energy	2,847	100	2,957	100	4,416	100	4,141	100
FCS energy out	1,680	59.0	1,744	59.0	2,605	59.0	2,442	59.0
FCS losses	1,167	41.0	1,213	41.0	1,811	41.0	1,699	41.0
Motor energy in	1,476	51.8	1,611	54.5	2,419	54.8	2,148	51.8
Total losses in power mode[b]	461	16.2	538	18.2	775	17.5	622	15.0
Total losses[c] in regenerative mode	84	3.0	144	4.9	202	4.6	39	0.9
Recovered energy from braking	194	6.8	310	10.5	452	10.2	86	2.1
Rolling resistance energy	297	9.4	326	11.0	482	10.9	448	10.8
Aerodynamic drag energy	444	15.4	293	9.9	508	11.5	953	23.0

[a] The % column refers to the percentage of energy with respect to the hydrogen energy.
[b] The total losses in power mode is the summation of the losses in the direct path from the FCS to the wheels.
[c] The total losses in regenerative mode is the summation of the losses in the path from the wheels to the ESS.

As mentioned in Sect. 3.1 of Chap. 1 the amount of energy that it is possible to recover from regenerative braking depends on the particular driving cycle. Our results show that for an FCHV according to the parameters of Table 2 (total mass of 1,380 kg), the maximum recoverable energy represents the 7.6% of the hydrogen energy that is spent to fulfill the NEDC cycle, whereas, in the case of UDDS and FTP, it represents the 11.7 and 11.4%, respectively. On the other hand, in a highway cycle like HWET, the regenerative braking energy only represents 2.3% of the hydrogen energy. Some works report greater ratios of energy recovery. For example, in [33], it is reported that in NEDC cycle with a similar vehicle, the braking energy is 35% of the traction energy, meanwhile with the UDDS cycle the braking energy ups to 50%. Nevertheless, these ratios only account for the energy available at the wheel. If the losses in the internal components where the energy passes through are taken into account, the braking energy available at the storage system is much lower. It is also important to note that all these ratios are calculated in the limit situation where no friction brake is used and assuming that in the ESS there is always sufficient capacity to store the available energy, allowing the maximum energy recuperation. Thus, these ratios only represent a superior limit and in real situations the available energy to store may be lower. Table 4 summarizes the main values of the energy flow analysis for the four cycles in study.

Table 5 Power demand to fulfill the traction power requirements for a 1,380 kg vehicle

Traction power requirements	Power [kW]
Sustained driving conditions	
Top speed (150 kph)	44.1
88.5 kph at 6.5% grade without overweight	37.3
88.5 kph at 6.5% grade with overweight (600 kg)	49.9
Transitory driving conditions	
Acceleration (0–96.5 kph in 10 s)	93.1

6 Determination of the Hybridization Degree

In order to be competitive with conventional vehicles, FCHVs must satisfy different types of driving conditions. This means that, both in transitory and sustained driving conditions, the power balance between the load and the power sources must be satisfied. In the case of time-sustained driving conditions, the fuel cell must be able to meet situations such as high-speed driving or overcoming a certain gradient. On the other hand, the combined FCS and storage system must be able to meet transitory conditions such as accelerations. In this sense, in the work of Ahluwalia et al. [33] it is stated that the hybrid system must satisfy the following requirements. First, the FCS by itself must be capable of supporting time-sustained driving conditions. This requirement includes two different tests: a top speed test where a specified speed is maintained throughout the time and a grade test where the vehicle must maintain a determined speed with a determined grade and a certain overweight. Second, with the assistance of ESS, the hybrid system must be able to fulfill a certain acceleration requirement.

Assuming the drivability requirements and exploiting the FCHV model developed in the ADVISOR environment, it is possible to quantify the amount of power that the powertrain needs to supply to fulfill the traction requirements. In Table 5, the power demands to fulfill each traction power requirement is presented. These powers correspond to a vehicle with a total mass of 1,380 kg. Besides, the power demand to maintain different sustained speeds and the power demand to maintain a sustained speed of 88.5 kph with a grade of 6.5% are shown in Fig. 11.

The hybridization degree analysis according to drivability conditions consists in determining the FCS rated power and the amount of SCs in the ESS necessary in order to fulfill the requirements stated in Table 5. Once these two design variables are determined, it is possible to calculate the total mass of the vehicle as the summation of the mass of the FCS, m_{FCS}, the mass of the ESS, m_{ESS}, and the mass of the basic vehicle, $m_{\mathrm{veh,basic}}$:

$$m_{\mathrm{veh,total}} = m_{\mathrm{ESS}} + m_{\mathrm{FCS}} + m_{\mathrm{veh,basic}}. \tag{16}$$

The mass of the basic vehicle is the mass of the vehicle without the FCS and the ESS (i.e., without the powertrain), but it is not a constant mass. It depends on the chosen FCS because the mass of the auxiliary system in the vehicle depends on the rated power of the FCS.

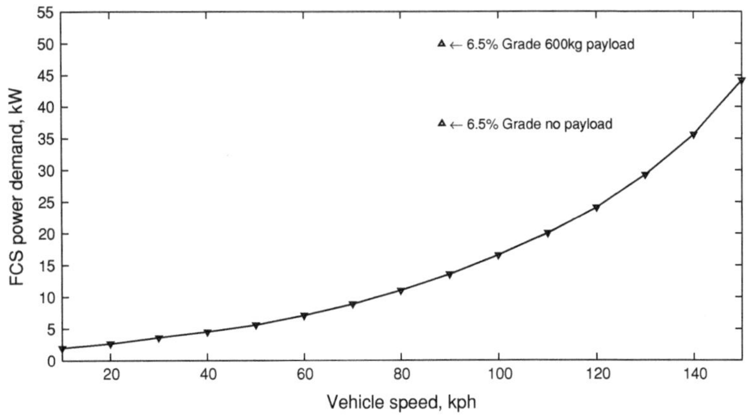

Fig. 11 Power demand versus vehicle speed for a vehicle of 1,380 kg

The procedure followed for dimensioning the hybridization degree is represented in Fig. 12. The first step of this dimensioning process concerns the time-sustained conditions. The dimensioning process starts with the calculation of the minimum necessary FCS that a vehicle of 1,380 kg, according to the case of study defined in Sect. 4 needs to meet the top sustained speed requirement. Then, the process continues with the calculation of the minimum necessary FCS that the vehicle needs to meet the grade requirement. Both estimations are done using only the FCS to power the vehicle because it is a sustained power and because it is sustained, it can only be supported by the FCS. The FCS that fulfills the two requirements is the maximum value between the two results. Then, the vehicle mass is calculated according to the mass of this FCS using (16). In this initial step, m_{ess} is assumed to be null.

The second step is related to the acceleration test and consists in the determination of the minimum amount of SCs in the ESS that is necessary to add to the previous FCS to reach the acceleration requirement. Once this step is done, the total mass is recalculated according to the determined number of SCs using Eq. 16.

The procedure is repeated from the first step using the vehicle mass calculated in the last step until the results converge in a combination of FCS and ESS such the vehicle is capable to meet both the sustained and the transitory driving conditions.

The results show that the minimum FCS is 35 kW and the minimum SC bank is 130 modules. Each module in the ADVISOR model has a specific power of 2.5 kW kg^{-1} and a specific energy of 6 W h kg^{-1} with a mass of 0.408 kg. Thus, the maximum energy in the bank is 318.2 W h and the final total mass of the vehicle results in 1,109 kg.

However, this is the minimum value of FCS and there is still a degree of freedom in the final selection between two extreme cases: the case with 35–kW FCS and 130 module bank, and the pure FCS case (with a larger FCS and no ESS) where all the power needed is supplied by the FCS. Therefore, an analysis is performed varying the ESS module number (i.e., varying the hybridization degree), and for different driving cycles, to find the combination with the lowest hydrogen consumption.

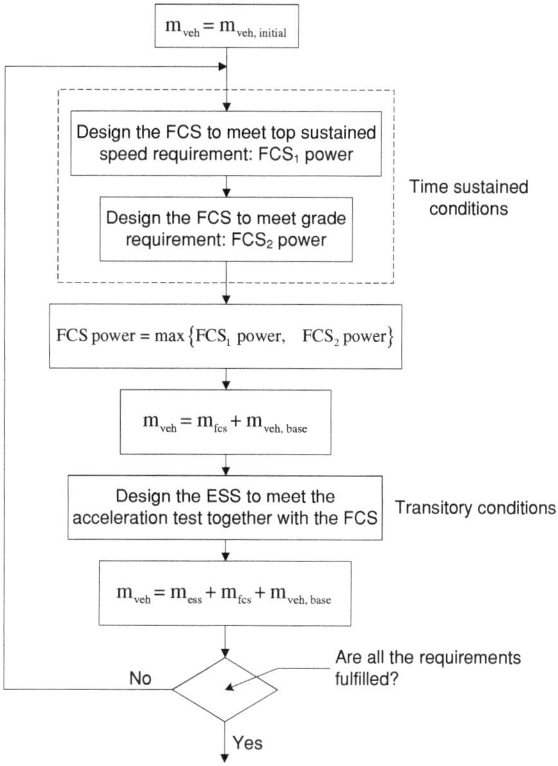

Fig. 12 Schematic diagram of the hybridization degree analysis according to drivability conditions

The results of hydrogen economy and total mass, shown in Figs. 13, 14, reveal that the best option, with lower mass and lower hydrogen consumption, is the vehicle consisting in an FCS of 35 kW and an ESS bank of 130 modules. In this case, the total vehicle mass results to be 1,109 kg and the hybridization degree is 79%. With regard to Fig. 13, the hydrogen consumption is the optimal. It is remarkable the relatively insensitive of the hydrogen consumption varying the ESS module number. This is because the increase of the optimal consumption in this analysis is only produced by the increase in the vehicle mass. On the contrary, in the case of ESS module number null, the increase in the optimal hydrogen consumption is high because there is no possibility of energy recovery. The computation of these hydrogen optimal consumptions are done using the analysis formulated in the following section.

7 Computation of the Optimal Hydrogen Consumption

In the previous section, the optimal hydrogen consumption is indicated for different profiles. In order to evaluate the performance of the EMS, which are developed in Sect. 1 of Chap. 8, in terms of hydrogen consumption, it is also useful to determine

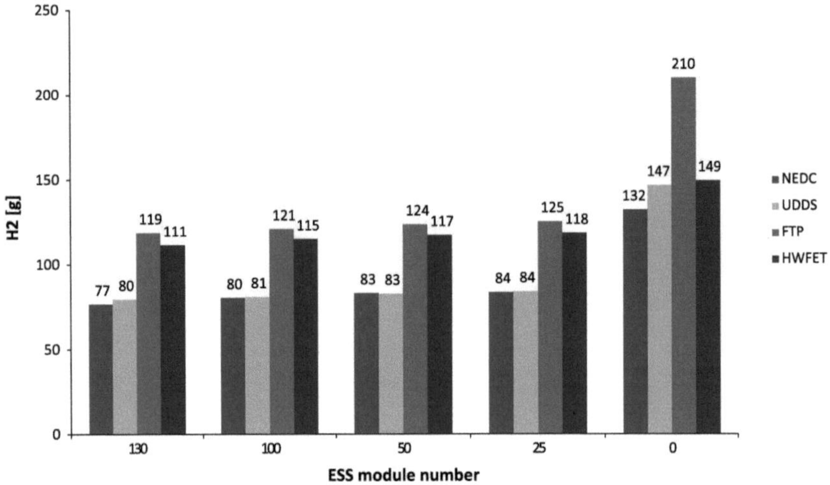

Fig. 13 Hydrogen consumption for different hybridization degrees (FCS power and ESS module number)

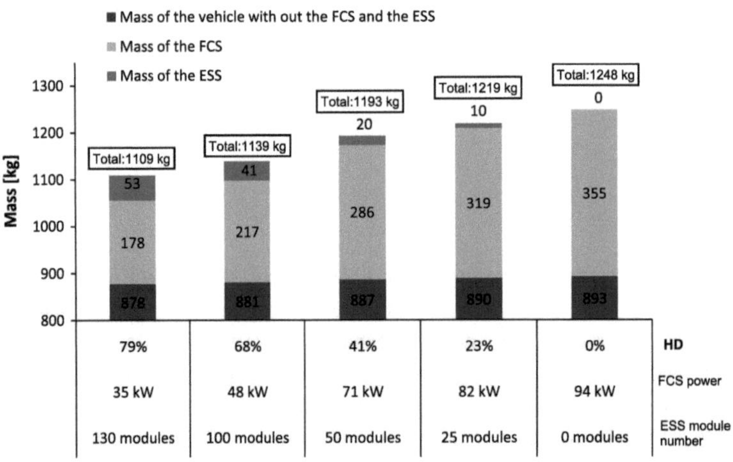

Fig. 14 Total vehicle mass for different hybridization degrees (FCS power and ESS module number)

the minimum amount of hydrogen that a specified vehicle needs to achieve a given driving speed profile. Thus, each EMS will be evaluated comparing its hydrogen consumption with respect to the consumption in optimal conditions, a reference context that represents the minimum amount of hydrogen necessary to achieve the load profile. In some works, e.g., [34–36], this analysis is performed through the use of the *dynamic programming* (DP) technique, a procedure that allows to determine a global optimal operation of the system for a given load profile by evaluating all possible control sequences in a systematic way [34]. However, a disadvantage of

Table 6 Results from the analysis of the optimal hydrogen economy for a vehicle of 1,109 kg

	NEDC	UDDS	FTP	HWFET
Hydrogen consumption [g]	77	80	119	111
Hydrogen consumption [g km^{-1}]	7.0	6.6	6.7	6.7
Braking recovery [W h]	194.2	310.4	452.4	85.8
Braking/hydrogen ratio[a] [%]	7.6	11.7	11.4	2.3

[a] The braking/hydrogen ratio is defined as the ratio between the energy recovered from braking in the ESS and the energy of the lower heating value energy of the hydrogen consumed in the FCS

this technique is the relative long computation time due to the large required grid density. The grid is a result of the discretization of time and the state variables at the appropriate grid levels. The grid density should be taken high, because it influences the accuracy of the result.

In this work, we propose a method to analyze the minimum hydrogen consumption with reduced computational cost because it avoids the discretization of the state variables. Only the time is discretized because the standard driving cycles in ADVISOR are defined every 1 s. The procedure is based on the following assumptions:

- The capacity of storage in the ESS is sufficient to recover all the available energy from regenerative braking.
- The friction brake is not employed during the entire cycle.
- When the power consumption is null the FCS is turned off.
- The FCS is already hot at the nominal operating temperature when the cycle starts.
- The total vehicle mass is known and in our case of study it is $m_{\text{veh}} = 1{,}109$ kg.

A simulation is performed for each driving cycle in such a way that the FCS works alternately in two operating points, namely *"On"* and *"Off"*, according to the actual SoE(k):

- When $\text{SoE}(k) \leq \text{SoE}(0)$, the FCS is operated at its point of maximum efficiency (the *"On"* point).
- When $\text{SoE(k)} > \text{SoE}(0)$, the FCS is turned off (the *"Off"* point).

As a result, the final state of energy SoE(N) is into a reduced range around the initial state of energy SoE(0).

Thus, the hydrogen consumption only accounts for the necessary hydrogen to fulfill the cycle of duration N. The main advantage of this method is that, unlike the usual methodology used in the literature, which is based on *dynamic programming*, it avoids the discretization of the state variables and only the time is discretized because the standard driving cycles that are utilized in ADVISOR are defined every 1 s. Therefore, this methodology allows the calculation of optimal hydrogen consumption in the FCHV with low computational time, maintaining a high accuracy. Figure 15 shows a scheme with the FCS operation to perform the calculation previously described and the results are collected in Table 6, where the *braking/hydrogen ratio* is defined as the ratio between the energy recovered from braking in the ESS

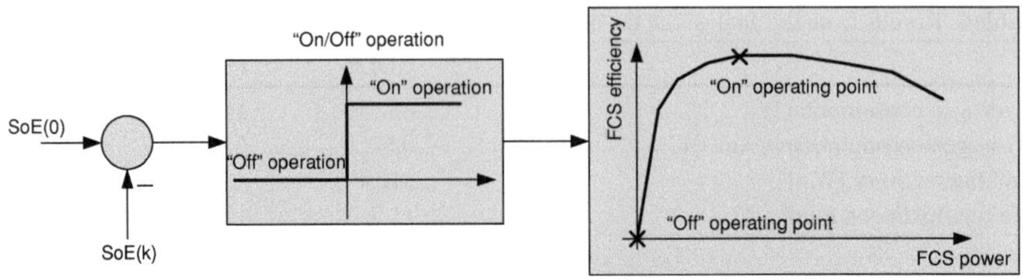

Fig. 15 Scheme showing the *On/Off* operation of the FCS model to perform the analysis of the minimum hydrogen consumption

and the energy of the hydrogen, calculated from the *lower heating value* (ΔH_{LHV}) of hydrogen, consumed in the FCS during the cycle. Since the FCS is either working at the point of maximum efficiency or is turned off (with the assumption that when the FCS is turned off the power consumption is null), the calculation of the minimum consumption is guaranteed.

8 Conclusion

In this chapter, the design and analysis of FCHS (fuel cell hybrid systems) oriented to automotive applications is approached. First, the advantages of these kinds of power systems are analyzed focusing on the mechanism that allows the improvement of the efficiency of FCHVs. Then, the electrical structures for FCHVs are addressed, analyzing the electrical topology, comparing two types of energy storage devices (i.e., batteries versus SCs) and concluding that SCs seem the best technical alternative nowadays, analyzing the requirements of HEVs concerning the storage system and finally, modeling the energy storage system. The main conclusions of this chapter are the following:

- Using a detailed model in ADVISOR to perform a precise study of the system, the result shows that through hybridization it is possible to improve the hydrogen economy in FCHVs significantly, compared with the pure fuel cell case without SCs (Fig. 13): 41.7 on NEDC, 45.6 on UDDS, 43.3 on FT and 22.5% on HWET. This is possible because of the reduction on the total mass of the vehicle, the more efficient operation of the FCS and the energy recovery from braking.
- In the determination of the hybridization degree there is a conflict between drivability and efficiency, which means that the most robust design to fulfill drivability requirements is not the most efficient in terms of hydrogen economy.
- The braking/hydrogen ratio defined in Sect. 7 is a more realistic indicator of other parameters used in the literature to analyze the reduction of the hydrogen consumption due to regenerative braking. It shows that it is possible to recover up to

7.6% of the hydrogen energy on NEDC, 11.7 on UDDS, 11.4 on FTP and 2.3% on HWET.

The main contributions stem from three methodologies proposed to design and analyze FCHS. First, an iterative and systematic methodology to determine the hybridization degree to fulfill drivability requirements was elaborated. Second, a new method to assess the optimal hydrogen consumption with low computational cost that avoids the discretization of the state variables was presented. This method demonstrated to be useful to evaluate the performance of any energy management strategy because it provides a benchmark to compare the hydrogen consumptions of each strategy. Third, an analysis of the energy flows inside the FCHV was done, emphasizing the calculation of the fraction of energy from regenerative braking.

Acknowledgment The author wish to thank to Maria Serra and Jordi Riera from the Institut de Robòtica i Informàtica Industrial (IRI-UPC-CSIC) for their contribution in this chapter.

References

1. Pukrushpan J (2003) Modelling and control of fuel cell Systems and fuel processors. PhD thesis, University of Michigan
2. Williams K, Keith W, Marcel M, Haskew T, Shepard , Todd B (2007) Experimental investigation of fuel cell dynamic response and control. J Power Sources 163(2):971–985
3. Larminie J, Dicks A (2003) Fuel Cell Systems Explained. 2nd edn. Wiley, New York
4. Caux S, Lachaize J, Fadel M, Shott P, Nicod L (2005) Energy management of fuel cell system and supercaps elements. In: Proceedings of the 16th IFAC World Congress, Prague
5. Ahluwalia R, Wang X, Rousseau A, Kumar R (2003) Fuel economy of hydrogen fuel cell vehicles. J Power Sources 130:192–201
6. Jeong K, Oh B (2002) Fuel economic and life-cycle cost analysis of a fuel cell hybrid vehicle. J Power Sources 105:58–65
7. Pukrushpan J, Stefanopoulou A, Peng H (2004) Control of fuel cell breathing: Initial results on the oxygen starvation problem. IEEE Control Syst Magazine 24:30–46
8. Drolia A, Jose P, Mohan N (2003) An approach to connect ultracapacitor to fuel cell powered electric vehicle and emulating fuel cell electrical characteristics using switched mode converter. Indus Electron Soc 1:897–901
9. Rajashekara K (2000) Propulsion system strategies for fuel cell vehicles. In:Fuel cell technology for vehicles, pp 179–187
10. Rajashekara K (2003) Power conversion and control strategies for fuel cell vehicles. Indus Electron Soc 3:2865–2870
11. Santi E, Franzoni D, Monti A, Patterson D, Barry N (2002) A fuel cell based domestic uninterruptible power supply. Applied Power Electronic Conference and Exposition, pages 605–613
12. Henson W (2008) Optimal battery/ultracapacitor storage combination. J Power Sources 1:417–423
13. Burke A (2000) Ultracapacitors: why, how and where is the technology. J Power Sources 91(1):37–50
14. Pede G, Iacobazzi A, Passerini S, Bobbio A, Botto G (2004) FC vehicle hybridisation: an affordable solution for an energy-efficient FC powered drive train. J Power Sources 125(2): 280–291
15. Maxwell Technologies (2009) Electrical Double Layer Capacitor: Boostcap ultracapacitor Series: BPAK. http://www.maxwell.com/

16. Arbizzani C, Biso M, Cericola D, Lazzari M, Soavi F, Mastragostino M (2008) Safe, high-energy SCs based on solvent-free ionic liquid electrolytes. J Power Sources
17. Mastragostino M, Soavi F (2007) Strategies for high-performance supercapacitors for HEV. J Power Sources 174(1):89–93
18. Gao W (2005) Performance comparison of a fuel cell-battery hybrid powertrain and a fuel cell-ultracapacitor hybrid powertrain. IEEE Trans Veh Technol 54(3):846–855
19. Kellaway M (2007) Hybrid buses—what their batteries really need to do. J Power Sources 168:95–98
20. Moseley P, Bonnet B, Cooper A, Kellaway M (2007) Lead-acid battery chemistry adapted for hybrid electric vehicle duty. J Power Sources 174(1):49–53
21. Lam L, Louey R, Haigh N, Lim O, Vella D, Phyland C, Vu L, Furukawa J, Takada T, Monma D, Kano T (2007) VRLA Ultrabattery for high-rate partial-state-of-charge operation. J Power Sources 174(1):16–29
22. Markel T, Brooker A, Hendricks T, Johnson V, Kelly K, Kramer B, O'Keefe M, Sprik S, Wipke K (2002) ADVISOR: a system analysis tool for advanced vehicle modeling. J Power Sources 110:255–266
23. Wipke K, Cuddy M, Burch S (1999) ADVISOR 2.1: A user-friendly advanced powertrain simulation using a combined backward/forward approach. IEEE Trans Veh Technol 48:1751–1761
24. Kim M, Peng H (2007) Power management and design optimization of fuel cell/battery hybrid vehicles. J Power Sources 165:819–832
25. Pukrushpan JT, Peng AGH (2002) Modeling and control for PEM fuel cell stack system. Am Control Conf. Proc 4
26. Correa J, Farret F, Canha L, Simoes M (2004) An electrochemical-based fuel-cell model suitable for electrical engineering automation approach. Indus Electron, IEEE Trans 51(5):1103–1112
27. Ahluwalia R, Wang X (2005) Direct hydrogen fuel cell systems for hybrid vehicles. J Power Sources 139:152–164
28. Pell WG, Conway BE, Adams WA, de Oliveira J (1999) Electrochemical efficiency in multiple discharge/recharge cycling of supercapacitors in hybrid EV applications. J Power Sources 80(1-2):134–141
29. Schiffer J, Bohlen O, de Doncker RW, Sauer DU (2005) Optimized energy management for fuelcell-superCap hybrid electric vehicles VPP Track 4: energy storage components/systems. vehicle power and propulsion, 2005 IEEE Conference, pp 716–723
30. Uzunoglu M, Alam MS (2006) Dynamic modeling, design, and simulation of a combined PEM fuel cell and ultracapacitor system for stand-alone residential applications. Energy Convers, IEEE Trans 21(3):767–775
31. Maxwell Technologies. Electrical Double Layer Capacitor: Boostcap ultracapacitor series: BPAK
32. DieselNet. Emission test cycles. Online, 2005. http://www.dieselnet.com/standards/cycles/
33. Ahluwalia R, Wang X, Rousseau A (2005) Fuel economy of hybrid fuel-cell vehicles. J Power Sources, 152:233–244
34. Kessels J (2007) Energy management for automotive power nets. PhD thesis, Technische Universiteit Eindhoven
35. Koot M, Kessels A, de Jager B, Heemels W, van den Bosch P, Steinbuch M (2005) Energy management strategies for vehicular electric power systems. IEEE Trans Veh Technol 54(3):771–782
36. Hofman T, Steinbuch M, van Druten RM, Serrarens AFA (2008) Rule-based equivalent fuel consumption minimization strategies for hybrid vehicles. In Proceedings of the 17th IFAC World Congress, Seoul pp 5652–5657

Energy Management Strategies for Fuel Cell Hybrid Systems

Diego Feroldi

Abstract In this chapter, the energy management strategies for fuel cell hybrid systems are addressed. To control the energy flow between the fuel cell, the energy storage system and the electrical load in fuel cell hybrid vehicles (FCHVs), three energy management strategies based on the fuel cell efficiency map are presented and validated through an experimental setup, which is developed to emulate FCHVs. The resulting hydrogen consumptions are compared with two references: the consumption of the pure fuel cell case, a vehicle without hybridization, and the optimal case with the minimum consumption. The optimal consumption for a given vehicle is determined through a methodology proposed that, unlike other previous methodologies, avoids the discretization of the state variables. The experimental results show that the operation in real time with high efficiency is possible.

1 Introduction

The energy management strategies (EMS) are algorithms that determine at each sampling time the power generation split between the fuel cell system (FCS) and the energy storage system (ESS) in order to fulfill the power balance between the load power and source power. Depending on how the power split is done a minimization of the hydrogen consumption can be obtained. To find a global optimal solution, control techniques where a minimization problem is resolved have been studied [1]. In general, these techniques do not offer a causal solution, and, in consequence, are not feasible because it is assumed that the future driving cycle is entirely known a priori. Nevertheless, their results can be used to evaluate the performance of new strategies in study. Some strategies perform an optimal management but they are not

D. Feroldi (✉)
CAPEG-CIFASIS-(CONICET-UNR-UPCAM) and DCC-FCEIA-UNR,
27 de Febrero 210 bis, S2000EZP Rosario, Argentina
e-mail: feroldi@cifasis-conicet.gov.ar

M. S. Basualdo et al. (eds.), *PEM Fuel Cells with Bio-Ethanol Processor Systems*, Green Energy and Technology, DOI: 10.1007/978-1-84996-184-4_8,

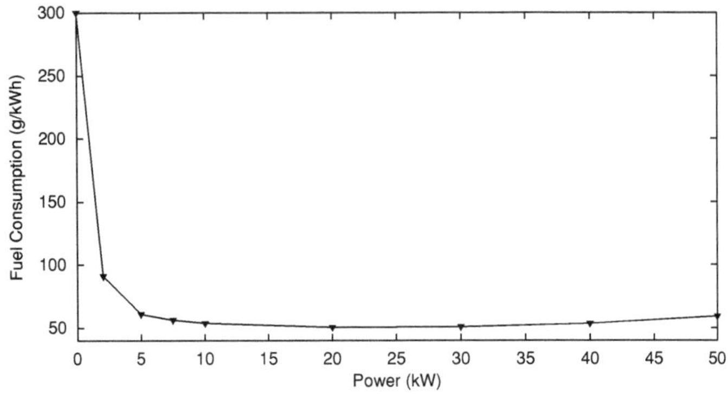

Fig. 1 Hydrogen consumption map as a function of the FCS power

suitable for real-time control. For instance, Murphey et al. [2] developed a power management strategy using machine learning and fuzzy logic. On the other hand, strategies which deal with local optimization are convenient for real implementation.

Another approach, particularly convenient to work in real-time operation, is the type of strategies where the system operation is based on heuristic rules from the knowledge of the performance of the components of the system. These strategies can achieve a comparable performance to those strategies where some optimization technique is utilized. In this way, two strategy based on the knowledge of the fuel-cell efficiency map are presented in this chapter.

In Sect. 3.5 of Chap. 1, the state of the art of EMS in FCHS was covered. In this chapter, we approach the EMS issue developing new strategies for FCHVs. Despite fuel cell hybrid systems having other applications, in this chapter we concentrate our attention on automotive applications because of the great potential of this application. The objective is to improve the performance of the hybrid vehicle in terms of hydrogen economy and transient response. The distribution of the chapter is as follows: in Sect. 2, the objective of the EMSs are formalized. Secondly, in Sect. 3, three new different strategies are proposed: two of them are based on the FCS efficiency map and the third is based on constrained nonlinear programming. In Sect. 4 the simulation results, showing the temporal response of the strategies and the performance comparative between them, are presented. In Sect. 5 the experimental results to validate the strategies are presented. Finally, in Sect. 6 the conclusions on EMS in FCHS close the chapter.

2 Objectives of Energy Management Strategies

The hydrogen consumption map of an FCS clearly reveals a zone, below a limit power, where the hydrogen consumption is very high. Meanwhile the zone above this limit power has a significative lower consumption as can be seen in Fig. 1. Thus, the principal objective of the energy management strategy is to minimize a cost function J, a mathematical expression that represents the cumulative hydrogen consumption during the time:

$$\min_{X} J(X) \quad \text{subject to} \quad H(X) = 0 \quad \text{and} \quad G(X) \leq 0, \tag{1}$$

where the vector X is

$$X = \begin{bmatrix} P_{\text{fcs}}(t) \\ P_{\text{ess}}(t) \end{bmatrix}. \tag{2}$$

This means that the EMS has to determine the optimal value of the FCS power, $P_{\text{fcs}}(t)$, and the ESS power, $P_{\text{ess}}(t)$, in order to minimize the cost function $J(X)$:

$$J(X) = \int_{0}^{t_c} F(X)\mathrm{d}t, \tag{3}$$

where t_c is the duration of driving cycle and $F(X)$ is the hydrogen consumption according to the hydrogen consumption map of the FCS:

$$F(X) = \text{Cons}_{H_2}\left(P_{\text{fcs}}(t)\right). \tag{4}$$

The constraint $H(X) = 0$ in Eq. 1 means that at each time the power balance in the DC bus must be satisfied:

$$P_{\text{fcs}}(t) \cdot \eta_B + P_{\text{ess}}(t) \cdot \eta_{B/B} \cdot \eta_{\text{ess}} = P_{\text{load}}(t) \quad \forall t \in [0, t_c]. \tag{5}$$

On the other hand, the constraint $G(X) \leq 0$ in Eq. 1 states the constraints in the FCS power and ESS power:

$$P_{\text{fcs,min}} \leq P_{\text{fcs}}(t) \leq P_{\text{fcs,max}}, \tag{6}$$

$$\Delta P_{\text{fcs,min}} = \Delta P_{\text{fcs,fall rate}} \leq \frac{dP_{\text{fcs}}(t)}{dt} \leq \Delta P_{\text{fcs,rise rate}} = \Delta P_{\text{fcs,max}}, \tag{7}$$

$$P_{\text{ess,min}}(\text{SoE}(t)) \leq P_{\text{ess}}(t) \leq P_{\text{ess,max}}(\text{SoE}(t)), \tag{8}$$

$$\text{SoE}_{\text{min}} \leq \text{SoE}(t) \leq \text{SoE}_{\text{max}}, \quad \forall t \in [0, t_c]. \tag{9}$$

The restrictions in FCS power are stated in Eq. 6: the maximum power is limited by the FCS rated power, whereas the minimum FCS power command must be limited to a value below which it is not suitable to operate because the parasitic load is too large, reducing the system net power.

There is a phenomenon of delay between the onset of the load on FCS and the response of the reactant supply system results which may lead to a "starvation" of reactants in FCS. This may be avoided by restricting the dynamics required by the load [3]. The same approach is followed in Rodatz et al. [3] and Philipps et al. [4]

where the maximum positive rise is limited to avoid damage in the stack. Therefore, in our work, the FCS power load is increased no faster than a certain power rise rate ($\Delta P_{\text{fcs,max}}$). Besides, we propose to operate FCS in such a way that the power is decreased no faster than a certain power fall rate ($\Delta P_{\text{fcs,min}}$) to prevent overpressure into the stack. Both situations can be managed by using the ESS as a power buffer. These restrictions are formulated in Eq. 7.

With regard to ESS, the maximum and minimum power flows are also limited, as formulated in Eq. 8. The maximum power that ESS can supply or store depends on the actual voltage in ESS, $V_{\text{ess}}(t)$, the maximum voltage $V_{\text{ess, max}}(t)$, and the minimum voltage $V_{\text{ess, min}}(t)$:

$$\begin{aligned} P_{\text{ess,chrg max}} &= \frac{(V_{\text{ess}}(t)-V_{\text{ess,max}})\cdot V_{\text{ess}}(t)}{R_d} \text{ during charging,} \\ P_{\text{ess,disch max}} &= \frac{(V_{\text{ess}}(t)-V_{\text{ess,min}})\cdot V_{\text{ess}}(t)}{R_d} \text{ during discharging,} \end{aligned} \tag{10}$$

where R_{d} is the ESS internal resistance. The charge voltage is related with the ESS energy, so that, it is possible to express Eq. 10 as a function of the *state of energy* SoE(t) in the ESS, (defined in Eq. 6 of Chap. 7):

$$P_{\text{ess}}(t) = \frac{k \cdot E_{\text{cap}}}{R_d \cdot C_R} \cdot \text{SoE}(t), \tag{11}$$

where E_{cap} is the maximum energy that ESS is able to store, R_{d} and C_{R} are the internal resistance and the capacitance of the supercapacitors, and k is a constant depending in the particular technology. Therefore,

$$P_{\text{ess, chrg max}}(t) = k_{\text{ess}} \cdot (\text{SoE}(t) - \text{SoE}_{\text{max}}), \tag{12}$$

$$P_{\text{ess, dischrg max}}(t) = k_{\text{ess}} \cdot (\text{SoE}(t) - \text{SoE}_{\text{min}}), \tag{13}$$

where

$$k_{\text{ess}} = \frac{k \cdot E_{\text{cap}}}{R_d \cdot C_R} \tag{14}$$

is the constant that relates the actual SoE and the power that the device can supply or store. Thus, $P_{\text{ess, chrg max}}(t)$ is the maximum power that ESS is able to store at the instant t (charging mode) and the $P_{\text{ess, dischrg max}}(t)$ is the maximum power that ESS is able to supply at instant t (discharging mode). According to the sign convention employed, the power is negative when ESS is in charging mode, and positive when ESS is discharging mode. The k_{ess} constant depends on the *DC* internal resistance R_{d}, the supercapacitor capacitance C_{R}, and the constant k depending on the particular supercapacitor technology. The internal resistance of supercapacitors is extremely low and the capacitance is exceptionally high, allowing a very fast operation both during charging and discharging. In the supercapacitors employed in this work, they are $R_d = 0.019\,\Omega$ and $C_R = 58\,\text{F}$, thus $k_{\text{ess}} = 479\,\text{kW}$.

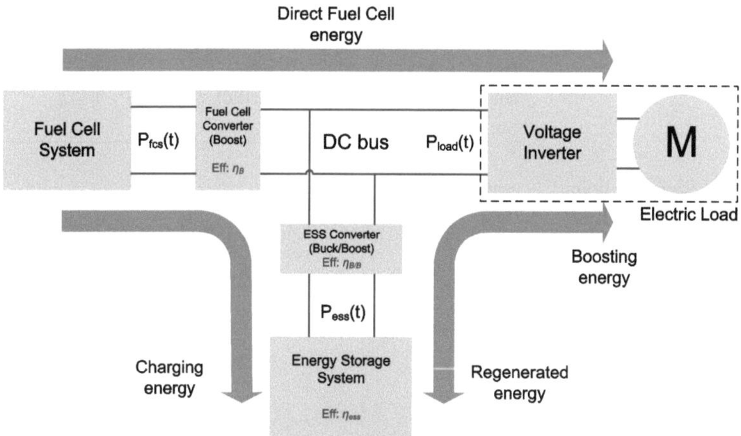

Fig. 2 Energy flow scheme in the strategy based on efficiency map

Besides the previous objective, another important objective of EMS is the improvement in the transient response of FCHS. As mentioned before in Sect. 2.3 of Chap. 3, the temporal behavior is fundamentally conditioned by the compressor dynamics. This drawback can be overcome by means of an adequate energy management strategy. In fact, it is possible to improve the transient response of FCHS by means of the ESS energy contribution. In that sense, the energy management strategy acts as follows: if the power requested to FCS exceeds a determined maximum rise rate, then the power rise rate that FCS actually gives is limited and the rest of the power is supplied by ESS if possible according to the present value of SoE(t). If, on the contrary, the power requested to FCS exceeds the maximum fall rate, then the actual power fall rate is limited and the surplus power is absorbed by ESS whenever there is sufficient storage capacity.

3 Energy Management Strategies

In this section, three energy management strategies for FCHVs are presented. Two are based on the knowledge of the fuel cell efficiency map and the third strategy is based on constrained nonlinear programming. The electrical topology previously established in Sect. 3.1 of Chap. 7 is represented in Fig. 2, which is common in the three strategies. In this figure, are also represented the energy flows in FCHS: the regenerated energy from load to the ESS bank, the charging energy from FCS to ESS, the boosting energy from ESS to the load, and the fuel cell energy that directly supplies the load from FCS.

The efficiency of the *Boost* power converter (η_B), that connects FCS with the power bus, the efficiency of the *Buck/Boost* converter ($\eta_{B/B}$), that connects ESS with

Table 1 Efficiency values assumed in hybrid-modelling process

Efficiency values		
Boost power converter efficiency	η_B	0.95
Buck/Boost power converter efficiency	$\eta_{B/B}$	0.95
Supercapacitor bank efficiency	η_{ess}	0.9

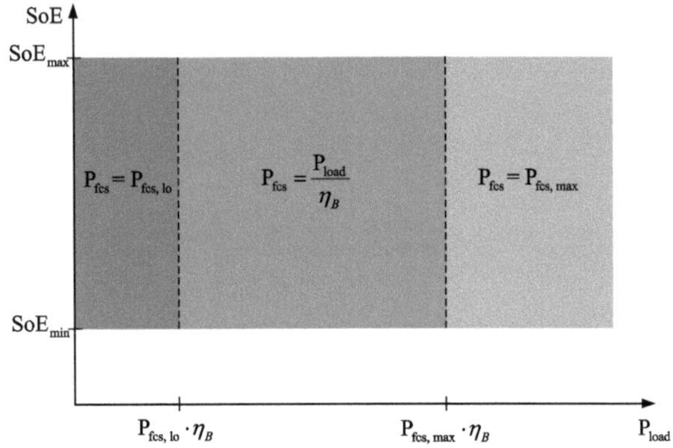

Fig. 3 FCS operating point as a function of the SoE in the ESS and the load power P_{load} for the strategy based on efficiency map

the bus and the efficiency of ESS itself, and the efficiency of ESS (the charging and discharging efficiency), are assumed as constants (see Table 1). The FCS efficiency is represented with an efficiency map as a function of the power. In this way, the power that FCS supplies to the bus is affected by the *Boost* converter efficiency; the power that ESS supplies is affected by the *Buck/Boost* efficiency and the ESS-discharging efficiency; whereas the power that ESS stores from the bus is affected by the *Buck/Boost* converter efficiency and the ESS-charging efficiency.

3.1 First Strategy Based on the FCS Efficiency Map

One of the most relevant characteristics of FCS is the efficiency map (see Fig. 3), a chart that shows how the efficiency changes with the load power and is related to the consumption map through Eq. 4 of Chap. 7. It is supposed that the FCS operating point is controlled and, thus, external parameters such as the ambient temperature have no influence on the efficiency map.

The first strategy proposed in this thesis work is a *quasi-load-following*[1] strategy where FCS is operated in an advantageous zone where the efficiency is high. In this case, the operating zone is constrained between an inferior limit ($P_{fcs, lo}$) and a superior limit ($P_{fcs, hi}$). The superior limit is imposed by the maximum power that the fuel cell can deliver (i.e., $P_{fcs, hi} = P_{fcs, max}$), whereas the inferior limit is determined according to the efficiency curve. As mentioned before, the efficiency of FCS at low

[1] A load-following strategy is a strategy that adjusts the power output of the fuel cell according to the load requirement.

power is very poor due to the parasitic power. Thanks to the inferior limit, the energy management strategy avoids this unfavorable zone (see Fig. 9 of Chap. 7).

Considering that the losses in the power converters (η_B and $\eta_{B/B}$) and the ESS efficiency η_{ess} are constants, the power balance in the DC bus previously defined in Eq. 5, can be rewritten as

$$P_{\mathrm{fcs}}(k) \cdot \eta_B + P_{\mathrm{ess}}(k) \cdot \eta_{B/B} \cdot \eta_{\mathrm{ess}} = P_{\mathrm{load}}(k) \qquad k = 1, 2, \ldots, N_c, \tag{15}$$

where $N_c = t_c/\Delta T$ is the cycle duration, assuming a constant sampling period $\Delta T = 1\,s$.

This EMS gives priority to the FCS-power supply since FCS is the primary power source and the direct power flow to the DC bus through the FCS power converter has lower losses than the indirectly way through the ESS power converter, the ESS itself and back to the load. Consequently, given the present $P_{\mathrm{load}}(k)$, if

$$P_{\mathrm{fcs,lo}} \cdot \eta_B \le P_{\mathrm{load}}(k) \le P_{\mathrm{fcs,\ max}} \cdot \eta_B \tag{16}$$

and

$$\Delta P_{\mathrm{fcs,\ fall\ rate}} \le \Delta P_{\mathrm{load}}(k) \le \Delta P_{\mathrm{fcs,\ rise\ rate}}, \tag{17}$$

where

$$\Delta P_{\mathrm{load}}(k) = P_{\mathrm{load}}(k) - P_{\mathrm{load}}(k-1). \tag{18}$$

Then, the FCS operates in load-following mode:

$$P_{\mathrm{fcs}}(k) = P_{\mathrm{load}}(k)/\eta_B \tag{19}$$

and

$$P_{\mathrm{ess}}(k) = 0. \tag{20}$$

If, on the contrary

$$P_{\mathrm{load}}(k) \le P_{\mathrm{fcs,\ lo}}(k) \cdot \eta_B, \tag{21}$$

then

$$P_{\mathrm{fcs}}(k) = P_{\mathrm{fcs,\ lo}}(k) \tag{22}$$

and, the ESS stores the rest of generated power, if the ESS is not too charged:

$$P_{\mathrm{ess}}(k) = -\min\left\{|P_{\mathrm{load}}(k) - P_{\mathrm{fcs}}(k) \cdot \eta_B| \cdot \eta_{\mathrm{ess}} \cdot \eta_{B/B},\ |\mathrm{SoE}(k) - \mathrm{SoE}_{\max}| \cdot k_{\mathrm{ess}}\right\}. \tag{23}$$

Conversely, if

$$P_{\text{load}}(k) \geq P_{\text{fcs, hi}}(k) \cdot \eta_B, \tag{24}$$

then

$$P_{\text{fcs}}(k) = P_{\text{fcs, max}} \tag{25}$$

and ESS supplies the rest of the load power, if there is enough energy in ESS:

$$P_{\text{ess}}(k) = \min\left\{\frac{(P_{\text{load}}(k) - P_{\text{fcs}}(k) \cdot \eta_B)}{\eta_{B/B} \cdot \eta_{\text{ess}}}, \quad (\text{SoE}(k) - \text{SoE}_{\min}) \cdot k_{\text{ess}}\right\}. \tag{26}$$

The representation of how the FCS operating point is determined is shown in Fig. 3. However, the transition between operating points is limited by the maximum rates, thus, finally

$$P_{\text{fcs}}(k) = \begin{cases} P_{\text{fcs}}(k-1) + \Delta P_{\text{fcs, fallrate}} \cdot \Delta T, & \text{if } \Delta P_{\text{fcs}}(k) \leq \Delta P_{\text{fcs, fallrate}}, \\ P_{\text{fcs}}(k-1) + \Delta P_{\text{fcs, riserate}} \cdot \Delta T, & \text{if } \Delta P_{\text{fcs}}(k) \geq \Delta P_{\text{fcs, riserate}}, \end{cases} \tag{27}$$

where $\Delta P_{\text{fcs}} = P_{\text{fcs}}(k) - P_{\text{fcs}}(k-1)$, with $P_{\text{fcs}}(k)$ as previously calculated in Eqs. 19, 22 or 25, according to the condition Eqs. 16, 21 or 24, respectively. The power $P_{\text{ess}}(k)$ is calculated as indicated in Eqs. 23 and 26.

3.2 Improved Strategy Based on the FCS Efficiency Map

The second strategy based on the FCS efficiency map is a strategy that operates FCS preferably in its point of maximum efficiency in order to improve the hydrogen economy. The operating point of FCS is determined based on the actual power demand and the state of energy of ESS. The FCS power command is determined according to the following rules. If the load power is

$$P_{\text{fcs,lo}} \cdot \eta_B \leq P_{\text{load}}(k) \leq P_{\text{fcs, hi}} \cdot \eta_B \tag{28}$$

and the SoE is

$$\text{SoE}_{\text{lo}} \leq \text{SoE}(k) \leq \text{SoE}_{\text{hi}}, \tag{29}$$

where $P_{\text{fcs, hi}}$ is

$$P_{\text{fcs, hi}} = P_{\text{fcs, max}} \cdot \eta_B \cdot X_{\text{fcs, hi}}, \tag{30}$$

and $X_{\text{fcs, hi}}$ is a fraction of the maximum FCS power; then, FCS is operated in its point of maximum efficiency:

$$P_{\text{fcs}}(k) = P_{\text{fcs, max eff}}. \tag{31}$$

The remaining power to achieve the load demand flows from or to ESS according to Eq. 26 if $P_{\mathrm{load}}(k) > P_{\mathrm{fcs,\ max\ eff}}$ (discharging mode), or Eq. 23 if $P_{\mathrm{load}}(k) < P_{\mathrm{fcs,\ max\ eff}}$ (charging mode).

If the load power is

$$P_{\mathrm{fcs,\ hi}} \cdot \eta_B \leq P_{\mathrm{load}}(k) \leq P_{\mathrm{fcs,\ max}} \cdot \eta_B, \tag{32}$$

and the SoE is

$$\mathrm{SoE}_{\mathrm{lo}} \leq \mathrm{SoE}(k) \leq \mathrm{SoE}_{\mathrm{hi}}, \tag{33}$$

then, FCS is operated in load following mode:

$$P_{\mathrm{fcs}}(k) = P_{\mathrm{load}}(k)/\eta_B \tag{34}$$

and $P_{\mathrm{ess}}(k)$ is as indicated in Eqs. 23 or 26.

On the other hand, if

$$P_{\mathrm{load}}(k) \geq P_{\mathrm{fcs,\ max}} \cdot \eta_B \quad \text{and} \quad \mathrm{SoE}(k) \leq \mathrm{SoE}_{\mathrm{hi}} \tag{35}$$

or

$$\mathrm{SoE}(k) \leq \mathrm{SoE}_{\mathrm{lo}}, \tag{36}$$

then, FCS is operated at its maximum power:

$$P_{\mathrm{fcs}}(k) = P_{\mathrm{fcs,\ max}}, \tag{37}$$

and $P_{\mathrm{ess}}(k)$ is as indicated in Eq. 26. If, on the contrary

$$P_{\mathrm{load}}(k) \leq P_{\mathrm{fcs,\ lo}} \cdot \eta_B \quad \text{and} \quad \mathrm{SoE}(k) \geq \mathrm{SoE}_{\mathrm{lo}}, \tag{38}$$

or

$$\mathrm{SoE}(k) \geq \mathrm{SoE}_{\mathrm{hi}} \tag{39}$$

then, FCS is operated at its lower operating point:

$$P_{\mathrm{fcs}}(k) = P_{\mathrm{fcs,\ lo}}, \tag{40}$$

and $P_{\mathrm{ess}}(k)$ is as in Eq. 23. Additionally, if $P_{\mathrm{load}}(k) = 0\ \forall\, t \in [k_1, k_2]$ with $(k_2 - k_1) > T_{\mathrm{off}}$, and, $\mathrm{SoE}(k) > \mathrm{SoE}_{\mathrm{hi}}$ with $k > k_2$, then, FCS is turned off to avoid unnecessary hydrogen consumption because of the parasitic losses in FCS. Figure 4 indicates the FCS operating point as a function of the SoE(k) in ESS and the load power $P_{\mathrm{load}}(k)$. In the same way as in the previous strategy, the transition between operating points is realized according to the constraints concerning the maximum fall power rate and the maximum power rate as indicated in Eq. 27.

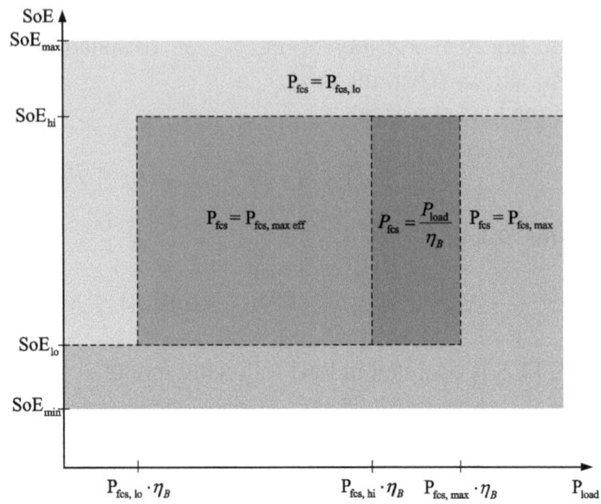

Fig. 4 FCS operating point as a function of the SoE in the ESS and the load power P_{load} for the improved strategy based on efficiency map

3.3 Strategy Based on Constrained Nonlinear Programming

In this strategy, an optimization problem with linear constraints is resolved at each sampling period ΔT where a nonlinear cost function, which represents the hydrogen consumption, is minimized. The problem can be put in the following form [5]:

$$\min_X J(X) \quad \text{subject to} \quad H(X) = 0 \quad \text{and} \quad G(X) \leq 0, \tag{41}$$

where the vector X is

$$X = \begin{bmatrix} P_{\text{fcs}} \\ P_{\text{ess}} \end{bmatrix}. \tag{42}$$

The cost function *J(X)* represents the hydrogen consumption in FCS:

$$J(X) = F(P_{\text{fcs}}(k)), \tag{43}$$

where

$$F(P_{\text{fcs}}(k)) = \text{Cons}H_2(P_{\text{fcs}}(k)) \cdot P_{\text{fcs}}(k) \cdot \Delta T, \tag{44}$$

and $\text{Cons}H_2(P_{\text{fcs}}(k))$ is the hydrogen consumption (g W^{-1}) as a function of $P_{\text{fcs}}(k)$. This relationship is shown in the form of a consumption map in Fig. 1.

The set of constraints $H(X) = 0$ represents the power balance in the *DC* bus:

$$P_{\text{fcs}}(k) \cdot \eta_B + P_{\text{ess}}(k) \cdot \eta_{B/B} \cdot \eta_{\text{ess}} = P_{\text{load}}(k) \quad k = 1, 2, \ldots, N_c. \tag{45}$$

On the other hand, the set of constraints $G(X) \leq 0$ represents the limitations in $P_{\text{fcs}}(k)$ and $P_{\text{ess}}(k)$. The $P_{\text{fcs}}(k)$ is limited in its maximum and minimum values, and in the maximum rise rate and fall rate:

$$P_{\text{fcs,lo}} \leq P_{\text{fcs}}(k) \leq P_{\text{fcs,hi}}, \tag{46}$$

and

$$P_{\mathrm{fcs}}(k-1)+\Delta P_{\mathrm{fcs,\,fall\,rate}}\cdot\Delta T \le P_{\mathrm{fcs}}(k) \le P_{\mathrm{fcs}}(k-1)+\Delta P_{\mathrm{fcs,\,rise\,rate}}\cdot\Delta T. \quad (47)$$

This can be rewritten as

$$P_{\mathrm{fcs,\,min}} \le P_{\mathrm{fcs}}(k) \le P_{\mathrm{fcs,\,max}}, \quad (48)$$

where

$$P_{\mathrm{fcs,\,max}}(k) = \max\left[P_{\mathrm{fcs,\,hi}};\quad P_{\mathrm{fcs}}(k-1)+\Delta P_{\mathrm{fcs,\,fall\,rate}}\cdot\Delta T\right], \quad (49)$$

$$P_{\mathrm{fcs,\,min}}(k) = \min\left[P_{\mathrm{fcs,\,lo}};\quad P_{\mathrm{fcs}}(k-1)+\Delta P_{\mathrm{fcs,\,rise\,rate}}\cdot\Delta T\right]. \quad (50)$$

On the other hand, $P_{\mathrm{ess}}(k)$ is limited to its maximum or minimum value depending on the actual $\mathrm{SoE}(k)$. The $\mathrm{SoE}(k)$ is also limited:

$$\mathrm{SoE}_{\mathrm{min}} \le \mathrm{SoE}(k) \le \mathrm{SoE}_{\mathrm{max}}, \quad (51)$$

thus,

$$k_{\mathrm{ess}}\cdot(\mathrm{SoE}(k)-\mathrm{SoE}_{\mathrm{max}}) \le P_{\mathrm{ess}}(k) \le k_{\mathrm{ess}}\cdot(\mathrm{SoE}(k)-\mathrm{SoE}_{\mathrm{min}}), \quad (52)$$

where k_{ess} is the constant defined in Eq. 14.

4 Simulation Results and Experimental Validation

The enunciated strategies were tested in a hybrid system consisting of the vehicle described in Table 2 of Chap. 7, provided with an FCS and a supercapacitors-based ESS with the principal parameters listed in Table 2. The supercapacitors bank is modeled according to the Sect. 4.2 of Chap. 7. The ESS can store a total of 612 *Wh*. The parameters $P_{\mathrm{fcs,\,lo}}$ and $P_{\mathrm{fcs,\,hi}}$ act as adjustment parameters, allowing to improve the fuel economy and performance according to the particular strategy and cycle. These parameters are shown in Table 3.

The strategies are tested on the four driving cycles described in Sect. 4.3 of Chap. 7: the *New European Driving Cycle* (NEDC), the *Urban Dynamometer Driving Schedule* (UDDS), the *Federal Test Procedure* (FTP) and the *Highway Fuel Economy Test* (HWFET). First, the simulation results corresponding to the three strategies are shown. Second, a comparison between the corresponding performances for the four cycles is done.

In Figs. 5, 6, 7, 8, the power split between the fuel cell system and the energy storage system, and the evolution of the SoE(t) in ESS are shown for the first strategy based on efficiency map. Figures 9, 10, 11, 12 show the power split between the fuel cell system and the energy storage system, and the evolution of the SoE(t) in ESS for

Table 2 Simulation parameters used to test the energy management strategies

Parameter	Symbol	Value	Unit
Number of supercapacitors	N_{cap}	250	–
ESS specific energy	$E_{ess,\ spec}$	6	$\mathrm{W\,h\,kg^{-1}}$
Maximum ESS energy	$E_{ess,\ max}$	612	W h
Maximum SoE	$\mathrm{SoE_{max}}$	1	–
Minimum SoE	$\mathrm{SoE_{min}}$	0.2	–
High limit of SoE	$\mathrm{SoE_{hi}}$	0.9	–
Low limit of SoE	$\mathrm{SoE_{lo}}$	0.3	–
Maximum ESS power	$P_{ess,\ max}$	255	kW
Maximum FCS power	$P_{fcs,\ max}$	15	kW
Maximum FCS rise rate power	$\Delta P_{fcs,\ rise\ rate}$	600	$\mathrm{W\,s^{-1}}$
Maximum FCS fall rate power	$\Delta P_{fcs,\ fall\ rate}$	−900	$\mathrm{W\,s^{-1}}$
Maximum FCS efficiency	$\eta_{fcs,\ max}$	0.6	–
FCS power of maximum efficiency	$P_{fcs,\ max\ eff}$	6	kW
Energy storage efficiency	η_{ess}	0.99	–
Boost converter efficiency	η_B	0.95	–
Buck/Boost converter efficiency	$\eta_{B/B}$	0.95	–

Table 3 Values of the parameters $P_{fcs,\ lo}$ and $P_{fcs,\ hi}$ used in the EMSs

	$P_{fcs,\ lo}$ [kW]	$P_{fcs,\ hi}$ [kW]
Strategy based on efficiency map	1	15
Improved strategy based on efficiency map	1	12
Strategy based on nonlinear programming	2	15

the second strategy based on efficiency map. Figures 13, 14, 15, 16 show the power split between the fuel cell system and the energy storage system, and the evolution of the SoE(t) in ESS for the third strategy, which is based on constrained nonlinear programming.

The comparison between the performance of the strategies, in terms of hydrogen consumption per kilometer, is done in Fig. 17. The comparison is done between the performance of the three strategies with respect to that corresponding to the optimal case where the consumption is minimum; the values in percentage indicate the increment in consumption with respect to the optimal case. The performance of the optimal case was estimated in Sect. 7 of Chap. 7 assuming that the entire cycle is known a priori; thus, it is possible to operate FCS during the entire cycle with maximum efficiency. In addition, the performance corresponding to the FCS pure case, where there is no hybridization, is included in Fig. 17. The analysis of the FCS pure case is performed with a 37.5 kW FCS, a power that is sufficient to fulfill the four cycles in study.

In general, the final state of charge SoE(N_c) is different from the initial state of charge SoE(0), resulting in a gain or a loss of energy in ESS over the driving cycle. Because of that, the results are corrected in order to compare the results correctly.

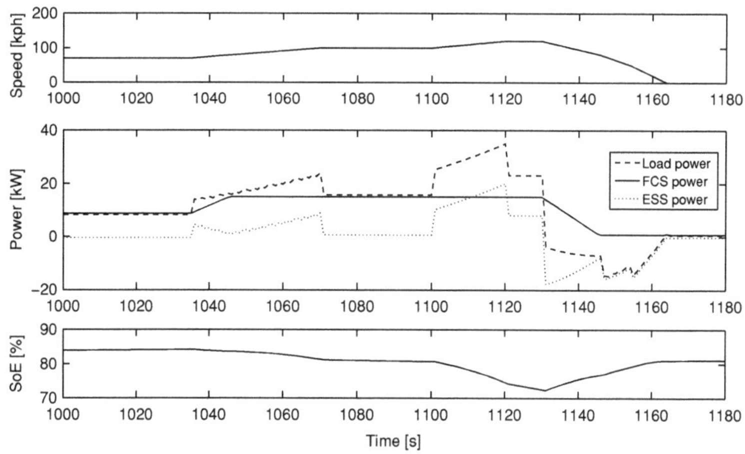

Fig. 5 Detail between 1000 and 1180 s of the results using the strategy based on efficiency map (NEDC)

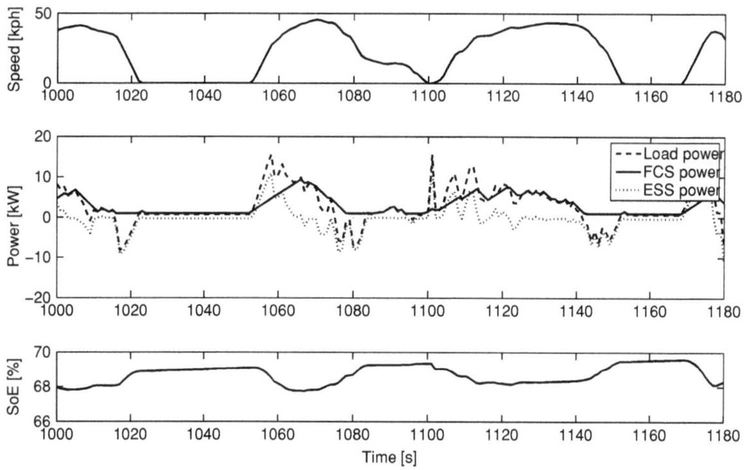

Fig. 6 Detail between 1000 and 1180 s of the results using the strategy based on efficiency map (UDDS)

The corrected consumption of hydrogen is based on the assumption that after the cycle, FCS continues to run in its point of maximum efficiency until the SoE again reaches the initial value. Thus, the corrected consumption results

$$\mathrm{Cons}_{\mathrm{H_2,corrected}} = \mathrm{Cons}_{\mathrm{H_2}} + \frac{\Delta E_{\mathrm{ess,stored}}}{\Delta H_{LHV} \cdot \eta_{\mathrm{fcs,\,max}} \cdot \eta_B \cdot \eta_{B/B} \cdot \eta_{\mathrm{ess}}}, \tag{53}$$

where, $\mathrm{Cons}_{\mathrm{H_2}}$ is the cumulative consumption of hydrogen over the cycle previous to the correction, $\Delta E_{\mathrm{ess,\,stored}}$ is the difference in the energy stored in ESS at the end of

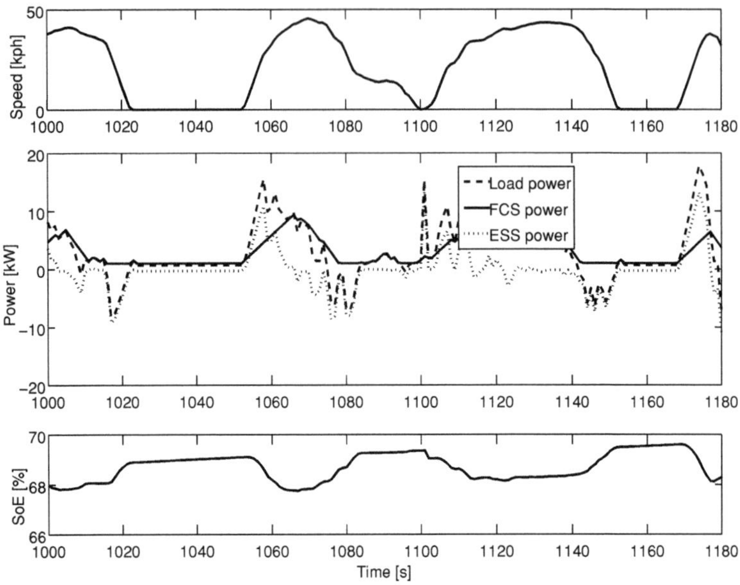

Fig. 7 Detail between 1000 and 1180 s of the Results using the strategy based on efficiency map (FTP)

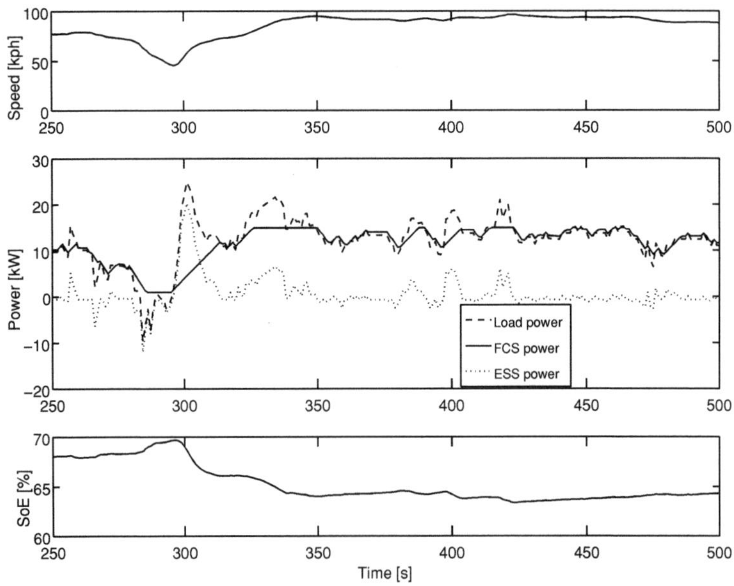

Fig. 8 Detail of the results using the strategy based on efficiency map (HWFET)

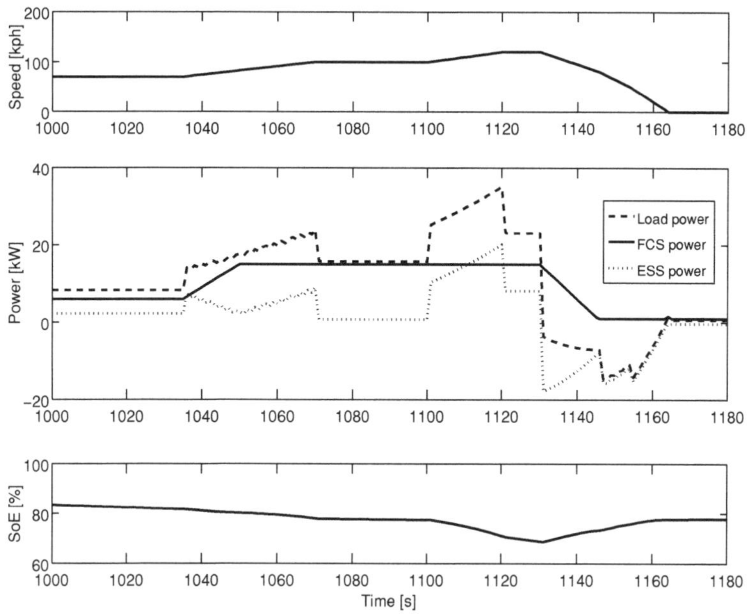

Fig. 9 Detail between 1000 and 1180 s of the power split running on NEDC, and the evolution of the SoE, using the improved strategy based on efficiency map

the cycle from the energy at the beginning of the cycle (positive if $\mathrm{SoE}(N_c) > \mathrm{SoE}(0)$ and negative in the opposite situation), ΔH_{LHV} is the lower heating value of hydrogen, $\eta_{\mathrm{fcs,\ max}}$ is the maximum efficiency of FCS and η_{ess} is the ESS efficiency.

4.1 Discussion

The analysis of the hydrogen economy in Fig. 17 shows that the three strategies have a good performance compared to the optimal case where the entire driving cycle is known a priori, which is not feasible in practice. In fact, the maximum deviation from the optimal case is 9.6% in the strategy based on efficiency map running on NEDC. The strategy based on constrained nonlinear programming gives the best performance in all the cases; however, the performance is similar to the two strategies based on efficiency map. On the other hand, compared to the pure fuel cell case in Fig. 18, the results show considerable hydrogen savings running on cycles NEDC, UDDS and FTP. On the contrary, running on HWFET the savings is exiguous. This cycle is a highway driving cycle and one of its characteristics is that the average deceleration is significantly lower than that corresponding to the other cycles (see Table 4). These results suggest that the strategies achieve the objectives satisfactorily when a significant energy is recovered from braking.

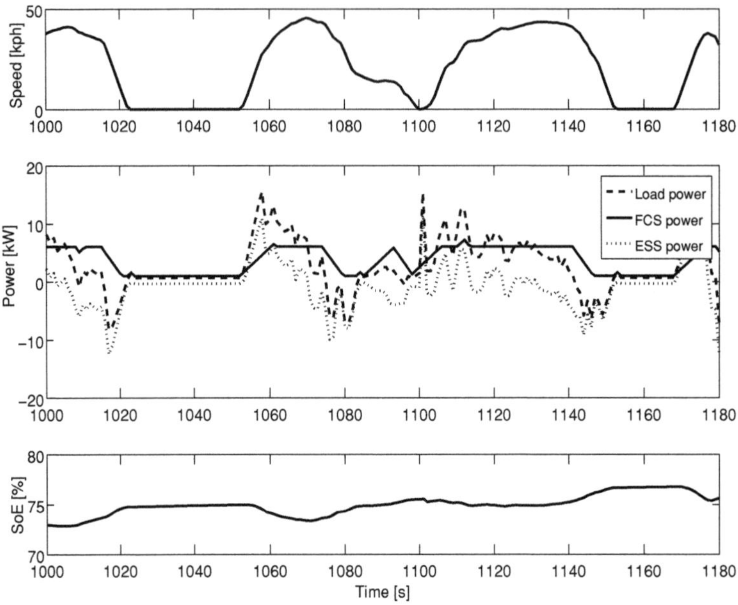

Fig. 10 Detail between 1000 and 1180 s of the power split running on UDDS, and the evolution of the SoE, using the improved strategy based on efficiency map

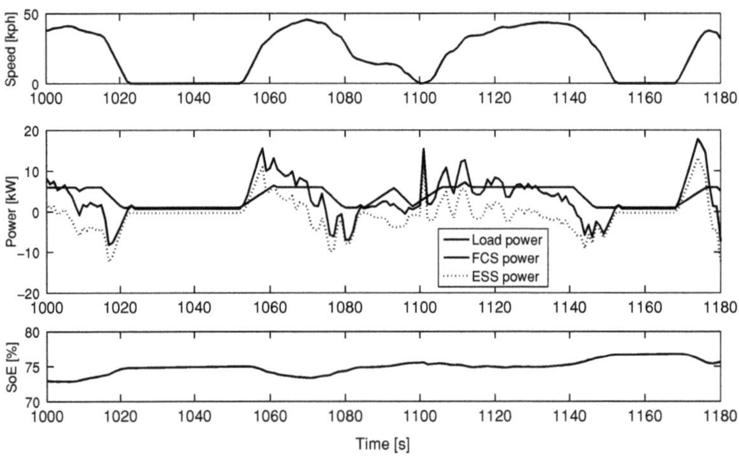

Fig. 11 Detail between 1000 and 1180 s of the power split running on FTP, and the evolution of the SoE, using the improved strategy based on efficiency map

It is remarkable that it is possible to meet the load power in the four driving cycles with a 15-kW FCS that is significantly lower to that corresponding in the pure fuel cell case with no hybridization (37.5 kW), an advantage that is translated into a

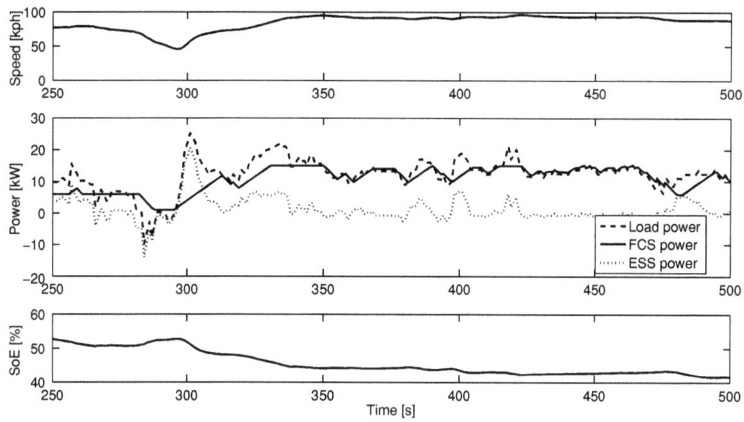

Fig. 12 Detail between 1000 and 1180 s of the power split running on HWFET, and the evolution of the SoE, using the improved strategy based on efficiency map

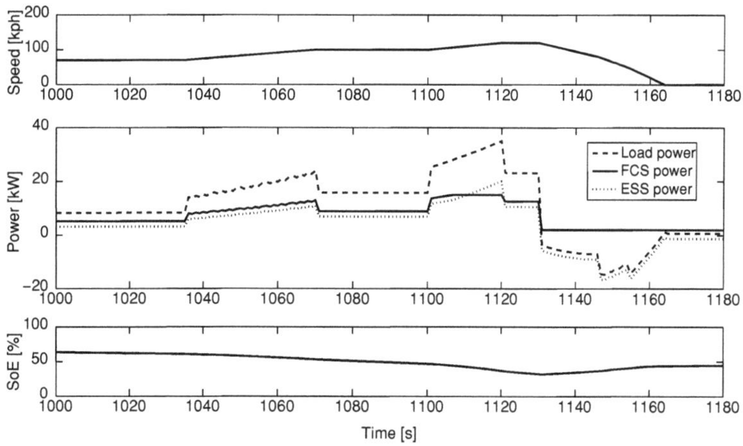

Fig. 13 Detail between 1000 and 1180 s of the power split running on NEDC, and the evolution of the SoE, using the strategy based on constrained nonlinear programming

Table 4 Average deceleration corresponding to the different standard driving cycles considered in this analysis

Cycle	Average deceleration ($m\,s^{-2}$)
NEDC	−0.79
UDDS	−0.58
FTP	−0.58
HWFET	−0.22

reduction in the production costs. This is possible thanks to the ESS power assistance in the proposed strategies.

The analysis of the influence of the initial SoE on the hydrogen consumption shows that the first and third strategies have minimal dependence, which denotes

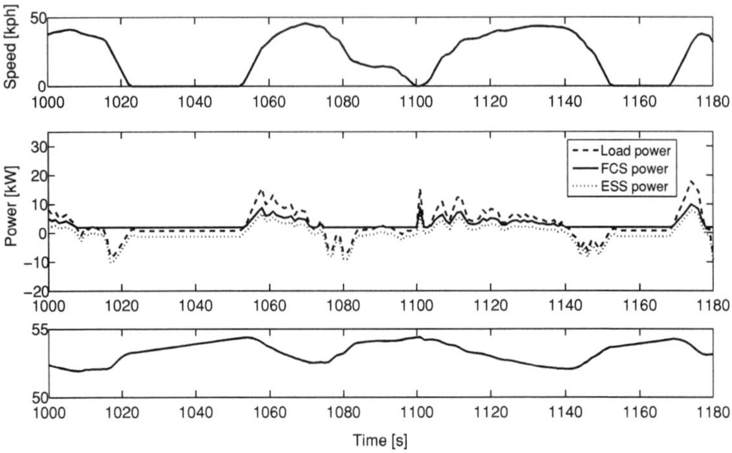

Fig. 14 Detail between 1000 and 1180 s of the power split running on UDDS, and the evolution of the SoE, using the strategy based on constrained nonlinear programming

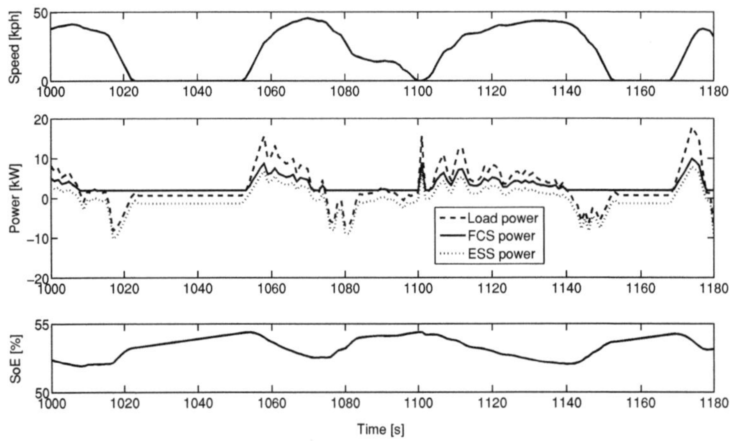

Fig. 15 Detail between 1000 and 1180 s of the power split running on FTP, and the evolution of the SoE, using the strategy based on constrained nonlinear programming

robustness to this parameter. On the other hand, the performance of the second strategy is dependent on the initial SoE, especially in the HWFET cycle. In this case, the performance is worse than in the first strategy if the initial SoE is lower than 60%. The results of the influence of the initial SoE in the three strategies running on the UDDS cycle and HWFET are shown in Fig. 19.

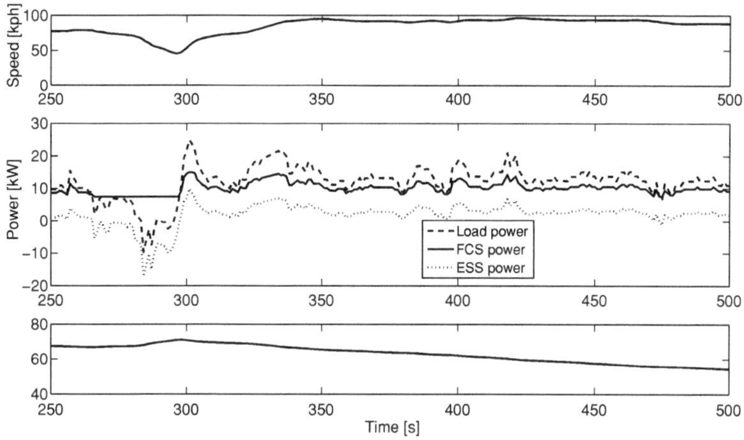

Fig. 16 Detail between 1000 and 1180 s of the power split running on HWFET, and the evolution of the SoE, using strategy based on constrained nonlinear programming

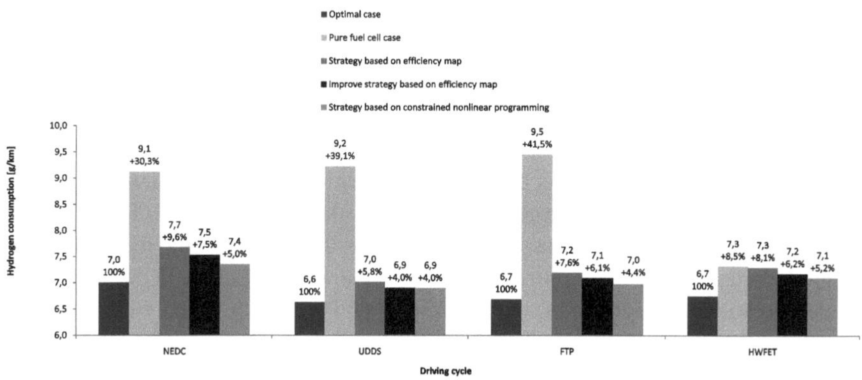

Fig. 17 Comparative between the hydrogen consumption of the proposed strategies, the optimum case and the FCS pure case. The percentages indicate the increment in consumption with respect to the optimal case

5 Experimental Validation

This section presents the implementation of the previously developed EMSs in Sect. 3 on an experimental test setup. The main objective is to show that through the use of these EMSs it is possible to operate FCS efficiently, reducing the hydrogen consumption, and thereby justifying the use of hybrid arrangements (i.e., hybridization) in fuel cell-based vehicles.

The arrangement of this section is as follows. First, the experimental setup is described. Then, the experimental results are presented showing the temporal

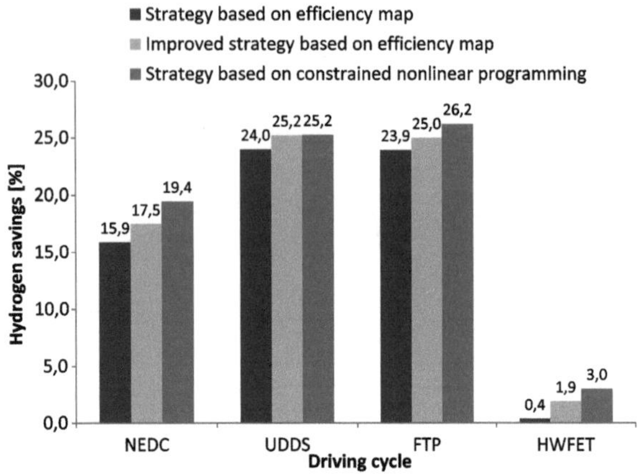

Fig. 18 Hydrogen savings with respect to the pure fuel cell case

behavior, the performance in terms of hydrogen consumption and the comparative between EMSs. Finally, the conclusions of this section are presented.

5.1 Experimental Setup

The experimental setup is a test environment in which some components of FCHV are actually present, some are emulated and some others are simulated. The objective of this setup is to reproduce, as close as possible, the FCHV behavior in order to validate the EMSs. In this experimental setup (shown in Fig. 20), FCS is a NEXA power module made by *Ballard* of 1.2 kW. The power module is fed with hydrogen from a pressurized tank and the hydrogen flow is measured with a mass flow meter (*Bronkhorst*, model *F-201AC*), an instrument device with a maximum flow capacity of 50 *SLPM*.

The vehicle power consumption is emulated with a programmable electronic load (*Höcherl&Hackl*, model *ZS1606*), a power device that can support up to 1600 W at 60 V, 150 A. This load is commanded by the *Host* computer in Fig. 20, which is in charge of carrying out EMS. The ESS is simulated with a model developed in LabVIEW[2] environment. This model is developed according to real supercapacitors from *Maxwell Technologies*. The main characteristics of these devices are listed in Table 5. The actual SoE is calculated according to this model, which is running in the *Host* computer.

The data acquisition and control system is composed of a *Host* computer and a computer running in real time, namely *RTOS* computer. The *RTOS* computer

[2] LabVIEW is the acronym for *Laboratory Virtual Instrumentation Engineering Workbench* and is a platform and development environment for a visual programming language from National Instruments. This platform is commonly used for data acquisition, instrument control and industrial automation. For more detail see [6].

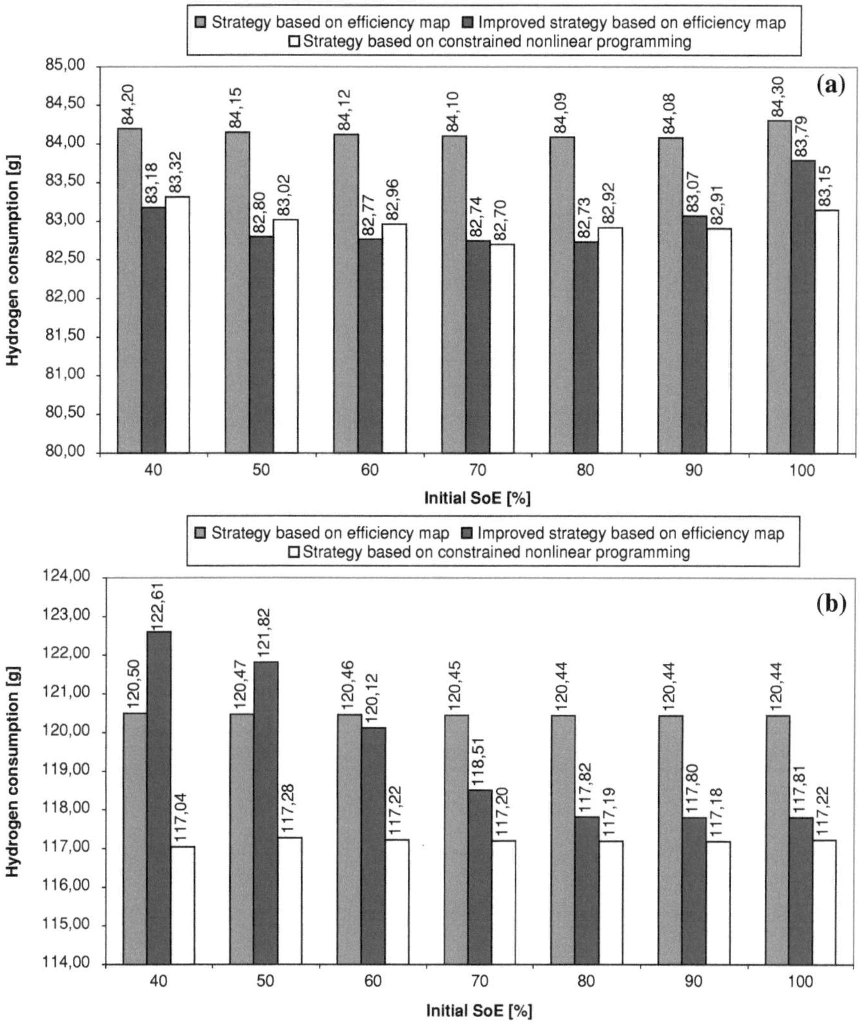

Fig. 19 Influence of the initial SoE on the hydrogen consumption in two driving cycles

communicates with the input/output (I/O) modules, made by *National Instrument*, throughout a *FPGA* target and a *PCI* bus. The two computers are connected via ethernet. The *Host* computer also allows monitoring the evolution of the system variables and commanding the system through a graphical interface developed in LabVIEW. The measured variables are the hydrogen flow, the load current and the stack voltage. These variables are measured every 200 ms.

The operation of the experimental setup is as follows. First, the FCS operation is determined according to the EMSs that were described in Sect. 3. To that end, the EMSs were programmed in the LabVIEW environment. These applications run in

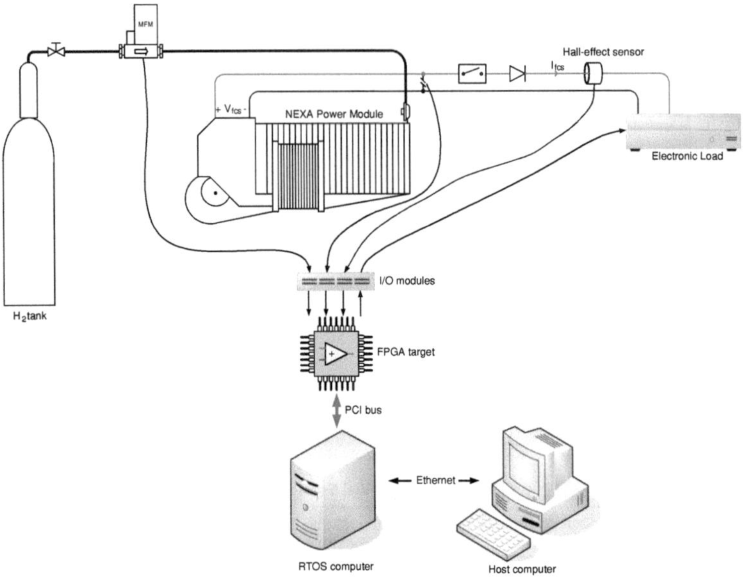

Fig. 20 Scheme showing the experimental setup to validate the EMSs

Table 5 Technical specifications of BOOSTCAP BPAK0350-15E supercapacitors from Maxwell technologies [7]

Specification	Value	Observations
Capacitance, C_R [F]	58	
Voltage, U_R [V]	15	
Maximum specific energy, [W h/kg]	3.63	Full discharge from nominal voltage ($U_R = 15$ V)
Specific power, [W/kg]	2850	Idem
Internal resistance, *DC* [Ω]	0.019	Discharging at constant current
Operating temperature, [°C]	−40 and 65	
Life time[a], $\Delta C/C_R$ [%]	≤20	From initial value after 10 years 25°C
Cycle life[b], $\Delta C/C_R$ [%]	≤20	From initial value after 500 K cycles 25°C (I = 5 A)

[a] Life time considers the aging degradation.
[b] Cycle life considers the degradation due to the amount of charge/discharge cycles.

the *RTOS* computer and assess a new request to FCS, $P_{fcs}(k)$, every 1 s. The $P_{ess}(k)$ value is determined from the power balance in the *DC* bus as in Eq. 15. Then, the $P_{fcs}(k)$ is required for the NEXA power module through the electronic load, which is commanded from the *RTOS* computer. The electronic load is operated in constant power mode, i.e., the load sinks a power in accordance with the external-programming signal from the *RTOS* computer. On the other hand, the $P_{ess}(k)$ is utilized in the ESS model, running in the *Host* computer, to actualize the actual SoE(k).

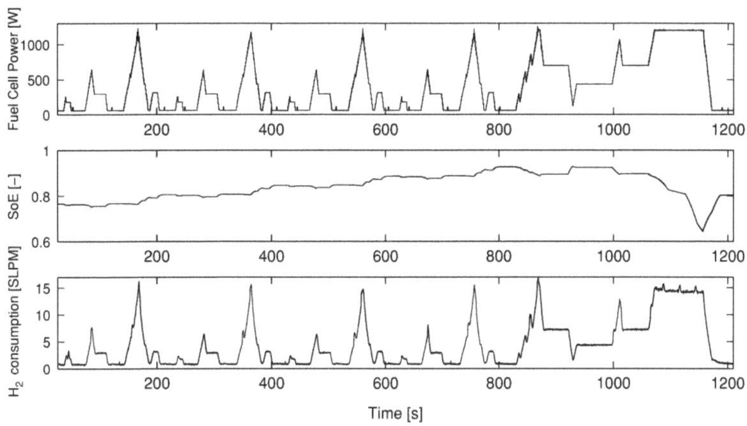

Fig. 21 Experimental results from the strategy based on efficiency map

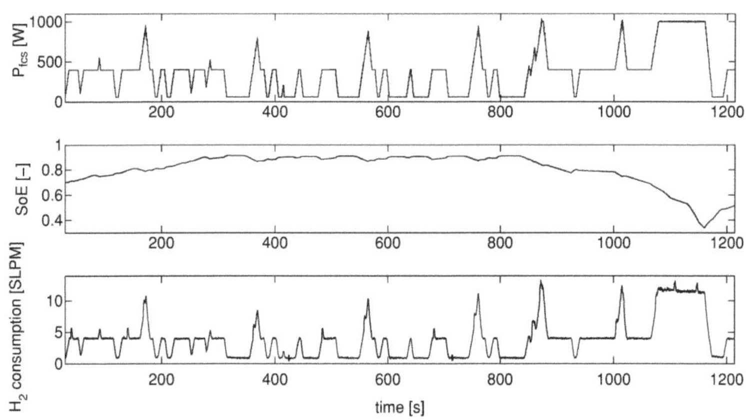

Fig. 22 Experimental results from the improved strategy based on efficiency map

5.2 Experimental Results

According to the previous explanation, the experimental setup was utilized to validate the enunciated EMSs running on the same four driving cycles as in Sect. 4. In the simulation stage, a 15-kW FCS was utilized to fulfill the driving cycles. In this stage, all the powers are scaled down with a factor of 12.5 to fit with the rated power of the NEXA power module (1.2 kW). In Figs. 21, 22, 23 are shown the experimental results showing the evolution of FCS power P_{fcs}, measuring the stack voltage and stack current; the evolution of SoE, from the ESS model; and the hydrogen consumption, measured with a mass flow meter.

The comparison of the hydrogen consumption between the strategies in the experimental setup and the strategies in the simulation environment is shown in Fig. 24.

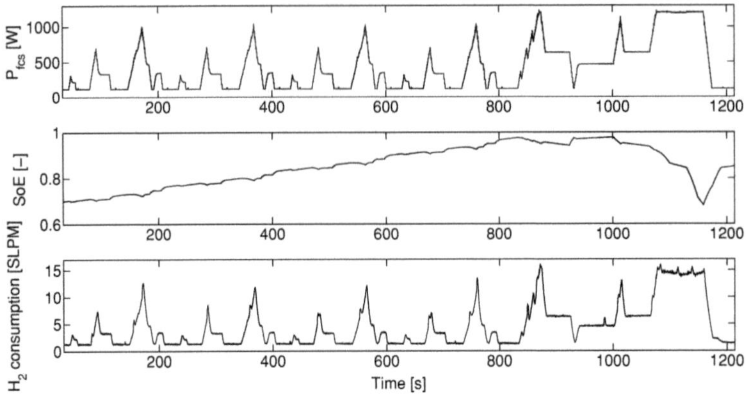

Fig. 23 Experimental results from the strategy based on constrained nonlinear programming

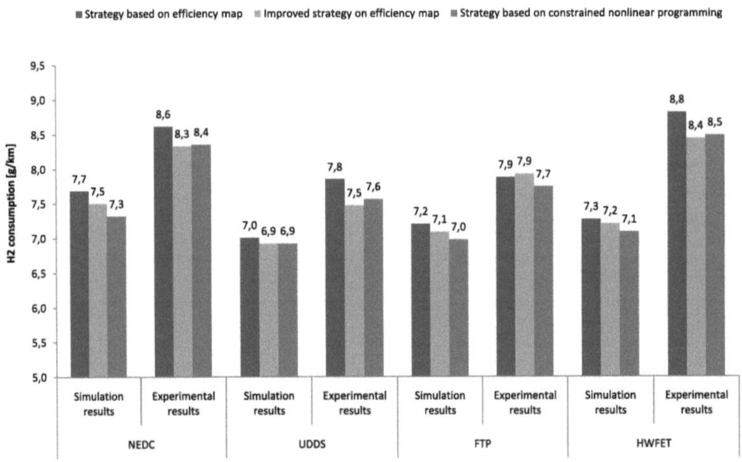

Fig. 24 Experimental validation of the proposed strategies

To do this comparison the experimental results are multiplied by the the same scale factor used to scale down the power. From this comparison, it is possible to draw some observations. First, the hydrogen consumption corresponding to the experimental setup is higher than that corresponding to the simulation setup. This is owing to the fact that the efficiency of NEXA is lower than the efficiency of the FCS model in the simulation setup as seen in Fig. 25, where it is shown the comparison between the experimental efficiency map of NEXA and the efficiency map of the FCS model in the *ADVISOR* simulator. The experimental map is obtained measuring the inlet hydrogen flow rate and using Eq. 4. Secondly, the relative consumptions in the experimental setup are similar to the experimental setup with some differences owing to discrepancies in the modeling and measurement errors.

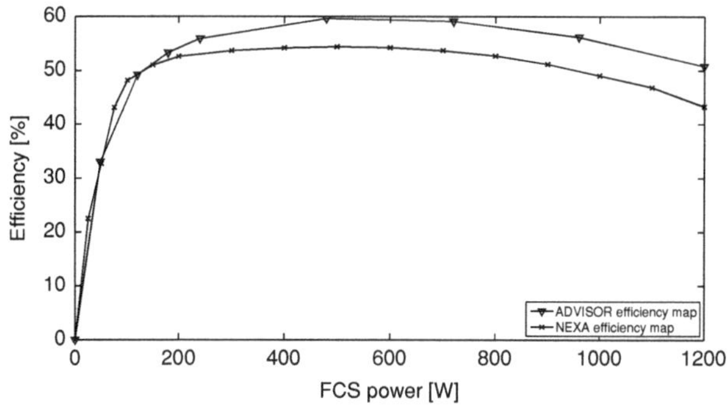

Fig. 25 Comparative between the experimental efficiency map of NEXA and the efficiency map of the FCS model in the *ADVISOR* simulator

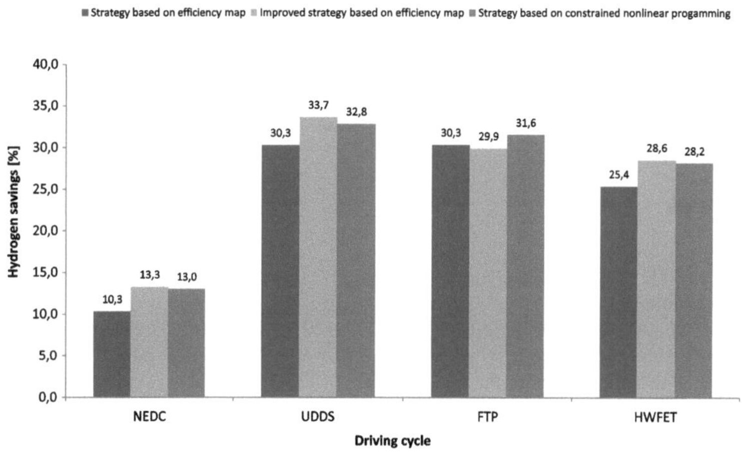

Fig. 26 Hydrogen savings achieved using the EMSs with respect to the pure fuel cell case

On the other hand, the hydrogen saving achieved using EMSs with respect to the pure fuel cell case with no hybridization is shown in Fig. 26. The results show the interest of hybridization, achieving hydrogen savings up to 33.7%.

6 Conclusions

In this chapter, the Energy Management Strategies for Fuel Cell Hybrid Vehicles were approached. The objectives of EMSs were enunciated and three strategies were proposed. These strategies are based on the knowledge of the efficiency map. The first strategy is a quasi-load-following strategy in which FCS is operated in an advantageous zone where the efficiency is high, whereas the second strategy operates

FCS preferably in its point of maximum efficiency in order to improve the hydrogen economy. In the third strategy, an optimization problem with linear constraints is resolved at each sampling period where a nonlinear cost function, which represents the hydrogen consumption, is minimized.

First, EMSs were tested in a simulation environment using four standard driving cycles and, then thet were tested in an experimental setup reproducing the FCVH behavior where some components are actually present, some are emulated and some others are simulated.

The results show that using the proposed EMSs it is possible to fulfill the load profiles, achieving a high reduction in the hydrogen consumption, a saving that is achieved by operating FCS properly and taking advantage of the energy from regenerative braking. To evaluate the savings, the hydrogen consumptions are compared with two references: the pure fuel cell case with no hybridization and the optimal case with minimum consumption where the driving cycle is known a priori.

Acknowledgments The author wishes to thank Maria Serra and Jordi Riera from the Institut de Robòtica i Informàtica Industrial (IRI-UPC-CSIC) for their contributions to this chapter.

References

1. Kessels J (2007) Energy management for automotive power nets. PhD thesis, Technische Universiteit Eindhoven
2. Murphey YL, Chen ZH, Kiliaris L, Masrur MA (2010) Intelligent power management in a vehicular system with multiple power sources. J Power Sources 2:835–846
3. Rodatz P, Paganelli G, Sciarretta A, Guzzella L (2004) Optimal power management of an experimental fuel cell/supercapacitors-powered hybrid vehicle. Control Eng Pract 13:41–53
4. Philipps F, Simons G, Schiefer K (2006) Dynamic investigation of PEFC stacks in interaction with the air supply system. J Power Sources 154(2):412–419
5. Gill P, Murray W, Wright M (1981) Practical optimization. Academic Press, London
6. National instrument. Introducing LabVIEW 8.6. Online, 2008. http://www.ni.com/labview/
7. Maxwell technologies (2009) Electrical double layer capacitor: boostcap ultracapacitor series: BPAK. http://www.maxwell.com/

Part II
PEM Fuel Cells in the Context of the Fuel Processor System with Bio-Ethanol

Design and Control of an Integrated Bio-Ethanol Processor with PEMFC

L. Nieto Degliuomini, S. Biset and M. Basualdo

Abstract The aim of this chapter is to present the important aspects of the synthesis of an integrated bio-ethanol processor plant for hydrogen production with proton exchange membrane fuel cell systems. It is based on performing a proper energy integration to determine the operating point of maximum efficiency. A large review about the different techniques for obtaining hydrogen from bio-ethanol is investigated to justify the selection of a process based on steam reforming, followed by high- and low-temperature shift reactors and preferential oxidation, coupled to a fuel cell. Applying simulation techniques with HYSYS commercial software and using its specific thermodynamic models, the performance of the complete system has been evaluated for a variety of operating conditions. These models involve mass and energy balances, chemical equilibrium and feasible heat transfer conditions. The main operating points of the variables were determined for those conditions where the endothermic nature of the reformer has a significant effect on the overall system efficiency. Then, a heuristic procedure for defining a preliminary control structure is applied via a sensitivity analysis, evaluating controllability aspects for the most critical disturbances and considering the main objectives of the process.

L. Nieto Degliuomini · M. Basualdo (✉)
CAPEG-CIFASIS-(CONICET-UNR-UPCAM), 27 de Febrero 210 bis,
S2000EZP Rosario, Argentina
e-mail: basualdo@cifasis-conicet.gov.ar

L. Nieto Degliuomini
e-mail: nieto@cifasis-conicet.gov.ar

S. Biset · M. Basualdo
CAPEG-Chemical Engineering Department, FRRo-UTN, Zeballos 1341,
S2000BQA Rosario, Argentina
e-mail: biset@frro.utn.edu.ar

M. S. Basualdo et al. (eds.), *PEM Fuel Cells with Bio-Ethanol Processor Systems*, Green Energy and Technology, DOI: 10.1007/978-1-84996-184-4_9,

1 Introduction

In an ideal project, dynamics and control strategies are considered during the process synthesis and design activities. This issue grows increasingly important as plants become more highly integrated with complex configurations, recycle streams and energy integration. Competitive economic pressures, safety issues and environmental concerns have all contributed to this. However, if a process control engineer becomes involved in the process design early enough, it would be better in the distant future. In this way it is most probable for the plant to provide more stable operation and less variability in the product quality.

Luyben et al. [1] remarked that a consideration of the dynamics and control strategies for new process designs has a much larger positive economic impact (when the design can potentially be modified) compared with control strategy upgrades on an existing process (with a fixed design). Most important, by working together with process engineers and plant engineers, the flowsheet can be changed until every objective is satisfied and the most profitable process, viewed over the entire lifetime of the project, can be developed. This involves making trade-offs between steady-state investment economics and dynamic performance measured in uptime, throughput, product quality and yield. Luyben et al. [1] considered that a close connection between process design and controllability exists since the way in which a process is designed fundamentally determines its inherent controllability. It means qualitatively how well the process rejects disturbances and how easily it moves from one operating condition to another.

Based on these concepts, through this chapter the interaction between process design and control plays the central role. In this context, the final flowsheet presented by Francesconi et al. [2] is studied in detail and adopted as a starting point for proposing a plantwide control strategy based on different kinds of philosophies for doing it. Then, according to the controllability properties of this flowsheet, the necessity of redesigning some parts of it may be concluded. Therefore, the bio-ethanol processor system (BPS) with proton exchange membrane fuel cell (PEMFC) constitutes a new plant where new technology is developing day by day to improve it. In Sect. 2 the reader will find an extensive review about several topics concerning the advances on this issue.

Since this flowsheet is one of the first presenting an overall integrated plant, till date, it means that there is no information about any successful plantwide control strategy to compare. Therefore, the real benefits of using particular techniques are evaluated through their capacity for achieving specific objectives.

Hence, this chapter begins with the steady-state model of the bio-ethanol processor for hydrogen production and fuel cell operation, provided by the synthesis stage given in Francesconi et al. [2]. The rigorous model, implemented in HYSYS flow sheeting software [3], consists of mass and energy balances, chemical equilibrium and feasible heat transfer condition elements. BPS converts a bio-ethanol fuel into a hydrogen (H_2) rich mixture that is directly fed to the PEMFC. The overall plant studied here includes an ethanol steam reforming (ESR), water gas shift (WGS) and preferential oxidation

(PrOx) reactors for the fuel processing, followed by a PEMFC and a post-combustion unit. This commercial simulator was used to compute the operating conditions for the process units under the most critical disturbances. Maximum heat integration within the system is necessary to achieve acceptable net electrical efficiency levels. The use of process integration techniques applying pinch analysis has already been reported for the design of a heat exchange network for integrated PEMFC [4]. In this chapter, the heat exchanger network involving the reformer, burner, gas cleaning units and the PEMFC stack was modeled using the LNG exchanger model which is a tool available in HYSYS. It allows to identify the most successful heat exchange opportunities, and to define the optimal operating conditions of the ethanol processor for obtaining the best global efficiency considering the plant balances. The steps followed during the conceptual design of this plant were reproduced so as to check whether the final operating point, proposed by Francesconi et al. [2], to achieve a highly heat integrated system can be controllable. The study presented here, about cost and performance requirements of the total power plant, is extremely useful for understanding the system-level interactions, implications on system performance and for defining a preliminary control structure synthesis for the most critical disturbances.

Initially, some more conventional control theoretic and heuristic tools, such as sensitivity analysis to critical perturbations for complete chemical plants, are accounted. They are taken from Luyben and Luyben [5] and Luyben et al. [1]. This methodology is characterized by the necessity of having an in-depth knowledge of the process. Therefore, for the implementation, some useful assumptions must be done. Hence, the recommendations given in Pukrushpan [6], were valuable for integrating to the synthesis the main objectives stated by Francesconi et al. [2]. Although in the work of Pukrushpan [6] the feed to the BPS was methane and did not consider any heat integration system, some analogous aspects were helpful for providing the preliminary considerations.

In this context, the problem to be solved is how to handle the integrated system efficiency accounting the best operating conditions of the reformer, the effect of the inlet temperature to the WGS and PrOx reactors, combustion preheating, expander unit, crude ethanol as fuel and the PEMFC specifications.

The analysis considered here is extended in the following chapter for constructing the process dynamic simulation working with the control structure chosen here. Meanwhile, in Chap. 12 new methodologies, highly systematic, for optimal sensor placement and plantwide control structure design will be tested with the dynamic model.

2 Description of the Ethanol Processor

The fuel processing plant has been built according to the recommendations of Francesconi et al. [2]. It chemically converts ethanol into hydrogen including a cleanup equipment for this last component, the fuel cell stack electrochemically converts the hydrogen energy into electric power. Additionally, associated

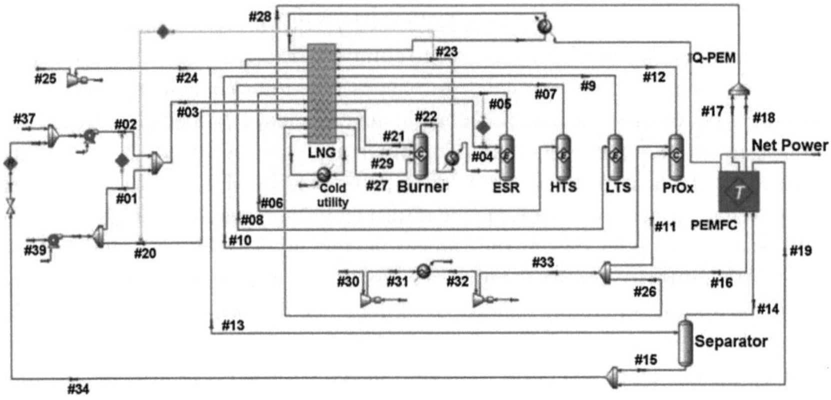

Fig. 1 Plant of the fuel processor system with PEMFC model in HYSYS

Table 1 Heating and cooling demands for the system streams

Stream number	#04	#05	#07	#09	#12	#14	#28
Temperature (°C)	709	709	538.9	236.7	406	80	80
Pressure (atm)	3	3	3	3	3	3	3
Mole flow (kmol h^{-1})	0.03670	0.06284	0.06284	0.06284	0.06284	0.06360	0.17497
Molar fraction hydrogen	0.8000	0.48602	0.52206	0.59209	0.54734	0.56642	0.04117
H_2O	0.0000	0.28037	0.24433	0.17430	0.18431	0.15588	0.15589
Methane	0.0000	0.02561	0.02561	0.02561	0.02445	0.02523	0.00919
CO	0.0000	0.11235	0.07631	0.06278	0.00002	0.00002	0.00001
CO_2	0.0000	9.56588	0.13169	0.20173	0.19856	0.20547	0.07467
Ethanol	0.2000	0.00000	0.00000	0.00000	0.00000	0.00000	0.00000
Oxygen	0.0000	0.00000	0.00000	0.00000	0.00008	0.00008	0.08233
Nitrogen	0.0000	0.00000	0.00000	0.00000	0.04525	0.04682	0.63667

equipment for heat, air and water management, and auxiliary equipment such as pumps and blowers are accounted. Figure 1 shows the main components: ethanol steam reformer, high (HTS) and low (LTS) temperature water gas shift reactors, a preferential oxidation reactor of CO, a proton exchange membrane fuel cell, a combustor, pumps, compressors and an expander. For the steady-state model the pressure drops are neglected and the operating pressure is fixed at 3 atm. The inlet flows depicted in Fig. 1 are as follows: Stream #01 is the ethanol flow provided at 25°C. Stream #02 is the water flow required for the steam reformer whose flow rate is controlled by the water to the ethanol molar ratio. Stream #30 is the airflow required for the operation of combustor, fuel cell and PrOx units. In the integrated system, the air compressor will be driven by a turbine expanding the post combustion gases. The compressor, expander and pumps isentropic efficiencies are 75%. The inlet air conditions are 1 atm, 25°C. The most important operating conditions are summarized in Tables 1 and 2.

Table 2 Fuel processor streams connected to the LNG unit

Pass		Input (°C)	Output (°C)	Q (kW)
#03–#04 (vaporizing)	Cold	61.32	126	0.45
#03–#04 (reheater)	Cold	127	709	0.32
#28–#29	Cold	80	500	0.67
#20–#21	Cold	25	300	0.09
#26–#27	Cold	127.2	300	0.00
Cold utility	Cold	20	25	2.36
ESR duty	Cold	709	709	0.41
#05–#06	Hot	709	500	0.13
#07–#08	Hot	538.9	150	0.23
#12–#13	Hot	406	80	0.23
#23–#24	Hot	850	287	0.95
#22–#23	Hot	1072	850	0.41
Q-PEMFC	Hot	65	55	2.27

The ideal unit cell voltage is calculated through Eq. 1, adopted from Godat and Marechal [7]:

$$E_{\text{rev}} = E_{\text{rev}}^{0}(T_{\text{cell}}) + \frac{R_g \cdot T_{\text{cell}}}{2F} \cdot \left[\ln\left(p_{\text{H}_2,a}^{*}\right) + \frac{1}{2}\ln\left(p_{\text{O}_2,c}^{*}\right)\right], \quad (1)$$

where E_{rev} is the voltage calculated for the cell temperature (T_{Cell}), and $p_{\text{H}_2,a}^{*}$ and $p_{\text{O}_2,c}^{*}$ are the partial pressure for H_2 and O_2 averaged (arithmetic mean) between the inlet and outlet conditions.

Useful work (electrical energy) is obtained from a fuel cell only when a current is drawn, but the actual cell voltage (V_{Cell}) is decreased from its equilibrium thermodynamic potential (E_{rev}) because of irreversible losses. When current flows, a deviation from the thermodynamic potential occurs corresponding to the electrical work performed by the cell. Therefore, the expression of the voltage of a single cell is

$$V_{\text{Cell}} = E_{\text{rev}} - \varphi, \quad (2)$$

where φ is the drop potential due to the operation irreversibility which represents the deviation from the equilibrium value. The actual electrical power generated by the cell (P_{Cell}) can then be calculated from

$$P_{\text{Cell}} = V_{\text{Cell}}\, i_{\text{Cell}}. \quad (3)$$

The operating voltage is defined as the power level at which unit cell voltage drops to 0.5 V from the ideal voltage [5]. The current i_{Cell} is related to the hydrogen molar flow rate at the anode:

$$i_{\text{Cell}} = 2F\left(f_{H_2,a}^{\text{in}} - f_{H_2,a}^{\text{out}}\right). \quad (4)$$

Table 3 Optimal values for the key system variables

Process variables	Description	Optimal value
T_{Ref}	Temperature reactor ESR	709°C
THTS	Temperature reactor HTS	500°C
TLTS	Temperature reactor LTS	150°C
TPrOx	Temperature reactor PrOx	237°C
$R_{w/e}$	Molar ratio H_2O–Ethanol	4
O_2 /CO	Feed relationship O_2–CO for the PrOx	2
O_2/H_2	Flor relationship O_2–H_2 at the PEMFC	4.76
P_{Cell}	Power cell	0.8027 kW
XCO	Molar Fraction de CO	10 ppm
P_{ESR}	Pressure at ESR	304 kPa
P_{HTS}	Pressure at HTS	304 kPa
P_{LTS}	Pressure at LTS	304 kPa
PPrOx	Pressure at PrOx	304 kPa
P_{Burner}	Pressure at the Burner	304 kPa
T_{Cell}	Cell Temperature	80°C
$P_{\#Cathode\ exhausted} - P_{\#Anode\ exhausted}$	Pressure difference at PEMFC	Min
H_{Mem}	Membrane humidity at PEMFC	80%

By doing an energy balance over the cell from the inlet to the outlet conditions, the heat produced by the cell which has to be removed is given by

$$Q_{PEMFC} = \sum_{i=1}^{inlets} f_i h_i \left(T_{Cell}\right) - \sum_{o=1}^{outlets} f_o h_o \left(T_{Cell}\right) - P_{Cell}. \tag{5}$$

It is assumed as isothermal and isobaric PEMFC.

2.1 Main Process Variables to be Accounted

Francesconi et al. [2] defined the main BPS variables and their optimal values so as to minimize the energy consumption. On the other hand, the most critical variables for PEMFC were explained in Pukrushpan [6]. In this chapter, both references were taken into account for listing the main process variables given in Table 3.

2.2 Ethanol Steam Reforming

There are numerous works that study the production of hydrogen through steam reforming of ethanol, focused on the development of new catalysts in order to obtain better conversions, be more selective and capable of working at low temperatures. A complete review of the advances made in the last few years can be found in references

[8–10]. On the other hand, there are very few publications boarding kinetic aspects, configuration and analysis of the reactor.

2.2.1 Reaction and Pathways

Despite the apparent simplicity of the stoichiometric reaction between ethanol and water for the production of hydrogen:

$$CH_3CH_2OH + 3H_2O \leftrightarrow 2CO_2 + 6H_2,$$

actually, steam reforming involves a complex system of reactions, which is why selectivity towards hydrogen is affected by a number of secondary unwanted reactions. Also, the efficiency depends on the process variables such as temperature, reactives ratios, speed, among others and, obviously, the activity and selectivity of the chosen catalysts.

Methane (CH_4), acetaldehyde (CH_3CHO), acetone (CH_3COCH_3), ethylene (C_2H_4), are all undesirable products, because they compete with the molecular hydrogen for the atomic hydrogen. The obtained distribution of products with different catalysts can be explained considering their activities towards the mechanism of the specific implied reactions. A possible sequence involved for hydrogen production could be the following, adapted from [11]:

- Ethanol dehydrogenation to acetaldehyde followed by decarbonilation or CH_3 CHO reforming:
  - Dehydrogenation:

$$C_2H_5OH \rightarrow CH_3CHO + H_2 \qquad \Delta H^0_{298} = 70.6\,\text{kJ mol}^{-1}$$

  - Decarbonilation:

$$CH_3CHO \rightarrow CH_4 + CO \qquad \Delta H^0_{298} = -21.1\,\text{mol}^{-1}$$

  - Reforming:

$$CH_3CHO + H_2O \rightarrow CO_2 + H_2 + CH_4 \qquad \Delta H^0_{298} = -63.3\,\text{kJ mol}^{-1}$$
$$CH_3CHO + H_2O \rightarrow 2 \cdot CO + 3 \cdot H_2 \qquad \Delta H^0_{298} = 184.2\,\text{kJ mol}^{-1}$$

- Decomposition or ethanol "cracking", followed by methane vapor reforming:
  - Decomposition:

$$C_2H_5OH \rightarrow CH_4 + CO + H_2 \qquad \Delta H^0_{298} = 49.5\,\text{kJ mol}^{-1}$$
$$2 \cdot C_2H_5OH \rightarrow 3 \cdot CH_4 + CO_2 \qquad \Delta H^0_{298} = -74.3\,\text{kJ mol}^{-1}$$

– Reforming:

$$CH_4 + 2 \cdot H_2O \leftrightarrow CO_2 + 4 \cdot H_2 \qquad \Delta H^0_{298} = 163.1\,\text{kJ mol}^{-1}$$
$$CH_4 + H_2O \leftrightarrow CO + 3 \cdot H_2 \qquad \Delta H^0_{298} = 206.14\,\text{kJ mol}^{-1}$$

- Dehydration to ethylene and water, followed by ethylene polymerization in the form of coke:
  – Dehydration:

$$C_2H_5OH \rightarrow C_2H_4 + H_2O \qquad \Delta H^0_{298} = 46.3\,\text{kJ mol}^{-1}$$

  – Polymerization:

$$C_2H_4 \rightarrow \text{coke}$$

- Decomposition of ethanol in acetone, followed by steam reforming:
  – Decomposition:

$$2 \cdot C_2H_5OH \rightarrow CH_3COCH_3 + CO + 3 \cdot H_2 \qquad \Delta H^0_{298} = 70.9\,\text{kJ mol}^{-1}$$

  – Reforming:

$$CH_3COCH_3 + 2 \cdot H_2O \leftrightarrow 3 \cdot CO + 5 \cdot H_2 \qquad \Delta H^0_{298} = 367.9\,\text{kJ mol}^{-1}$$

- Vapor ethanol reforming to synthesis gas ($CO + H_2$):

$$C_2H_5OH + H_2O \rightarrow 2 \cdot CO + 4 \cdot H_2 \qquad \Delta H^0_{298} = 254.8\,\text{kJ mol}^{-1}$$

- Water gas shift reaction:

$$CO + H_2O \rightarrow CO_2 + H_2 \qquad \Delta H^0_{298} = -41.1\,\text{kJ mol}^{-1}$$

- Methanation:

$$CO + 3 \cdot H_2 \rightarrow CH_4 + H_2O \qquad \Delta H^0_{298} = -205.3\,\text{kJ mol}^{-1}$$
$$CO_2 + 4 \cdot H_2 \leftrightarrow CH_4 + 2 \cdot H_2O \qquad \Delta H^0_{298} = -163.1\,\text{kJ mol}^{-1}$$

- Coke from methane decomposition:

$$CH_4 \rightarrow 2 \cdot H_2 + C \qquad \Delta H^0_{298} = 74.9\,\text{kJ mol}^{-1}$$

- Acetic acid formation:

$$C_2H_5OH + H_2O \rightarrow CH_3COOH + 2 \cdot H_2 \qquad \Delta H^0_{298} = 40.9\,\text{kJ mol}^{-1}$$

The predominant reaction is the ethanol dehydrogenation to acetaldehyde, which is an intermediate product in the ethanol steam reforming stage.

2.2.2 Thermodynamic Aspects

Thermodynamics of ethanol reforming has been extensively studied [12–16] showing that the ethanol reforming is feasible for temperatures above 203°C, being methane, carbon oxides and hydrogen the main products.

The effect of operating parameters over the equilibrium composition can be determined by stoichiometric methods [16] or non-stoichiometric [12, 13]. By the non-stoichiometric technique, the equilibrium composition of the system is found with direct minimization of the Gibbs free energy for a set of chemical species without specifying the possible reactions that can occur in the system. On the other hand, the stoichiometric technique describes the system by a set of reactions stoichiometrically independent, which are chosen arbitrarily from a set of possible reactions. Fishtik et al. [14] suggested that studying the system choosing arbitrarily the set of chemical reactions can lead to erroneous conclusions and propose an analysis with response answers, independent methodology from the initial set of reactions. This systematic allows to detect the dominant reactions to be used in thermodynamic studies or subsequent kinetic analysis.

2.2.3 Catalysts

There are many utilized catalysts for the production of hydrogen by ethanol steam reforming. Nowadays, the main references are centered in the search for better catalysts able to improve conversions and to work at low temperatures [17–25] and presenting more selectivity. The ethanol steam reforming over Nickel (Ni), Cobalt (Co), Nickel/Copper (Ni/Cu), and noble metals [Palladium (Pd), Platinum (Pt), Rhodium (Rh)] has been extensively studied. The main concern is to develop an active catalyzer that inhibits the formation of coke and the production of CO. A short description and analysis is presented for the different groups of catalysts:

- Cobalt catalysts

Cobalt-based catalysts are considered appropriate systems to reform ethanol. Llorca et al. [20] obtained hydrogen free of CO from ethanol reforming using a mix with molar ratio 13:1 Water/ethanol in the rank of 300–500°C over Co-based catalysts in various supports. They observed a minor conversion over the Co/Al_2O_3 catalyst. Ethanol dehydration to ethylene takes place in a big extent due to the acid behavior of Al_2O_3. The Co/MgO based catalyzer shows a low conversion around 30%, and the main reaction was the dehydrogenation of ethanol to acetaldehyde. At low temperatures, 100% of conversion of ethanol was obtained over CO/V_2O_5, nearly 84% of ethanol was converted by means of dehydrogenation into acetaldehyde, while the rest was by the ethanol reforming reaction. Co/ZnO exhibited the highest performance of all the utilized catalysts. 100% of conversion of ethanol was reached and the highest selectivity towards H_2, and CO_2 by mole of ethanol were obtained without deactivating the catalyst. Llorca et al. [26] proposed Co catalyzers supported over ZnO. The use of $Co(CO)_8$ as a precursor produces a catalyzer highly stable that allows the production of H_2 free from CO at low temperatures (350°C).

They concluded that the method of preparation of the catalyzer affects its functioning and its structural characteristics.

Benito et al. [24] analyzed different active phases (Ni, Cu and Co) supported over Zirconium. The catalyst based on Co/ZrO_2 shows the greatest selectivity and a good resistance to carbon deposition. According to reported results it is possible to reach total conversion at temperatures higher than 600°C obtaining a high selectivity towards H_2 (70%), being CO_2, CO and CH_4 the only by-products obtained. The reaction gives 4.25 hydrogen moles by mole of fresh ethanol fed.

- Nickel based catalyzers

Galvita et al. [18] worked with water/ethanol molar ratios of 3 and 8, and they obtained the greatest conversion over a catalyst of Ni/MgO at a ratio of 8 (28% v/v ethanol). They investigated the reforming of ethanol for the production of synthesis gas in a fixed bed catalytic reactor of two stages. The reaction conditions were as follows, temperature in the segment of 210 to 380°C; 1 atm pressure; and water/ethanol ratio of 1.04 and 8.1. In the first bed, ethanol was converted into a mix of methane, carbon oxides and hydrogen over a Pd/C catalyzer (Pd supported over Sibunit, a carbon porous special material), then the mix was converted to synthesis gas over a Ni catalyzer for the reforming of methane. They observed that the conversion of ethanol rises with temperature, which reaches 100% at 330 and 360°C for a ratio of 8.1 and 1.04, respectively. They concluded that the use of a fixed bed reactor of two stages prevents the coke formation and provides a close to equilibrium conversion.

Auprete et al. [19] also studied the effects of different metals (Rh, Pt, Ni, Cu, Zn and Fe) and the role of the supports (γ–Al_2O_3, 12% CeO_2–Al_2O_3, CeO_2 and $Ce_{0.63}Zn_{0.37}O_2$) on the reforming of ethanol. The authors proposed a catalytic formulation free of any promotor of the water gas shift reaction. The experiments took place at water/ethanol ratios of 3 and constant temperatures between 500 and 800°C and 1 atm. At 700°C the Rh and Ni catalyzers supported in γ-Al_2O_3 rise as the most actives and selectives in the reforming reaction of ethanol. The Ni/Al_2O_3 reached the higher conversion but with low selectivity to CO_2 compared with the Rh/Al_2O_3. They also investigated the role of the oxides as catalyzers support based on Rh and Ni. The results obtained at 600°C shows that the activity of the Rh based catalyzer follows the next descending order: $Rh/Ce_{0.63}Zn_{0.37}O_2 > Rh/CeO_2–Al_2O_3 > Rh/CeO_2 > Rh/\gamma\text{-}Al_2O_3$. They observed a similar tendency for the nickel case.

Comas et al. [23] examined the steam reforming over a Ni/Al_2O_3 catalyzer in the rank of 300 to 500°C. They concluded that at high temperatures, over 500°C, and high water/ethanol ratios (6) the conversion of hydrogen Eq. 2 and selectivity (91%) is augmented. The excess of water in the feed increases the methane reforming and suppresses the carbon deposition.

- Noble metal catalysts

Ethanol reforming over Rh-based catalysts has been exhaustively studied. Liguras et al. [27] found that Rh-supported catalyzers are acceptable under severe conditions (800°C) and can be used for the production of hydrogen in fuel cell applications.

Cavallaro et al. [21] reported that impregnated Rh supported over γ-Al_2O_3 is highly appropriate for ethanol steam reforming. The evaluation of Rh/γ-Al_2O_3 as a catalyzer at 650°C shown that the main products of the reaction are CO_2, CO, CH_4 and CH_3CHO. The stability of the catalyzer was investigated with and without oxygen, they observed that the catalyzer is quickly deactivated without oxygen, the presence of O_2 increases stability and only 10% of the activity is lost after 95 h of operation.

Cavallaro and Freni [17] investigated the ethanol steam reforming over a CuO/ZnO/ Al_2O_3catalyzer and others. The experiments were made in a fixed bed reactor at 1 atm pressure, in the rank of temperatures from 357 to 477°C and with water/ethanol molar ratios of 6 and 10. At temperatures around 377°C they did not obtain the remaining intermediate products such as acetic acid, acetaldehyde and ethyl acetate. These compounds were present at temperatures closer to 325°C where the production of H_2 and CO_2 is low. The production of H_2, CO_2 and CO increased as temperature risen.

Duan and Senkan [25] analyzed the conversion and selectivity of several catalyzers testing various metals as reaction promoters. They evaluate nearly 840 different materials combining 42 elements of the periodic table, impregnated in porous particles of γ-Al_2O_3, SiO_2, TiO_2, CeO_2, and ZrO_2/Y_2O_3. The catalyzers were evaluated under identical operating conditions at 300°C and 1 atm pressure, using a gaseous mix of 2% C_2H_5 OH and 12% H_2 O within a carrier gas (Helium). By means of this systematic investigation they found that among all the explored materials, the catalysts based on Pt/TiO_2 and Pt/CeO_2 obtain the higher ethanol conversion (>90%) and hydrogen selectivity (≈30%).

2.2.4 Kinetic Aspects

Despite the number of studies focusing on the development of active catalysts for the steam reforming of ethanol, only few of them give information about the kinetic expressions. Only recently some publications have appeared offering some starting point to understand the aspects of chemical kinetics and use them for reactor design purposes. This is due to the complexity of the reformer and the multiple reactions that may occur simultaneously if they are not properly driven through the operational conditions.

Akande et al. [28] determined a kinetic expression for the crude ethanol steam reforming over a 15% Ni–Al_2O_3 catalyzer prepared by co-precipitation. The kinetic experiments were carried out under atmospheric conditions in a fixed bed tubular reactor in the rank of 320–520°C. They assumed a Rideal–Eley mechanism for the obtainment of the kinetic expression considering that the limiting step of the reaction is the dissociation of absorbed crude ethanol.

Sahoo et al. [29] made a kinetic study of the ethanol reforming over a Co/Al_2O_3 obtained by the impregnation method. Using the Langmuir–Hinshelwood mechanism, they developed a kinetic expression considering the superficial reaction mechanisms of the ethanol steam-reforming reaction, water gas shift reaction and ethanol decomposition. The proposed mechanism is based on the analysis of the distribution of obtained products working at temperatures between 400 and 700°C and with the water/ethanol ratio varying between 1 and 8. They indicated that the formation of acetaldehyde from the ethoxi group (CH_3CH_2O–) is the limiting step of the reaction.

Akpan et al. [30] formulated kinetic models under the assumption of the Langmuir-Hinshelwood–Hougen–Watson and Eley–Rideal mechanisms to describe the kinetics of the production of hydrogen for the reforming of crude concentrated ethanol. They used an Ni- based commercial catalyst under atmospheric pressure in the rank of temperature from 400°C to 590°C. They simulated the reforming process with a 2D pseudo-homogeneous model.

2.3 Water Gas Shift Reactors

The carbon monoxide produced from the reforming reaction must be reduced to a concentration of about 10 parts per million in volume (ppm) due to the poisoning that the CO adsorption produces over the catalyzer of the fuel cell. The primary elimination of the CO is produced in the WGS reactors, these units produce the WGS exothermic reaction supposed to be in equilibrium. This stage also generates additional H_2. The WGS reaction habitually takes place in two adiabatic reactors in series with a heat exchanger-cooler between them.

The first reactor, high temperature water gas shift, which operates typically at 350–550°C. The HTS reactor uses a ferric oxide chrome-promoted catalyzer. The second reactor, low temperature water gas shift, operates at lower temperature, between 150–250°C, using a copper–zinc catalyzer supported over aluminium. The LTS is capable of achieving a CO residual concentration of about 0.5–1.5% in dry volume. For the integration of the fuel cell, the gaseous current exiting the WGS stages needs to be processed in additional modules. It is considered the WGS reactors' configuration proposed by [31] with two adiabatic reactors, HTS and LTS, being the inlet temperature of the first reactor 500°C and 150°C for the second.

2.3.1 Reaction

The water gas shift reaction is extensively applied at the industrial stage. It is used in the synthesis of ammonia and production of hydrogen by means of hydrocarbon reforming. The CO reacts with steam according to the following scheme:

$$CO + H_2O \rightarrow CO_2 + H_2 \qquad \Delta H^0_{298} = -41.1\,\text{kJ}\,\text{mol}^{-1}$$

Industrially, the main objective of the WGS is to increase and adjust the molar ratio H_2/CO in the synthesis gas and to remove the CO from the exit gases.

The reaction is moderately exothermic and its equilibrium constant decreases with temperature; high conversions are favored by low temperatures. The equilibrium position is not virtually affected by pressure. Under adiabatic conditions the conversion in a single-bed catalyst is limited thermodynamically, the reaction proceeds and the heat of reaction increases the operation temperature restricting the conversion. The thermodynamic limitation over the reaction can be reduced using two or more catalytic beds with intermediate cooling.

Although WGS is promoted by many materials, industrially, it takes place in two adiabatic stages with two different catalyzers, with a heat exchanger-cooler between them. The first stage (HTS) uses an iron-based catalyzer (Fe/Cr) that operates between 300–550°C and converts most parts of the CO. The typical values of CO reached at the exit of a single reactor are in the rank of 2–4%. The second reactor operates at lower temperatures, between 180 and 230°C, using a copper–zinc catalyzer supported over alumina. The LTS reactor is thermodynamically favored and can achieve a residual CO concentration of about 0.1–0.3%. The composition of the catalyzer is usually $CuO/ZnO/Al_2O_3$. Such catalysts are pyrophosphoric, they spontaneously generate heat increasing their temperature when exposed to air after activation.

In the last few years, a renewed interest was produced in the study of this reaction for its application to power generating systems with fuel cells. WGS is the biggest and heaviest component. The reaction is relatively slower than the others and is inhibited at high temperatures because of its thermodynamic equilibrium [32].

Campbell [33] established that the reaction is strongly controlled by the diffusion between pores near the 200°C and besides mentions that the catalyzers that contain copper are prone to lose activity by sintering, which is why the rank of use must be limited to 150–250°C.

The efforts for improving the performance of the WGS reactor have been distributed in a wide range of subjects including the development of more active catalysts than the traditional ones [34, 35], experimental and theoretical studies to obtain accurate kinetic expressions for different catalysts [36, 37], and theoretical studies analyzing the heat and mass transfer in the catalyst particle [38]. Other research areas are centered in the evaluation of unconventional reactors such as microreactors [39], monolithic [40, 41] or membrane reactors [42].

2.3.2 Kinetic Aspects

There are many works that determine the kinetic expression of the water–gas-shift reaction for both types of catalyzers, of high and low temperatures.

Keiski et al. [43] adjusted the reaction data for an iron-based catalyzer to a power law expression. The adjusted data corresponded to industrial conditions.

The most popular kinetic expression used in the design of industrial reactors for low temperature catalysts is the expression proposed by Moe [44]. They presented kinetic data obtained using a commercial catalyzer ($CuO/ZnO/Al_2O_3$) under industrial operating conditions of the reactor. The expression is represented by a Langmuir–Hinshelwood type model.

The effect of pressure is significant for industrial reactors operated at 15–20 atm. In fuel processors the pressures are closer to atmospheric (1–5 atm).

Choi and Stenger [36] determined the kinetics under the most likely conditions for the reaction in a small-scale methanol reformer, which are similar in pressure and composition to the ethanol reformer. They used a commercial catalyzer based on $Cu/ZnO/Al_2O_3$ from Süd-Chemie.

2.4 Preferential Oxidation Reactor

The preferential oxidation reactor is used to eliminate the unconverted CO in the WGS reactors. In order to avoid catalyst poisoning in the PEMFC, this reactor is required to reach very low levels of CO in the exit flow. The preferential oxidation is one of the possible technologies for this task. It uses oxygen to proceed with the following reaction:

$$CO + \frac{1}{2}O_2 \rightarrow CO_2 \qquad \Delta H^0_{298} = -283\,\text{kJ mol}^{-1}$$

Unfortunately, the selectivity of the catalyzer does not avoid the combustion of a fraction of the hydrogen in the gaseous flow through the following reaction:

$$H_2 + \frac{1}{2}O_2 \rightarrow H_2O \qquad \Delta H^0_{298} = -242\,\text{kJ mol}^{-1}$$

The carbon monoxide oxidation to obtain carbon dioxide in a hydrogen reach stream was investigated for the first time in the ammonia industry in the 1960s. As a result of these first works, a highly active and selective catalyzer was obtained, called Selectoxo [45]. Some authors found that, depending on the operating conditions and the used catalyzer, the inverse water gas shift reaction may also occur [46, 47]:

$$CO + 3H_2 \rightarrow CH_4 + H_2O \qquad \Delta H^0_{298} = -205.3\,\text{kJ mol}^{-1}$$

Excessive consumption of H_2 is undesired because it decreases the efficiency of the system and complicates the temperature control which is why the catalyzer must strongly favor the CO oxidation over the H_2 oxidation. Ideally the catalyzer should oxidate near 1% (10,000 ppm) of CO to less than 5 ppm without consuming part of the 30–70% of present H_2. For instance, with a O_2 /CO ratio equal to 2 and using a Pt/alumina catalyst, a concentration of 100 ppm is obtained at the exit of the reactor. However, with the same catalyzer and using a ratio of 3, the CO depletion is complete from a mix containing 1% of carbon monoxide. But the excess of O_2 consumes also the H_2 decreasing the efficiency of the system. Besides, if excess of air is aggregated, the temperature control becomes difficult due to the exothermic nature of the reaction [48]. These are the reasons why a highly selective catalyzer is needed, that allows to work with O_2 amounts near to stoichiometry.

Therefore, the choice of a highly selective catalyst, the design of the reactor and the definition of the operating conditions are critical aspects. The catalyst, the

Table 4 Heats of adsorption over Pt

Substance	Heat of adsorption (kJ mol^{-1})
H_2	77
CO	126
O_2	230

temperature, O_2/CO ratio, the type of operation of the reactor and the amount of stages to use are the key variables of the process.

Several catalysts had been proposed to make the preferential oxidation reaction of CO, most of them based on noble metals. Several researchers have reported the use of catalysts based on Iridium, Palladium, Platinum, Gold, Rhodium, Ruthenium and Platinum alloys.

With the objective of reducing costs and improving selectivity, recently, non noble metal catalysts had been considered. Some works reported the use of Copper, Ceria, Nickel, Cobalt and different combinations of them. Such metals have demonstrated a high selectivity for the oxidation of carbon monoxide in a H_2 rich stream.

Catalysts based on Pt (over alumina or silica) are the most used and investigated as conventional catalysts in the preferential oxidation of CO. The oxidation speed of CO is several orders of magnitude higher than that of the CO over a Pt surface and the concentration of it in the gaseous flow entering the CO–PrOx unit is two orders of magnitude higher than the CO, the preferential oxidation is possible due to the relative values of the heats of adsorption of the involved substances over a surface of Pt, as shown in Table 4 [49]:

Most of the analyses presented in the literature agree that, at low temperatures, the surface of the catalyst is covered with adsorbed CO, the fraction of which decreases with the increase of temperature, and free spots appear for the adsorption of O_2 and the oxidation is produced between the adsorbed substances to give CO_2. Above a certain temperature, the fraction of surface covered by CO decreases even more and the H_2 begins its chemisorption and reaction with adsorbed O_2 to give water. Therefore, a common fact for all these catalytic systems is that there is a range (window) of operating temperature, such that it is operated at a temperature high enough for the speed of reaction to be elevated and minimize the size of the reactor, and low enough to avoid simultaneous oxidation of H_2. The range of temperatures that satisfy these conditions is so tight that the catalysts cannot allow small perturbations in the system. Meaning that the aforementioned types are useless without a precise temperature control.

Undesired reactions take higher importance operating at higher temperatures where the inverse WGS reaction takes place consuming H_2 as the methylation reaction. Zalc and Loffler [32] pointed out that the reaction kinetics reveal the importance of reactors' temperature control, due to, precisely, the tight range of temperatures in which it can work in order to maintain a high selectivity of the desired reaction. So, an adiabatic reactor in a single stage would have a poor performance; an optimal design of the reactor must contemplate an effective control of the temperature. Because of this reason, they propose to support the catalyst on heat exchangers or use water or air injection between stages.

2.4.1 Kinetic Aspects

Despite the number of publications that suggest kinetic expressions for the carbon monoxide oxidation, very few of them consider the hydrogen oxidation simultaneously. Choi and Stenger [46] derived the kinetic expressions for both, the oxidation of CO and H_2 considering also the water gas shift reaction. Such expressions were obtained for a Pt-Fe/Al_2O_3 catalyst. Bissett et al. [50] determined the superficial kinetics for the simultaneous oxidations of H_2 and CO considering a catalyst of 0.5%w of Pt/Al_2O_3 within a range of temperatures of 80–200°C.

2.5 Proton Exchange Membrane Fuel Cell Stack Model

The fuel cell is modeled as described in Sect. 1.8 of Chap. 2. It is supposed to be isothermal and isobaric. The output voltage of a single cell (v_{fc}) is determined from the reversible voltage, which is decreased by the irreversible losses as shown in Eq. 17.

In Eq. 17, E is the thermodynamic potential of the cell; v_{act} is the voltage drop related to the activation of the anode and cathode; v_{ohm} is the ohmic voltage drop related to the resistances of the conduction of protons through the solid electrolyte and the electrons through their path; and v_{conc} represents the voltage drop related to the concentration reduction of the reactant gases. The voltage drop associated to the internal currents and/or the fuel crossover is taken into account, using a fixed current density even at "no-load" operation condition. The voltage of a stack (v_{st}) of n cells connected in series can be calculated by Eq. 16.

The reversible thermodynamic potential (E) is given by the Nernst equation, as shown in Eq. 18. The activation overpotential (v_{act}) is determined from the semi-empirical Eq. 19. The ohmic overpotential (v_{ohm}) is expressed using the Ohm's law Eq. 23. The total internal resistance is a complex function of temperature and current. The expression for resistance includes all the important membrane parameters in Eq. 24. The diffusion overpotential (v_{conc}) is directly related to the concentration drop of the reactant gases. Oxygen and hydrogen pressure reduction depends on the electrical current and physical characteristics of the system. To determine this voltage drop, a maximum current density, i_{max}, is defined under which the fuel is being used at the maximum supply rate (Eq. 26).

2.6 Post-Combustion System

The depleted fuel of the PEMFC, formed by cathode and anode outlets, is burnt off in the post combustion system. The generated heat will be used to balance the energy requirement of the fuel processing section. Supplementary amount of ethanol will be considered if the energy content of the depleted fuel is not sufficient to satisfy the balance. The supplementary fuel requirement will be computed in order to achieve a set of feasibility specifications. A minimum approach temperature of $\Delta T = 100°C$ in the cold side of the reformer is first imposed. Complete and stoichiometric combustion for the burner unit has been assumed.

After the heat exchange with the LNG unit, the exhaust gases can be expanded in a turbine coupled to the air compressor. Additional ethanol burning will be considered, if necessary, to balance the air compression work requirement.

2.7 Heat Exchange Model

The heat exchangers are not explicitly considered in the simulation model described because the heat recovery system is unknown a priori. Therefore, the LNG unit of the commercial process simulator HYSYS has been used to model the integrated heat exchange system. It is useful because it allows to know the maximum energy recovery of the process streams without the need for design at this stage of the heat exchanger networks. The most probable fact is that the further incorporation of those equipments would decrease the efficiency magnitude.

The LNG exchanger model is a HYSYS operation that solves heat and material balances for multi-stream heat exchangers and heat exchanger networks. The LNG calculations are based on energy balances for hot and cold fluids. Each stream is divided using 20 intervals, and the occurrence of phase changes are always verified. In this approach, the results of the simulation model will characterize the hot and cold streams of the system.

The LNG unit allows analyzing the system energy integration by means of the process integration method (or pinch technology) [51]. Process integration studies start with the definition of a list of hot and cold streams. Hence, heat sources and heat requirements of the system that are characterized by a heat-temperature diagram, usually specified by a heat load, an inlet and a target temperature. The heat sources are then composed to compute the hot composite curve that represents the heat availability in the system as a function of the temperature. The same procedure is applied for the cold streams to obtain the cold composite curve. Considering that the heat exchange will be technically feasible if the temperature difference between the hot and cold streams are always superior to a predefined ΔT_{min}, the maximum heat recovery by heat exchange between the hot and the cold streams will be obtained when the ΔT_{min} constraint is activated. This point is called the system pinch point. By means of energy balance, one may then compute the minimum energy requirement of the system and the minimum heat to be evacuated from the system. Using the composite curve calculation together with a simulation model, the simulation of the heat exchangers network is defined without knowing about its configuration. By this approach, it is possible to compute the influence of the decision variables so that the best system configuration can be specified.

2.8 Definition of Efficiencies

The overall efficiency of BPS is defined here as the net energy output of the system obtained from the gross output by subtracting the electrical energy needed to operate FCS auxiliaries such as pumps and compressors divided by the heating value of the ethanol consumed in the fuel processor for reforming ($f_{\text{EtOL}}^{\text{BP}}$) and burning ($f_{\text{EtOL}}^{\text{Burn}}$):

$$\eta_{\text{BPS}}^{\text{HV}} = \frac{P_{\text{System}}}{\left(HV_{\text{EtOL}}\left(f_{\text{EtOL}}^{\text{BP}} + f_{\text{EtOL}}^{\text{Burn}}\right)\right)}. \tag{6}$$

The HV factor can be the higher heating value (HHV) or the lower heating value (LHV) of the ethanol. HHV represents the actual amount of chemical energy in the fuel (relative to standard conditions), while LHV neglects heat below 150°C. Therefore, HHV should be compared with the mechanical and electrical energy produced by the power system. However, as reported in numerous works LHV value is useful to present both values. The net power of the system is computed as follows:

$$P_{\text{System}} = P_{\text{Cell}} - W_{\text{Pump}}^{\text{Water}} - W_{\text{Pump}}^{\text{EtOL}} - \frac{Q_{\text{Cooler}}}{\eta_c} - \left(P_{\text{Comp}} - P_{\text{Tur}}\right), \tag{7}$$

where $W_{\text{Pump}}^{\text{Water}}$ and $W_{\text{Pump}}^{\text{EtOL}}$ are the work of the water and ethanol pumps, respectively. Q_{Cooler} is the heat removal from the fuel cell system; it represents the cold utility of the process. This heat is formed by the heat generated by the fuel cell and extra cold utilities such as heat removal from PrOx outlet stream in the condenser separation unit in order to meet the fuel cell operating temperature. In the above expression, η_c is the cooling system efficiency defined as the rate of heat removal over the electrical power consumed by the pumps and fans associated to the refrigeration system. A cooling system efficiency of 25% is adopted.

The processor efficiency is defined as the ratio of the heating value of the gas stream incoming to the anode divided by the sum of the heating value of the ethanol fed to the processor and the remainder gas purged from the stack. Again, the processor efficiency can be computed based on the lower or higher heating values.

3 Definition of the Plant Operating Point Based on Energy Efficiency

The integrated process model has been used to determine the optimal operating conditions to be considered in the system. The influence of the water-to-ethanol molar ratio and reforming temperature has been studied. Furthermore, the remaining reactors temperatures were considered as variables, but the changes in efficiency due to these variables are marginal. The procedure of calculation to close the energy balance is as follows: fixing the water-to-ethanol molar ratio and the reforming temperature,

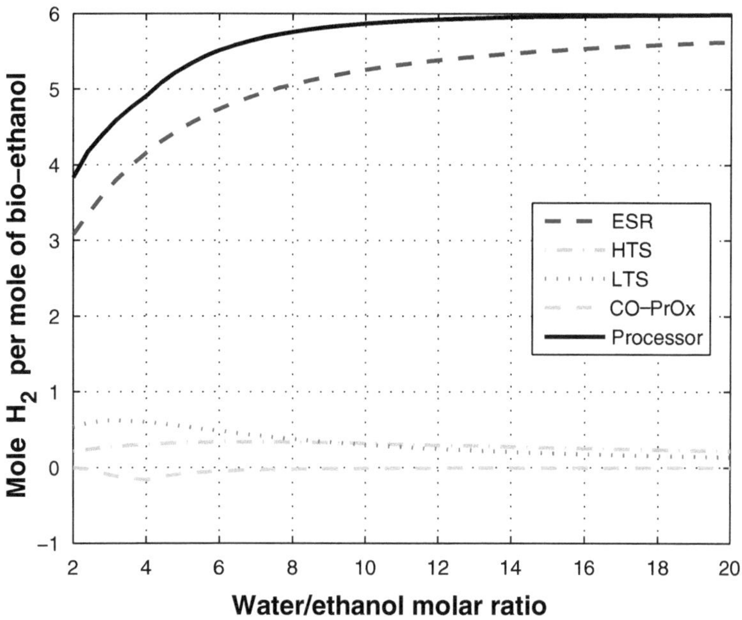

Fig. 2 Reactor yields versus water/ethanol molar ratio

the simulation model determines the temperature of combustion gases (T_{cg}) which leave the reformer unit. If this temperature is lower than $T_{Ref} + \Delta T$, an extra amount of ethanol is burned, which is computed in order to achieve this thermal requirement (ΔT = 100°C). Flue gases exchange heat with the other units, by means of the LNG operation, before being expanded in a turbine. If the temperature of the flue gases after this heat exchange is lower than the inlet temperature in the turbine (T_{Tur}^{in}) necessary to balance the air compression work, a supplementary burning of ethanol is computed.

3.1 Influence of the Water-to-Ethanol Ratio

In this section, the effect of the water-to-ethanol ratio (R) on the net efficiency of the system will be analyzed. Figure 2 presents the reactors yields versus the total performance of the processor. These are computed as the ratio of mole of hydrogen produced by mole of ethanol incoming to the reformer.

Although a high reagents molar relation presents a good yield of hydrogen, the water quantity demands extra energy in the vaporization and reheating processes before entering the reformer. This extra energy is partially recovered from the hot streams, which increase their flow rate, but heat exchanges occur at finite temperature differences and consequently the efficiency decreases. Figure 3 shows the global

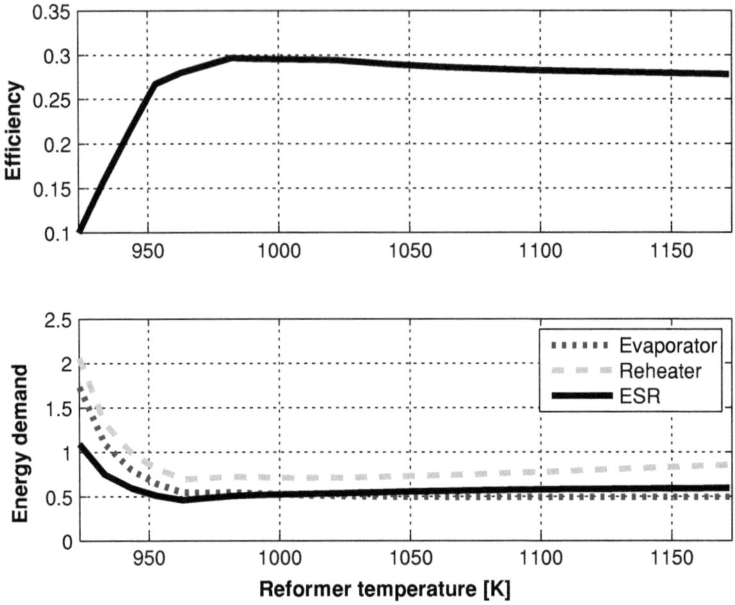

Fig. 3 Efficiency and energy demand versus water/ethanol molar ratio

efficiency and energy demand of the evaporator, reheater and reforming reactor versus the water-to-ethanol molar ratio at constant reforming temperature ($T_{Ref} = 709°C$).

Smaller molar ratios present lower efficiencies and the PrOx unit consumes more H_2 to reduce the level of CO to the required values. At higher water-to-ethanol molar ratios, the water excess must be evaporated and re-heated consuming additional fuel in the reformer, and then diminishing the system efficiency. Due to excess water, thermodynamic equilibrium of the water gas shift reaction on both HTS and LTS reactors is shifted to hydrogen formation diminishing the CO outlet concentration. This improves the performance of PrOx reactor, consuming less H_2 at higher R values.

The system composite curves were drawn in order to give a good view of the possibilities for a heat exchanger network (Fig. 4). The composite curves are two for describing the total system cooling and heating demand, respectively, as a function of temperature intervals. Using the LNG operation the composite curves from the system can be built as seen in Fig. 4 for the specific case of the system at the optimal operating conditions.

Figure 5 compares the composite curves for two different water-to-ethanol molar ratios. It shows the previously analyzed case, corresponding to $R = 4$ (maximum efficiency case), and for $R = 10$ at the same operating reforming temperature. The extra water quantity increases the energy demand in the evaporator, and the cold utility. These changes move the composite curves producing a net efficiency of 34%.

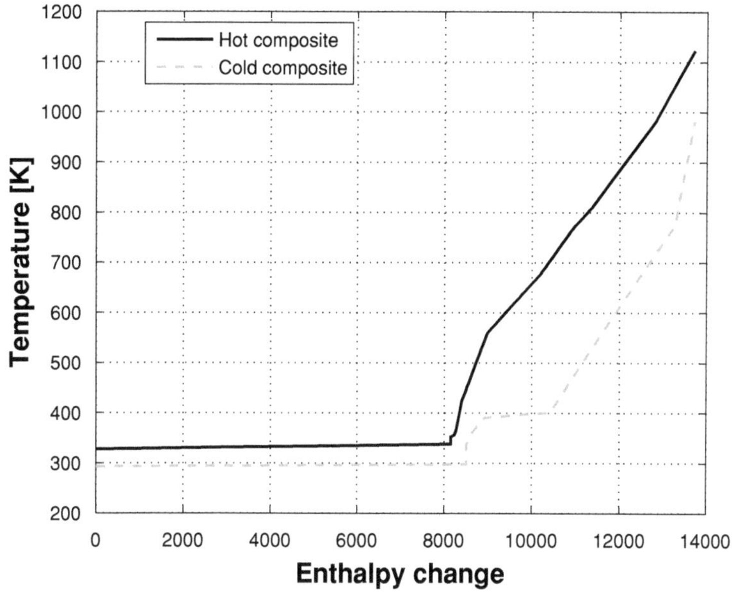

Fig. 4 Composite curves for optimal $T_{ref} = 982$ K and $R = 4$

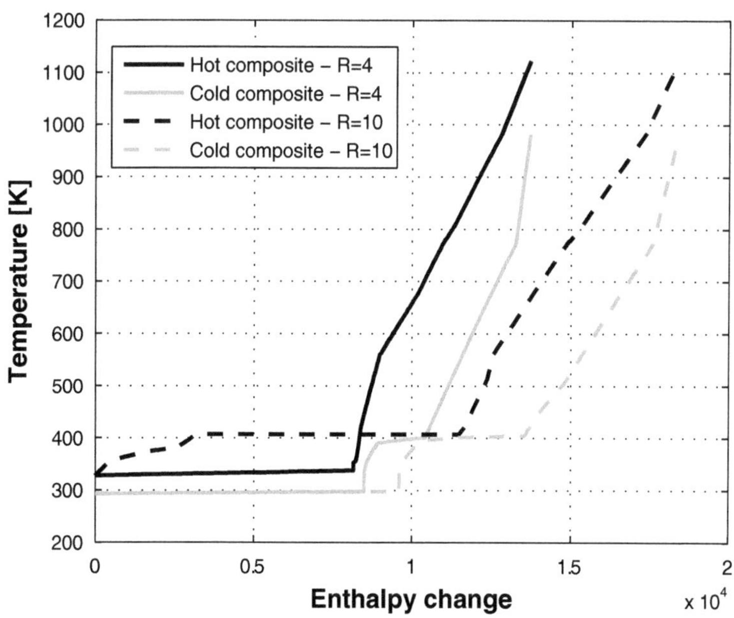

Fig. 5 Comparative composite curves for $R = 4$ and $R = 10$

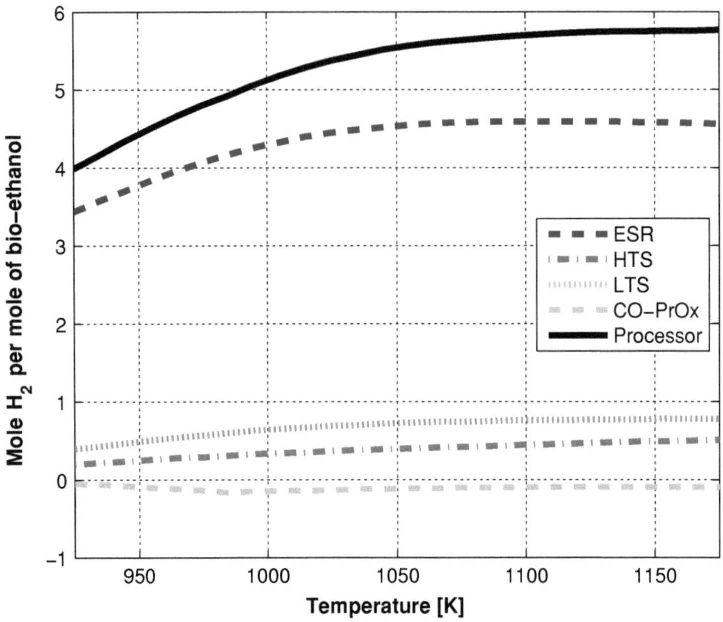

Fig. 6 Reactor yields versus temperature of the reformer

3.2 Influence of Reformer Temperature

The effect of the reforming temperature is analyzed considering the water-to-ethanol molar ratio of 4. The yield of the reactors (Fig. 6) show that increasing the temperature increases the hydrogen production in the reformer, HTS and LTS units, although the hydrogen consumed by the PrOx unit is also increased .

Figure 7 shows the variation of the energy demand with the temperature. The demand of the evaporator decreases, the reactor increases and the reheater presents a minimum.

Operating at lower temperatures, the poor yield of hydrogen implies a greater flow of reactive mixture needed to achieve the target power. This superior flow increases the energy demand of the evaporator.

The energy demand of the reformer increases because the conversion of methane is enhanced with the temperature, producing greater energy requirements. The sum of these effects produces a maximum net efficiency of $\eta_{FCS}^{LHV} = 38\%$ below 709°C.

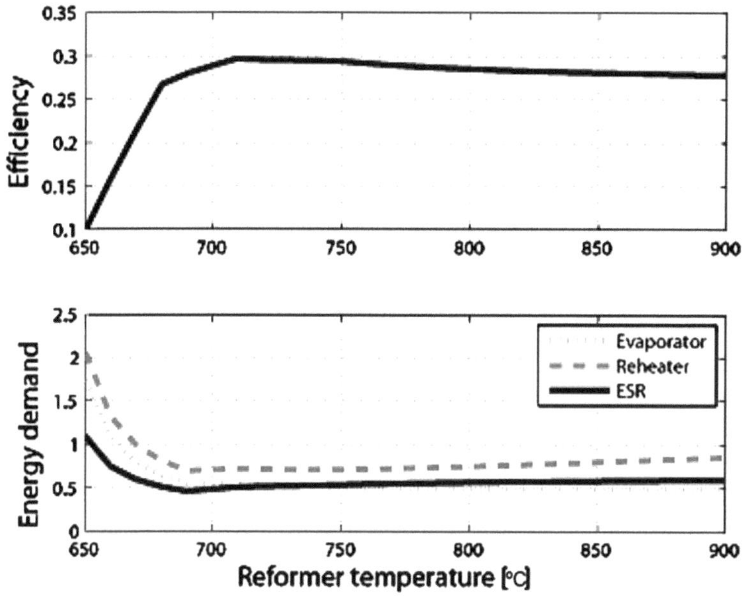

Fig. 7 Efficiency and energy demand versus reforming temperature

3.3 The Self-Sufficient Limit

In the late 1970s Pinch technology emerged as a tool for the design of heat exchanger networks [51]. One of the most significant features of pinch analysis is that it can be used to set performance targets for a process before a detailed design. Most importantly, these design options are evaluated in a whole process context to ensure that they give a global improvement. Once a process configuration and conditions that give satisfactory targets have been established, a heat exchanger network (HEN) is designed. The pinch design method makes this design task relatively simple.

The system composite curves were drawn in order to give a good view of the possibilities for a heat exchanger network (Fig. 4). Using the LNG operation the composite curves can be built from the data given in HYSYS.

For the LNG operation, heating and cooling demands between reactors were calculated: all internal streams were forced to their working temperatures, and all heating or cooling demands for all streams were recorded. Note that every single inlet stream to the fuel cell system was preheated up to its working temperature.

The net electric efficiency based on HHV is approximately 35%. The remaining 65% is accounted for: dissipation as heat in the PEMFC cooling system (38%), energy in the flue gases (10%) and irreversibility in compression and expansion of gases. The tasks that demand more energy are vaporizing and reheating of the reactive mixture (0.79 kW), the reformer reactor (0.41 kW) and the preheating of the exhaust gas to the burner (0.67 kW). These values represent the energy demand (in kilowatt)

Table 5 Effect of input temperatures of the clean up system reactors

η_{FCS}^{LHV} (%)	η_{FCS}^{HHV} (%)	HTS (°C)	LTS (°C)	CO–PrOx (°C)	T_{Ref} (°C)	*R* molar ratio
38.3	34.9	500	150	237	709	4.0
38.3	34.9	500	150	150	704	3.8
38.8	35.3	350	150	237	709	3.6
38.8	35.3	350	150	150	709	3.6
39.3	35.8	150	150	150	714	3.3

per net kW produced by the system. The requirement for the recycle conformed by the streams passing from #28 to #29, shown in Fig. 1, is comparatively higher due to the more significant flow rate because of the presence of N_2.

Respect to the necessary energy to drive the FCS auxiliaries, mainly pumps and blowers for water, ethanol, air and heat management, the largest load is the air compressor (0.16 kW), which delivers air to the cathode compartments of the stack and to the PrOx reactor.

3.4 Effect of Input Temperatures of the Clean-up System Reactors

In previous analysis, input temperatures to HTS, LTS and PrOx reactors were fixed following the scheme proposed by Little [31]. HTS and LTS units are modeled as equilibrium reactors and in the PrOx the conversion is fixed. These units operate at adiabatic conditions, so the input temperature variation affects the output temperature and the CO conversion in the WGS.

Table 5 shows the new values of operating reformer variables that maximize the efficiency when the input temperatures of the train of purification reactors are varied.

The impact of a wrong selected temperature could drive to be outside the reasonable operating parameters and may change the processor design. Besides, PrOx operating temperature variation could vary the selectivity to burning hydrogen (which is not accounted in the model). However, at this stage, the goal is to analyze the sensitivity of the net system efficiency on the input temperature. Then, according to the results, the design aspects of each reactor could be revisited later if some controllability problems are detected. In the table, T_{Ref} and R represent the new optimal conditions calculated for each case.

The variation of the input temperatures does not affect sensitively the net efficiency of the system even though the operating conditions of the reformer are modified. The net efficiency differs slightly. However, smaller molar optimal relations are found when the temperatures of the input flows of the reactor decrease.

Table 6 Effect of input temperatures of the clean up system reactors

T (°C)	η_{FCS}^{LHV} (%)	η_{FCS}^{HHV} (%)	Ethanol to burn (mole h^{-1})
80	30.4	27.7	21.5
200	32.3	29.4	16.2
300	34.0	31.0	11.7
400	36.1	32.8	7.1
500	38.3	34.9	2.5
600	38.3	34.9	2.5
700	38.3	34.9	2.5

Table 7 Influence of the turbine on the net efficiency of the system

	η_{FCS}^{LHV} (%)	R	T_{Ref} (°C)	P_{Comp} (kW)	P_{Tur}	Power stack (kW)	Net power (kW)
$P_{Tur} = 0$	34.4	5.5	679	1.7	–	12.4	10
$P_{Tur} = P_{Comp}$	38.3	4.0	709	1.6	1.6	10.4	10
$P_{Tur} \leq P_{Comp}$	38.9	4.9	687	1.6	1.3	10.8	10

3.5 Influence of the Combustion Preheating

The burner is fed with three streams: pure ethanol, air and depleted gases from PEMFC. In the cases under study, the oxygen from de cathode exhaust is enough to burn the fuel (H_2, CH_4, Bio-ethanol) in stoichiometric form. Consequently, there is no need of extra air. These streams are previously heated (combustion preheating) allowing a heat recover from below the reforming temperature to make it available above the reforming temperature.

In Table 6, efficiency values for several preheated temperatures of gases to burn are shown. The results reveal that by increasing the input temperature of the burner the required amount of additional fuel is reduced. Efficiency based on HHV is about 35% with preheating while it is 28% without preheating.

3.6 Efficiency Analysis Without Considering Expander Unit

In the previous sections, the fuel cell system was analyzed considering that the required compression work was balanced with a turbine. The goal here is to study the effect on the net efficiency when the expander unit is not present. Table 7 presents the results obtained considering the following cases: without turbine; the obtained work from the turbine is equal to the compression work; partial work recuperation with the turbine. The second case was utilized in all previous analysis considering extra ethanol burning to balance both works. The last case considers that the turbine is included, but when compression work is superior to that of the expander unit, the difference is compensated by means of the fuel cell power.

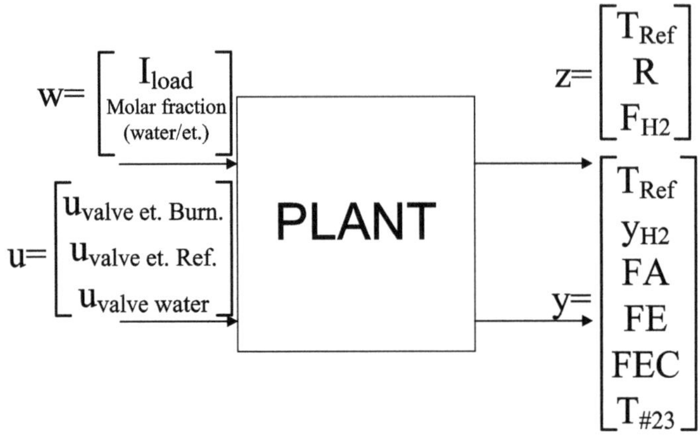

Fig. 8 Generalized block diagram for the plant control structure

This analysis shows that the best option is working with a partial recovery of work. Net efficiency is slightly improved when it is compared with total recovery, but the operating conditions vary significantly.

4 Preliminary Control Problem Formulation

As previously discussed, one of the key requirements of the BPS controller is to quickly replenish the hydrogen that is consumed in the fuel cell anode during current (load) changes. On the other hand, the BPS controller needs to reduce the H_2 generation when there is a step-down in the current drawn from the fuel cell, so H_2 must not be wasted. This H_2 on demand operation involves the following objectives (1) to protect the stack from damage due to H_2 starvation (2) to protect ESR from overheating and (3) to keep overall system efficiency high, which includes high stack H_2 utilization and high BPS Ethanol-to-H_2 conversion. Objectives (1) and (2) are important during transient operations while objective (3) can be viewed as a steady-state goal. Objectives (2) and (3) are also related since keeping the desired ESR temperature during steady-state implies proper regulation of the oxygen-to-carbon ratio which corresponds to high BPS conversion efficiency.

The stack current, I_{load}, is considered as an exogenous input that can be measured. The preliminary control problem is formulated using the general scheme configuration shown in Fig. 8.

The PEMFC control is formulated following the analysis given on Sect. 2.2 of Chap. 3 (Feroldi et al. [52]) remarked on the advantages of using a regulating valve for the cathode outlet flow in combination with the compressor motor voltage as manipulated variables in a fuel cell system. Then, at a given current load, the cathode input and output flow rates determine its pressure and stoichiometry, and con-

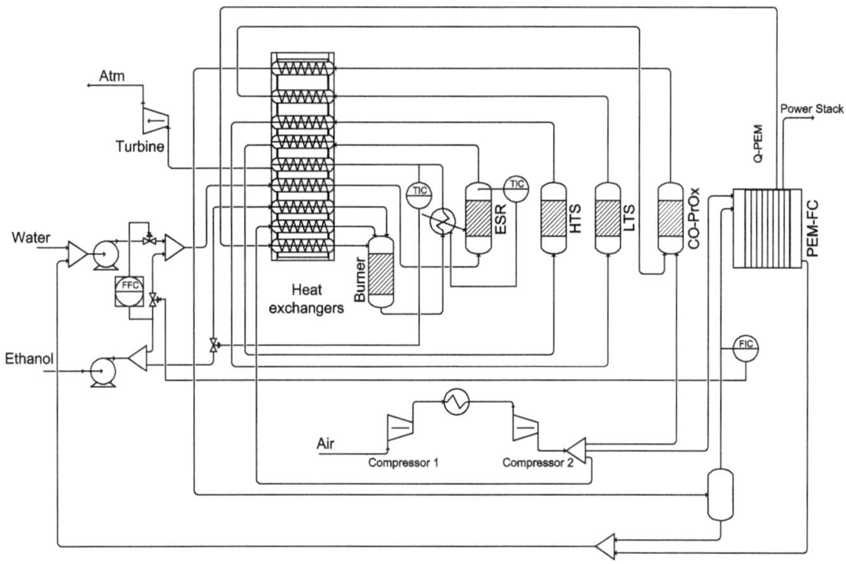

Fig. 9 Control structure no. 1

sequently set the oxygen partial pressure, the generated voltage and the compressor power consumption. Therefore, in order to maintain a high efficiency during the operation, the cathode output regulating valve has to be adjusted to the operating conditions, specially marked by the current drawn from the stack. In addition, in [53] was demonstrated that the appropriate valve manipulation produces an improvement in the system transient response. The implementation was done with a predictive control strategy based on dynamic matrix control (DMC), using the compressor voltage and the cathode output regulating valve as manipulated variables. Hence, both the fuel cell voltage and oxygen excess ratio in the cathode, are regulated and thus, the system performance can be improved. Based on this previous study in the following sections two control structures are analyzed based on a sensitivity evaluation to the most critical perturbations for the overall system.

4.1 Structure Number 1

This structure can be seen in Fig. 9, which consists of controlling the molar relationship between ethanol and water through a ratio control, the ESR temperature through the heat duty given from the exhausted gases of the Burner. The ethanol feed flow rate to the ESR is manipulated according to the H_2 requirements for the PEMFC stated by the current load.

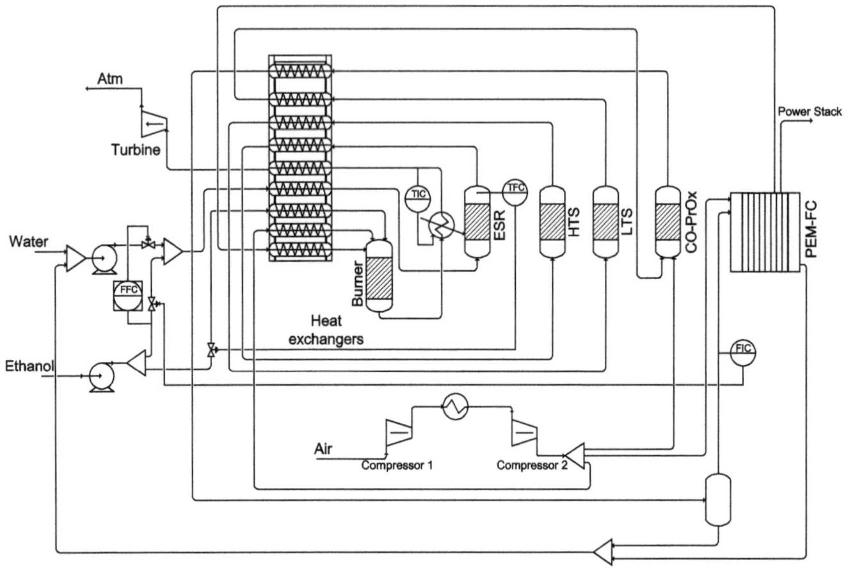

Fig. 10 Control structure no. 2

4.2 Structure Number 2

This structure can be seen in Fig. 10 which consists of controlling the molar relationship between ethanol-water through a ratio control, the ESR temperature through ethanol feed flow rate to the Burner and its temperature through the heat duty given by the exhausted gases to the ESR (T#23). The ethanol feed flow to the ESR is manipulated according to the H_2 requirements for PEMFC.

4.3 Assumptions for Control Problem Simplification

Several assumptions are made in order to simplify the plantwide control of BPS. The three cleaning reactors (two water gas shift, WGS, of high and low temperatures and the preferential oxidation, PrOx) are lumped together as one volume assuming that they are under perfect control. Therefore, the desired values of the reactants are supplied to these equipments. It is considered that the composition of the air entering the blower is constant. Additionally, all the temperatures, except those of ESR, are assumed to be determined optimally by the LNG HYSYS tool, which represents a perfectly controlled heat exchange network. Finally, all gases obey the ideal gas law and are perfectly mixed. Based on all these considerations, it is possible to take into consideration the ESR temperature, the relationship between H_2O and Ethanol (feed plant) and the feed flow rate of H_2 to the PEMFC as the most critical process variables to be controlled. In Fig. 8 can be seen the generalized block diagram representing the main plant control variables accounted:

- *W*: (perturbations) feed molar fraction (H_2O and Ethanol) and current demands;
- *U*: (manipulated variables) the pre-heated ethanol flow rate valve, the ethanol feed flow rate to the Burner and the water recycle feed flow rate from the PEMFC;
- *Z*: (performance variables) T_{Ref}, $R_{w/e}$ and H_2 feed flow rate to the PEMFC;
- *Y*: controlled process variables

Therefore, the control objectives are to keep the performance variables under optimal values, given by the synthesis stage, through variations on the manipulated variables even though perturbations occurrence.

4.4 Steady-State Sensitivity Analysis

BPS, as any other chemical plant, has a certain (sometimes large) number of units with multiple recycle streams. Many different control strategies are possible, and it would be impractical to perform a detailed dynamic study for each alternative. Therefore, a disturbances sensitivity analysis can be done to screen out poor control structures. This procedure is performed by accounting with a steady-state process model trying to deal with problems such as snowball effect, produced sometimes by material recycle. The principal objective is to use a steady-state analysis that can reveal structures that require large changes in manipulated variables when load disturbances occur or when a change throughput is made. These structures can be eliminated from further study. The idea is to specify a control structure (fix the variables that are held constant in the control scheme) and specify a disturbance,then solve the nonlinear algebraic equations to determine the values of all variables at the new steady-state condition. For realistically complex processes, analytical solution is out of the question and numerical methods must be used. The use of modern software tools (such as SPEEDUP, HYSYS or GAMS) which make these calculations relatively easy to perform is recommended by Luyben [5]. Hence, in this chapter, two control structures are proposed based on the critical control objectives detailed above. Hence, in order to justify the selection of the control structures the BPS modeled in HYSYS is tested.

Following this methodology and assuming as critical disturbances the water molar fraction in the feed flow rate and the current load demand for both structures are analyzed. From Tables 8, 9, 10, 11 can be seen how these perturbations impact the manipulated variables FA, FE and FEC and the efficiency. Here, the overall efficiency of the FCS was analyzed at Sect. 3 as the net energy output of the system obtained from the gross output by subtracting the electrical energy needed to operate FCS auxiliaries such as pumps and compressors divided by the heating value of the consumed ethanol in the fuel processor for reforming and burning,

$$\eta_{\text{BPS}}^{\text{HHV}} = \frac{P_{\text{system}}}{\text{HHV}_{\text{et.}} \times (\text{FE} + \text{FEC})}. \tag{8}$$

The HV factor can be the higher heating value (HHV) or the lower heating value (LHV) of the ethanol. HHV represents the actual amount of chemical energy in

Table 8 Sensitivity analysis for control structure no. 1

	Base case	XE (−1%)	Variation (%)	XE (−5%)	Variation (%)	XE (−10%)	Variation (%)
XE (Ethanol molar fraction in %)	1	0.99	−1.000	0.95	−5.000	0.9	−10.000
FA (kmole h^{-1})	2.936E–02	2.955E–02	0.647	3.039E –02	3.508	3.156E–02	7.493
FE (kmole h^{-1})	7.340E–03	7.388E–03	0.654	7.597E–03	3.501	7.890E–03	7.493
FEC (kmole h^{-1})	7.581E–04	7.878E–04	3.918	9.159E–04	20.815	1.086E–03	43.253
Efficiency	0.2881	0.2881	0.000	0.2880	−0.035	0.2878	−0.104

Table 9 Sensitivity analysis for control structure no. 2

	Base case	XE (−1%)	Variation (%)	XE (−5%)	Variation (%)	XE (−10%)	Variation (%)
XE (Ethanol molar fraction in %)	1	0.99	−1.000	0.95	−5.000	0.9	−10.000
FA (kmole h^{-1})	2.936E–02	2.968E–02	1.090	3.094E–02	5.381	3.220E–02	9.673
FE (kmole h^{-1})	7.340E–03	7.420E–03	1.090	7.734E–03	5.368	8.049E–03	9.659
FEC (kmole h^{-1})	7.566E–04	7.566E–04	0.000	7.566E–04	0.000	9.060E–04	19.746
Efficiency	0.2881	0.2880	−0.035	0.2888	0.243	0.2885	0.139

Table 10 Sensitivity analysis for control structure no. 1

	Base case	*P* (−1%)	Variation (%)	*P* (−5%)	Variation (%)	*P* (−10%)	Variation (%)
XE (Ethanol molar fraction in %)	1	1	0.000	1	0.000	1	0.000
FA (kmole h^{-1})	2.936E–02	2.906E–02	−1.022	2.788E–02	−5.041	2.642E–02	−10.014
FE (kmole h^{-1})	7.340E–03	7.265E–03	−1.022	6.970E–03	−5.041	6.605E–03	−10.014
FEC (kmole h^{-1})	7.581E–04	7.511E–04	−0.923	7.222E–04	−4.736	6.875E–04	−9.313
Efficiency	0.2881	2.880E–01	−0.035	2.880E–01	−0.035	2.879E–01	−0.069

Table 11 Sensitivity analysis for control structure no. 2

	Base case	*P* (−1%)	Variation (%)	*P* (−5%)	Variation (%)	*P* (−10%)	Variation (%)
XE (Ethanol molar fraction in %)	1	1	0.000	1	0.000	1	0.000
FA (kmole h^{-1})	2.936E–02	2.905E–02	−1.056	2.778E–02	−5.381	2.618E–02	−10.831
FE (kmole h^{-1})	7.340E–03	7.263E–03	−1.049	6.945E–03	−5.381	6.545E–03	−10.831
FEC (kmole h^{-1})	7.581E–04	7.566E–04	0.000	7.566E–04	0.000	7.566E–04	0.000
Efficiency	0.2881	2.879E–01	−0.069	2.877E–01	−0.139	2.875E–01	−0.208

the fuel (relative to standard conditions), while LHV neglects heat below 150°C. In Tables 5, 6, 7, 8, 9, the HHV values are included.

From the simulation results can be concluded that structure no. 1 does not present significant variation on the efficiency, however, the percentage variations on the manipulated variables are more important than those produced using structure no. 2. In the latter the efficiency is kept close to the same value as the base case. The stronger effect is detected for the FEC when a disturbance at molar fraction of about 10% occurs, the FEC reaches the 40% variation. It indicates that structure no. 1 is the most sensitive since actuators' saturation is probable to occur during the operation plant and, as a consequence, a loss of controllability. For structure no. 2 the variation is less than that produced in structure no. 2 even though its magnitude indicates to be careful on the controller to be chosen for this specific case.

5 Conclusions

In this chapter, a large review of several research works for improving the different aspects of a bio-ethanol processor with PEMFC was accounted. It allowed to perform a complete study about the best operation conditions for the overall system considering mass and energy integration. The construction of the steady-state model using a valuable dedicated software for chemical plants, like HYSYS, was also given. It was helpful for doing the hard calculations related to the search for a good operating point for this complex and new process. In addition, it was remarked that the bio-ethanol processor efficiency depends on the ability to keep the process at the given operating conditions mainly for the reformer and the PEMFC in the context of a well-designed energetic integration. Then, a plantwide control heuristic procedure was applied. It is mainly based on the large experience on other analogous systems and in the requirements of the isolated units' operation. As a result, two important control loops could be formulated accounting for their capacity for rejecting the influence of some critical perturbations in the context of the overall plant. The effect of these disturbances was quantified through the sensitivity analysis which gives an idea about the rate of variation for the selected manipulated variables. Hence, this methodology helps to screen between different control structures. However, it needs to have a good process and control experience for reducing the problem dimensionality. Anyway, this preliminary exploration gave a first acceptable result that can be further investigated with the assistance of a well done control-oriented dynamic model which will be presented in the following chapter. It will be useful to show a closer behavior of the real plant. In Chap. 12 a more detailed study for defining more systematically the final control structure and the controller synthesis by applying new methodologies will be presented.

References

1. Luyben WL, Tyreus BD, Luyben ML (1999) Plantwide process control. McGraw-Hill Professional Publishing, New York
2. Francesconi JA, Mussati MC, Mato RO, Aguirre PA (2007) Analysis of the energy efficiency of an integrated ethanol processor for PEM fuel cell systems. J Power Source 167(1):151–161
3. Aspen Technology (2006) HYSYS User Manual
4. Boland P Hewitt G.F, Thomas B.E.A, Guy A.R, Marsland R.H, Linnhoff B, Townsend D (1985) A user guide on process integration for the efficient use of energy. Institute of Chemical Engineers, Rugby, UK
5. Luyben ML, Luyben WL (1997) Essentials of process control. Chemical Engineering Series. McGraw-Hill, International Editions, New York
6. Pukrushpan J, Stefanopoulou A, Varigonda S, Eborn J, Haugstetter C (2006) Control-oriented model of fuel processor for hydrogen generation in fuel cell applications. Control Eng Practice 14(3):277–293
7. Godat J, Marechal F (2003) Optimization of a fuel cell system using process integration techniques. J Power Source 118(1–2):411–423
8. Vaidya P, Rodrigues A (2006) Insight into steam reforming of ethanol to produce hydrogen for fuel cells. Chem Eng J 117(1):39–49
9. Haryanto A, Fernando S, Murali N, Adhikari S (2005) Current status of hydrogen production techniques by steam reforming of ethanol: a review. Energy Fuels 19(5):2098–2106
10. Ni M, Leung D, Leung M (2007) A review on reforming bio-ethanol for hydrogen production. Int. J. Hydrogen Energy 32(15):3238–3247
11. Haryanto A, Fernando S, Murali N, Adhikari S (2005) Current status of hydrogen production techniques by steam reforming of ethanol: A review. Energy Fuels 19(5):2098–2106
12. Garcia E, Laborde M (1991) Hydrogen production by the steam reforming of ethanol: thermodynamic analysis. Int. J. Hydrogen Energy 16(5):307–312
13. Vasudeva P, Mitra N, Umasankar P, Dhingra S (1996) Steam reforming of ethanol for hydrogen production: thermodynamic analysis. Int. J. Hydrogen Energy 21(1):13–18
14. Fishtik I, Alexander A, Datta R, Geana D (2000) A thermodynamic analysis of hydrogen production by steam reforming of ethanol via response reactions. Int. J. Hydrogen Energy 25(1):439–452
15. Comas J, Laborde M, Amadeo N (2004) Thermodynamic analysis of hydrogen from ethanol using cao as a co2 sorbent. J. Power Source 138(1–2):61–67
16. Mas V, Kipreos R, Amadeo N, Laborde M (2006) Thermodynamic analysis of ethanol/water system with the stoichiometric method. Int. J. Hydrogen Energy 31(1):21–28
17. Cavallaro S, Freni S (1996) Ethanol steam reforming on rh/al2o3 catalysts. Energy Fuels 14(3):119–128
18. Galvita V, Semin G, Belyaev V, Semikolenov V, Tsiakaras P, Sobyanin V (2001) Synthesis gas production by steam reforming of ethanol. Appl Catal A 220(1–2):123–127
19. Auprete F, Descorme C, Duprez D (2002) Bio-ethanol catalytic steam reforming over supported metal catalysts. Catal Commun 3(6):263–267
20. Llorca J, Homs N, Sales J, DeLa Piscina P (2002) Efficient production of hydrogen over supported cobalt catalysts from ethanol steam reforming. J Catal 209(2):306–317
21. Cavallaro S, Chiodo V, Freni D, Mondello N, Frusteri F (2003) Performance of rh/al2o3 catalyst in the steam reforming of ethanol: H2 production for mcfc. Appl Catal A 249(1):119–128
22. Luo R, Misra M, Himmelblau D (2005) Sensor fault detection via multiscale analysis and dynamic PCA. Ind Eng Chem Res 38:1489–1495
23. Comas J, Marino F, Laborde M, Amadeo N (2004) Bio-ethanol steam reforming on ni/al2o3 catalyst. Chem Eng J 98(1–2):61–68
24. Benito M, Sanz J, Isabel R, Padilla R, Arjona R, Daza L (2005) Bio-ethanol steamreforming: insights on the mechanism for hydrogen production. J Power Source 151:11–17

25. Duan S, Senkan S (2005) Catalytic conversion of ethanol to hydrogen using combinatorial methods. Ind Eng Chem Res 44(16):6381–6386
26. Llorca J, DeLa Piscina P, Dalmon J, Sales J, Homs N (2003) Co-free hydrogen from steam reforming of bioethanol over zno-supported cobalt catalysts: effect of the metallic precursor. Appl Catal B 43(3):355–369
27. Liguras D, Kondarides D, Verykios X (2003) Production of hydrogen for fuel cells by steam reforming of ethanol over supported noble metal catalysts. Appl Catal B 43(4):345–354
28. Akande A, Aboudheir A, Idem R, Dalai A (2006) Kinetic modeling of hydrogen production by the catalytic reforming of crude ethanol over a co-precipitated ni-al2o3 catalyst in a packed bed tubular reactor. Int J Hydrogen Energy 31(12):1707–1715
29. Sahoo D, Vajpai S, Patel S, Pant K (2007) Kinetic modeling of steam reforming of ethanol for the production of hydrogen over co/al2o3 catalyst. Chem Eng J 125(3):139–147
30. Akpan E, Akande A, Aboudheir A, Ibrahim H, Idem R (2007) Experimental, kinetic and 2-d reactor modeling for simulation of the production of hydrogen by the catalytic reforming of concentrated crude ethanol (crcce) over a ni-based commercial catalyst in a packed-bed tubular reactor. Chem Eng Sci 62(12):3112–3126
31. Little A (1994) Multi-fuel reformers for fuel cells used in transportation-multi-fuel reformers: phase i. Technical report, Cambridge Arthur D. Little
32. Zalc J, Loffler D (2002) Fuel processing for pem fuel cells: transport and kinetic issues of system design. J Power Source 111(1):58–64
33. Campbell J (1970) Influences of catalyst formulation and poisoning on the activity and die-off of low temprature shift catalysts. Ind Eng Chem Proc Des Dev 9(4):588–595
34. Loffler D, McDermott S, Renn C (2003) Activity and durability of water-gas shift catalysts used for the steam reforming of methanol. J Power Source 114(1):15–20
35. Ruettinger W, Ilinich O, Farrauto R (2003) A new generation of water gas shift catalysts for fuel cell applications. J Power Source 118(12):61–65
36. Choi Y, Stenger H (2003) Water gas shift reaction kinetics and reactor modeling for fuel cell grade hydrogen. J Power Source 124(2):432–439
37. Ayastuy J, Gutierrez-Ortiz M, Gonzalez-Marcos J, Aranzabal A, Gonzalez-Velasco J (2005) Kinetics of the low temperature wgs reaction over a cuo/zno/al2o3 catalyst. Ind Eng Chem Res 44(1):41–50
38. Levent M (2001) Water-gas shift reaction over porous catalyst: temperature and reactant concentration distribution. Int. J. Hydrogen Energy 26(6):551–558
39. Kim D, Mayor J, Ni J (2005) Parametric study of microreactor design for water gas shift reactor using an integrated reaction and heat exchange model. Chem Eng J 110(1-3):1–10
40. Pasel J, Samsun R, Schmitt D, Peters R, Stolten D (2005) Test of a water-gas-shift reactor on a 3 kwe-scale–design points for high- and low-temperature shift reaction. J Power Source 152:189–195
41. Quiney A, Germani G, Schuurman Y (2006) Optimization of a water-gas shift reactor over a pt/ceria/alumina monolith. J Power Source 160(2):1163–1169
42. Basile A, Chiappetta G, Tosti S, Violante V (2001) Experimental and simulation of both pd and pd/ag for a water gas shift membrane reactor. Sep Purif Technol 25(1–3):549–571
43. Keiski R, Salmi T, Pohjola V (1992) Development and verification of a simulation model for a non-isothermal water-gas shift reactor. Chem Eng J 48(1):17–29
44. Amadeo N, Laborde M (1995) Hydrogen production from the low-temperature water-gas shift reaction: kinetics and simulation of the industrial reactor. Int J Hydrogen Energy 20(12):949–956
45. Hulteberg P, Brandin J, Silversand F, Lundberg M (2005) Preferential oxidation of carbon monoxide on mounted and unmounted noble-metal catalysts in hidrogen-rich streams. Int J Hydrogen Energy 30(11):1235–1242
46. Choi Y, Stenger H (2004) Kinetics, simulation and insights for co selective oxidation in fuel cell applications. J Power Source 129(2):246–254

47. Echigo M, Tabata T (2004) Development of novel ru catalyst of preferential co oxidation for residential polymer electrolyte fuel cell systems. Catal Today 90(3–4):269–275
48. Echigo M, Tabata T (2003) A study of co removal on an activated ru catalyst for polymer electrolyte fuel cell applications. Appl Catal A 251(1):157–166
49. Toyoshima I, Somorjai G (1979) Heat of chemisorption of o_2, h_2, co, co_2 and n_2 on polycristalline and single crystal transition metal surfaces. Catal Rev - Sci Eng 19(1):105–159
50. Bissett E, Oh S, Sinkevitch R (2005) Pt surface kinetics for a prox reactor for fuel cell feedstream processing. Chem Eng Sci 60(17):4709–4721
51. Linnhoff B, Townsend P, Boland P, Hewitt G.F, Thomas B.E.A, Guy A.R, Marsland R.H (1994) A user guide on process integration for the efficient use of energy. Institute of chemical engineers, Rugby, UK, rev sub edition
52. Feroldi D, Serra M, Riera J (2007) Performance improvement of a pemfc system controlling the cathode outlet air flow. J Power Source 169(1):205–212
53. Qi A, Peppley B, Karan K (2007) Integrated fuel processors for fuel cell application: a review. Fuel Process Technol 88(1):3–22

Control Oriented Dynamic Model of the Bio-Ethanol Processor System

L. Nieto Degliuomini, S. Biset, P. Luppi and M. Basualdo

Abstract In this chapter the pseudo-dynamic rigorous model of the bio-ethanol processor system (BPS) and proton exchange membrane fuel cell (PEMFC), working at the optimal efficiency point, is presented. It is based on the steady state HYSYS model given in the previous chapter and is able to deal with this kind of integrated plant for proposing and testing a suitable control structure. The model can be implemented thanks to the use of a specific communication protocol between the programs HYSYS and MATLAB to coordinate the calculations. The heat integration configuration is the same given in Chap. 9. The optimal heat exchangers network is initially considered with a faster dynamic than the rest of units of the plant so as to simplify the solution of the problem. It allows assuming that the stream temperatures are instantaneously at their optimum values. Moreover, the dynamic behavior of the reactors and the PEMFC are modeled in MATLAB. The principal objectives given previously for the BPS and PEMFC can be tested here through the dynamic simulations. They are performed using the critical control loops defined through the sensitivity analysis presented in the previous chapter. The controllers tuning for doing the preliminary tests are done thanks to the obtention of black box models, using the simulated plant excited with step changes and recording the reaction curves data.

L. Nieto Degliuomini · P. Luppi · M. Basualdo (✉)
CIFASIS-(CONICET-UNR-UPCAM), 27 de Febrero 210 bis,
S2000EZP Rosario, Argentina
e-mail: basualdo@cifasis-conicet.gov.ar

L. Nieto Degliuomini
e-mail: nieto@cifasis-conicet.gov.ar

S. Biset · M. Basualdo
Department of FRRo-UTN, CAPEG-Chem. Eng., Zeballos 1341,
S2000BQA Rosario, Argentine
e-mail: biset@frro.utn.edu.ar

P. Luppi
e-mail: luppi@cifasis-conicet.gov.ar

M. S. Basualdo et al. (eds.), *PEM Fuel Cells with Bio-Ethanol Processor Systems*,
Green Energy and Technology, DOI: 10.1007/978-1-84996-184-4_10,

1 Introduction

Dynamic rigorous models of fuel cell power plants are extremely important in understanding the variables interactions, implications on system performance and transient behaviors, hardly analyzed nowadays. The benefit of having representative dynamic models is because they reveal much information about reactors performance, and they are specially needed when transients during process disturbances are simulated. The development of a library models set is described in Eborn et al. [1] for unique component designs, building upon publicly and commercially available tools. They gave examples of models and showed some simulation results. Designs for the catalytic partial oxidation (CPO) and water gas shift reactors in the context of fuel cell power systems were presented in Görgün et al. [2, 3] using natural gas (methane) as fuel, nonlinear observer and control-oriented physics-based designs respectively. However, these designs were based on the reactors invariants and did not take into account reaction rate expressions. A control-oriented model of a catalytic partial oxidation of natural gas processor using physics-based principles was presented in Pukrushpan et al. [4]. They gave a guide about the control design of the bio-ethanol processor system (BPS) combined with a proton exchange membrane fuel cell (PEMFC) employing related theoretic tools. Varigonda et al. [5] highlighted major issues of BPS control, such as how regulating the amount of hydrogen in the fuel cell anode and the temperature of the CPO reactor during transient power demands from the fuel cell (FC). The limitations of a decentralized controller and the benefits of a multivariable controller were identified by Pukrushpan et al. [6] using linear and feedback analysis. A mathematical model of the reformer was developed by Chuang et al. [7] and the selected parameters of the model were fit to match experimental data from the dynamic response of the process. Then, they designed a reformer which had the minimum possible combined volume of the steam reformer and WGS reactors. There are few publications regarding bio-fuel processors dynamic behavior, some of them using methanol as raw material, such as Chuang et al. [7], and very recently some publications on using bio-ethanol to produce hydrogen. Some of these are Aicher et al. and García et al. [8, 9]. Eventhough they present dynamic models or experimental results, and use them to synthesize control structures, they do not contemplate the energy integration which produces strong interaction effects. The plant presented here is specifically designed to be useful for a combination of the fuel cell system with energy storage devices, which makes this technology economically feasible.

According to the review given above, the dynamic model of the BPS and PEMFC presented here is one of the first that deals with this energetically integrated process to achieve advantageous efficiencies. It is developed by using three commercial softwares. As a starting point some parts of the steady-state (SS) model, described in Chap. 9, implemented in the program HYSYS are useful for constructing the pseudo-dynamic model. The term 'pseudo' is included because those parts of the plant that present faster dynamic behavior can be taken from the original SS model. Then, even though this assumption is valid and does not produce serious difficulties for the final results, in a rigorous sense, the model is 'pseudo' dynamic. To complete the plant simulation, the set of reactors and PEMFC are modeled in MATLAB SIMULINK. The data transference between HYSYS and MATLAB can be performed thanks to specific protocol which will be detailed at Chap. 11. The third well-known program for mechanical engineers, named ADVISOR, is connected to simulate realistic disturbances. They are provided from information about standard vehicles driven through different scenarios such as urban or highway circuits. They generate different load profiles to the PEMFC. The heat integration considered in Chap. 9 to improve the efficiency of the hydrogen production remains in the pseudo-dynamic model. The heat exchanger network dynamic effect is neglected. So, the LNG tool can be used here for simulating this part of the process. This assumption helps to focus the attention in the main equipments of the plant, the reactors.

Therefore, the core of this chapter is the simulation of the plug flow reactors' dynamic behavior. They have been modeled as stage-wise unit operations, ideally mixed portions. General concepts of catalytic reactor modeling are presented in many textbooks of chemical reaction engineering, for example the book of Fogler [10].

Therefore, this chapter is organized in such a way that the reader would be able to follow the different steps of the model construction. In Sect. 2 a deep description of the fuel processor system with its reactors together with the main involved differential equations are given. In addition, a brief description of the MATLAB/HYSYS interface, the heat integration approximation solved by the 'LNG' tool and a short discussion about the main considerations for the PEMFC model are included. In Sect. 3 a transfer function matrix is given, from the simulation results with the plant in open loop. It is configured with the reduced order black box models set via simple rules of identification techniques. In Sect. 4 the best control structure found in the previous chapter is implemented. The controllers tuning via internal model control methodology is performed using the black box models. In Sect. 5 the results with the closed loop plant, simulated with the dynamic rigorous model are presented and evaluated. Finally, the conclusions together with some proposed improvements are presented in Sect. 5.

2 Bio-ethanol Processor System

The liquid fuel conversion into useful power in the context of the automobile industry reports its maximum efficiency for the internal combustion (IC) engines of around 35%. In the last decades the alternative use of the fuel cells has been growing up specially because they can achieve up to 83% of global efficiency, [11]. The main reason for this difference is that FCs do not have mobile parts so the energy losses because of friction are avoided.

The great difference of efficiencies should be enough reason to consider the fuel cells as excellent alternatives to the internal combustion engines, however, there are more advantages. Traditionally, engines burn fossil fuels and their residues are dumped to the atmosphere. It has been stated in Agarwal [12] that traditional fuels can be mixed with renewable ones for internal combustion engines without a considerable loss of efficiency. Fuel cells can be completely fueled with renewable sources. Particularly, proton exchange membrane fuel cells, require hydrogen to produce electricity but, considering the difficulties and risks of its storage and transportation, they can be provided with an onboard processing system to produce H_2 from renewable materials, such as bio-ethanol. It is relatively easy and cheap to produce, and can be obtained from agricultural or sugar industry residues. In such a way, the CO_2 produced in gas reforming will be reabsorbed by crops growth for the following reaps, presenting a nearly closed carbon loop and without net greenhouse gasses emissions.

The primary methods used in reforming hydrocarbons to produce hydrogen for use in PEM fuel cells are ethanol steam reforming (ESR), Partial oxidation, and a combination of both called autothermal reforming. Ethanol steam reforming is considered in this work mainly because of its ability to produce higher H_2 molar fractions. A very important issue for this reaction scheme is the endothermic nature of reforming. Therefore, it needs an auxiliary combustion reactor to fulfill the thermal requirements of the main reactor. One of the byproducts of the reforming reaction is Carbon monoxide which must be kept in a very low concentration of the hydrogen rich stream. The reason for this is because the membrane of the fuel cell is sensitive to poisoning by CO, so additional processing is needed in order to keep up to the requirements of the PEMFC.

2.1 Plant Configuration

A diagram of the bio-ethanol processor plant is shown in the Fig. 1. The main reactor of the system is the ESR, in which the most part of the conversion of bio-ethanol to H_2 is made. The group of series-parallel reaction occurring in this reactor gives rise to the appearance of other intermediate substances. One of this substances is the carbon monoxide, of particular interest and concern because it turns out to be poisonous to the catalyzer utilized in the PEMFC. Due to this sensitiveness, it is necessary to strip the stream of gasses from this compound, for which a reactors set is utilized as cleaning system. The first reaction to achieve this goal is the so called water gas shift where the equilibrium is driven to the products at low temperatures. However, the speeds of the reactions are strongly dependant on this variable, so implementing a reactor that operates at temperature close to that wanted for the equilibrium, would need a reactor of extremely large dimension. Due to this situation, it is advantageous to consider a two-stage water gas shift reaction. The first stage of high temperature, oriented to produce a quick reaction, and the second at low temperature in such a way that the resulting stream is found in an equilibrium state that favors the conversion of the reactives. Intermediate cooling will be necessary to obtain the desired characteristics.

The hydrogen production is thought for feeding the PEMFC for vehicles, so it is necessary to take into account the requirement of CO which concentration must be less than ten parts per million in volume. So it requires a third stage of cleaning given by means of the oxidation of the remaining monoxide. Hence, the last reactor named

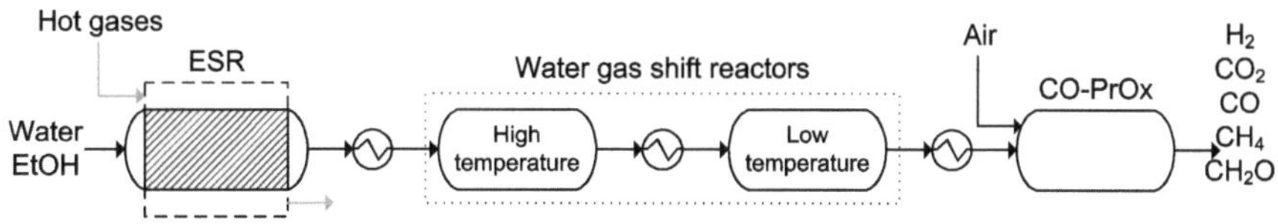

Fig. 1 Bio-ethanol processor system

Preferential Oxidation of carbon monoxide is in charge of achieving a hydrogen adequate for PEM usage.

The bio-ethanol is mixed with vaporized water to feed the ESR, producing an output stream rich in hydrogen, but containing water, carbon dioxide (CO_2), CO, methane, acetaldehyde and unreacted bio-ethanol. The overall reaction is endothermic, and the heat requirement is supplied by a burner, which is fed with compressed air and fresh ethanol. In the burner reactor, the complete combustion of ethanol is produced and the exit flow of hot gasses are passed through the heat exchanger jacket of the ESR. The hydrogen rich synthesis gas is conditioned, and finally, the resulting stream is fitted to feed a PEMFC designed to supply the load requirements of a small vehicle. A Fuel Cell is defined as an electrochemical device that converts the chemical energy of a reaction directly to electrical energy. It can generate electricity combining hydrogen and oxygen electrochemically without passing through a combustion stage. Theoretically, the only produced emissions are clean water and heat. The operating point for this plant is obtained from Benito et al. [14], where several details about its design were given and it is intended to feed a PEMFC with stringent mobility requirements.

2.2 *Mathematical Model Implementation*

The main concepts of catalytic reactor modeling are presented in many textbooks of chemical reaction engineering, for example in Larminie et al. [11]. In the case analyzed here, the considered reactors are plug flow reactor (PFR). Their conditions vary according to the location of the point under consideration within the vessel, and they should be strictly solved by partial differential equations. An alternative way of solving them is to discretize the model, considering a lumped configuration. The reactor is divided in several slices of smaller size, and each one of them is considered as a continuous stirred tank reactor (CSTR), as can be seen in Fig. 2. Hence, a series of a specific number of reactors, space independent, is assumed with homogeneous conditions. It simplifies the resolution by means of ordinary differential equations instead of using parcial differential equations.

Under this assumption, the total mass balance is considered in the following manner:

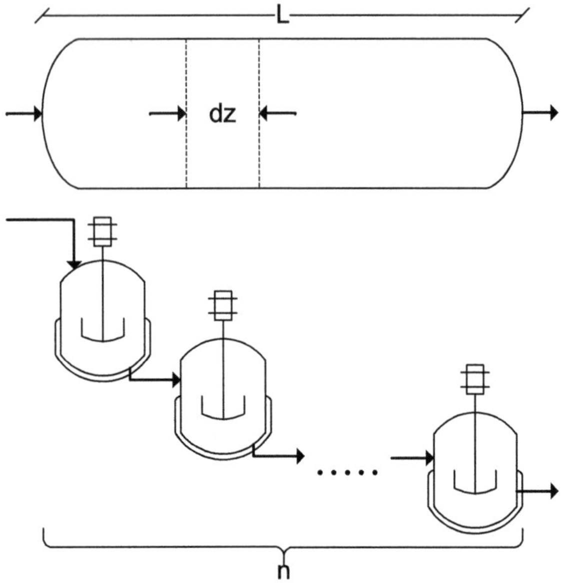

Fig. 2 A plug flow reactor can be modeled as a number of continuous stirred tank reactors in series

$$\frac{\mathrm{d}\,(n_\mathrm{t})}{\mathrm{d}t} = \mathrm{Fr_{in}} - \mathrm{Fr_{out}} + \left[\sum_{j,i} \nu_{j,i} \cdot (-r_i)\right] \cdot w_\mathrm{cat}, \tag{1}$$

where n_t is the total number of moles for each slice, $\mathrm{Fr_{in}}$ and $\mathrm{Fr_{out}}$ are the entering and exiting molar flow, $\nu_{j,i}$ is the stoichiometric coefficient for the component j in reaction i, $(-r_i)$ is the reaction rate and w_cat is the total weight of the utilized catalyst. Additionally, the pressure variation for each slice ΔP is neglected, ideal gasses behavior is considered, then $\mathrm{d}\,(n_\mathrm{t})/\mathrm{d}t$ is equal to zero, hence the exit flow can be determined. So, considering the component mass balance, the term $\mathrm{d}(n_\mathrm{t} \cdot y_j)/\mathrm{d}t$, can be written as $n_\mathrm{t} \cdot \mathrm{d}(y_j)/\mathrm{d}t$ and can be calculated through the following general expression

$$\frac{\mathrm{d}\,(y_j)}{\mathrm{d}t} = \frac{\mathrm{Fr_{in}} \cdot y_{j,\mathrm{in}} - \mathrm{Fr_{out}} \cdot y_j + \left[\sum_i \nu_{j,i} \cdot (-r_i)\right] \cdot w_\mathrm{cat}}{n_t}, \tag{2}$$

where y_j is the molar fraction of component j. Each of the involved reactions assume the following general kinetic expression:

$$(-r_i) = k_i \cdot \exp\left(-\frac{E_i}{R_\mathrm{g} \cdot T}\right) \cdot \left(\prod_{j=1}^{Rc} y_j^{\nu_j^i} - \frac{\prod_{j=1}^{Pr} y_j^{\nu_j^i}}{K\mathrm{eq}_i}\right). \tag{3}$$

where $(-r_i)$ stands for the reaction rate. It is based in the molar fractions of the substances (y_j) where α_j^i represents the stoichiometric coefficient of substance

j in reaction i. Pr represents the products and Rc the reactants, k_i is calculated from Arrhenius law as a temperature function. Parameters determined for each reaction are the frequency factor (k_i) and activation energy (E_i). The expression of the equilibrium constant for each reaction ($K\text{eq}_i$) is determined with the variation of the Gibbs free energy.

From the energy balance the temperature variation is obtained as follows:

$$\frac{\mathrm{d}t(T)}{\mathrm{d}t} = \frac{\mathrm{Fr}_{\mathrm{in}} \cdot cp_{\mathrm{in}} \cdot T_{\mathrm{in}} - \mathrm{Fr}_{\mathrm{out}} \cdot cp \cdot T + \left[\sum_i \nu_{j,i} \cdot (-\Delta H_i) \cdot (-r_i)\right] \cdot w_{\mathrm{cat}} + Q_i}{n_t \cdot cp}, \tag{4}$$

where T is the temperature of each slice, cp is the specific heat of the stream, $(-\Delta H)$ is the heat of reaction and Q_i is the exchanged heat, neglected in all the reactors considered as adiabatic. Only in the case of the ESR Q_i must be accounted. The heat generated by the different reactions is supposed to vary with temperature according to

$$\Delta H_i = \sum_j \nu_{j,i} \cdot \Delta H^0_{f,j} + \int_{\mathrm{T}^0}^{\mathrm{T}} \left(\sum_j \nu_{j,i} \cdot cp^0_j \right) \mathrm{d}T, \tag{5}$$

where cp^0_j is the specific heat of component j and $\Delta H^0_{f,j}$ the formation heat of component j at reference temperature T^0.

2.3 Ethanol Steam Reforming

Despite the apparent simplicity of the stoichiometric reaction for the production of hydrogen,

$$\mathrm{(ESR)CH_3CH_2OH + 3H_2O \leftrightarrow 2CO_2 + 6H_2},$$

actually, this reaction involves a complex system of reactions, and the selectivity towards H_2 is affected by a number of undesired secondary reactions. The reforming of ethanol is feasible for temperatures higher than 230°C.

The analysis made by Francesconi [13] adjusting experimental data of a cobalt catalyzer supported in zircon (CO/ZrO_2), leads to the reaction scheme of acetaldehyde decomposition, according to the mechanism proposed by Benito et al. [14]:

Reaction A:	$\mathrm{CH_3CH_2OH \leftrightarrow CH_3CHO + H_2}$	$\Delta H^{\mathrm{A}}_{298\,\mathrm{K}} = 70.6\,\mathrm{kJ/mol}$
Reaction B:	$\mathrm{CH_3CHO \leftrightarrow CO + CH_4}$	$\Delta H^{\mathrm{B}}_{298\,\mathrm{K}} = -21.9\,\mathrm{kJ/mol}$
Reaction C:	$\mathrm{CH_4 + H_2O \leftrightarrow CO + 3H_2}$	$\Delta H^{\mathrm{C}}_{298\,\mathrm{K}} = 205.9\,\mathrm{kJ/mol}$
Reaction D:	$\mathrm{CO + H_2O \leftrightarrow CO_2 + H_2}$	$\Delta H^{\mathrm{D}}_{298\,\mathrm{K}} = -41.18\,\mathrm{kJ/mol}$

To achieve a good yield it is necessary to operate at a high temperature (>550°C).

The total mass balance for the ESR is

$$\frac{dn_{t,i}}{dt} = \mathrm{Fr}_{i-1} - \mathrm{Fr}_i + \left[\left(-r_{A,i}\right) + \left(-r_{B,i}\right) + 2 \cdot \left(-r_{C,i}\right)\right] \cdot w_{\mathrm{cat}} = 0 \tag{6}$$

and the mass balances for each component are the following:

$$\frac{\mathrm{d}ya_i}{\mathrm{d}t} = \frac{\left(\mathrm{Fr}_{i-1} \cdot ya_{i-1} - \mathrm{Fr}_i \cdot ya_i - \left(-r_{\mathrm{A,i}}\right) \cdot w_{\mathrm{cat}}\right)}{n_{\mathrm{t,i}}}$$

$$\frac{\mathrm{d}yb_i}{\mathrm{d}t} = \frac{\left(\mathrm{Fr}_{i-1} \cdot yb_{i-1} - \mathrm{Fr}_i \cdot yb_i + \left[-\left(-r_{\mathrm{C,i}}\right) - \left(-r_{\mathrm{D,i}}\right)\right] \cdot w_{\mathrm{cat}}\right)}{n_{\mathrm{t,i}}}$$

$$\frac{\mathrm{d}yc_i}{\mathrm{d}t} = \frac{\left(\mathrm{Fr}_{i-1} \cdot yc_{i-1} - \mathrm{Fr}_i \cdot yc_i + \left[\left(-r_{\mathrm{B,i}}\right) - \left(-r_{\mathrm{C,i}}\right)\right] \cdot w_{\mathrm{cat}}\right)}{n_{\mathrm{t,i}}}$$

$$\frac{\mathrm{d}yd_i}{\mathrm{d}t} = \frac{\left(\mathrm{Fr}_{i-1} \cdot yd_{i-1} - \mathrm{Fr}_i \cdot yd_i + \left[\left(-r_{\mathrm{B,i}}\right) + \left(-r_{\mathrm{C,i}}\right) - \left(-r_{\mathrm{D,i}}\right)\right] \cdot w_{\mathrm{cat}}\right)}{n_{\mathrm{t,i}}}$$

$$\frac{dye_i}{dt} = \frac{\left(\mathrm{Fr}_{i-1} \cdot ye_{i-1} - \mathrm{Fr}_i \cdot ye_i + \left(-r_{\mathrm{D,i}}\right) \cdot w_{\mathrm{cat}}\right)}{n_{\mathrm{t,i}}}$$

$$\frac{\mathrm{d}yf_i}{\mathrm{d}t} = \frac{\left(\mathrm{Fr}_{i-1} \cdot yf_{i-1} - \mathrm{Fr}_i \cdot yf_i + \left[\left(-r_{\mathrm{A,i}}\right) + 3\left(-r_{\mathrm{C,i}}\right) + \left(-r_{\mathrm{D,i}}\right)\right] \cdot w_{\mathrm{cat}}\right)}{n_{\mathrm{t,i}}}$$

$$\frac{\mathrm{d}yg_i}{\mathrm{d}t} = \frac{\left(\mathrm{Fr}_{i-1} \cdot yg_{i-1} - \mathrm{Fr}_i \cdot yg_i + \left[\left(-r_{\mathrm{A,i}}\right) - \left(-r_{\mathrm{B,i}}\right)\right] \cdot w_{\mathrm{cat}}\right)}{n_{\mathrm{t,i}}}$$

From the energy balance can be obtained the expression for temperature variation given by the following:

$$\frac{\mathrm{d}T_i}{\mathrm{d}t} = \frac{\begin{pmatrix} \mathrm{Fr}_{i-1} \cdot \mathrm{cp}_{i-1} \cdot T_{i-1} - \mathrm{Fr}_i \cdot \mathrm{cp}_i \cdot T_i + \left[\left(-r_{\mathrm{A,i}}\right) \cdot \left(-\Delta H_{\mathrm{A}}\right) + \left(-r_{\mathrm{B,i}}\right) \right. \\ \left. \cdot \left(-\Delta H_{\mathrm{B}}\right) + \left(-r_{\mathrm{C,i}}\right) \cdot \left(-\Delta H_C\right) + \left(-r_{\mathrm{D,i}}\right) \cdot \left(-\Delta H_{\mathrm{D}}\right)\right] \cdot w_{\mathrm{cat}} + Q_i \end{pmatrix}}{n_{\mathrm{t,i}} \cdot \mathrm{cp}_i}.$$

To fulfill the heat requirements of the reactions, the ESR is provided with a hot gasses chamber, it collects the products of combustion from the ethanol burner, and exchanges heat with the main reactor through the jacket. Since there is no reaction in this volume, only the total mass balance is needed.

Therefore, the exchanged heat Q_i is calculated by the following algebraic expression:

$$Q_i = \frac{4}{Dt} \cdot \left(h_1 \cdot D\right) \cdot V_{\mathrm{sl}} \cdot \left(T_{\mathrm{g}} - T\right), \tag{7}$$

where Dt is the tube inner diameter, V_{sl} is the volume of the chamber, T_{g} is the temperature of the hot gasses and $(h_1 \cdot D)_i$ is the warming correlation, given by Keiski et al. [16]

$$\left(h_1 \cdot D\right)_i = 0.813 \cdot \mathrm{Re}^{0.9} \cdot \exp\left(-6\frac{Dp}{Dt}\right) \cdot k_{\mathrm{mix}}, \tag{8}$$

Table 1 Design conditions for the reformer

Reactants		Hot gasses	
Molar ratio	4	Temperature [K]	1,331
H_2O [mole · h^{-1}]	302.4	Flow [mole · h^{-1}]	2,342
EtOH [mole · h^{-1}]	75.6	H_2O	22.4%
Temperature [K]	982	CO_2	8.8%
Pressure [atm]	3	O_2	4.0%
		N_2	64.8%

Table 2 Results for annular reactor optimization

Particle diameter [cm]	0.05
Inerts fraction	0.20
Tube length [cm]	43.61
Tube diameter [cm]	10.93
Catalytic fixed bed thickness [cm]	0.50
First isolation thickness [cm]	1.03
Second isolation thickness [cm]	13.20
Fixed bed volume [cm^3]	869 (1.4%)
Tubes volume [cm^3]	2,304 (3.6%)
Gasses chamber volume [cm^3]	4,091 (6.5%)
Isolations volume [cm^3]	56,139 (88.5%)
Total volume [cm^3]	63,404

where k_{mix} is the gaseous mix thermal conductivity, Re is the Reynolds number, Dp is the particle diameter.

As a design condition is chosen a water/ethanol molar ratio of 4 and an inlet temperature to the reactor of 982 K, as stated in Chap. 9. From preliminary calculations, considering a production of 260 mole · h^{-1} of H_2 to feed a PEMFC of 10 kW, a total flow of reactants of 378 mole · h^{-1} is needed.

From a thermodynamic point of view under these conditions, an isothermal reactor can reach a production rate of 328 mole · h^{-1} of H_2. But given that in an actual reactor the conditions will be away from the equilibrium, for the determination, this excess of flow will be considered.

Table 1 shows the conditions for the design of the reactor. The compositions and hot gasses flows leaving the burner that supply the energy to the reactor were obtained from the mass and energy balances of the integrated system from Chap. 9. Fixing the operating conditions of the reactor in such a system, the composition, temperature and flow are determined, considering the combustion of a mix of alcohol along with the remaining gasses from the fuel cell.

From the analysis made by Francesconi [13], it is concluded that the annular reactor presented better performance. So, the reactor considered has an annular geometry and a rated power of 10 kW. The design obtained for the reactor by optimization techniques is shown in Table 2.

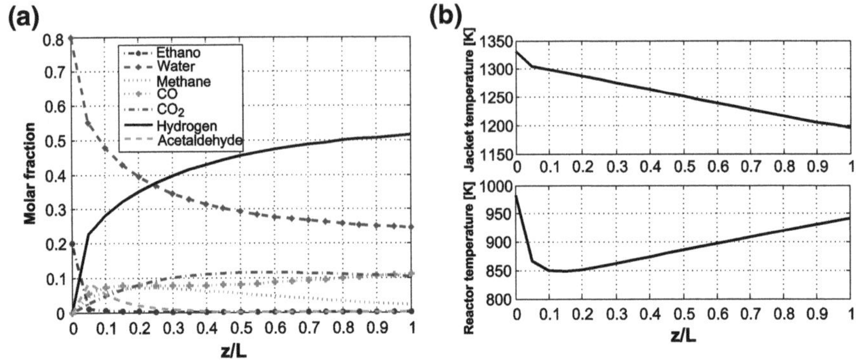

Fig. 3 Steady state compositions and temperature profiles for the ESR. **a** ESR molar composition profile, **b** Temperatures profile for ESR

Through the simulation according to the assumed hypothesis and the operating conditions described previously, the profiles obtained for the steady state operation of the reforming reactor are shown in Fig. 3. The first detail that can be noticed is the fast consumption of ethanol in the entrance of the reactor, producing a maximum in the concentration of acetaldehyde, and a sudden decrease in the temperature inside the reactive bed. After the acetaldehyde reaches its peak, it decreases gradually to almost disappear at the exit of the reactor. The slow diminution of the amount of water inside the reactor is directly accompanied with the increase in hydrogen concentration, the main product. Methane also presents a maximum, and then decreases little by little. Carbon oxides present a sudden increase in the concentration and then keep nearly constant along the rest of the reactor. According to the endothermic reaction of the reformer, in the first section there is a relatively abrupt decrease of the temperature in the hot gasses inside the jacket and then continue decreasing gradually, as they apport their available heat. The internal temperature decrease of the reactor is much more clear at the beginning and then rises constantly to reach a value very close to the original one.

2.4 Water Gas Shift

In these reactors, the only reaction that takes place is the WGS, widely used in the industry in ammonia synthesis and hydrogen production by means of hydrocarbon reforming. The reaction scheme is the following:

$$\text{Reaction WGS: } CO + H_2O \leftrightarrow CO_2 + H_2 \quad \Delta H^{WGS}_{298\,K} = -41.18\,\text{kJ/mol}.$$

The main objectives of the WGS reaction are to remove the CO from streams and adjust the H_2/CO molar ratio. This reaction is moderately exothermic and its equilibrium constant decreases as temperature rises, so lower temperature favors high conversions. The reactors that belong to the cleaning system are considered adiabatic, and this condition limits their performance, this is the reason why more than one WGS reactor with intermediate cooling is included.

Since both reactors have the same reaction scheme, the balances are equal for both of them, the total mass balance is

$$\frac{dn_{t,i}}{dt} = Fr_{i-1} - Fr_i = 0 \tag{9}$$

and the component mass balances can be expressed as

$$\frac{dya_i}{dt} = \frac{(Fr_{i-1} \cdot ya_{i-1} - Fr_i \cdot ya_i)}{n_{t,i}} \tag{10}$$

$$\frac{dyb_i}{dt} = \frac{\left(Fr_{i-1} \cdot yb_{i-1} - Fr_i \cdot yb_i - \left(-r_{WGS,i}\right) \cdot w_{cat}\right)}{n_{t,i}} \tag{11}$$

$$\frac{dyc_i}{dt} = \frac{(Fr_{i-1} \cdot yc_{i-1} - Fr_i \cdot yc_i)}{n_{t,i}} \tag{12}$$

$$\frac{dyd_i}{dt} = \frac{\left(Fr_{i-1} \cdot yd_{i-1} - Fr_i \cdot yd_i - \left(-r_{WGS,i}\right) \cdot w_{cat}\right)}{n_{t,i}} \tag{13}$$

$$\frac{dye_i}{dt} = \frac{\left(Fr_{i-1} \cdot ye_{i-1} - Fr_i \cdot ye_i + \left(-r_{WGS,i}\right) \cdot w_{cat}\right)}{n_{t,i}} \tag{14}$$

$$\frac{dyf_i}{dt} = \frac{\left(Fr_{i-1} \cdot yf_{i-1} - Fr_i \cdot yf_i + \left(-r_{WGS,i}\right) \cdot w_{cat}\right)}{n_{t,i}} \tag{15}$$

$$\frac{dyg_i}{dt} = \frac{(Fr_{i-1} \cdot yg_{i-1} - Fr_i \cdot yg_i)}{n_{t,i}} \tag{16}$$

From the energy balance the temperature variations for both reactors is

$$\frac{dT_i}{dt} = \frac{Fr_{i-1} \cdot cp_{i-1} \cdot T_{i-1} - Fr_i \cdot cp_i \cdot T_i + \left(-r_{WGS,i}\right) \cdot (-\Delta H_{WGS}) \cdot w_{cat}}{n_{t,i} \cdot cp_i}. \tag{17}$$

Because the considered reactors are supposed to be adiabatic, there is no heat exchange, and all the heat produced in the reaction is used to raise the temperature of the stream.

The exit flows of the reforming reactor are considered the entering values at the inlet of the WGS reactors. A heat exchanger is necessary to properly conditioning the inlet temperature, this value is an operating parameter that comes from the resolution of the optimization problem. The molar flows and inlet compositions to the WGS reactors are shown in Table 3.

Table 3 Inlet conditions to WGS reactors

Molar flows	$(\text{mole} \cdot h^{-1})$	Molar	Fraction
F_{CH_4}	30	y_{CH_4}	0.045
F_{H_2}	324	y_{H_2}	0.491
F_{CO}	64	y_{CO}	0.096
F_{H_2O}	167	y_{H_2O}	0.252
F_{CO_2}	76	y_{CO_2}	0.114

Table 4 Design for the HTS reactor

Particle diameter [cm]	0.05
Tube length [cm]	6.7
Tube diameter [cm]	5.0
Isolation thickness [cm]	5.6
Temperature [K]	700
Fixed bed volume [cm^3]	131 (3%)
Tubes volume [cm^3]	47 (1%)
Isolation volume [cm^3]	1,353 (31%)
Heads volume [cm^3]	2,609 (60%)
Inert inlet [cm^3]	229 (5%)
Unit volume [cm^3]	4,370

2.4.1 High Temperature Shift

The high temperature water gas shift (HTS) reactor that uses an iron based catalyzer (Fe/Cr) and operates between 300°C and 550°C is the first for the cleaning stage and makes most of the conversion of CO.

Kinetics parameters used in this work for the HTS reactor are given by Keiski et al. [16]. They adjust reaction data (industrial conditions) for an iron based catalyzer and propose a power law expression which is

$$\begin{aligned} r_{WGS} = {} & 1.197 \times 10^3 \exp\left(-\frac{78{,}213}{R_g T}\right) C_{CO}^{0.73} C_{H_2O}^{0.55} \\ & - 9.0919 \times 10^4 \exp\left(-\frac{116{,}270}{R_g T}\right) C_{CO}^{-0.27} C_{H_2O}^{-0.45} C_{CO_2} C_{H_2}, \end{aligned} \tag{18}$$

where C_j is the concentration of component j (CO, H_2O, CO_2 and H_2).

Considering the joint optimization of the cleaning reactors set, with the total volume as the objective variable, the design obtained for the HTS is shown in Table 4.

It can be seen in Fig. 4 the compositions and temperature profiles for the axial direction of the reactor. Here, carbon dioxide and hydrogen concentrations rise significantly, simultaneously with the decrease in the carbon monoxide and water compositions. In terms of the objectives of the processor, the reactions are very favorable. Variations of the components are even, due to the fact that they all have unitary coefficients in the stoichiometry of the only reaction considered. The temperature

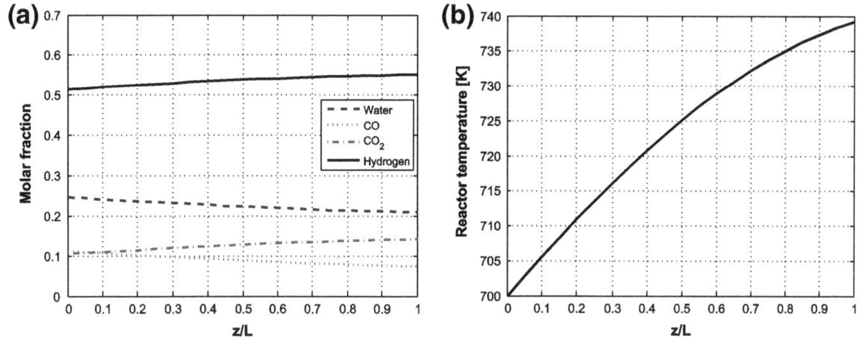

Fig. 4 Steady state compositions and temperature profiles for the HTS. **a** HTS molar composition profile, **b** Temperatures profile for HTS

Table 5 Design for the LTS reactor

Particle diameter [cm]	0.05
Tube length [cm]	11.1
Tube diameter [cm]	9.8
Isolation thickness [cm]	2.2
Temperature [K]	421.5
Fixed bed volume [cm^3]	836 (20%)
Tubes volume [cm^3]	216 (5%)
Isolation volume [cm^3]	1,004 (24%)
Heads volume [cm^3]	1,886 (46%)
Inert inlet [cm^3]	185 (4%)
Unit volume [cm^3]	4,126

rises along the reactor, according to the enthalpy change that the reaction presents, and the lack of heat dissipation, as it was described earlier. In the figures only some of the components are shown, they are the ones that are involved in the reaction. Also, the reaction does not present volume variation, so the concentrations of the other substances remain constant.

2.4.2 Low Temperature Shift

The low temperature water gas shift (LTS) reactor operates at lower temperatures (150–230°C) and uses a Copper–Zinc catalyzer supported over Alumina, the composition is usually $CuO/ZnO/Al_2O_3$. It is thermodynamically favored, so higher conversions rates are obtained because the equilibrium constant is low.

The kinetic expression obtained by Choi et al. [17] is used. They determine the parameters under the most likely conditions for a methanol reforming reactor, that is composition and pressures, which are similar to the ethanol processor in study. They used a Süd-Chemie commercially available catalyzer and provide the following expression:

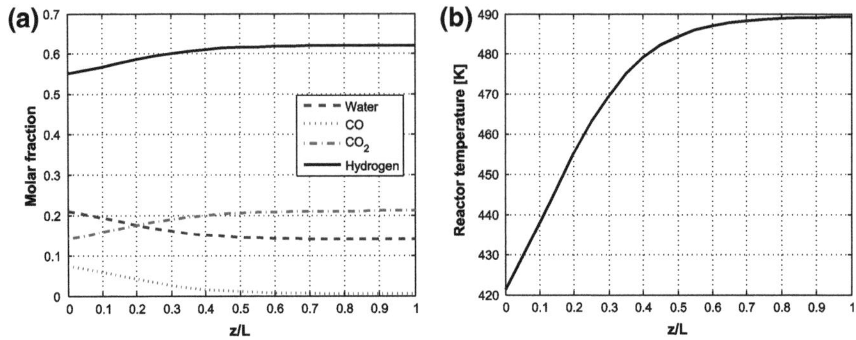

Fig. 5 Steady state compositions and temperature profiles for the LTS. **a** LTS molar composition profile, **b** Temperatures profile for LTS

$$r_{WGS} = 82.2 \exp\left(-\frac{47{,}400}{R_g T}\right)\left(P_{CO}P_{H_2O} - \frac{P_{CO_2}P_{H_2}}{K_{eq}}\right), \tag{19}$$

where K_{eq} represents the equilibrium constant, which can be calculated with the following equation provided by Choi et al. [18]:

$$ln(K_{eq}) = \frac{5{,}693.5}{T} + 1.077\ ln(T) + 5.44 \times 10^{-4}T - \frac{49{,}170}{T^2} - 13.148. \tag{20}$$

Considering the joint optimization of the cleaning reactors set, the design obtained for the LTS is shown in Table 5.

Composition and temperature profiles for the low temperature reactor are presented in Fig. 5. The variations of the compositions allow an analogous analysis to the HTS reactor. The main difference of this reactor is that it reaches an equilibrium value, the variation of concentrations decreases up to achieve the end of the reactor. The final part of this reactor appears to be unnecessary, that is why it could be erroneously concluded that the reactor is oversized. However, the final diminution in the concentration of CO is the necessary to reach the designed operating point of the CO–PrOx, and that determines its size and the selectivity towards CO of the last combustion reaction.

2.5 Preferential Oxidation of Carbon Monoxide

The last reactor needed to achieve the hydrogen quality objective to feed a PEMFC is the preferential oxidation of carbon monoxide (CO–PrOx). It conforms the cleaning system too. The main (desired) reaction given in this stage is the oxidation of CO into CO_2. However, the undesired oxidation of H_2 is produced too, it reduces the efficiency of the BPS and increases the temperature of the stream. It has been reported by Choi et al. [18] that WGS reaction should be considered at this stage. The reaction scheme for the CO–PrOx is the following:

$$\begin{array}{lll} \text{Reaction WGS:} & CO + H_2O \leftrightarrow CO_2 + H_2 & \Delta H^{WGS}_{298\,K} = -41.18\,\text{kJ/mol} \\ \text{Reaction E:} & CO + \frac{1}{2}O_2 \leftrightarrow CO_2 & \Delta H^{E}_{298\,K} = -283\,\text{kJ/mol} \\ \text{Reaction F:} & H_2 + \frac{1}{2}O_2 \leftrightarrow H_2O & \Delta H^{F}_{298\,K} = -242\,\text{kJ/mol} \end{array}$$

Oxygen is injected into the CO–PrOx, the amount needed is about twice the stoichiometric relation to have a good selectivity and satisfy the requirements of the FC.

The total mass balance for this reactor can be expressed as follows:

$$\frac{dn_{t,i}}{dt} = \text{Fr}_{i-1} - \text{Fr}_i + \frac{1}{2}\left[-\left(-r_{E,i}\right) - \left(-r_{F,i}\right)\right] w_{cat} = 0. \tag{21}$$

Because the reaction rates are based in CO and H_2, the mass balances by component are the following:

$$\frac{dya_i}{dt} = \frac{(\text{Fr}_{i-1} \cdot ya_{i-1} - \text{Fr}_i \cdot ya_i)}{n_{t,i}} \tag{22}$$

$$\frac{dyb_i}{dt} = \frac{\left(\text{Fr}_{i-1} \cdot yb_{i-1} - \text{Fr}_i \cdot yb_i + \left[\left(-r_{F,i}\right) - \left(-r_{WGS,i}\right)\right] w_{cat}\right)}{n_{t,i}} \tag{23}$$

$$\frac{dyc_i}{dt} = \frac{(\text{Fr}_{i-1} \cdot yc_{i-1} - \text{Fr}_i \cdot yc_i)}{n_{t,i}} \tag{24}$$

$$\frac{dyd_i}{dt} = \frac{\left(\text{Fr}_{i-1} \cdot yd_{i-1} - \text{Fr}_i \cdot yd_i + \left[-\left(-r_{E,i}\right) - \left(-r_{WGS,i}\right)\right] w_{cat}\right)}{n_{t,i}} \tag{25}$$

$$\frac{dye_i}{dt} = \frac{\left(\text{Fr}_{i-1} \cdot ye_{i-1} - \text{Fr}_i \cdot ye_i + \left[\left(-r_{E,i}\right) + \left(-r_{WGS,i}\right)\right] w_{cat}\right)}{n_{t,i}} \tag{26}$$

$$\frac{dyf_i}{dt} = \frac{\left(\text{Fr}_{i-1} \cdot yf_{i-1} - \text{Fr}_i \cdot yf_i + \left[-\left(-r_{F,i}\right) + \left(-r_{WGS,i}\right)\right] w_{cat}\right)}{n_{t,i}} \tag{27}$$

$$\frac{dyg_i}{dt} = \frac{(\text{Fr}_{i-1} \cdot yg_{i-1} - \text{Fr}_i \cdot yg_i)}{n_{t,i}} \tag{28}$$

$$\frac{dyh_i}{dt} = \frac{\left(\text{Fr}_{i-1} \cdot yh_{i-1} - \text{Fr}_i \cdot yh_i + \frac{1}{2}\left[-\left(-r_{E}\right) - \left(-r_{F,i,i}\right)\right] w_{cat}\right)}{n_{t,i}} \tag{29}$$

The energy balance is expressed:

$$\frac{dT_i}{dt} = \frac{\begin{pmatrix} \text{Fr}_{i-1} \cdot cp_{i-1} \cdot T_{i-1} - \text{Fr}_i \cdot cp_i \cdot T_i + [(-r_{WGS}) \cdot (-\Delta H_{WGS}) \\ + (-r_E) \cdot (-\Delta H_E) + (-r_F) \cdot (-\Delta H_F)] \cdot w_{cat} \cdot T_i \end{pmatrix}}{n_{t,i} \cdot cp_i}. \tag{30}$$

Again, adiabatic reaction is considered so all the heat generated from the reaction increases the temperature of the stream.

Although many publications present kinetic expressions for CO oxidation, only a few consider the simultaneous H_2 oxidation. Linnhoff et al. [19] gives equations for both reactions and consider the WGS, with a Pt–Fe/Al_2O_3 catalyzer:

$$r_{CO} = 0.098 \cdot \exp\left(-\frac{33{,}092}{R_g \cdot T}\right) \cdot P_{CO}^{-0.1} \cdot P_{O_2}^{0.5} \tag{31}$$

$$r_{H_2} = 0.005703 \cdot \exp\left(-\frac{18{,}742}{R_g \cdot T}\right) \cdot P_{O_2}^{0.5} \tag{32}$$

$$r_{WGS} = 1.2227778 \cdot \exp\left(-\frac{34{,}104}{R_g \cdot T}\right) \cdot \left(P_{CO} \cdot P_{H_2O} - \frac{P_{CO_2} \cdot P_{H_2}}{K_{eq}}\right) \tag{33}$$

where K_{eq} can be calculated by Eq. 20, and P_j is the partial pressure of component j (CO, H_2O, CO_2 and H_2).

Analyzing the original kinetic expression, it can be noticed that the rate of reaction tends to infinite as the partial pressures become closer to zero. Such behavior brings numerical convergence problems, and besides it represents an unreal limit behavior, given that a null concentration would give an infinite rate of reaction.

Due to the values that reach the partial pressures of the substances (O_2 and CO), in the case under investigation are around 1×10^{-6}. It is necessary to modify the original expressions correcting the limit behavior to introduce no numerical convergence problems. The corrected expressions are the following:

$$r_{CO} = 0.098 \cdot \exp\left(-\frac{33{,}092}{R_g \cdot T}\right)\left(\left(P_{O_2} + \mathbf{C}\right)^{0.5} - \mathbf{C}^{0.5}\right)\left(\frac{P_{CO}}{\mathbf{C}} + P_{CO}^{1.1}\right), \tag{34}$$

$$r_{H_2} = 0.005703 \cdot \exp\left(-\frac{18{,}742}{R_g \cdot T}\right)\left(\left(P_{O_2} + \mathbf{C}\right)^{0.5} - \mathbf{C}^{0.5}\right). \tag{35}$$

By the incorporation of a constant (**C**) the behavior of the kinetics is corrected in such a way that the rate of reaction tends to zero when the concentrations of the reactants tend to zero. On the other hand the incorporation of this constant does not modify significantly the values at higher concentrations. In Fig. 6, the behavior of the kinetic expression can be seen in its original version compared to the modified expression for two different values of **C** (1×10^{-5} and 1×10^{-6}). It can be concluded that with the value of 1×10^{-6} the error produced is not significant.

In Table 6, the inlet conditions to the reactor are resumed.

Considering the joint optimization of the cleaning reactors set, the design obtained for the CO–PrOx is shown in Table 7.

Figure 7 shows the compositions and temperature profiles for the last reactor of the cleaning system. Only the concentration of the substances involved in the reactions taking place are shown. As it can be noticed, CO and hydrogen concentrations decrease, while the concentrations of water and carbon dioxide rise, the water in least amount, because the WGS reaction is produced in this reactor too. Despite the extremely small variation in the compositions, the temperature variation inside the

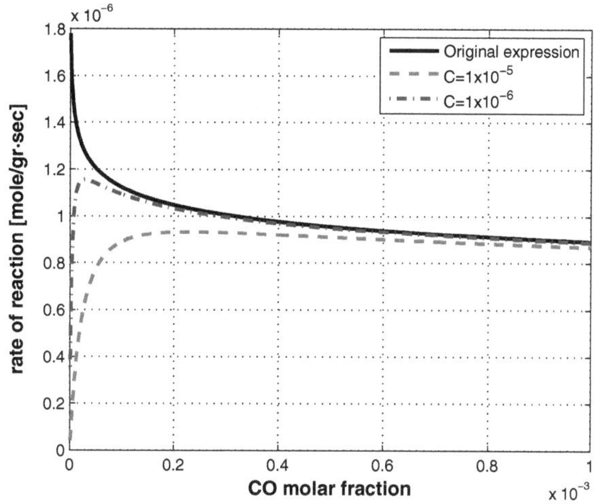

Fig. 6 Comparison of the kinetic expressions for different values of **C**

Table 6 Inlet conditions to WGS reactors

Molar flows	$(\text{mole}\cdot h^{-1})$	Molar	Fraction
F_{CH_4}	30	y_{CH_4}	0.045
F_{H_2}	385	y_{H_2}	0.583
F_{CO}	2.6	y_{CO}	0.004
F_{H_2O}	105	y_{H_2O}	0.160
F_{CO_2}	136	y_{CO_2}	0.206

Table 7 Design for the CO–PrOx reactor

Particle diameter [cm]	0.05
Tube length [cm]	8.6
Tube diameter [cm]	7.1
Isolation thickness [cm]	3.0
Temperature [K]	436.5
Fixed bed volume [cm^3]	342 (11%)
Tubes volume [cm^3]	103 (3%)
Isolation volume [cm^3]	910 (30%)
Heads volume [cm^3]	1,490 (50%)
Inert inlet [cm^3]	158 (5%)
Unit volume [cm^3]	3,002

reactor is significant, according to the strongly exothermic nature of the reactions taking place, mainly the combustion. In Fig. 7, it is shown in detail the molar fraction of carbon monoxide along the axial direction of the reactor, entering within the specified design conditions, and exiting the reactor with the required concentration to avoid catalyst poisoning of the fuel cell, that is below 10 ppm.

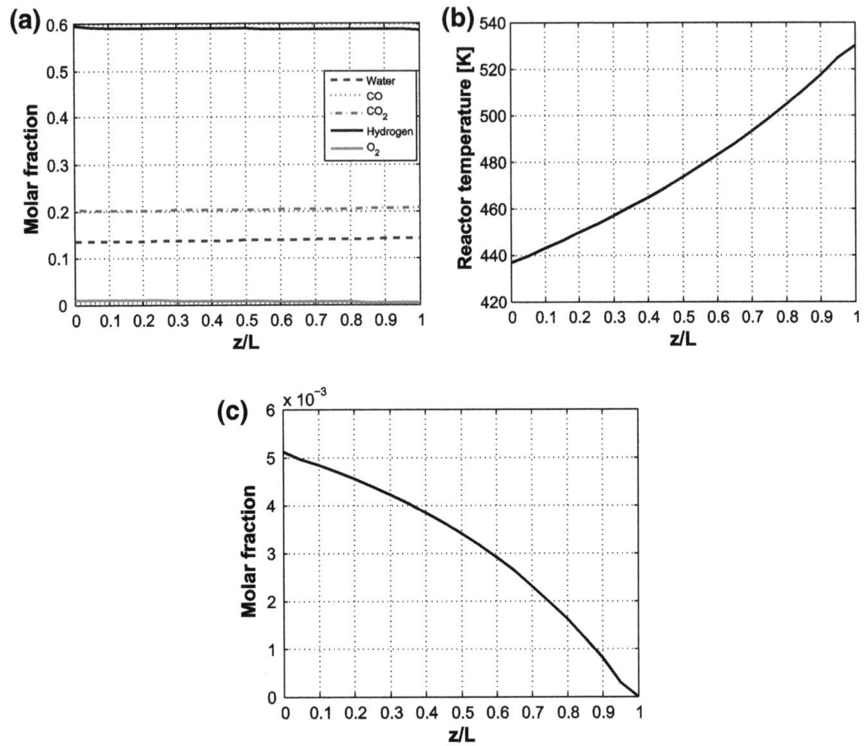

Fig. 7 Steady state compositions and temperature profiles for the CO–PrOx. **a** COP molar composition profile, **b** Temperatures profile for COP, **c** Molar composition profile of CO

2.6 Burner

A stream of ethanol mixed with air is fed into this reactor to burn the ethanol and obtain hot gasses that will go to the jacket of the ESR to provide the necessary heat to reach good yields. Complete combustion of reactants is considered, if the stoichiometric amount of oxygen is available. This reactor is modeled as an adiabatic CSTR. The reaction taking place is the following:

$$\mathrm{CH_3CH_2OH} + 3 \cdot \mathrm{O_2} \rightarrow 2 \cdot \mathrm{CO_2} + 3 \cdot \mathrm{H_2O} \Delta H^{\mathrm{BRN}}_{298\,\mathrm{K}} = -1{,}279\ \mathrm{kJ/mol}$$

and the total mass balance is the following:

$$\frac{dn_t}{dt} = \mathrm{Fr_{in}} - \mathrm{Fr_{out}} + \mathrm{EtOH_{reacted}} - \frac{1}{2}\mathrm{H_{2reacted}} = 0, \qquad (36)$$

meanwhile the mass balances by component can be expressed as

$$\frac{dya}{dt} = \frac{(Fr_{in} \cdot ya_{in} - Fr_{out} \cdot ya_{out} - EtOH_{reacted})}{n_t}$$
$$\frac{dyb}{dt} = \frac{(Fr_{in} \cdot yb_{in} - Fr_{out} \cdot yb_{out} + 3 \cdot EtOH_{reacted} + 2 \cdot CH_{4reacted} + H_{2reacted})}{n_t}$$
$$\frac{dyc}{dt} = \frac{(Fr_{in} \cdot yc_{in} - Fr_{out} \cdot yc_{out} - CH_{4reacted})}{n_t}$$
$$\frac{dyd}{dt} = \frac{(Fr_{in} \cdot yd_{in} - Fr_{out} \cdot yd_{out})}{n_t}$$
$$\frac{dye}{dt} = \frac{(Fr_{in} \cdot ye_{in} - Fr_{out} \cdot ye_{out} + 2 \cdot EtOH_{reacted} + CH_{4reacted})}{n_t}$$
$$\frac{dyf}{dt} = \frac{(Fr_{in} \cdot yf_{in} - Fr_{out} \cdot yf_{out} - H_{2reacted})}{n_t}$$
$$\frac{dyg}{dt} = \frac{(Fr_{in} \cdot yg_{in} - Fr_{out} \cdot yg_{out})}{n_t}$$
$$\frac{dyh}{dt} = \frac{\left(Fr_{in} \cdot yg_{in} - Fr_{out} \cdot yg_{out} - 3 \cdot EtOH_{reacted} - 2 \cdot CH_{4reacted} - \frac{1}{2} CH_{2reacted}\right)}{n_t}$$

The energy balance for both reactors is

$$\frac{dT}{dt} = \frac{Fr_{in} \cdot cp_{in} \cdot T_{in} - Fr_{out} \cdot cp_{out} \cdot T_{out} + (-r_{WGS}) \cdot (-\Delta H_{WGS}) \cdot w_{cat}}{n_t \cdot cp_{out}} \tag{37}$$

The amount of ethanol that reacts ($EtOH_{reacted}$) is equal to the entering flow if the quantity of oxygen is enough to produce the complete combustion.

2.7 Heat Integration

To obtain a high efficiency in the bio-ethanol processor system, it is necessary to use the heat of all the intermediate and product streams, in order to minimize the use of auxiliary services. To achieve the maximum possible heat exchange, it is not possible to simply take the available heat and provide it where is needed. Then a well designed heat exchangers network is needed to achieve an optimal energy integrated system. In the Table 8 can be seen the thermal requisites of the process streams connected to the LNG unit.

The Pinch analysis, as described by Linnhoff et al. [19], gives an idea of the possibility of the heat exchanges between flows and allows to compute the needed minimum of cooling water and steam required by the system to reach the desired operating points. The process integration study begins with the definition of a list of hot and cold streams, which defines heat sources and heat requirements of the

Table 8 BPS streams connected to the LNG unit

Type	Description	In (°C)	Out (°C)
Cold	Water and ethanol mix	80	709
Hot	Cooling from ESR exit to HTS inlet	672	427
Hot	Cooling from HTS exit to LTS inlet	466	150
Hot	Cooling from LTS exit to CO–PrOx inlet	216	165
Hot	Cooling from CO–PrOx exit to PEMFC inlet	267	80
Cold	Preheating of cathode and anode gasses	80	500
Cold	Preheating of air for combustion	134	300
Hot	Combustion gasses from ESR jacket exit to turbine inlet	938	350
Cold	Preheating of extra ethanol for combustion	25	300

Fig. 8 Composite curves of the system

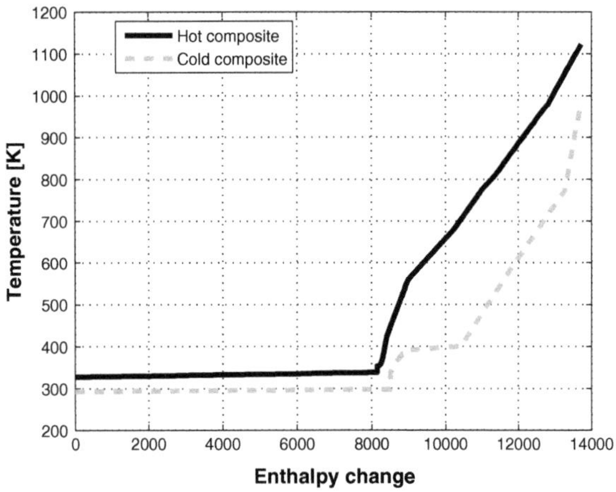

system. The heat exchange is technically feasible only if the temperature difference between the hot and cold streams is always superior to a predefined $\Delta T_{\min}$, and the maximum heat recovery is obtained when the $\Delta T_{\min}$ constraint is activated. This point is called the Pinch point. The hot Pinch temperature is 323.17 K, the cold Pinch temperature is 298.23 K and therefore the minimal approach is 24.94 K.

The interest of the process integration techniques is based in the fact that they can identify the bottle necks of the heat exchangers network in the system (Pinch points) and propose modifications of the process that will increase its global efficiency. With the LNG operation of the simulator, the composite curves of the system can be seen in Fig. 8 and essentially are the same considered in this chapter.

Cost and performance requirements of the power plant lead to a highly integrated designs and stringent control objectives. The heat integration needed for the model is achieved by the LNG tool in HYSYS. This operation was explained in Chap. 9 that solves material and heat balances for multi-stream exchangers and heat exchangers network. Each stream is divided in a number of intervals, in this case, ten intervals are considered, and the phase change is verified.

For constructing the dynamic model it was taken into account that heat exchangers have fast dynamics compared to other unit operations in a process. Normally the time constant is measured in seconds but could be up to a few minutes for large exchangers. Process-to-process exchangers should be modeled rigorously by partial differential equations since they are distributed systems. This introduces the correct amount of dead time and time constant in the exit stream temperatures, but the models are inconvenient to solve. It was found that for the purpose of plantwide control studies it is not necessary to have such detailed descriptions of the exchanger dynamics, since these units rarely dominate the process response. Luyben et al. [20] considered that another alternative is to use an effectiveness method to calculate the steady-state exchanger exit temperatures and then delay these temperatures by first order time constants to capture the dynamics. Here, with the LNG tool, the simulation of the heat exchangers network is defined without actually knowing its configuration. The minimum heat requirement of the system and the minimum heat to be evacuated can be computed for each operating point or with the system under different disturbances. Even though this option gives a good solution for this stage, in the future this point will be analyzed more exhaustively.

2.8 Proton Exchange Membrane Fuel Cell

The PEMFC dynamic model used in this chapter corresponds to the same described in Chap. 2, hence only the most relevant equations are given here. Then, the considered fuel cell stack is constituted by an anode, where the fuel is injected, and a cathode, where the oxidant, normally oxygen or air, is injected. The electrodes are separated by a semipermeable membrane that allows the proton exchange and makes the oxidation reaction to produce electrical power. The cell generates an open-circuit voltage which is affected by a number of losses (activation, concentration and ohmic) that leads to a useful actual voltage. The rigorous dynamic model of a PEMFC presented by Pukrushpan et al. [21] was specifically adapted here to produce a maximum power of 10 kW. In that work the anode was fed with a hydrogen tank and a BPS based on partial oxidation of natural gas was analyzed.

Remembering, the overall reaction at the fuel cell to convert chemical energy directly into electrical energy is the following:

$$H_2 + \frac{1}{2}O_2 \rightarrow H_2O + \text{Heat} + \text{Electrical power}.$$

According to Pukrushpan et al. [21], the reversible open circuit voltage of a fuel cell can be written as

$$\begin{aligned} E^0 &= 1.229 - 0.85 \times 10^{-3}\left(T_{\text{fc}} - 298.15\right) \\ &\quad + 4.3085 \times 10^{-5} T_{\text{fc}}\left[\ln\left(p_{\text{H}_2}\right) + \frac{1}{2}\ln\left(p_{\text{O}_2}\right)\right], \end{aligned} \tag{38}$$

where T_{fc} is the operation temperature of the FC expressed in Kelvin, and p_{H_2} and p_{O_2} are the partial pressure of hydrogen in the anode and oxygen in the cathode respectively, in atmospheres. Both fuel and oxidant are considered humidified, and the vapor pressure from the Nernst equation is assumed equal to saturated vapor pressure. A typical voltage is 1.2 volts, although when the FC operates, the actual voltage is less than the value calculated. This difference is a result of losses or irreversibilities:

$$\begin{aligned} &\text{Activation loss:} && v_{act} = a \ln\left(\frac{i}{i_0}\right) \\ &\text{Ohmic loss:} && v_{ohm} = i \cdot R_{ohm} \\ &\text{Concentration loss:} && v_{conc} = i \cdot \left(c_2 \frac{i}{i_{max}}\right)^{c_3} \end{aligned}$$

where a is a constant, and i_0, the exchange current density is also a constant. R_{ohm} is the internal electrical resistance with units of $\Omega\ \text{cm}^2$. c_2, c_3 and i_{max} are constants that depend on the temperature and the reactant partial pressure, taken from the literature, Pukrushpan et al. [21]. The parameters were obtained adjusting the presented equations to the data of a fuel cell operating at steady state and at designed operating conditions.

Transient behavior of manifold filling, of membrane hydration, of the air compressor and of the heat management are included in the model. Interaction between processes are also included. The necessary hydrogen production is calculated as a function of the stack current:

$$W_{H_2} = \frac{n \cdot I_{st}}{2 \cdot F}, \tag{39}$$

where W_{H_2} refers to H_2 production (taken as setpoint), n is the number of cells in the stack and F the Faraday constant. Actual hydrogen production setpoint should be determined directly from H_2 partial pressure in the anode, but this is a variable that cannot be measured in a FC. Complete utilization of fuel is assumed in the stack, meaning the purge rate is equal to zero, so the steady-state consumption of fuel constitutes an acceptable approximation for the H_2 flow demanded by the stack.

To adapt the fuel cell to the rated power needed, the number of cells in series is set to $n = 80$. The new working point requires the optimization of the operation of the FC. As stated in Sect. 3 of Chap. 2, there is an optimal oxygen excess, $\lambda_{O_2}^{opt}$, for each demanded current. That condition also determines an optimal voltage to the compressor, net power obtained and pressure of the supply manifold. Using the model and making an analysis as recommended by Grujicic et al. [22], the new optimal operating conditions can be obtained. The net power obtained for different currents, under optimal compressor voltage can be seen in Fig. 9a. Under the same conditions, the ideal voltage for the compressor can be obtained, as shown in Fig. 9b, c shows the optimal oxygen excess under the new operating conditions.

As can be seen in Fig. 9, the differences in the behavior of the FC under the new operating condition can not be neglected. When it is compared to Fig. 5 it is

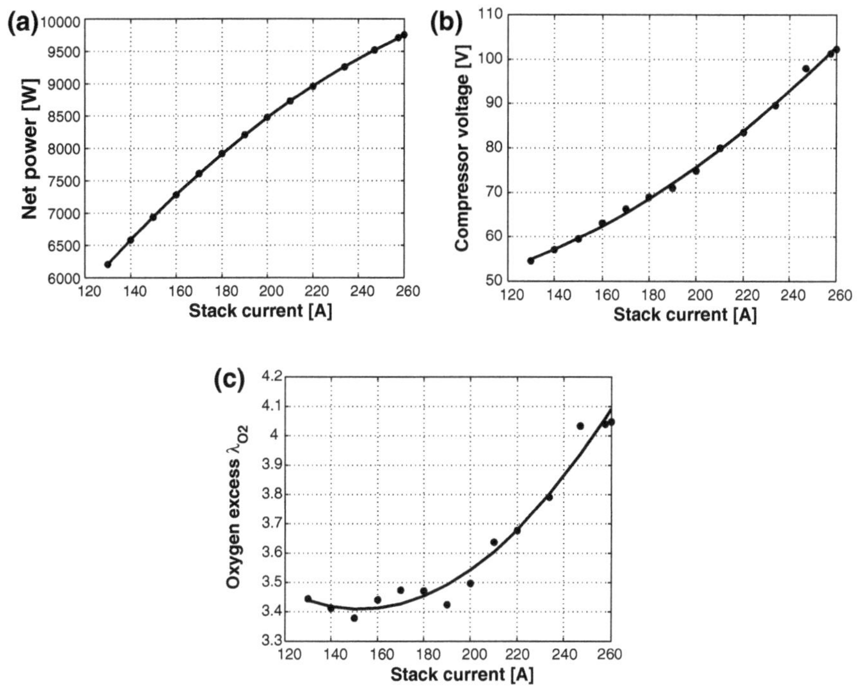

Fig. 9 Optimal values for different current loads. **a** Maximum net power, **b** Optimal compressor motor voltage, **c** Optimal oxygen excess

noticeable that not only the relations have changed, but also the tendencies of the optimal operating conditions against the changes in the demanded current. In order to illustrate the differences, the equations for the approximation to a quadratic function are presented:

$$P_{\max}^{\text{net}} = -8.57 \times 10^{-2} I_{\text{st}}^2 + 60.64 I_{\text{st}} - 223.97 \tag{40}$$

$$v_{\text{CM}}^{\text{opt}} = 1.19 \times 10^{-3} I_{\text{st}}^2 - 9.63 \times 10^{-2} I_{\text{st}} + 47.29 \tag{41}$$

$$\lambda_{\text{O}_2}^{\text{opt}} = 5.90 \times 10^{-5} I_{\text{st}}^2 - 1.80 \times 10^{-2} I_{\text{st}} + 4.78 \tag{42}$$

The net power obtained has a similar behavior in both cases. The compressor motor voltage needed is also growing in the interval analyzed, but presents a different concavity, growing quadratically in the new case, while in the other case it would have grown to an upper bound, for then decreasing. The behavior of the oxygen excess on the other hand, is radically different. It has a minimum in the interval, so it is not monotonous like the others, first decreasing, and then increasing. Considering a fixed setpoint for this variable, or a re-scaling, would have been a huge mistake, so this analysis turns out a deal of great importance. Then, the quadratic function obtained by this step will state for the variable setpoint of the oxygen excess.

2.9 Computing Implementation

The pressure requirements are satisfied with compressors and turbines from HYSYS. Also, it supports the important data bank information for the different components, and the 'LNG' operation, which calculates the optimum heat exchange for the system, minimizing auxiliary services. In addition, the dynamic behaviors of the reactors are modeled in MATLAB, and it works as the integrator, supporting the differential equations and the data obtained from the steady-state case in HYSYS.

In order to successfully coordinate calculations between these programs, an activex-com controller for HYSYS is needed, which is in a directory named *hysyslib* that should be installed in the toolboxes and addressed to the MATLAB path. This interface is sustained by the use of the spreadsheets of HYSYS for the transference and actualization of the information. The spreadsheet is a grid that contains all the variables that the user is interested in, and MATLAB is able to connect to this, read and write all the data that is available.

3 Linear Model Development

For design and testing any control structure is very useful to obtain a reduced order model of the complete process. The first consideration is to determine those key variables that are responsible for the full plant stabilization. In this case, the pressure of each reactor (ESR, HTS, LTS, CO–PrOx) must be controlled, by manipulating their corresponding exit flow. The pressures are assumed under perfect control. These five stabilizing control loops are indicated in the Table 9 with (*). The H_2 production to feed the anode of the fuel cell stack, is controlled with the fresh ethanol flow entering to the bio-ethanol steam reforming (ESR).

Once the process is stabilized, some identification technique, independently of which methodology is chosen, can be performed so as to obtain from the rigorous model a reduced order dynamic model of the plant. In this chapter, the identification was done according to simple rules of reaction curves, accounting in many textbooks of process control such as of Seborg et al. [23]. In Fig. 10 can be seen the most important dynamic responses excited by the specific step changes detailed below. For the Fig. 10a step change of -5% was done at time $= 2.5$ s , in the fresh ethanol feed to the processor. The response of the H_2 production variable was recorded, and used for identification of transfer function. Fig. 10b shows the response of the temperature of the ESR exit flow, when a step change of -5% is applied in the fresh ethanol feed to the burner at time $= 2.5$ s. Other important variable is shown in Fig. 10c, the temperature of the exit gasses from the jacket of the ESR, for a step change of -5%

Table 9 Variables in the BPS+FC process

Measured		Manipulated		Disturbances	
y_1	ESR exit temperature	u_1	Water to ESR inlet	d_1	Ethanol purity
y_2	Jacket exit gasses temperature	u_2	Exchanged heat Q	d_2	Stack current
y_3	Burner exit temperature	u_3	Ethanol to Burner		
y_4	Burner entering molar flow	u_4	Oxygen to Burner		
y_5	Molar ratio H_2O/Ethanol	u_5	Oxygen to CO–PrOx		
y_6	HTS exit temperature	u_6	CM voltage		
y_7	LTS exit temperature	u_7	ESR exit flow(*)		
y_8	CO–PrOx exit temperature	u_8	HTS exit flow(*)		
y_9	Molar ratio O_2/CO	u_9	LTS exit flow(*)		
y_{10}	Burner exit molar flow	u_{10}	CO–PrOx exit flow(*)		
y_{11}	CO–PrOx CO exit concentration	u_{11}	Bio-ethanol flow(*)		
y_{12}	Net Power				
y_{13}	Oxygen excess				
y_{14}	Stack voltage				
y_{15}	ESR pressure(*)				
y_{16}	HTS pressure(*)				
y_{17}	LTS pressure(*)				
y_{18}	CO–PrOx pressure(*)				
y_{19}	H_2 production rate(*)				

Table 10 Parameters of the transfer functions

	Loop 1	Loop 2	Loop 3	Loop 4	Loop 5	Loop 6	Loop 7
K	0.033	−116.5	1.87×10^3	1,127	1.3×10^5	5.071	0.029
τ	0.100	1.610	1.155	0.05	0.05	0.140	0.100
θ	0	0.200	0.705	0.05	0.05	0.140	0

applied in the exchanged heat Q at the time = 2.5 s. A tangent is drawn at the point of inflection of the reaction curves. The intersection of the tangent line and the time axis is the time delay. By extending the tangent to intersect the steady state value of the response line, the point of intersection corresponds to the sum of time delay and the time constant. The obtained parameters for the approximation to a first order with time delay transfer function following the methodology given below are given in Table 10.

In Chap. 12 a system identification experiment, based on subspace state-space methods named "n4sid", developed by Van Overschee et al. [24] will be implemented in the Matlab environment [25, 26]. It is based on exciting the inputs of the process with rich signals and collect the output data periodically with a suitable sample time, T_s. The input signals are selected as random steps whose amplitudes varying is such a way to capture a closely linear behavior of the selected variables.This strategy allows to estimate the system matrices of the space–state linear model, $\hat{\mathbf{A}}, \hat{\mathbf{B}}, \hat{\mathbf{C}}, \hat{\mathbf{D}}$ and the model order using singular value decomposition from the impulse response Hankel matrix. The linear model structure and the matrix dimension can be observed in (12.30) and (12.31).

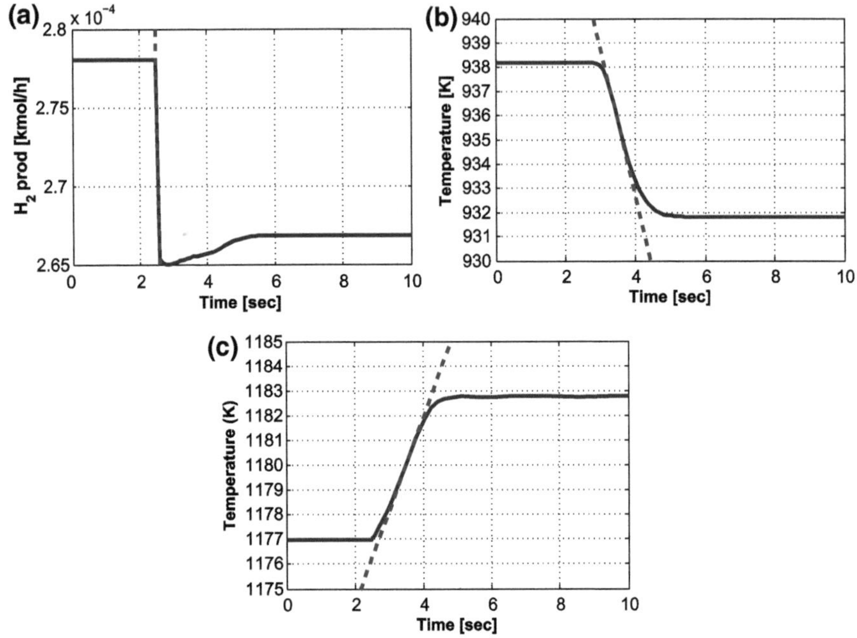

Fig. 10 Transfer function identification. **a** H_2 to fuel cell anode, **b** ESR product flow, **c** ESR jacket exit gases

4 Control Structure

The control structure must be generated based on the main objectives of the BPS. They are to maintain H_2 levels on the anode of the FC, because starvation can cause permanent damage, and overfeeding will lead to H_2 waste, the CO levels of the anode inlet stream must be under 10 parts per million (volume), and keeping the temperatures of the reactors set and FC prevent damages, maintaining the system efficiency.

The development of the control oriented dynamic model presented here allows testing more efficiently the adopted control structure based on a previous analysis, completely performed in steady state model described in detail in Chap. 9. In addition, thanks to the analysis performed with the dynamic model two more control loops were added which are described in Fig. 11 and detailed in Table 11. The most critical controlled variables were selected taking into account the objectives mentioned above. They are directly related to the optimal operating conditions of the BPS and the less sensitivity structure.

The final test can be done by defining the tuning parameters. The linear model around the operating point of interest, obtained through the reaction curves, is very useful. It is adopted as the nominal model for implementing internal model control (IMC) method. It was selected because it has proven good performance in control

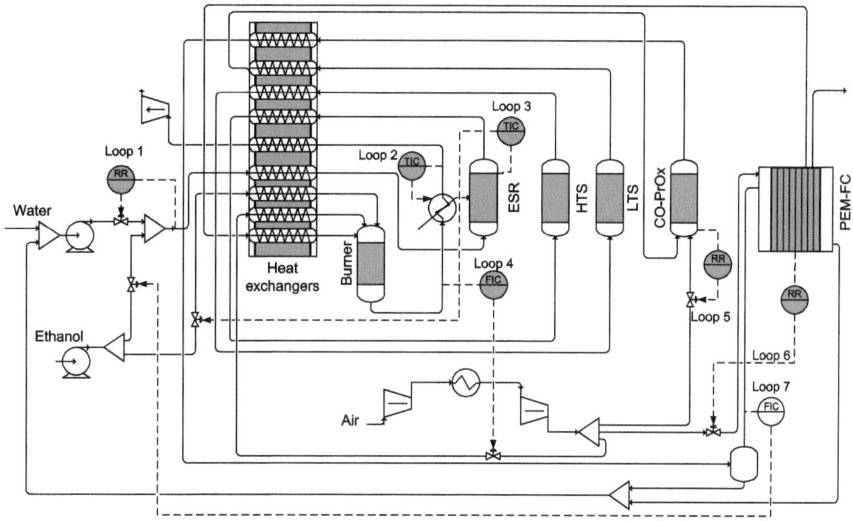

Fig. 11 Implemented control structure

Table 11 Detail of the control loops

	Controlled	Manipulated
Loop 1	y_5	u_1
Loop 2	y_2	u_2
Loop 3	y_1	u_3
Loop 4	y_{10}	u_4
Loop 5	y_9	u_5
Loop 6	y_{13}	u_6
Loop 7	y_{19}	u_{11}

schemes. It will be further used in Chap. 12 over this dynamic model, both for theoretical approaches and practical implementation. The tuning parameters were chosen following the recommendations given in the work of Rivera [27], and they are shown in Table 12.

5 Simulation Results with the Dynamic Model

The energy balance taking into account the PEM fuel cell for generating electric power gave an optimal operating point for the BPS. Then, it was considered that having a control oriented model at this point would be useful for obtaining a preliminary plantwide control structure able to keep the plant very close to it. In addition, an iterative methodology between synthesis and control could be done. It represents a valuable tool for developing an efficient hydrogen production via bio ethanol to respond to PEM requirements for using in vehicles. However this kind of process is

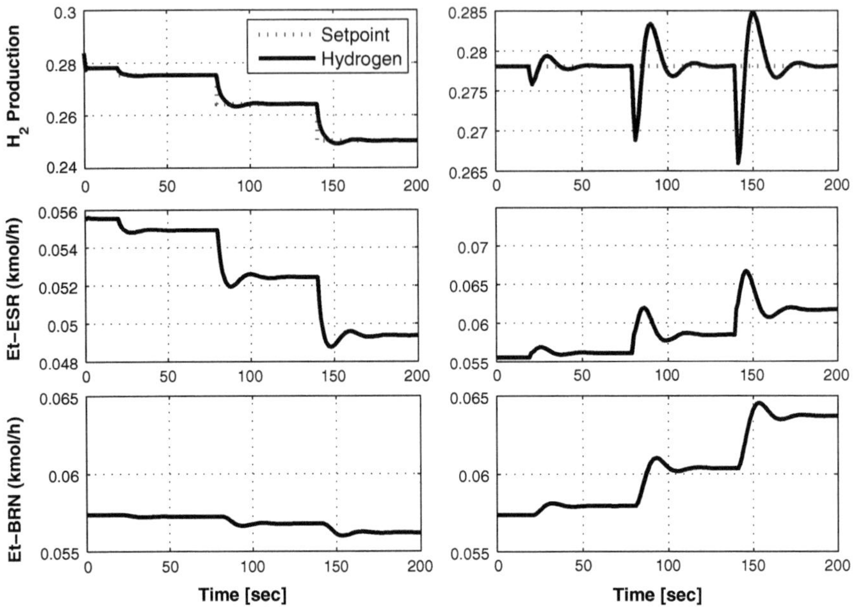

Fig. 12 Dynamic simulations of Hydrogen production, controlled variable, and manipulated variables, feed flow rate ethanol to ESR and burner reactors under perturbation

Table 12 Controllers tuning

	Loop 1	Loop 2	Loop 3	Loop 4	Loop 5	Loop 6	Loop 7
k_c	318.2	−0.0736	0.0011	0.001	1×10^{-6}	5.071	51.55
τ_i^c	0.105	1.710	1.5	0.047	0.05	0.140	0.015

Table 13 Abbreviations of used components

a	Ethanol	CH_3CH_2OH
b	Water	H_2O
c	Methane	CH_4
d	Carbon monoxide	CO
e	Carbon dioxide	CO_2
f	Hydrogen	H_2
g	Acetaldehyde	CH_3CHO
h	Oxygen	O_2

very incipient so the authors of this work didn't find any reference of other dynamic model similar to the one presented here. Therefore, the resulting responses were compared to the stationary states reported at Universidad Nacional del Litoral [13] for the 10 KW FC. A good agreement with those points were found.

Then, the same step changes considered at Biset et al. [28] were performed and shown in Fig. 12. Hence, the steps applied in fresh ethanol molar fraction and the current load were of about −1, −5 and −10% at 20, 80 and 140 s, respectively.

The first group of figures, at the left, corresponds to a step change disturbance in the current I_{st}, and the second group, at the right, in fresh ethanol molar fraction. For the first group of figures, the current demands from the PEM, considered as a disturbance, require a specific hydrogen production and that is the reason of the set point variations accordingly to Eq. 39. Therefore as can be seen from the simulation results the main objectives of the plant can be achieved satisfactory.

6 Conclusion

In this chapter, the main characteristics of the rigorous dynamic model of a fuel processor system along with a fuel cell stack, highly integrated to obtain good efficiencies and maximum heat recovery is presented. In addition, it is enough flexible to analyze other possible kinetics, sizing, etc., accounting different scenarios. However, to achieve a successful operation when the typical disturbances occur is necessary to account with a well designed plant-wide control structure. Focusing on this objective the rigorous model allowed obtaining a linearized and reduced order of a dynamic model via system identification techniques. In this context, it was possible to obtain a preliminary control policy since the dynamic interactions are well understood with it. As future works are considered the use of this dynamic model for testing new methodologies of plant-wide control structure and optimal sensor location and check them under standard requirements (Table 13).

References

1. Eborn J, Pedersen L, Haugstetter C, Ghosh S. (2003) System level dynamic modeling of fuel cell power plants. Proceedings of the American Control Conference. pp 2024–2029
2. Görgün H, Arcak M, Varigonda S, Bortoff SA. Nonlinear observer design for fuel processing reactors in fuel cell power systems. Proceedings of the American Control Conference, pp 845–850
3. Görgün H (2005) Control-oriented modeling of fuel processing reactors in fuel cell power systems. J Electr Electron Eng 5(1):1279–1286
4. Pukrushpan JT, Stefanopoulou A, Varigonda S, Eborn J, Haugstetter C (2003) Control-oriented model of fuel processor for hydrogen generation in fuel cell applications. Control Eng Pract 14(3):277–293
5. Varigonda S, Pukrushpan JT, Stefanopoulou A (2003) Challenges in fuel cell power plant control: the role of system level dynamic models. IEEE Trans Control Sys Technol 13(1):3–14
6. Pukrushpan JT, Stefanopoulou A, Varigonda S, Pedersen LM, Ghosh S, Peng H (2003) Control of natural gas catalytic partial oxidation for hydrogen generation in fuel cell applications. IEEE Trans Control Sys Technol 13(1):3–14
7. Chuang CC, Cheng YH, Ward JD, Yu CC, Liu YC, Lee CH (2008) Optimal design of an experimental methanol fuel reformer. Int J Hydrogen Energy 33:7062–7073
8. Aicher T, Full J, Schaadt A (2009) A portable fuel processor for hydrogen production from ethanol in a 250wel fuel cell system. Int J Hydrogen Energy 34:8006–8015
9. García VM, López E, Serra M, Llorca J, Riera J (2009) Dynamic modeling and controllability analysis of an ethanol reformer for fuel cell application. Int J Hydrogen Energy 35(18):9768–9775

10. Fogler HS (2005) Elements of chemical reaction engineering. Prentice-Hall PTR, 4th edn
11. Larminie J, Dicks A (2003) Fuel cell systems explained. John Wiley, New York
12. Agarwal AK (2007) Biofuels (alcohol and biodiesels) applications as fuels for internal combustion engines. Prog Energy Combus Sci 33(3):233–271
13. Francesconi J (2008) Modelado, Síntesis y Optimización del Proceso de Reformado de Bioetanol para la Producción de Hidrógeno grado PEM. PhD thesis, Universidad Nacional del Litoral
14. Benito M, Sanz JL, Isabel R, Padilla R, Arjona R, Daza L (2005) Bio-ethanol steam reforming: insights on the mechanism for hydrogen production. J Power Sources 151:11–17
15. Leva M, Weintraub M, Grummer M, Clarke EL (1948) Cooling of gasses flowing in turbulent motion through packed tubes. Ind Eng Chem 40:747–752
16. Keiski R, Salmi T, Pohjola V (1992) Development and verification of a simulation model for a non-isothermal water-gas shift reactor. Chem Eng J 48(1):17–29
17. Choi Y, Stenger H (2003) Water gas shift reaction kinetics and reactor modeling for fuel cell grade hydrogen. J Power Sources 124(2):432–439
18. Choi Y, Stenger HG (2004) Kinetics simulation and insights for CO selective oxidation in fuel cell applications. J Power Sources 129(2):246–254
19. Linnhoff B, Townsend P, Boland P, Hewitt GF, Thomas BEA, Guy AR, Marsland RH (1994) A user guide on process integration for the efficient use of energy. Institute of Chemical Engineers, Rev sub edn. Rugby, UK.
20. Luyben ML, Luyben WL (1997) Essentials Of Process Control. McGraw-Hill, Int edn. Chemical Engineering Series.
21. Pukrushpan JT, Stefanopoulou AG, Peng H (2004) Control of Fuel Cell Power Systems: Principles, Modeling, Analysis and Feedback Design. Springer, London
22. Grujicic M, Chittajallu KM, Law EH, Pukrushpan JT (2004) Model-based control strategies in the dynamic interaction of air supply and fuel cell. Proc Inst Mech Eng. Part A: J Power Energy 218(7):487–499
23. Seborg DE, Edgar TF, Mellichamp DA (2004) Process Dynamic and Control. Wiley, New York
24. Van Overschee P, De Moor B (1994) N4SID Subspace algorithms for the identification of combined deterministic-stochastic systems. Automatica 30:75–93
25. Ljung L. (1999) System identification (Theory for the user). 2nd edn
26. Ljung L (2002) System identification toolbox (user's guide version 5). Technical report. The Math Works Inc
27. Rivera DE (2007) Una metodología para la identificación integrada con el diseño de controladores IMC-PID. Revista Iberoamericana de Automática e Informática Industrial 4(4): 129–134
28. Biset S, Nieto Degliuomini L, Basualdo M, Garcia VM, Serra M (2009) Analysis of the control structures for an integrated ethanol processor for proton exchange membrane fuel cell systems. J Power Sources 192(1):107–113

Mathematical Model Implementation for Simulation Purposes

L. Nieto Degliuomini, S. Biset, D. Feroldi and M. Basualdo

Abstract In this chapter the reader can find a guide for developing a realistic computational model of BPS with PEMFC for both steady state and dynamic modes. It is performed thanks to the use of the potentiality of three very well-known softwares for the engineering community. It is assumed that the reader has some knowledge about programming in the MATLAB environment. The other two softwares are HYSYS and ADVISOR. Therefore, the main details about the engineering steps for moving from steady state to the dynamic simulation of this complex system are given. The guide includes several figures which capture the screens of the different softwares and the chosen options for each one to complete the model. The code for the communication protocol between HYSYS and MATLAB is also included. In addition, some useful references to help the readers in getting more insight into the use of the three softwares are presented.

L. Nieto Degliuomini · D. Feroldi · M. Basualdo (✉)
CAPEG-CIFASIS-(CONICET-UNR-UPCAM),
27 de Febrero 210 bis,
S2000EZP Rosario,
Argentina
e-mail: basualdo@cifasis-conicet.gov.ar

L. Nieto Degliuomini
e-mail: nieto@cifasis-conicet.gov.ar

M. Basualdo · S. Biset
CAPEG-Chemical Engineering Department, FRRo-UTN,
Zeballos 1341,
S2000BQA Rosario,
Argentina
e-mail: biset@frro.utn.edu.ar

D. Feroldi
DCC-FCEIA-UNR, Pellegrini 250,
S2000BTP Rosario,
Argentina
e-mail: feroldi@cifasis-conicet.gov.ar

M. S. Basualdo et al. (eds.), *PEM Fuel Cells with Bio-Ethanol Processor Systems*,
Green Energy and Technology, DOI: 10.1007/978-1-84996-184-4_11,

1 Introduction

In this chapter, the main aspects of the procedure for moving from steady state to the dynamic simulation of the bio-ethanol processor with PEMFC, using commercial softwares, are presented. Initially, the steady-state (SS) simulation is described by using the program HYSYS. Based on this, some parts of the SS model are useful for constructing the pseudo-dynamic model of the complete plant thanks to the connection with the software MATLAB-SIMULINK. Both programs offer the valuable opportunity of writing the codes of specific equipments, mainly in the cases of reactors and PEMFC. Additionally, some realistic assumptions are adopted so as to get a complete prototype of this complex process. Once the control structure is designed, the dynamic model helps in testing the performance of the closed-loop system. Another well-known program for mechanical engineers, named ADvanced VehIcle SimulatOR (ADVISOR), is used to provide standard disturbances to PEMFC. It allows taking into account different kinds of vehicles, for distinct scenarios, according to the velocity and sequences of starts and stops. The aim of this chapter is to provide a good working knowledge about the development of a representative computational model for testing several operating conditions.

2 Development of the Steady-State Model in HYSYS

In this section the main information about constructing the steady-state model of the bio-ethanol processor, shown in Fig. 1, is given. It is performed in HYSYS version 3.2, hence prior knowledge of this software is assumed. In case the reader is unfamiliar with the software, the book of Luyben [1] is recommended.

2.1 Basis Environment Configuration

At the beginning of the model construction in the HYSYS environment a new case must be defined, and all the components to be used in the simulation must be added. Further details about the simulation basis configuration can be seen in Hyprotech HYSYS 3.2 simulation basis [2]. For the case of BPS they are as follows:

- Hydrogen
- Water (H_2O)
- Methane
- Carbon monoxide (CO)
- Carbon dioxide (CO_2)
- Ethanol
- Oxygen
- Nitrogen

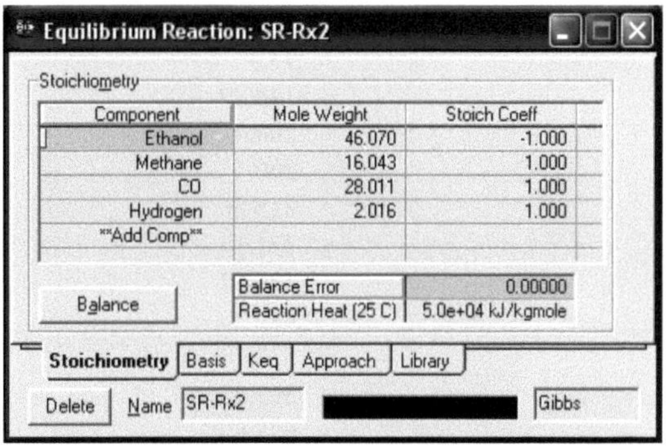

Fig. 1 Stoichiometry of the ethanol decomposition reaction

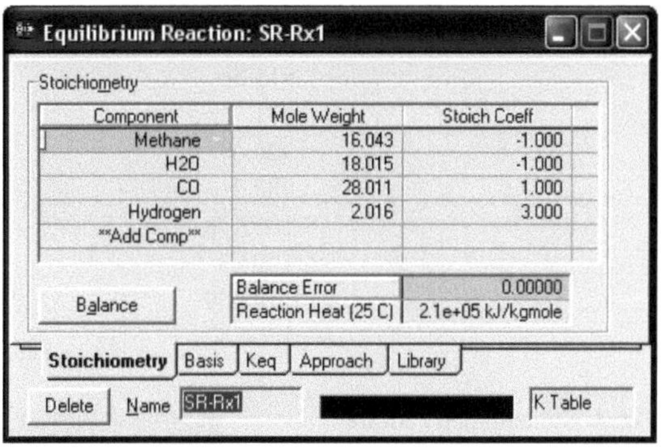

Fig. 2 Stoichiometry of the methane reforming reaction

The next step needs to define which would be the most adequate fluid package for these components. Here, the Wilson fluid package is chosen, assuming the vapor as ideal gas. Then, the reactions set that will be produced at each reactor is available to be chosen. There are several reactions to be considered:

- Ethanol dehydrogenation and acetaldehyde decomposition:

$$C_2H_5OH \leftrightarrow CH_3CHO + H_2$$
$$CH_3CHO \rightarrow CH_4 + CO$$

A complete acetaldehyde conversion is considered, hence these reactions are grouped into one. In Fig. 1 is shown how the equilibrium reaction is loaded in HYSYS.

The specific bases used in this case are the activity, vapor phase and the source for the K_{eq} is the Gibbs free energy with auto detect option.

Methane reforming reaction:

$$CH_4 + H_2O \leftrightarrow CO + 3 \cdot H_2$$

The stoichiometry of the reaction can be set as shown in Fig. 2. An equilibrium reaction is chosen.

Table 1 K_{eq} versus T_{ESR}

T_{ESR} [°C]	K_{eq}
93.3	7.813×10^{-19}
148.89	6.839×10^{-15}
204.4	7.793×10^{-12}
260	2.173×10^{-9}
315.56	2.186×10^{-7}
371.1	1.024×10^{-5}
426.67	2.659×10^{-4}
482.2	4.338×10^{-3}
537.78	4.9×10^{-2}
593.3	0.4098
648.89	2.679
704.4	14.26
760	63.43
815.56	242.6
871. 1	816.6
926.67	2464
982.2	6755
1037.78	17010
1093.3	39670
1148.89	86640
1204.4	178400

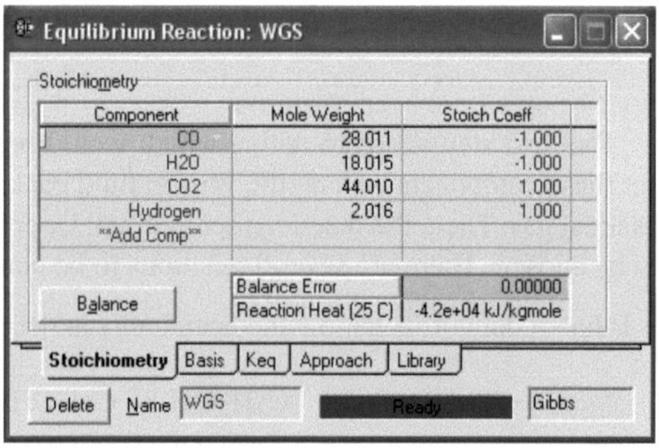

Fig. 3 Stoichiometry for the WGS reaction

The basis used in this case is partial press, vapor phase, the basis unit is atmospheric pressure (atm), the source for estimating the K_{eq} is obtained from the data presented in Table 1 where K_{eq} versus T_{ESR} are given.
Water gas shift reaction:

$$CO + H_2O \leftrightarrow CO_2 + H_2$$

The stoichiometry of the equilibrium reaction can be set in HYSYS as shown in Fig. 3.

The basis used are the activity, vapor phase, the source for the K_{eq} is the Gibbs free energy with auto detect.

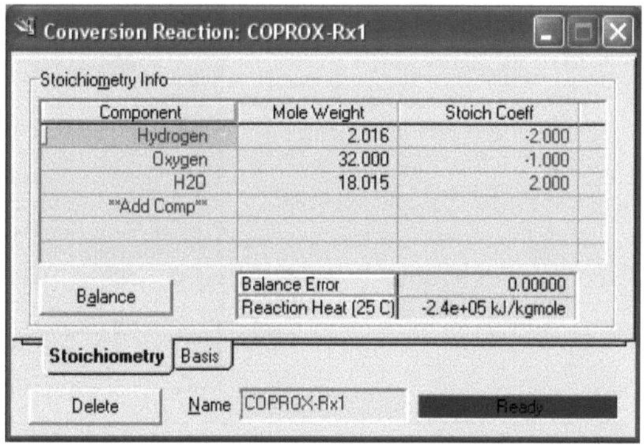

Fig. 4 Stoichiometry for hydrogen combustion

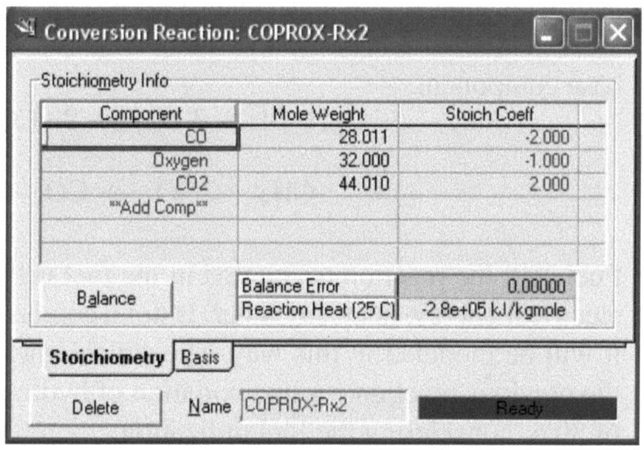

Fig. 5 Stoichiometry for CO combustion

- Hydrogen oxidation:

$$H_2 + \frac{1}{2}O_2 \rightarrow H_2O$$

  This is a conversion reaction. The stoichiometry can be configured in the same way as the previous one, as shown in Fig. 4.

  The reaction is based on the overall conversion of hydrogen adjusted with $C_0 = 75\%$.

- Carbon monoxide oxidation:

$$CO + \frac{1}{2}O_2 \rightarrow CO_2$$

  This reaction is the conversion type. The stoichiometry should be set in the configuration window as shown in Fig. 5.

  The base component for the reaction is the overall conversion of CO, with $C_0 = 25\%$.

Combustion reactions: for the post-combustion reactor, the combustion reactions of the components need to be configured. The four substances that are burned are

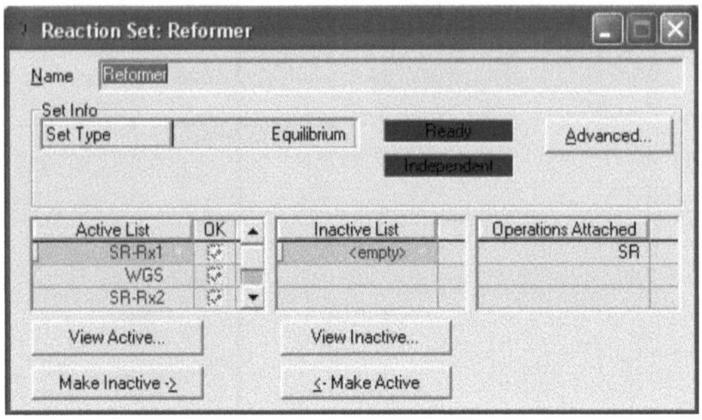

Fig. 6 Set of reactions for the ESR

hydrogen, ethanol, methane and CO. The stoichiometry must be configured as the aforementioned reactions, and the overall conversion set in $C_0 = 100\%$ for each base component.

$$C_2H_5OH + 3 \cdot O_2 \rightarrow 2 \cdot CO_2 + 3 \cdot H_2O$$

$$CH_4 + 2 \cdot O_2 \rightarrow CO_2 + 2 \cdot H_2O$$

- Fuel cell: the reaction accounted in the fuel cell is the conversion of hydrogen to electrical energy, heat and water. Is not a combustion, but for simulation purposes it will be modeled in this way. The stoichiometry has the same configuration as the previous reaction for the oxidation of hydrogen, with a H_2 overall conversion of 80%, considering the fuel utilization.

It is also recommended to group the reactions in sets, so that there is a set for each reactor. There are five different sets of reactions:

- Ethanol steam reformer: the steam reforming reaction comprises the methane reforming, water gas shift, ethanol dehydrogenation and acetaldehyde decomposition. To configure a set of reactions, it needs to be added first, by pressing the `Add Set...` button and choosing the corresponding reaction selecting from the active list, as shown in Fig. 6.
- Water gas shift: in both reactors occurs only one reaction, which is the same for both.
- Preferential oxidation of CO: it is configured with the water gas shift reaction and the oxidations of hydrogen and CO.
- Post-combustion: this reactor accounts for the four combustion reactions, that is ethanol, methane, hydrogen and CO; all set to 100% conversion for complete combustion consideration.
- Fuel cell: this set only has the fuel cell reaction, that comprises oxidation of hydrogen for simulation purposes.

Finally, the 'Reactions' label for the basis manager window will be as shown in Fig. 7.

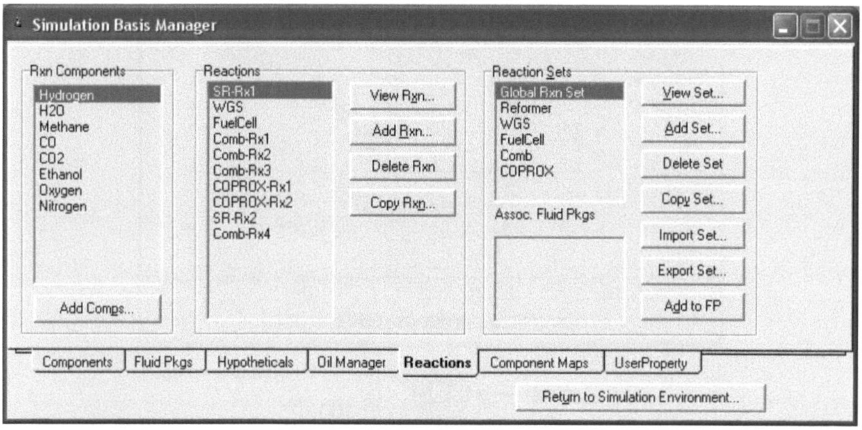

Fig. 7 Configuration window for the reactions

2.2 Inlet Streams and Conditioning

Once the basis environment is properly configured, the simulation environment is ready to work. Detailed characteristics and working conditions for the operations and objects in HYSYS can be seen in Hyprotech. HYSYS 3.2 operations guide [3]. The first thing to provide to construct the flow sheet through the tool "PFD" are the material inlet streams of ethanol, water and air. Since they come from environmental conditions, they will be set in 25°C and 1 atm of pressure. Initially, the actual flows are uncertain, but a detailed study of the most efficient conditions is already described in the previous chapters. The three inlet streams need to be reconditioned for the bio-ethanol processor. The liquid flows require pumps to increase pressure to 3 atm. For the configuration of each pump, an inlet stream is needed, which is the fresh feed, an outlet stream that has to be added and a new energy stream. The adiabatic efficiency is considered as 75% and the configuration window can be seen in Fig. 8.

The air stream needs to be compressed in two stages with intermediate cooling. The first of the compressors takes the air from ambient pressure and it must be increased up to 1.75 atm. The cooler between the compressors cools the air to 50°C and the second compressor is responsible for achieving the pressure of 3 atm needed for the processor. The adiabatic efficiency considered for both compressors is 70%.

The heat exchanger needs both material streams, and a new energy stream, the pressure drop is neglected. In Fig. 9, the schemes for the air compressing section can be seen.

Ethanol and air have to be fed to multiple parts of the bio-ethanol processor, hence splitters are needed to divide the streams. The operation needed is called tee in HYSYS; it needs the inlet and outlet material streams and also some of their conditions to be properly configured. The compressed ethanol flow needs to be divided into two streams, one for the ESR reactor and the other for the post-combustion

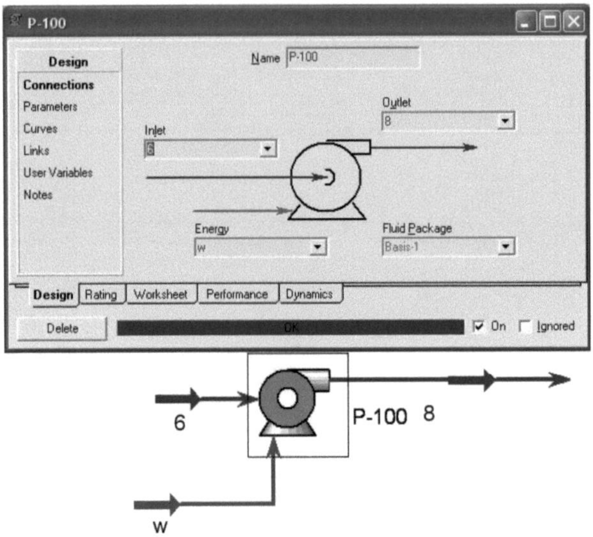

Fig. 8 Configuration window for the pumps

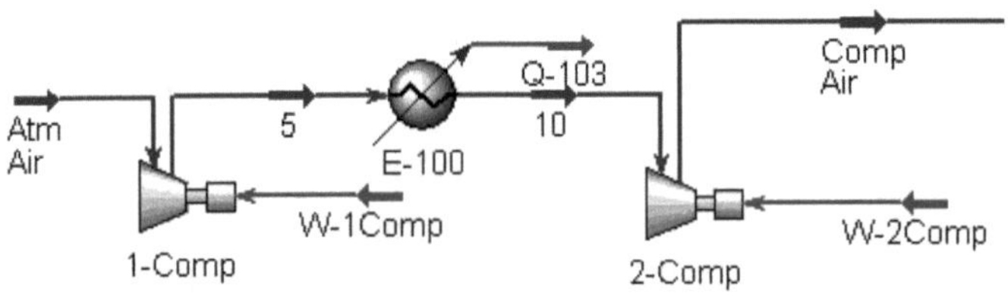

Fig. 9 Configuration window for the air compression

reactor. The air has three divisions, for the burner reactor, the CO-PrOx reactor, and the PEMFC. The conditions in both splitters are to keep the temperature and pressure of the streams by modifying only the molar (or mass) flows.

The next step needed is the mixing of the ethanol and water before entering the processor. The unit required for this operation is the mixer from HYSYS, which needs the two material streams, the water and the ethanol to ESR. If both inlets are properly determined, the exiting stream does not need extra specifications.

2.3 Heat Exchangers Network

As stated in the previous chapters, a heat exchangers network is needed to integrate the energy of the streams in order to maximize the efficiency of the system. The operation of HYSYS used to do so, is the LNG tool. It allows to compute the maximum heat exchange between the streams passing through it, minimizing the consumption of

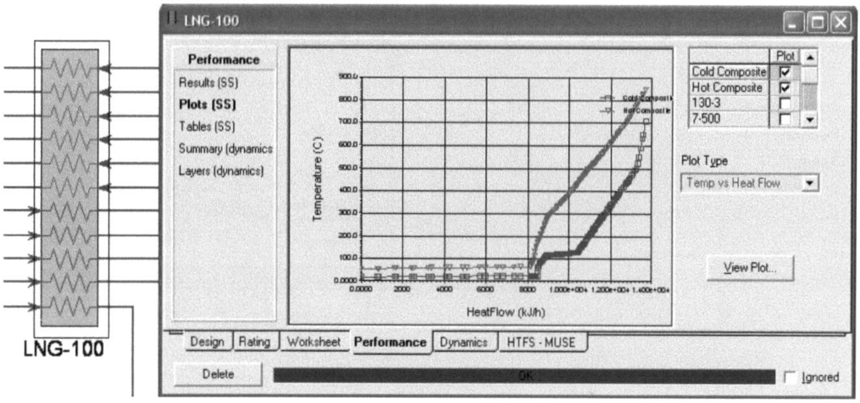

Fig. 10 LNG tool and configuration window in HYSYS

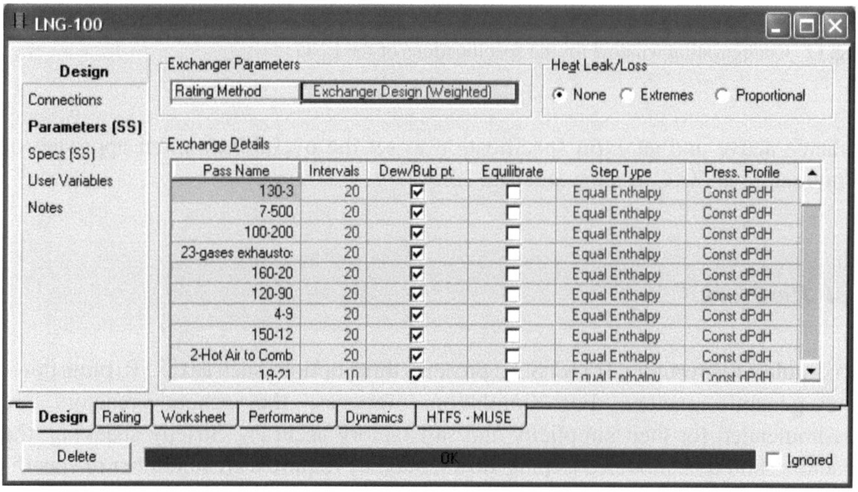

Fig. 11 Configuration window for the parameters of the LNG

auxiliary services such as steam and cooling water. Figure 10 shows the LNG and the configuration window with the plot of the composite curves.

To configure the operation, the inlet and outlet streams are needed, and also the pressure drop, which in this case is set to zero. The type of stream (hot/cold) is defined according to the difference in temperature. The configuration parameters used are: no heat leak/loss; weighted rating method; twenty intervals; equal enthalpy for step type and constdPdH for `press.profile`. In Fig. 11 can be seen the configuration window for the mentioned parameters.

The specifications for the solver are 1×10^{-4} for tolerance; 100 maximum iterations; one unknown variable; one constraint; zero degrees of freedom; the heat

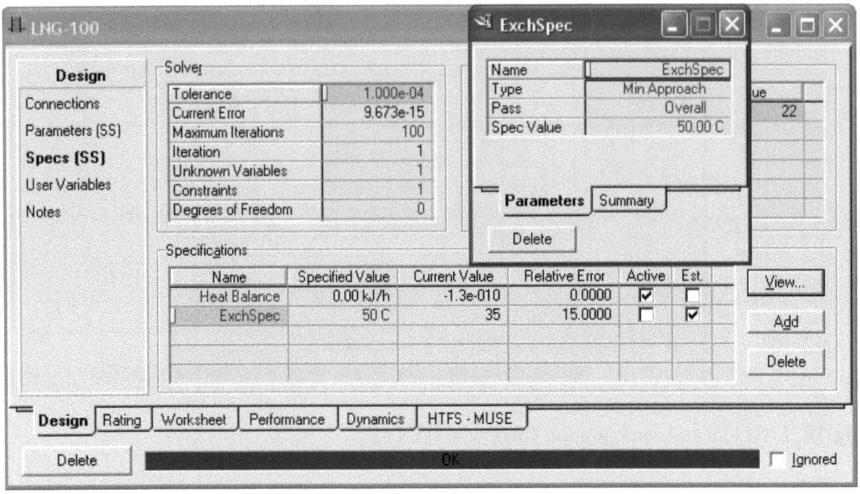

Fig. 12 Configuration window for the specifications of the LNG

balance active and an extra specification to set the overall minimum approach of temperature at 50°C as shown in Fig. 12.

2.4 Reactors

To simulate the reactors, HYSYS 3.2 presents three options such as CSTR, plug-flows and general reactors. For simulation purposes, the general reactors are recommended for their simplicity and satisfactory accuracy. Strictly speaking, the reactors should be modeled as plug-flows, but they require a great number of specifications to be properly operated, which exceeds the requirements for the preliminary efficiency analysis. Within the general reactors, there are three possibilities to be chosen: Gibbs, equilibrium and conversion types.

2.4.1 Post-Combustion Reactor

The simulation of the burner reactor that fulfills the thermal requirements of the main reactor, the ESR, is performed by choosing the conversion type. It needs the entering flows, and two outlets, one for liquids, and the other for gaseous products as can be seen in Fig. 13.

The parameters needed to configure it are the pressure drop and the duty which are set to zero, because the reactor is assumed adiabatic. The used reaction set corresponds to that previously detailed in Sect. 2.1 for the combustion reactor. The three inlets to this equipment are the exhaust gases from the anode to burn the remaining

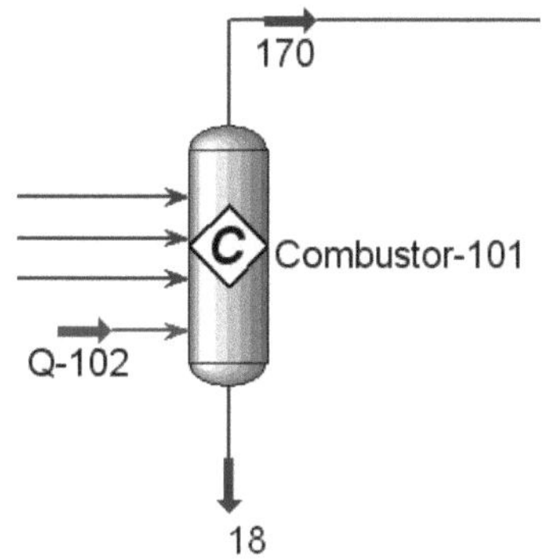

Fig. 13 Conversion reactor for the Burner

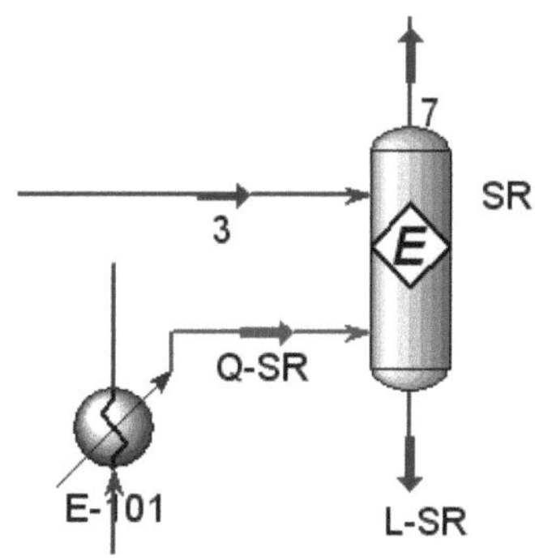

Fig. 14 Equilibrium reactor for the ESR

compounds, the fresh ethanol to supply the demand of the PEMFC and the compressed air, all working at 3 atm and pre-heated to 773 K.

2.4.2 Ethanol Steam Reforming Reactor

For the main reaction, an equilibrium reactor is recommended. It can be used since the reactions are considered to reach equilibrium. Figure 14 shows the reactor used in process flow diagram of HYSYS.

It needs the inlet flow (the mix of water and ethanol), the heat stream, a liquid and a gaseous outlet. The pressure drop is neglected. The reaction set is that previously configured for the ESR. The heat required is supplied by the hot gases from the burner. These gases are passed through a cooler, which can be seen in Fig. 14, and the obtained energy stream is connected to the ESR. At this stage, the amount of fresh feed of ethanol to the burner can be obtained with the ADJUST tool from HYSYS, as shown in Fig. 15.

This tool is useful for determining the required molar flow of the fresh feed ethanol to the burner as the temperature of the gases (the target variable) after the exchange of heat in the ESR. The desired value is specified by the user which in this case is

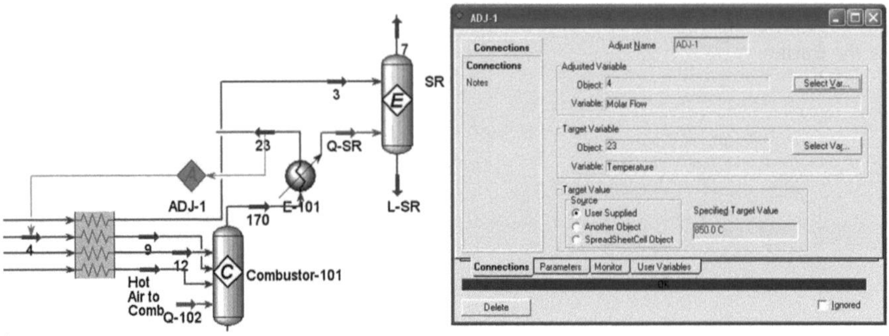

Fig. 15 ESR and thermodynamic conditioning

850°C. At the beginning, the molar flow of the ethanol should be higher, but will be re-adjusted when the recycle stream is connected to the reactor.

In addition, as the isothermal condition is assumed, the tool named SET from HYSYS is included in the model. It allows that the inlet and outlet temperatures be equal. It is done by multiplying the inlet temperature (source) by 1.0 in this case and with zero offset. These values are user-defined.

2.4.3 High Temperature Water Gas Shift Reactor

For modeling the HTS reactor, a separator tool from HYSYS is used for simulation purposes. It can be configured to consider specific reactions. It needs the inlet stream, which is the outlet of the ESR after passing through the LNG, an energy stream that will be null because it is considered adiabatic, and two outlets, liquid and gaseous. The pressure drop and duty are set to zero. The reaction set is the one configured for the WGS reactors, with a conversion of 32.07% based on CO.

2.4.4 Low Temperature Water Gas Shift Reactor

For the LTS reactor a general equilibrium reactor is chosen. It needs to do proper connections for inlets and outlets, including the energy stream, although adiabatic operation is considered. The pressure drop is considered zero. The reaction set is the same as selected for the HTS.

2.4.5 Preferential Oxidation Reactor

The following reactor of the processor is that responsible for the last cleaning stage for producing the PEMFC required quality of hydrogen. It is intended for the preferential oxidation of CO. For this purpose, a conversion reactor is chosen and Fig. 16 shows the appearance in the process flow diagram.

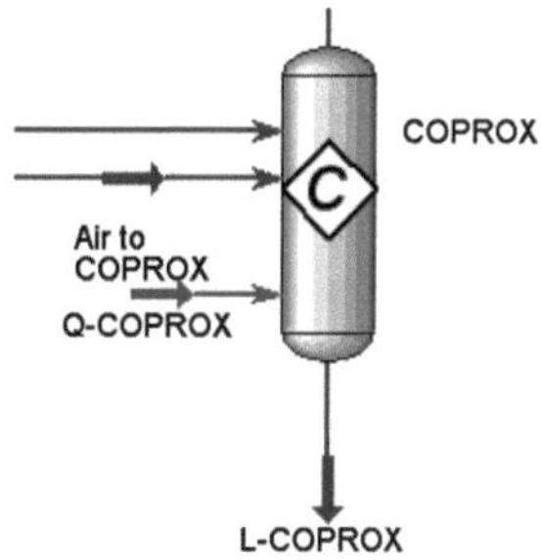

Fig. 16 Conversion reactor for the CO-PrOx

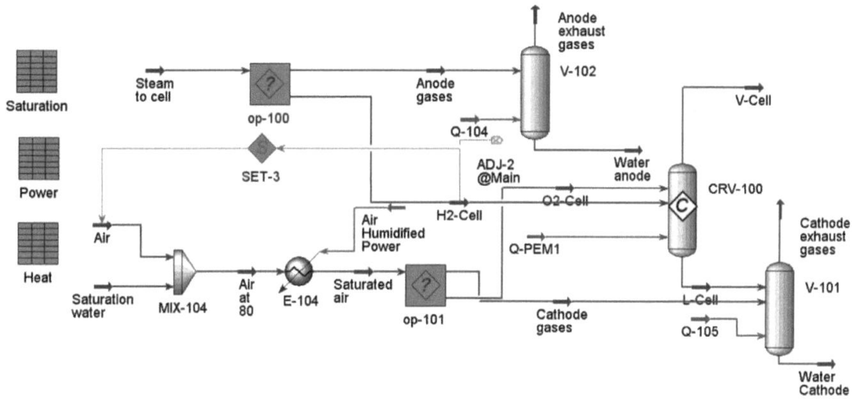

Fig. 17 Sub-flowsheet for the PEMFC

Analogous to the previous reactors, all the energy and material streams considered adiabatic, must be specified, so that the duty is zero. The pressure drop is neglected, and the reaction set used is that configured previously for this reactor, including the WGS reaction and the oxidation of CO and hydrogen.

2.5 Proton Exchange Membrane Fuel Cell

The implementation of the model in HYSYS was done by the construction of a "sub-flowsheet" where, through the selection of basic operations (reaction, separation), the cell can be modeled. A conversion reactor is accounted to specify the reaction, two separators to split the condensed water and two blocks, programmed in visual basic language as a virtual operation unit, to separate the reactants and feed the reaction section. By the use of spreadsheets available in HYSYS it is possible to calculate both the power and heat generated by the cell. In Fig. 17 can be seen the sub-flowsheet environment for simulating the fuel cell operation.

The implemented model is assumed isothermal and isobaric. The oxidizing agent (air) is fed to the cathode with 80% humidity, while the applied current to the anode is considered for saturation conditions. If the feed to the reformer—consisting of a specific ratio of water/ethanol—is low, the current at the anode is not saturated but reaches higher humidity levels than 80%. The supplied oxygen is twice the quantity corresponding to the stoichiometric magnitude.

The amount of water needed to humidify the air stream is determined from Eq. 1.

$$N_{H_2O}^{Hum} = \frac{N_{Total}^{Air} \phi \frac{p_{s,v}}{P_{Total}}}{1 - \phi \frac{p_{s,v}}{P_{Total}}} \tag{1}$$

where $N_{H_2O}^{Hum}$ is the molar flow needed to reach the specified relative humidity (ϕ), P_{Total} the pressure in atm, and $p_{s,v}$ the saturation pressure of the water at the operating temperature. N_{Total}^{Air} represents the total flow of dry air (without humidity).

A water mass balance is done to determine the excess (or defect) of this component. In case of excess, it is recycled to the processor. To calculate the generated heat by the fuel cell an energy balance is performed between the inlets and outlets of the system. The heat generated by the cell that must be dissipated can be computed by the Eq. 2.

$$Q_{Cell} = \sum_{i=1}^{In} N_i h_i \left(T_{Cell}, P_{Cell}\right) - \sum_{o=1}^{Out} N_o h_o \left(T_{Cell}, P_{Cell}\right) - W_{St} \tag{2}$$

This balance accounts for the heat for water vaporization for the humidification and the latent heat corresponding to water condensation in the anode and cathode streams. Once the cell is properly modeled, the amount of ethanol to the fuel processor can be obtained by the ADJUST tool from HYSYS. The adjusted variable is the molar flow of the pre-heated ethanol (before it is mixed with water), and the target variable is the molar flow of hydrogen to the fuel cell, specified by the user in 0.0333 kmole/h. The amount of water will be defined with the help of the SET tool, with the pre-heated ethanol as source and the water molar flow (the objective), with a factor of 4 and selecting no offset option.

The inlet gases must be cooled down to 80°C, and the liquids split before entering in the sub-flowsheet. The water leaving both anode and cathode is mixed with that separated from the inlet gases, and fed back to the fuel processor. The exiting gases are mixed together and fed to be burned once in the post-combustion reactor.

2.6 Expander

Finally, the exhaust gases from the jacket of the reformer still have some useful energy. It is used in the heat exchangers, so this stream is passed through the LNG. Its last stage is to be expanded to atmospheric pressure. Hence, a turbine must be

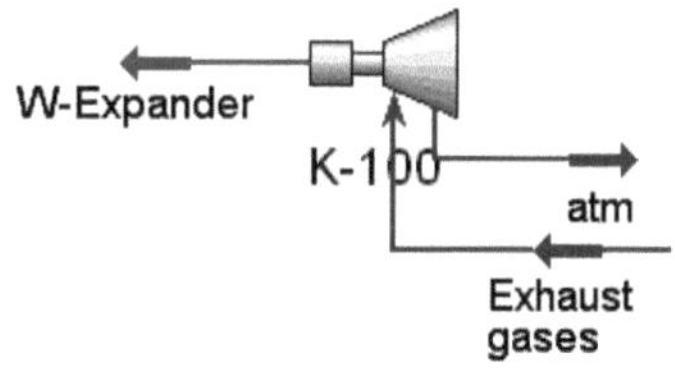

Fig. 18 Schematics for the expansion turbine

included to take advantage of this useful power, before releasing the gases to the environment. The turbine operation from HYSYS requires the connection of the material and energy streams, as shown in Fig. 18, where the adiabatic efficiency of 70% is adopted.

3 Converting from Steady State to Dynamic Simulation

From this section up to the end of this chapter, a detailed description of the pseudo-dynamic model of BPS with PEMFC is given. It is considered "pseudo" because some parts of the overall system present a very fast dynamic behavior. This fact allows to assume that many variables involved can instantaneously achieve new values during the process dynamic simulation run time. Analyzing the complete system of BPS and PEMFC in the HYSYS environment, it is possible to assume that the dynamic simulation could be done starting from the SS model, described above, and introducing all the information needed for doing the transformation. However, it was considered a better option to implement the mathematical models of the plug flow reactors and PEMFC in the MATLAB-SIMULINK environment. In addition, several valuable tools in HYSYS, already selected for the SS model, could be a part of the dynamic model. It is possible only for those subsystems that present faster dynamics as it was remarked previously. Therefore, the connection between MATLAB-SIMULINK and HYSYS, with the help of a proper communication protocol, allows to perform the implementation of this very complex system. Finally, realistic disturbances, corresponding to different requirements from vehicles under standard trajectories, could be included thanks to the use of the program ADVISOR. Then, a description of how the models of BPS and PEMFC are embedded in this software as a subsystem is given at the end of this chapter.

3.1 The Use of the Communication Protocol for Commercial Softwares

In this section, the communication protocol between MATLAB and HYSYS will be detailed. The main reason to use this protocol is because the dynamic behavior of the reactors in HYSYS is more difficult to include due to the many conditions as presented in this work. In addition, more resiliency can be obtained when its own code can be developed with specific purposes. The use of the connection with MATLAB-SIMULINK allows the reactors from the HYSYS SS model to be directly replaced by sub-flowsheet objects, sharing the inputs and outputs. Inside the sub-flowsheet environment, material streams are placed, isolated between them and just connected to the corresponding inputs and outputs. These streams are linked to the corresponding cells of the spreadsheets, which collect the desired variable or parameter from the PFD. So, the rest of the elements of the SS of the HYSYS case, such as basis environment and auxiliary equipments still work in it, but it will be periodically updated according to the imposed conditions from the reactors working dynamically in MATLAB. HYSYS acts with MATLAB in a relationship between client and server. The differential equations integration is performed by MATLAB. Therefore, the process dynamics are mainly governed by the reactors, with auxiliary equipments such as compressors and turbines assumed faster, justifying the validity of this pseudo-dynamic behavior.

In order to connect and coordinate the calculations between MATLAB and HYSYS properly, a special communication protocol is needed. It is publicly available under the terms of the GNU Library General Public License version 2 as published by the Free Software Foundation in http://www.pvv.org/~olafb/software/hysyslib/. It connects MATLAB as a controller for HYSYS via ActiveX/COM.

Several steps are needed to connect the softwares. MATLAB can read the values from the HYSYS spreadsheets. So, the first stage is to connect MATLAB to the case in HYSYS by means of running the file hyconnect.m. The following is the code to be saved as a .m file in the path of MATLAB:

```
function hyapp = hyconnect(FileNameString,
                                         VisibleBoolean)

hy = feval('actxserver','Hysys.Application');

if nargin <= 1
 hy.Visible = 1;
else
 hy.Visible = VisibleBoolean;
end

hyapp = hy;
```

Since the activeX-library is not included in MATLAB at compilation, a feval or eval syntax must be used. `hyconnect(FileNameString,VisibleBoolean)` connects to the Hysys case-file specified by FileNameString. If the arguments are not given, the current and open Hysys-case will be used by default. The optional boolean variable VisibleBoolean controls whether the application window will be shown (takes values 0 and 1, 1 by default). Then, it returns the activeX application server object.

With the connection ready to be used, where the spreadsheet is to be read from, must be specified. The file hyspread.m makes the bonding to the object in HYSYS:

```
function SpreadsheetObject =
     hyspread(ApplicationObject,SpreadsheetNameString)
SpreadsheetObject = get(ApplicationObject.
           ActiveDocument.Flowsheet.Operations,'Item',
                                   SpreadsheetNameString);.
```

The function hyspread returns the activeX object for a spreadsheet in HYSYS. It connects to the spreadsheet specified by SpreadsheetNameString.

Once the connection to the spreadsheet is ready, it is needed to read the cells inside the object. The file hycell.m is used to specify the cells, or an array of them to be read:

```
function CellObject = hycell(SpreadsheetObject,
                                          CellNameString)
 if isa(CellNameString,'cell')
  for n1 = 1:size(CellNameString,1)
   for n2 = 1:size(CellNameString,2)
    CellObject{n1,n2}=get(SpreadsheetObject,'Cell',
                                   CellNameString{n1,n2});
   end
  end
 else
  CellObject{1} = get (SpreadsheetObject,'Cell',
                                          CellNameString);
 end.
```

The function hycell returns the column cell array of activeX objects of cells in Hysys. hycell (SpreadsheetObject, CellNameString) returns the object for the cell name given by CellNameString. CellNameString may be a cell array.

Once the connections to the cells are done, the final step is to read and store them in a useful way for MATLAB operations. The file hyvalue.m is required for this task.

The function hyvalue returns the value of a HYSYS spreadsheet cell. hyvalue (CellObject) - CellObject can be a single object or a cell array of objects.

The hyset function is used to change a value of a spreadsheet cell in HYSYS, it is a useful function because it allows to write from MATLAB in HYSYS cells:

```
function CellValue = hyvalue(CellObject)
if isa(CellObject,'cell')
 for n1 = 1:size(CellObject,1)
  for n2 = 1:size(CellObject,2)
   if isa(CellObjectn1,n2,'activex')
    CellValue(n1,n2) = CellObjectn1,n2.CellValue;
   end
  end
 end
else
 CellValue = CellObject.CellValue;
end.
```

```
function hyset(CellObjects,Value)
if isa(CellObjects,'cell')
 for n1 = 1:size(CellObjects,1)
  for n2 = 1:size(CellObjects,2)
   if isa(CellObjectsn1,n2,'activex')
    CellObjectsn1,n2.CellValue = Value(n1,n2);
   end
  end
 end
else
 CellObjects.CellValue = Value;
end
```

Therefore, hyset (CellObjects, Value) changes the value of CellObjects to Value. CellObjects and Value must be matrices of the same dimension. CellObjects can contain a single object or a cell array of objects.

4 Programming of the Reactors

In this section, the main details of the developed code for simulating the reactors dynamic behavior are given. It is assumed that the reader has some knowledge about programming in MATLAB. More information and insights into it can be seen in the MATLAB 6 user's guide [4]. All the reactors share a common structure of the program, having adaptations that are ideal to each reaction. Each reactor considers a set of reactions different from the others, implying other auxiliary calculations and parameters. Since they can be programmed in a very similar manner, the ethanol steam reforming reactor is used as an example, with detailed explanations. In addition, this reactor comprises a unique feature, that is the heat exchange jacket, which needs extra calculations for the heat exchanged in it, while the others are considered adiabatic.

4.1 Header and Initialization

The reactors were programmed as S-functions in MATLAB. The first part of the code is to properly configure the function and the variables. As an example, the Ethanol Steam Reforming reactor is analyzed:

```
function [sys,x0,str,ts] = ESR(t,x,u,flag)
global np V_rodesr Dpesr Dtesr...
switch flag
case 0
 sizes = simsizes;
 sizes.NumContStates= 9*20;
 sizes.NumDiscStates= 0;
 sizes.NumOutputs= 10*20;
 sizes.NumInputs= 21;
 sizes.DirFeedthrough=20;
 sizes.NumSampleTimes=1;
 sys = simsizes(sizes);
```

The first sentence declares the function, the input and the output variables. The file must be named like the function with the extension .m, in this case ESR.m. The sentence `global` states the variables that will be considered as global, otherwise they will be considered local and will be available only for the subprograms which need them for specific calculations. The switch flag is to solve each part of the algorithm, because it is divided into three main parts. In this case, the algorithm has a first stage for initializing the differentiable variables `case 0`, a set of differential equations to be integrated by the selected numerical method. Finally, the integrated variables to be available for graphics and tables. The definition of the vectors for the inputs and outputs must also be included.

The following lines of the code are meant to make the connection with HYSYS, as explained above:

```
hy=hyconnect;
spread=hyspread(hy,'ESR-sprd');
cells_01=hycell(spread,'A1','A2',...,'A20');
```

These sentences make the connection with the HYSYS spreadsheet and with the specified cells. In this case, only the column A is shown, but there is one for each component and each temperature (the reactor and the jacket temperatures).

Twenty rows are considered because the plug flow reactor is modeled as 20 lined up continuous stirred tank reactors, and there is a differential equation for each slice. The next part is to assign the initial values:

```
cells_value_01 = hyvalue(cells_01);

x0 = [cells_value_01 cells_value_02 ...];

str = [];
ts = [-1 0];
```

First, the value of the cells must be read, and then arranged in the vector of the initial values. The following lines declare that there is no state ordering. Before starting to solve the equations, the constants defined as global are specified, simply declaring their values. As examples of global constants can be mentioned the number of discretization points, the length and diameter of the reactor, calculating the volumes and stating the constants for the reaction.

4.2 Solving the Differential Equations

The calculation is done by means of a solver configured in Simulink. In this case the fixed-step solver ode1(Euler) is used, with a step size of 5×10^{-5}. To call the solver, a different function is used, with all the differential equations explicitly stated, referred from the main file:

```
case 1

  sys=dyn_esr(t,x,u);
```

As stated earlier, there are different cases for each part of the resolution, case 1 is meant to solve the differential equations. It is directly referred to as the file/function `dyn_esr.m` with the required calculations.

4.2.1 Explicitness of the Differential Equations

The differential equations and the auxiliary calculations needed are implemented in a different file. The program `din_esr` is a function of time, differential and manipulated variables, giving as result the derivatives for the equations. The header of the file is as follows:

```
function [dfdt]=din_esr(t,x,u)

global np V_rodesr Dpesr Dtesr...
```

The first sentence is to declare the function, the following needs to be equal to the previous file in order to work properly. Next, the variables are reassigned:

```
ya=[x(1:np)'];
```

Table 2 Distinctions between reactors

	ESR	WGS	CO-PrOx	Burner
Reactions	Ethanol dehydrogenation Acetaldehyde decomposition Methane reforming Water gas shift	WGS	CO combustion H_2 combustion WGS-inverse	Combustions: Ethanol Methane
Heat exchange	Endothermic	Adiabatic	Adiabatic	Adiabatic
K_{eq} calculation	All	Yes	Only WGS	No
Molar flow	Vary	Constant	Vary	Vary

The vector `x` of initial conditions needs to be separated in the corresponding variables to be treated properly by the calculations. Then, the inlet conditions to the reactor must be accounted:

```
F_esr = u(1);
ya0_esr = u(2);
```

The process independent variables of the reactor are read as the input vector named `u`. The manipulated and disturbance variables such as molar flows and the entering temperatures, the process variables and the compositions, belong to this vector. The pressure drop must be calculated, arranging a vector with the inlet and outlet pressures, with different pressure drops for each slice, since it is assumed constant for each one. For the calculation of the molar flow entering in each slice an additional equation must be considered:

```
na = F_esr*ya0_esr;
```

Several auxiliary calculations are needed before solving the differences. First, the *cp* for all the components and for each slice (according to the temperature). In the case of ESR, because it has a jacket for heat exchange, it needs the viscosity and thermal conductivities.

For all the reactors, the ΔH of the reactions are needed for each slice, as a function of temperature. The equilibrium of reactions when it is concerned, and the kinetic expressions for the reactions.

Another necessary calculation is the molar flow going out of each slice of the reactor. It is equal to the difference between the entering flow and the overall mole change for each reaction. The outlet of the first slice can be solved directly, but the others are dependent on the formers, so they need to be calculated in the specific order by a loop `for`:

```
for i=2:np
  F_out_esr(i) = F_out_esr(i-1)-(ra(i) + rb(i)+
                                   +2*rc(i))*mass_cat;
end
```

`ra (i)`, `rb (i)` and `rc (i)` are the reaction kinetics for the reactions of ESR. `mass_cat` is the mass of the catalyst involved in the slice. They include a change in the number of moles. The number of moles involved in the WGS reaction remains constant. In addition, the Leva warming correlation and the heat exchanged Q must be calculated for this specific reactor. There are a few distinctions between the calculations needed for the reactors. Distinctions are highlighted in Table 2. The most important is the reactions set considered, which is unique for each reactor, except for both water gas shift reactors. The thermodynamic condition is adiabatic for most reactors except for ESR, which needs the calculation of the mean *cp* for the flows and mean viscosity to get the exchanged heat between jacket and fixed bed. Depending on the nature of the reactions, the calculation of K_{eq} has to be done. All the reactions of ESR are based on equilibrium, so the WGS reactor also needs this parameter. The combustion reactions occurring in the CO-PrOx are conversion reactions, so they do not need K_{eq}. The burner only accounts for combustion so no equilibrium is applied. As stated earlier, the molar flow through each slice is needed, the WGS reaction has no change in the number of moles, so for this reactors, the molar flow remains constant in each slice, including input and output.

The rate of reaction is needed for all reactions. To calculate the temperature change, the *cp* of all the components in all the slices is required, also the ΔH for all reactions. Finally, the balances are calculated according to the equations stated in Chap. 10. As an example, the derivative for the composition of one of the components is shown:

```
dyadt(2:np) = (F_out_esr(1:(np-1)).*ya(1:(np-1))-
          F_out_esr(2:np).*ya(2:np)+ra(2:np).*mass_cat)
                                         ./ntesr(2:np);
```

The `ntesr` is the total moles in each slice (holdup), and `dyadt` is the difference calculated to be integrated. Once the differential equations for all the components and temperatures for each slice are calculated, they need to be arranged in the vector that MATLAB is able to integrate:

```
dfdt=[dyadt dybdt ... dTdt]';
```

`dfdt` is the output vector of the function analyzed. `dTdt` id the difference calculated for the temperature.

4.3 Results

To bring the calculated integrations to Simulink again, the `sys` vector is used. The results from the calculation of the differential equations must be called and the outcoming vector of the S-function will have as much components as configured earlier with the NumOutputs:

The `case 3` is the section for printing the results. The second line corresponds to the FeedThrough defined previously, in this particular case where the molar flow

```
case 3
  sys(1:np)=F_out_esr;
  sys(np+1:10*np)=x(1:9*np);
```

changes along the reactor. The last line is just taking the values of the solution vector to export it to Simulink. Following, the update of the values to HYSYS is done. The first part is to make the connection, as in Sect. 4.1, with the program, the spreadsheet and the cells. Instead of reading the values of the cells, they are written with the updated values:

```
hyset(cells_01,x(1:np,:)');
```

The `hyset` function is used in this case, instead of the `hyvalue` used previously to take the values of the cells. Once the results are printed, the calculations for the reactors are done. The cases 2, 4 and 9 of the S-function are unused.

5 Incorporating the Reactors in Simulink

In the previous section, the development of the code to model the dynamic behavior of the reactor was presented. This still needs an external solver to make the integrations needed to obtain the desired results. This integration can be done in Simulink, a graphical block diagramming tool integrated with MATLAB [5]. The code as presented, can be used in an S-function block, which reads and executes the instructions from a program written in a specific format. It needs a vector of inputs and an output, that is another vector composed of all the results obtained. The selected solver do the integrations. In this case, the ode1(Euler) solver is used, with a fixed step size of 5×10^{-5}. The schematics of the inclusion of the reactor in Simulink can be seen in Fig. 19.

The input vector and components can be arranged in one input vector with the 'Mux' tool. It can have as many inputs as needed. The S-function must call the proper .m file that makes the calculations for each reactor. The outlet vector can be de-convoluted into the required outputs as a vector to feed the next reactors, or as each individual component, if needed. Some conditioning of the vector may be needed between reactors, mainly when other components are added. Any variable of any part of the process can be saved or just watched. The exit and entering flows are interconnected between reactors according to the scheme ESR-HTS-LTS-COPrOx, the resulting stream from the last cleaning reactor is fed to PEMFC, and the exhaust gases fed back to the post-combustion reactor. The hot gases from the burner are connected to the inlet of the jacket of ESR, therefore completing the recycle stream. The set of reactors that conform to the processor can be seen in Fig. 20.

For the sake of clarity, each reactor was included in a subsystem, each of these subsystems forming a set within another subsystem. So the processor will be represented by a single subsystem in the parent level. This plant has a number of inputs, the manipulated variables plus the disturbances, allocated in the main flowsheet. The

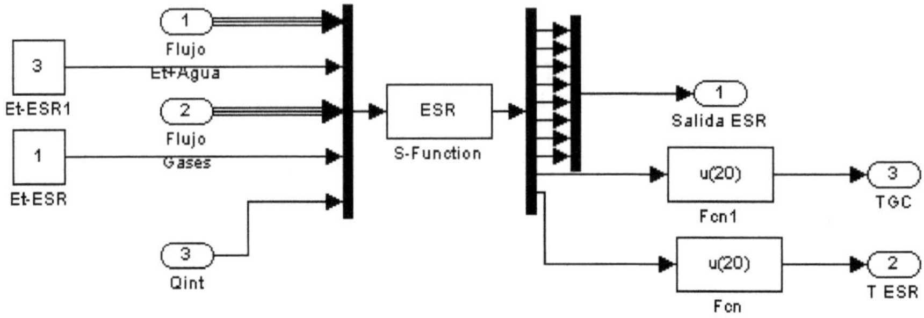

Fig. 19 S-function for the ESR in Simulink

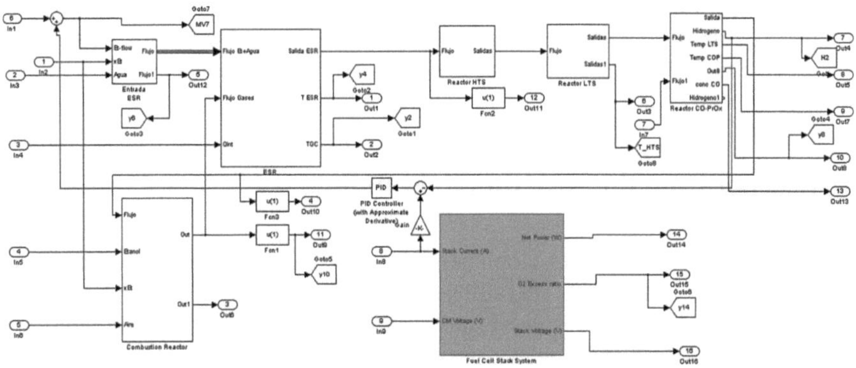

Fig. 20 Sub-flowsheet for the BPS in Simulink

controlled variables are connected to 'Goto' tools visible globally, which can be collected by 'From' tools in any part of the model. The 'From' with the corresponding variable feeds a 'Sum' block that calculates the difference with the set point, to give the current error to a PID controller block, set with the proper parameters. The master flowsheet in Simulink is shown in Fig. 21.

6 Using ADVISOR for Vehicle Modeling

In Sect. 4 of Chap. 7, ADVISOR was introduced in the context of the vehicle modeling and the main characteristics and advantages of this simulation tool were stated. In this section, a more detailed discussion about the benefits of this tool is performed. Besides, this section provides an overview of how to use ADVISOR. The information provided in this section is based on the documentation of the simulation tool [6].

ADVISOR is a tool for vehicle modeling and analysis composed of a set of model, data and script text files for use with MATLAB and Simulink. It is designed for rapid

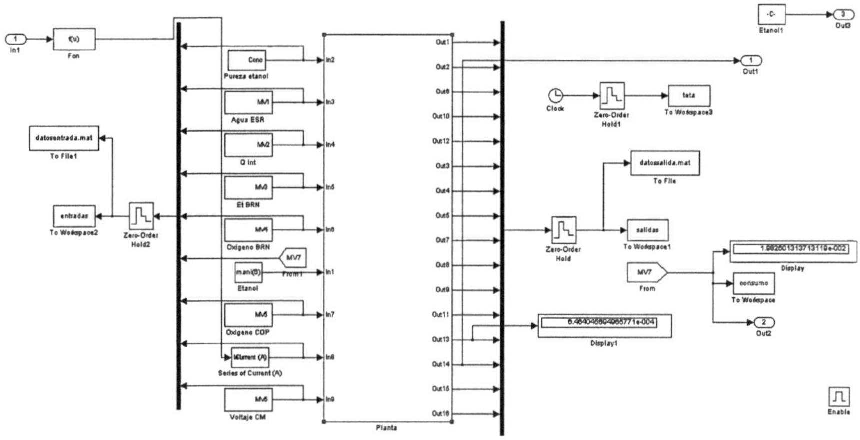

Fig. 21 Master flowsheet of the plant in Simulink

analysis of the performance and fuel economy of conventional, electric and hybrid vehicles. Some of the utilities of this tool are as follows:

- estimating the fuel economy of vehicles;
- accounting for the energy losses in conventional, hybrid or electric vehicles;
- comparing tailpipe emissions;
- evaluating energy management strategies in hybrid vehicles;
- optimizing the gear ratios in the transmission to minimize fuel use or maximize performance.

The models in ADVISOR are mostly empirical, relying on drivetrain component input/output relationships measured in the laboratory, and quasi-static, using data collected in steady-state tests and correcting them for transient effects such as the rotational inertia of drivetrain components. ADVISOR uses basic physics and measured component performance to models existing or future vehicles. The main advantage is the prediction of the performance of vehicles that have not yet been built, predicting fuel use, tailpipe emissions, acceleration performance and gradeability. In general, the analysis takes two steps:

1. Defining a vehicle using measured or estimated component and overall vehicle data.
2. Prescribing a speed versus time trace, along with road grade that the vehicle must follow.

With this analysis, it is possible to determine whether the vehicle is able to follow the defined speed trace, the fuel and/or electric energy required, the fluctuation of the state-of-charge of the energy storage system throughout a cycle, the peak powers delivered by the drivetrain components, the distribution of torques and speeds and the average efficiency of the transmission. Moreover, following an iterative procedure

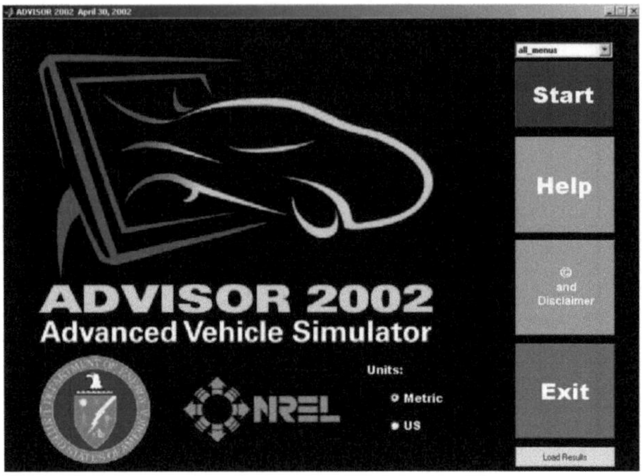

Fig. 22 Startup ADVISOR window

where the vehicle definition and/or driving cycle is changed, it is possible to determine several issues such as the maximum road grade that the vehicle can maintain at a given speed indefinitely, the smallest engine for a given acceleration, the final drive ratio that minimizes the fuel use, and the fuel economy sensitivity to mass, aerodynamic drag, or other vehicle or component variations. Some of these issues are determined automatically using ADVISOR, while others require some custom programming.

The principal limitation of ADVISOR consists in that the component models are quasi-static, and cannot be used to predict the phenomena with a time scale of less than a tenth of a second, approximately. Phenomena such as physical vibrations, electric field oscillations and other dynamics cannot be captured. However, recent linkages with other tools such as Saber, Simplorer and Sinda/Fluint allow a detailed study of these transients.

ADVISOR uses the required speed as an input, and thus determines what drivetrain torques, speeds and powers would be required to meet that vehicle speed. Because of this flow of information back through the drivetrain, from tire to axle to gearbox and so on, ADVISOR is what is called a backward-facing vehicle simulation. Forward-facing vehicle simulations include a model of a driver, who senses the required speed and responds with an accelerator or brake position, to which the drivetrain responds with a torque.

Once the tool using the command `ADVISOR` in the MATLAB command window is opened, Fig. 22 appears, which is the start-up figure. This initial window presents the options to select US or metric units, start using ADVISOR, click help to go to a local ADVISOR web page, exit ADVISOR or read the copyright and disclaimer.

Clicking on *Start*, it drives to the input figure (see Fig. 23). This figure shows the default values for a specific vehicle and then it is possible to select the desired drivetrain configuration for the vehicle (series, parallel, etc.) from a popup menu. After selecting the drivetrain configuration, all the components of the vehicle can be

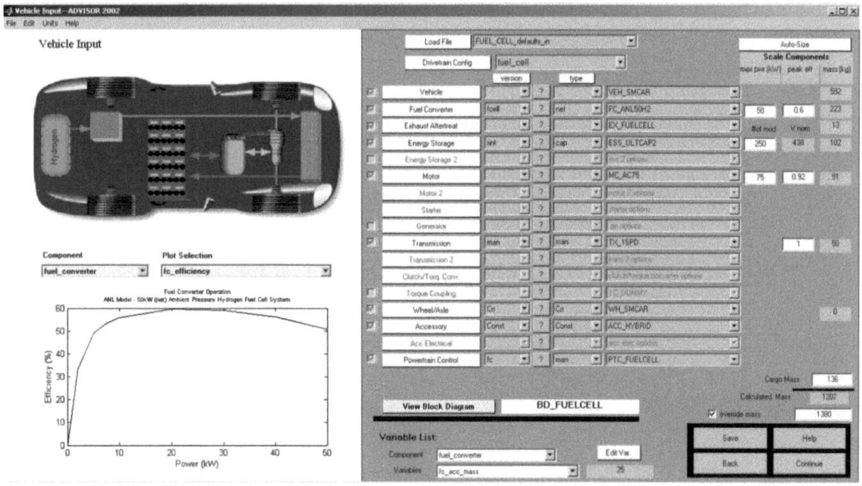

Fig. 23 ADVISOR vehicle configuration window

selected using the popup menus, or by clicking on the component in the picture. To the left of the component popup menus is a *push* button that allows to add or delete components by selecting their corresponding listed m-files. The m-file of a specific component can be accessed for viewing or modifying from either the component pushbutton or by clicking on the component of the picture.

After selecting all the desired components for the vehicle, scalar input variables can be modified. One way this can be done is with the variable list at the bottom of the figure and the *Edit Var.* button. First, the variable to change selected and then click the edit button to change its value. The default value is always shown for reference. The *View All* button allows to see all the altered variables. Clicking on the *Help* button, it is possible to see a brief description and the units used for the input variables. A second way in which it is possible to edit variables is by typing in a desired value in the edit boxes next to the component. A final way to edit the mass of the vehicle is to use the override mass button. In this way, the calculated mass is ignored and the value input into the box is used instead.

To load or save a particular vehicle configuration click on the *Load Vehicle* button on the top of the figure or click on *Save* at the bottom of the figure, respectively. The file is saved in the format 'filename_in.m' and then a saved vehicle can be accessed by pushing the *Load* button. At the bottom left portion of the figure there is a popup menu and axes with the ability to view information about components such as their efficiency maps, emissions maps, fuel use maps, etc. These are plotted along with their maximum torque envelopes where appropriate. Besides, any component m-file can be viewed by clicking the component buttons.

The *Auto-Size* button takes the selected vehicle and adjusts vehicle parameters until it meets acceleration and gradeability goals. The parameters that are modified

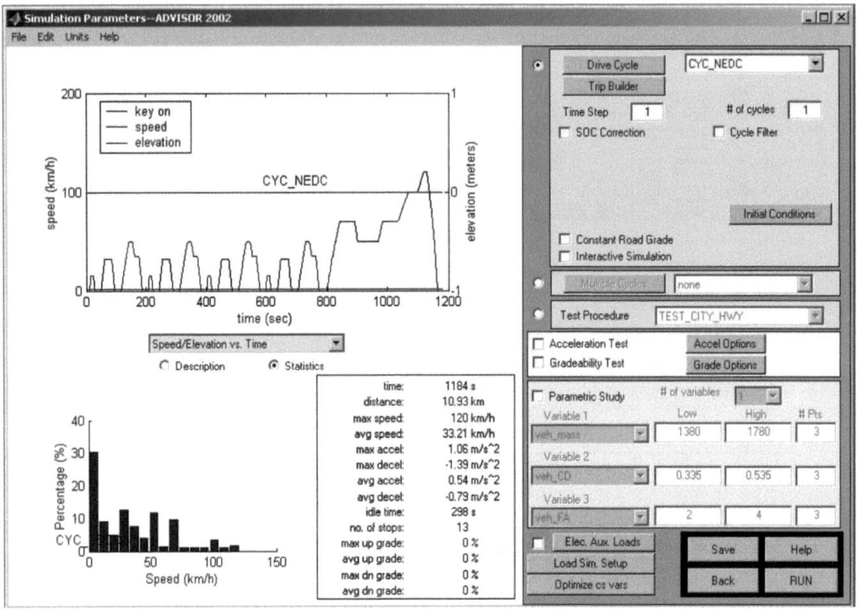

Fig. 24 ADVISOR simulation setup window

are the fuel converter torque scale, the motor controller torque scale, the number of energy storage system modules and the vehicle mass. The minimum torque scale is set so that its peak power output is 45 kW and the number of battery modules is limited to a maximum nominal voltage of 480 V. The default performance targets are maintaining at least a 6% grade at 55 mph, and obtaining less than a 12 s 0–60 mph time, 23.4 s 0–85 mph time, and 5.3 s 40–60 mph time.

The *Back* button drives to the opening window, losing all unsaved information, and the *Continue* drives to the simulation setup figure. The simulation setup figure (see Fig. 24) provides several options to test the previously defined vehicle. If the *Drive Cycle* radio button is selected, it is possible to select the driving cycle from a list of available options. Also, it is possible to select the number of times the cycle is repeated as well as if the SoC is corrected. The initial conditions can also be set from this window and the *Cycle Filter* option allows to smooth out the selected drive cycle. Besides, a cycle can be created by combining many different cycles back to back and this new cycle can be saved in the normal cycle format.

Selecting the *Auxiliary Loads* button invokes a graphical interface to select different auxiliary loads and their on/off times as related to the drive cycle.

With respect to the SoC correction, there are two options: linear- and zero-delta. The Linear SoC correction routine runs two simulations: one that gives a positive change in the state of charge and one that gives a negative change in SOC. The corrected value of the variables of interest (e.g. miles per gallon and emissions) are then interpolated from the zero change in SoC from a linear fit to the two data points. The zero-delta correction routine adjusts the initial SOC until the simulation run yields a zero change in SoC around a tolerance band of $\pm 0.5\%$.

On the other hand, it is possible to simulate the vehicle under different situations and test procedures:

- Selecting the *Constant Road Grade* checkbox, it is possible to run the drive cycle using a constant road grade.
- Selecting the *Interactive Simulation* checkbox causes a real-time interactive simulation interface to activate while the simulation is running.
- Selecting *Multiple Cycles* checkbox, it is possible to run many different cycles with the same initial conditions. This procedure saves the setup information including initial conditions and all the results.
- Selecting the *Test Procedure* checkbox, an acceleration test is run in addition to the selected cycle. Acceleration times, maximum accelerations and distanced traveled in 5 s are displayed in the results figure.
- Selecting the *Gradeability* checkbox, a gradeability test is run in addition to the chosen cycle. The grade displayed in the results figure is the maximum grade maintainable at the input speed.
- Selecting *Parametric Study* checkbox, it is possible to see the effect that up to three variables have on the vehicle, select a parametric study. The low and high values may be set, as well as the number of points desired for that variable. A parametric study runs a set of simulations to cover the matrix of input points, such that if three variables are selected with 3 points each, 27 simulations are run.
- Clicking *Load Sim. Setup* button, the control strategy optimization setup window is opened. Checking the radio boxes, it is possible to select the design variables used to optimize for the chosen objectives and constraints below.

Finally, the *Save* button allows to save the simulation setup and the *Run* button proceeds to run the simulation. Figure 25 shows the results of the simulation. This window presents some summary results (fuel economy, emissions, total distance, etc.) and allows to plot up to four time series plots by selecting a variable from the popup menu. If the acceleration and gradeability checkboxes were selected in the simulation setup screen, appropriate results are also displayed.

By clicking the *Energy Use Figure* button, a new window is opened showing how energy was used and transferred for the vehicle during the simulation. The *Output Check Plots* button pulls up plots that show the vehicle's performance, some of which are not available under the time series plots. The fuel used over the simulation run and the operation points of the fuel converter can be viewed as a function of speed and torque of the fuel converter by clicking on the 'tools' menu and selecting 'FC operation'.

When a simulated vehicle deviates from the requested drive cycle, the trace miss analysis is automatically called. This analysis displays statistical information about the severity of cycle trace-miss in the "Warnings/Messages" window of the Results screen. Information also includes the absolute average difference between the speed requested and speed achieved, the percent of simulated time with trace miss greater

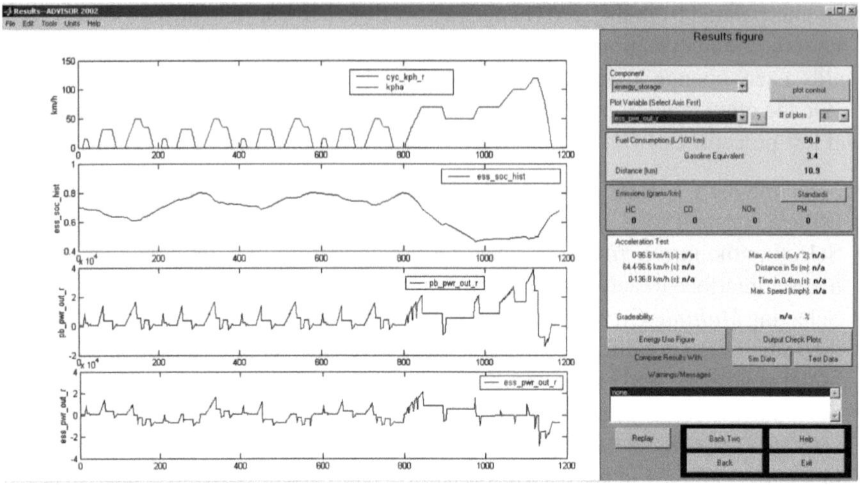

Fig. 25 ADVISOR result windows

than 3.2 kph (absolute), the greatest percent difference based on the maximum cycle speed, the greatest percent difference based upon local cycle speed, and the greatest absolute difference.

7 Inclusion of BPS in the Simulation Environment

For the simulation of the bio-ethanol processor, against a standardized drive cycle, in a fuel cell hybrid system, ADVISOR is used. Instead of using the entire system, only the model and parameters are utilized. The first step for including the processor, is to configure the vehicle, as previously described. The parameters are set in the input figure (Fig. 23). Instead of clicking in the *Continue* button, the *View block diagram* is pushed. A regular Simulink blocks window is shown with the model as presented for simulation purposes. The main window from ADVISOR, the input figure, can be discarded, as it will not be necessary. The model can be executed and simulated as a common Simulink model (.mdl).

There are two main changes to make to this .mdl file. The first is to replace the original energy management strategy with the described in Chap. 8. For this purpose, the original must be deleted and replaced by a subsystem containing the required instructions, including the working zone definitions, and the decision variable to keep the cell working or to turn it off.

Then, the original fuel cell model must be deleted, and again, replaced by a new subsystem. This new subsystem must contain the bio-ethanol processor model, as described in Sect. 5. The processor model was the main system, but now it is working

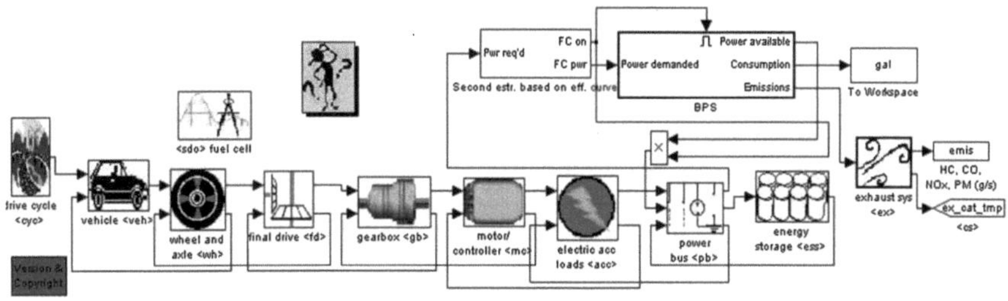

Fig. 26 ADVISOR model with the BPS

within a subsystem. The parent system is formed by the components of the vehicle model, i. e. drive cycle, processor, power bus, energy storage system, etc. The main model window is displayed in Fig. 26. It can be noted the inclusion of the processor and energy management systems.

Several other modifications need to be done to have a properly working model. An `Enable` block must be inserted in the subsystem. This block creates an enabled subsystem, meaning that the subsystem is working as long as there is a 1 in the binary input for this purpose, and stops with a 0. The output that indicates the working condition from EMS must be connected to the input port of the processor subsystem.

The required input to the subsystem is the power demanded to the fuel cell. It is determined by EMS, and since the processor is commanded by the current demand, the power needs to be translated into I_{st} according to Chap. 10. A function block is ideal for the task, with a quadratic function of the input variable. The output determines the current demand and must be connected to the processor and fuel cell.

The outputs that the fuel cell from ADVISOR presented were the power obtained, the consumption and the emissions. The emissions were stated as ideally water and heat, so for analysis purposes they will be neglected. Actually the goal of that sector was to compare harmful emissions from the vehicle, and the fuel cell hybrid vehicle has none, so this simplification does not affect the validity of the results. The power obtained is connected to the actual power obtained from the fuel cell. The consumption states for the sum of the ethanol fed to the burner and ESR reactors. It is not useful to compare with different types of vehicles, but can be used for the same vehicle with different configurations, sizes of capacitors, management strategies, drive cycles, etc. The power obtained must be multiplied by the working condition of the cell before the input to the power bus in order to obtain realistic measures, because when the fuel cell is off, the subsystem freezes the output values, and does not reflect the real behavior of the system. The same happens with the fuel consumption.

Another fundamental issue is the solver configuration. For the analysis presented, the fixed-step ode1(Euler) solver was used, with a step size of 5×10^{-5} and SingleTasking mode. It shows stable performance for several cases, and acceptable simulation times (compared to other type of solvers). As given, the model can run up to 12 simulation seconds per computing hour in a Pentium Dual-Core, 2×250 GHz CPU and 2 GB of RAM, in closed loop and with a drive cycle. There are many other configurations that can be set to the preference of the user, but those suggested are fully tested and working. Executing the Simulink file, the results for the drive cycles can be obtained directly.

Finally, the model can be saved, and it will be done within the ADVISOR environment. It cannot be opened externally to access the system, but must be done from the input figure, via the *View block diagram* button for the fuel cell hybrid vehicle. The drive cycles can be switched or modified, or any other substantial changes can be done.

References

1. Luyben ML, Luyben WL (1997) Essentials of process control. McGraw-Hill chemical engineering series. International Editions, Singapore
2. Hyprotech. HYSYS 3.2 simulation basis (2003) Technical report, Aspen Technology Inc
3. Hyprotech. HYSYS 3.2 operations guide (2003) Technical report, Aspen Technology Inc
4. The MathWorks MATLAB 6 user's guide (2002) Technical report, The Math Works Inc
5. The MathWorks Simulink 4 user's guide (2002) Technical report, The Math Works Inc
6. Markel T, Brooker A, Hendricks T, Johnson V, Kelly K, Kramer B, O'Keefe M, Sprik S, Wipke K (2002) ADVISOR: a system analysis tool for advanced vehicle modeling. J Power Sources 110:255–266

Plant-Wide Control for Fuel Processor System with PEMFC: Control Structure Selection and Optimal Sensor Location

L. Nieto Degliuomini, D. Zumoffen and M. Basualdo

Abstract The synthesis and a plant-wide control strategy based on the process knowledge for the fuel processor with fuel cell were detailed in Chaps. 9. There, the unit dimensions and the equipment interconnections were defined to conform the flow sheet of the complete process at the more efficient operating point. The knowledge-based plant-wide control strategy was tested through the control-oriented dynamic model in Chaps. 10. In this chapter a new systematic and generalized procedure is applied to simultaneously solve the optimal sensor placement integrated to the plant-wide control design. This methodology allows to configure the loops pairing by considering the trade-off between servo and regulator behavior. It can be done thanks to defining a proper function, named net load effect, accounting for both set point and disturbance effects. Although some concepts used in this approach are not new, the main contribution of this method is the selection of the adequate objective function. It is mathematically expressed in a new way, in terms of Frobenius norm of specific matrices related with the reduced models of the plant and very useful for evaluating the process interaction. Then, it drives the search supported by genetic algorithms (GA), which evaluates all the possible combinations of input–output variables. It allows to solve successfully and with less computational effort the combinatorial optimization problem, in spite of the high dimension usually involved in large-scale chemical plants. It must be emphasized that this approach is developed for working in cases where only steady-state plant information is available. However, if a

L. Nieto Degliuomini · D. Zumoffen · M. Basualdo (✉)
CAPEG-CIFASIS-(CONICET-UNR-UPCAM), 27 de Febrero 210 bis,
S2000EZP Rosario, Argentina
e-mail: basualdo@cifasis-conicet.gov.ar

L. Nieto Degliuomini
e-mail: nieto@cifasis-conicet.gov.ar

D. Zumoffen · M. Basualdo
CAPEG-FRRo-UTN, Zeballos 1341, S2000BQA Rosario, Argentina
e-mail: zumoffen@cifasis-conicet.gov.ar

M. S. Basualdo et al. (eds.), *PEM Fuel Cells with Bio-Ethanol Processor Systems*,
Green Energy and Technology, DOI: 10.1007/978-1-84996-184-4_12,

dynamic model is also disposable, the algorithm is extended to use it. In addition, a mathematical demonstration is presented so as to understand why it is possible through the proposed sequence of calculations to find a well-conditioned control structure. This methodology has been successfully tested on several well-known benchmark cases of chemical plants. Hence, in this chapter it is applied to the challenging and novel case of the fuel processor with fuel cell. Thus, through a set of dynamic simulations for different scenarios the high capacity of this approach can be shown.

1 Introduction

The process synthesis stage generally defines some partial control objectives such as product quality, product rate, stabilization of some inherent unstable behavior, active constraints, etc. But there may be hundreds or thousands of additional variables to decide that could/should be controlled to improve the overall process behavior (either for set point changes or disturbances), which is not a trivial problem. The selected plant-wide control structure affects the investment cost, since it is related to the number of sensors, actuators, controllers to be installed and the final dynamic control performance. Therefore, having a systematic methodology to obtain an efficient plant-wide control structure, able to reduce both the heuristic load and the design complexity at the synthesis stage, is helpful from the starting point of the project and during the lifetime of the process.

An important number of process control researchers have developed many systematic plant-wide control methodologies and applied them to chemical processes. According to Larsson and Skogestad [1], these methodologies can be classified based on heuristic and mathematic tools.

On the heuristic approaches the pioneering work of Buckley [2] is the first report on plant-wide control. The main issues in this area were introduced there and presented, it is still the industrial approach to plant-wide control nowadays. Other important authors in this category are Luyben and co-workers [3] with a nine-step approach, which was developed based on the process experience of the group. More recently, Konda et al. [4] proposed to obtain the plant-wide control structure using an integrated framework of simulation and heuristics which uses steady state and dynamic simulation to support the decisions taken. All the proposed methodologies mentioned above are systematic in nature and address many of the major issues involved in the plant-wide control problem, such as the effects of recycles and energy integration. The main result from these methodologies are the different ways to achieve a decentralized plant-wide control structure.

The second category relies on a rigorous mathematical framework of dynamic theory, constrained optimization and systems analysis. The mathematical approaches, although rigorous, are often very difficult to formulate for large-scale systems and, in many cases, they are computationally very intensive. However, several authors suggest approaches in this category. Cao and Saha [5] developed an improved and more efficient algorithm of the "branch and bound" that provided a global ranking to all possible input and output combinations. Based on this ranking an efficient control structure with least complexity for stabilizing control is detected which leads to a decentralized proportional controller. In Cao and Kariwala [6] is presented a bidirectional branch and bound algorithm, i.e. branch pruning is considered in both upward (gradually increasing subset size) and downward directions simultaneously. Thus, the total number of subsets evaluated is reduced dramatically for efficient handling of large-scale processes. In Robinson et al. [7] is presented an approach based on splitting the optimal controller gain matrix that results from solving an output optimal control problem into feedback and feedforward parts, and from these parts can be obtained the information about designing decentralized plant-wide control system architectures, or if it is preferred, a model predictive control (MPC). In Jorgensen and Jorgensen [8] was presented the problem of selecting the control structure formulating a special mixed integer linear program (MILP) employing cost coefficients computed using Parseval's theorem combined with relative gain arrays (RGA) and internal model control (IMC) concepts.

As an example of a combination between mathematical and heuristic approaches can be cited Skogestad [9, 10]. In those works, the focus was on the selection of controlled variables and for keeping them constant, the process is maintained close to the optimum when disturbances and control errors are present. The control accomplished with this goal is called self-optimizing. Alstad and Skogestad [11] showed how to select the CVs as a linear combination of measurements in order to obtain a minimum deviation from the optimal value, i.e. a control structure with better self-optimizing characteristics. More recently, Suraj Vasudevan et al. [12] found the control structures following three of the approaches cited before for a vinyl acetate monomer plant. The analysis of the results indicated that while all the procedures gave stable control structures, the integrated framework and self-optimizing in Konda et al. [4] and Skogestad [10] respectively, gave more robust control structures than that resulting from the practical experience and engineering judgement approach proposed by Luyben et al. [3].

It is remarkable that in the majority of the works cited above the use of the pioneering work of Bristol [13], the relative gain array (RGA)-based techniques for control-loop configuration is reported. Although, it is well known and has widespread industry applications, some limitations are recognized. Based on these, several new extensions of the RGA have appeared in the literature. Particularly, in this work, the non-square relative gain array (NRG), introduced by Chang and Yu [14], the generalized relative disturbance gain (GRDG) presented in Chang and Yu [15] and the recently developed relative normalize gain array (RNGA) together with the RGA have been considered to generate the most critical stages of the new systematic approach. The stage dedicated to determine the best group of controlled variables (CVs) is implemented on the basis of an extension of the NRG theoretical concepts. In the work presented by Chang and Yu [14] the NRG calculation is limited to the problem dimensionality. In this work, the objective function to be minimized for defining the CVs is the sum of the square deviations in steady state of the uncontrolled output variables (SSD_{yr}).The optimal solution can be found by the use of genetic algorithms (GA) independent of the problem dimension. It allows to solve successfully the combinatorial optimization problem, in spite of the high dimension involved in large-scale chemical plants. On the other side, the GRDG theoretical concepts presented in Chang and Yu [15] were limited to consider only disturbance effects, by time, and did not consider the set point changes effect. Thus, the methodology implemented in the fuel processor system and fuel cell given here, allows to configure the loops pairing by considering the trade-off between changing operating points (servo) and regulator behavior. It can be done thanks to defining a proper function, named net load effect, containing both set point and critical disturbance effects. The relative importance between them is handled through specific weighting parameters. For selecting the best pairs between CVs and the manipulated variables (MVs) the minimization of the net load effect is done subjected to find a proper plant model, in the context of IMC design.

The complete methodology used here can work satisfactorily under two possibilities: the availability or not of the process dynamic model. For the first option it is recommended to check the control structure suggested by the RGA with that proposed by the relative normalize gain array (RNGA), recently introduced by He et al. [16]. The second one corresponds to the procedure detailed above. Therefore, in this chapter the reader can find a step-by-step application of this systematic methodology that integrates and extends the results for plant-wide control which has been tested previously, by the group of Basualdo, in several academic cases of study [17–22].

2 Systematic and Generalized Methodology for Plant-wide Control

In this section the generalized procedure for plant-wide control, called minimum square deviation (MSD), is presented. The main steps of the algorithm are schematically given in Fig. 1. In this block diagram a direct connection with process synthesis area is assumed, which allows to perform an intensive interchange of information and by successive iterations a controllable process is achievable. Simultaneously, the main control objectives, derived from the synthesis stage, can be effectively incorporated within this procedure.

The MSD methodology shown in Fig. 1 can be split into three groups acting sequentially:

1. Optimization and stabilization: in this stage, the operating point can be selected by optimizing the original process conditions (if necessary) considering, for instance, operational costs minimization. Simultaneously, it defines active constraints and the minimal number of loops to guarantee process stability.
2. Optimal sensor location (OSL): according to the availability of the degrees of freedom from the previous stage, the OSL is defined using an evaluation index of the sum of square deviations (SSD). The combinatorial problem is solved with genetic algorithms (GA), as a mixed-integer programming tool. In this manner, the controlled variables of the plant are defined together with the sensor location.

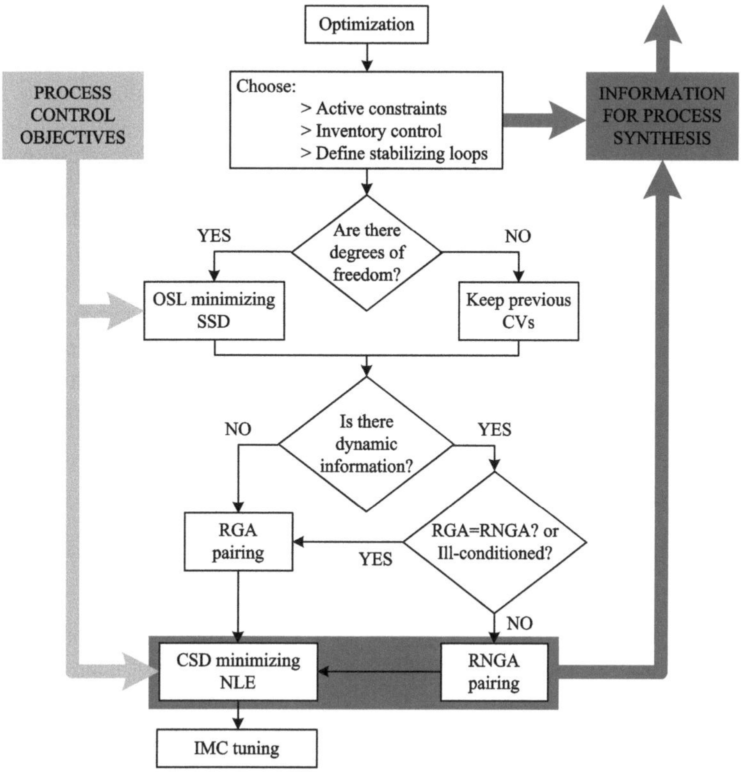

Fig. 1 Minimum square deviation (MSD) procedure for plant-wide control

3. Control structure design: the control structure based on the controlled variables defined previously. In this case the approach is different if dynamic information is available. In the affirmative case, the structure is preferred to be defined by the relative normalized gain array (RNGA) [23]. In the negative case or when the structure suggested by the RGA is the same as suggested by the RNGA, the CSD is defined using a new index called net load evaluation (NLE). This procedure presents again a combinatorial problem that can be effectively solved by means of GA.

An extended discussion about these new concepts and the way they interact within the generalized proposal is presented next.

2.1 Optimization, Stabilization and Identification

The main objective of all industrial processes is to obtain the most possible benefits. Generally, process operation must be optimized accounting for economic objectives. This procedure allows to obtain important information regarding which variables should be controlled in this context. As a first recommendation it is suggested to control those variables found in their active constraints (equalities or inequalities). They must remain at that point in order to guarantee the optimization results.

In some cases, industrial processes could present an unstable behavior in open-loop operation. Then, it is necessary to define the minimal number of control loops needed for stabilizing the plant. These loops include, for instance, the level control in tank, given the pure integrative nature of its behavior. In the proposed strategy it is suggested to define these loops as a starting point. In the literature there are several procedures to do this, as an example can be mentioned the works of [24] and [25].

Once the process is stabilized, an extra stage consisting of system identification, using different strategies, can be done in order to obtain dynamic information about the plant. This reduced dynamic model can be very useful for supporting the controllers' tuning stage, the uncertainty analysis and the control structure selection via RNGA.

2.2 Optimal Sensor Location: Controlled Variables Selection

After finishing the previous stage it must be considered whether additional variables can be controlled by those manipulated variables that have not been used yet. It means that there is the same number of degrees of freedom in the system. In this case, the problem is to decide which variables must be selected to achieve a specific goal. This could become a really complicated and tedious problem especially when working with large industrial plants.

In order to solve this problem in this section the well-known control structure, named Internal Model Control (IMC), presented in Fig. 2, is useful for giving support for the calculations. Here $\mathbf{G}(s) = [\mathbf{G}_s(s)^T, \mathbf{G}_r(s)^T]^T$ represents the transfer function matrix (TFM) of the process with n inputs and m outputs, being $m > n$, then

$$\mathbf{y}(s) = \mathbf{G}(s)\mathbf{u}(s) + \mathbf{D}(s)\mathbf{d}_*(s) \tag{1}$$

where $\mathbf{D}(s) = [\mathbf{D}_s(s)^T, \mathbf{D}_r(s)^T]^T$ is the TFM corresponding to the effect of output disturbances with dimension $m \times p$. Respectively, $\mathbf{y}(s)$, $\mathbf{u}(s)$ and $\mathbf{d}_*(s)$ are output, input and disturbance vectors with dimension $m \times 1$, $n \times 1$ and $p \times 1$ respectively.

The process can be divided into two parts: one square, $\mathbf{G}_s(s)$, that includes the n output variables to be controlled and the other generally not square, $\mathbf{G}_r(s)$ that contains the m-n uncontrolled variables. In this way, the process can be rewritten as

$$\mathbf{y}_s(s) = \mathbf{G}_s(s)\mathbf{u}(s) + \mathbf{D}_s(s)\mathbf{d}_*(s) \tag{2}$$

$$\mathbf{y}_r(s) = \mathbf{G}_r(s)\mathbf{u}(s) + \mathbf{D}_r(s)\mathbf{d}_*(s), \tag{3}$$

where $\mathbf{G}_s(s)$, $\mathbf{D}_s(s)$, $\mathbf{G}_r(s)$ and $\mathbf{D}_r(s)$ are the TFM of $n \times n$, $n \times p$, $(m-n) \times n$ and $(m-n) \times p$ respectively. Being $\widetilde{\mathbf{G}}_s(s)$ an approximated model of $\mathbf{G}_s(s)$.

Fig. 2 Generalized IMC structure

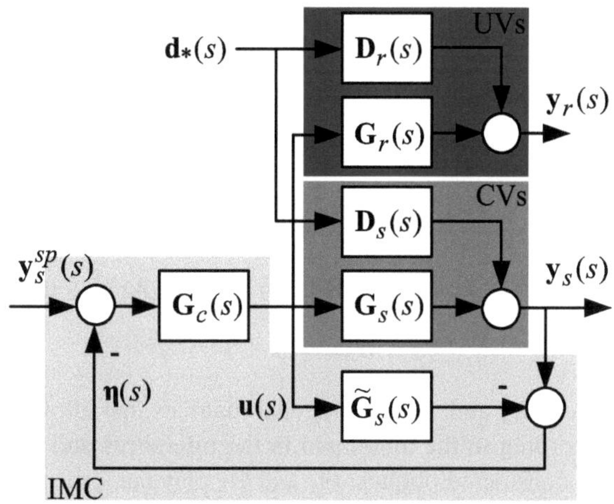

Once the process is stabilized, a steady-state model can be obtained or, in some cases, a simplified dynamic model. Generally, as a result of the process synthesis stage a good steady-state model is achievable. Using this information, a novel procedure for the selection of the CVs is presented here, also called OSL with control structure information. The main features of this methodology is that it tries to avoid the use of heuristics or expert knowledge, so it can be applied independently of the specific process. Second, the OSL is performed only with steady-state information. And third, the partial control objectives are introduced directly through the weighting matrices.

This methodology is based on assuming perfect control, in the least square sense, considering that n variables are in their desired operating point. Hence, the sum of the square deviations (SSD) of the m-n uncontrolled variables (UVs) can be minimized. This situation generates a steady-state index for the selection of the CVs. The chosen variables in this stage guarantee that the plant will remain as close as possible to its nominal operating point when it is affected by set point modifications or disturbances.

The process outputs in steady state, ($s = 0$), can be stated as

$$\mathbf{y}_s = \mathbf{y}_s^{sp} \tag{4}$$

$$\mathbf{y}_r = \mathbf{G}_r\mathbf{G}_s^{-1}\mathbf{y}_s^{sp} - \left(\mathbf{G}_r\mathbf{G}_s^{-1}\mathbf{D}_s - \mathbf{D}_r\right)\mathbf{d}_* \tag{5}$$

when IMC theory is considered, $\mathbf{G}_c = \widetilde{\mathbf{G}}_s^{-1}$. So, Eq. 5 can be rewritten as

$$\mathbf{y}_r = \mathbf{S}_{sp}\mathbf{y}_s^{set} + \mathbf{S}_d\mathbf{d}_* \tag{6}$$

where

$$\mathbf{S}_{sp} = \left[\mathbf{G}_r\mathbf{G}_s^{-1}\right], \quad \text{and} \quad \mathbf{S}_d = \left[\mathbf{D}_r - \mathbf{G}_r\mathbf{G}_s^{-1}\mathbf{D}_s\right] \tag{7}$$

As can be seen from Eq. 6, the deviation of UVs about their nominal operating point depend specifically on the CVs selection. Therefore, the idea is to choose those

CVs that minimize the next criterion or index

$$\begin{aligned} SSD_{yr} &= \sum_{i=1}^{n} \|\mathbf{e}_{set}(i)\|_2^2 + \sum_{j=1}^{p} \|\mathbf{e}_d(j)\|_2^2 \\ &= \sum_{i=1}^{n} \|\Lambda_2 \mathbf{S}_{sp} \Lambda_1 \mathbf{y}_{set}^n(i)\|_2^2 + \sum_{j=1}^{p} \|\Theta_2 \mathbf{S}_d \Theta_1 \mathbf{d}_*^p(j)\|_2^2 \\ &= \mathrm{tr}\left[\left(\Lambda_2 \mathbf{S}_{sp} \Lambda_1\right)^T \left(\Lambda_2 \mathbf{S}_{sp} \Lambda_1\right)\right] + \mathrm{tr}\left[\left(\Theta_2 \mathbf{S}_d \Theta_1\right)^T \left(\Theta_2 \mathbf{S}_d \Theta_1\right)\right] \\ &= \|\Lambda_2 \mathbf{S}_{sp} \Lambda_1\|_F^2 + \|\Theta_2 \mathbf{S}_d \Theta_1\|_F^2 \end{aligned} \tag{8}$$

where Λ_1 and Λ_2 are diagonal matrices that allow to incorporate relative weights according to the amplitude in the references and the relative importance between the considered outputs. Θ_1 and Θ_2 play the same role as Λ_1 and Λ_2 but for disturbances. Therefore, the original control objectives of the process can be included in the analysis in a simple and effective way. Where $\mathbf{e}_{set}$ and $\mathbf{e}_d$ refer to the steady-state errors occurring for the outputs $\mathbf{y}_r$ when set point and disturbance changes happen respectively. The vectors $\mathbf{y}_{set}^n(i)$ and $\mathbf{d}_*^p(j)$ of dimension $n \times 1$ and $p \times 1$ have an unitary entry in the positions i and j respectively and zero elsewhere. tr $[\cdot]$ means the trace of the matrix and $\|\cdot\|_F^2$ represents the square Frobenius norm.

Considering m output variables and n potential CVs (measurements) the problem is to select the set of n variables from m such that the criterion in (8) is minimized. This kind of problem requires a great number of combinations for finding the best solution. For example, in this particular case the total amount of combinations is $m!/(n!(m-n)!)$. For instance, in a medium-scale process with $m=50$ and $n=12$ the combinatorial dimension results as 1.2140×10^{11}. In this case an exhaustive search solution is unpractical. Currently, several proposed techniques for solving plant-wide control try to reduce the problem dimension via heuristics and engineering knowledge. This procedure can give, in some occasions, acceptable solutions. However, normally they are suboptimal and sometimes could be unfeasible ones. To avoid this result and generalize the treatment of this combinatorial problems, in this chapter is proposed a procedure based on stochastic global search. Particularly, mixed-integer optimization based on genetic algorithms (GA). In the Appendix, a brief description about this type of searching tool is presented.

Basically, the original problem can be parameterized to a particular CVs selection by using the methodology proposed in GA. In other words, a particular individual or chromosome $\mathbf{C}_i = [c_1, \ldots, c_m]$, which belongs to a binary alphabet, parameterizes all the potential measurement points (sensors locations). Thus, $c_j = 1$ indicates that a sensor in the location j is needed and $c_j = 0$ the opposite situation. In this context, the combinatorial problem can be stated as shown in Eq. 9, with $\mathbf{C}_i$ as the searching argument.

$$\min_{\mathbf{C}_i} SSE_{yr}(\mathbf{C}_i) = \min_{\mathbf{C}_i} \left[\|\Lambda_2 \mathbf{S}_{sp}(\mathbf{C}_i)\Lambda_1\|_F^2 + \|\Theta_2 \mathbf{S}_d(\mathbf{C}_i)\Theta_1\|_F^2\right] \tag{9}$$

subject to

$$\det\left(\mathbf{G}_s(\mathbf{C}_i)\right) \neq 0 \tag{10}$$

the constraint in (10) guarantees that the optimal solution for (9) is also a feasible one. Remembering (7) and Fig. 2 it can be observed that the optimal solution $\mathbf{C}_i$, parameterizes $\mathbf{G}_s$ and this particular selection of the CVs eventually affects the feasibility of the future IMC control design.

From the optimization stated in (9) the following properties and conclusions can be obtained,

1. The $SSD(\mathbf{y}_r)$ minimization decreases the steady-state effect on $\mathbf{y}_r$ (UVs) produced by set points and disturbance modifications. Specifically, it reduces the multivariable gains of $\mathbf{S}_{sp}$ and $\mathbf{S}_d$.
2. The properties of the matrix $\mathbf{G}_s$ are improved by increasing its minimum singular value, $\underline{\sigma}(\mathbf{G}_s)$.

   The $\underline{\sigma}(\mathbf{G}_s)$ increase produces that $\mathbf{G}_s$ be away from singularity, so its inverse $\mathbf{G}_s^{-1}$ existence is guaranteed.

   The condition number of $\mathbf{G}_s$, $\gamma(\mathbf{G}_s) = \overline{\sigma}(\mathbf{G}_s)/\underline{\sigma}(\mathbf{G}_s)$, decreases and tends to the minimum. This occurs because the maximum singular value $\overline{\sigma}(\mathbf{G}_s)$ is bounded.

   This searching methodology drives to obtain a good process matrix, $\mathbf{G}_s$ which guarantees that the elements in $RGA = \mathbf{G}_s \otimes (\mathbf{G}_s^{-1})^T$ do not have extremely high values.

3. Processes with high values in the RGA elements are very difficult or impossible to control [26, 27], in addition these kinds of processes are very sensitive to both modeling errors and multiplicative input uncertainties.
4. The maximization of $\underline{\sigma}(\mathbf{G}_s)$ does not necessarily lead to a gain minimization of $\mathbf{S}_{sp}$ and $\mathbf{S}_d$.
5. The minimization of $||\mathbf{S}_{sp}||_F^2$ produces that $\underline{\sigma}(\mathbf{G}_s)$ increases and tends to the maximum allowed value based on the characteristics of the process, $\underline{\sigma}(\mathbf{G}_s)_{\text{max}}$. Eventually it may happen that $\underline{\sigma}(\mathbf{G}_s) = \underline{\sigma}(\mathbf{G}_s)_{\text{max}}$.
6. The minimization of $||\mathbf{S}_d||_F^2$ produces that $\underline{\sigma}(\mathbf{G}_s)$ increases, but in this case the maximum reachable value is bounded by the disturbance model characteristics, $\underline{\sigma}\left(\mathbf{G}_s\right) \leq \overline{\sigma}\left(\mathbf{G}_r\right)\overline{\sigma}\left(\mathbf{D}_s\right)/\overline{\sigma}\left(\mathbf{D}_r\right)$.

A complete mathematical analysis of these properties can be found in the Matrix Properties Section of the Appendix.

2.3 Control Structure Design

Once the previous stage is finalized, there are no more degrees of freedom as shown in Fig. 1. In other words, the n CVs were selected from the set of m potential measurement points. This selection was performed via optimization by minimizing the

SSE_{yr} index. When the stabilization stage uses all the process degrees of freedom the optimization presented in the Sect. 2.2 is avoided. In both cases the new problem is to decide the input–output pairing, $n \times n$, and the controller structure. The latter can be decentralized (diagonal), full or sparse

In this context, and remembering Fig. 1, different ways are proposed. If there is only steady-state information about the process, then the initial input–output pairing is proposed via RGA. On the other hand, if there is some dynamic process information (although approximately) as time constants and dead times,then the preferred pairing tool is the RNGA. In the Appendix, a brief description is given about these tools. Anyway, both strategies can suggest the same pairing solution or eventually the normalization used in the RNGA approach can produce ill-conditioned matrices. In these cases, the RGA pairing solution is adopted.

The approach for obtaining the control structure is based on the steady-state index called net load evaluation (NLE) which allows to quantify the benefits of using a particular control structure. In this sense, the approach applied here allows to improve the dynamic behavior of the closed-loop plant response [17–19, 21, 22]. The net load, $\mathbf{y}_s^{net}(s)$, arises from Fig. 2 by considering plant-model mismatch $\mathbf{G}_s(s) \neq \widetilde{\mathbf{G}}_s(s)$ and both set point and disturbance changes. This net load, basically, introduces unwanted transient effects on the CVs. The previous works [15, 28] analyzed this effect via relative disturbance gain array (RDGA). A brief description of the RDGA methodology is included in the Appendix.

Considering Fig. 2, the IMC theory and plant-model mismatch of the process CVs can be written as displayed in (11), independently of the proposed IMC structure.

$$\mathbf{y}_s(s) = \widetilde{\mathbf{G}}_s(s)\mathbf{G}_c(s)\mathbf{y}_s^{sp}(s) + \left(\mathbf{I} - \widetilde{\mathbf{G}}_s(s)\mathbf{G}_c(s)\right)\mathbf{y}_s^{net}(s), \tag{11}$$

where

$$\mathbf{y}_s^{net}(s) = \mathbf{A}(s)\mathbf{y}_s^{sp}(s) + \mathbf{B}(s)\mathbf{d}_*(s) \tag{12}$$

$$\mathbf{A}(s) = \left[\mathbf{I} + \left(\mathbf{G}_s(s) - \widetilde{\mathbf{G}}_s(s)\right)\mathbf{G}_c(s)\right]^{-1}\left(\mathbf{G}_s(s) - \widetilde{\mathbf{G}}_s(s)\right)\mathbf{G}_c(s) \tag{13}$$

$$\mathbf{B}(s) = \left[\mathbf{I} + \left(\mathbf{G}_s(s) - \widetilde{\mathbf{G}}_s(s)\right)\mathbf{G}_c(s)\right]^{-1}\mathbf{D}_s(s), \tag{14}$$

being $\mathbf{y}_s^{net}(s)$ the net load which affects the CVs during the transient response. In the context of the IMC structure $\mathbf{G}_c(s) = \widetilde{\mathbf{G}}_s^{-1}(s)\mathbf{F}(s)$, where $\mathbf{F}(s)$ is the low-pass filter matrix with $\mathbf{F}(0) = \mathbf{I}$, the Eq. 11 can be written as,

$$\mathbf{y}_s(s) = \mathbf{F}(s)\mathbf{y}_s^{sp}(s) + (\mathbf{I} - \mathbf{F}(s))\,\mathbf{y}_s^{net}(s) \tag{15}$$

and from Eqs. 12–15 the following conclusions can be obtained:

1. The steady-state gain of $\mathbf{y}_s^{net}(s)$ is not necessarily null.
2. The effect of $\mathbf{y}_s^{net}(s)$ on the CVs in steady state is null due to the integrator behavior of $\left(\mathbf{I} - \widetilde{\mathbf{G}}_s(s)\mathbf{G}_c(s)\right)$.
3. At steady state the unwanted effects due to set points and disturbances are perfectly rejected, it is an IMC property.

4. Basically, $\mathbf{y}_s^{net}(s)$ affects the CVs in a transitory way. The severity of this effect depends on both the multivariable steady-state gain of $\mathbf{y}_s^{net}(s)$ and the time constant of the integrator behavior. This last tuning parameter is set by the user, so as to determine the closed loop time response velocity.

Note that it depends on the control requirements such as stability and the process characteristics. This topic will be addressed opportunely at the controller tuning section. Working in steady-state and accounting that $\mathbf{G}_c = \widetilde{\mathbf{G}}_s^{-1}$, the Eqs. 13 and 14 can be expressed as,

$$\mathbf{A} = \mathbf{I} - \widetilde{\mathbf{G}}_s \mathbf{G}_s^{-1} \tag{16}$$

$$\mathbf{B} = \widetilde{\mathbf{G}}_s \mathbf{G}_s^{-1} \mathbf{D}_s \tag{17}$$

from these equations can be observed that when a perfect model is available, $\widetilde{\mathbf{G}}_s = \mathbf{G}_s$,the set point changes do not affect the CVs, $\mathbf{A} = \mathbf{0}$. On the other hand, from disturbances point of view, there is no attenuation in this case, being $\mathbf{B} = \mathbf{D}_s$.

The basic idea is that, by accounting both the process control objectives and a suitable model parametrization, a trade-off solution between servo and regulator controller behaviors can be obtained. In other words, it can be achieved through the net load effect minimization, $\mathbf{y}_s^{net}$. Parameterizing the model as,

$$\widetilde{\mathbf{G}}_{s\Gamma} = \mathbf{G}_s \otimes \Gamma, \quad \text{con} \quad \Gamma = \begin{bmatrix} \gamma_{11} & \cdots & \gamma_{1n} \\ \vdots & \ddots & \vdots \\ \gamma_{n1} & \cdots & \gamma_{nn} \end{bmatrix} \tag{18}$$

where $\otimes$ is the element-by-element product and Γ is a binary matrix with dimension $n \times n$, allows to select a specific model structure that minimizes the net load evaluation, NLE(Γ), in a sum of square deviation (SSD) sense,

$$\mathrm{NLE}(\Gamma) = \sum_{i=1}^{n} \left\| \Delta_1 \mathbf{A}_\Gamma \Delta_2 \mathbf{y}_{set}^n(i) \right\|_2^2 + \sum_{j=1}^{p} \left\| \Xi_1 \mathbf{B}_\Gamma \Xi_2 \mathbf{d}_*^p(j) \right\|_2^2 \tag{19}$$

by analogy with the Eq. 8 can be expressed as

$$\begin{aligned} \mathrm{NLE}(\Gamma) &= \mathrm{tr}\left[(\Delta_1 \mathbf{A}_\Gamma \Delta_2)^T (\Delta_1 \mathbf{A}_\Gamma \Delta_2) \right] + \mathrm{tr}\left[(\Xi_1 \mathbf{B}_\Gamma \Xi_2)^T (\Xi_1 \mathbf{B}_\Gamma \Xi_2) \right] \\ &= \|\Delta_1 \mathbf{A}_\Gamma \Delta_2\|_F^2 + \|\Xi_1 \mathbf{B}_\Gamma \Xi_2\|_F^2 \end{aligned} \tag{20}$$

where Δ_1, Δ_2, Ξ_1 and Ξ_2 are matrices that allow to incorporate the relative importance of both, set points and disturbance changes and, on the other hand, sort the outputs according to the plant objectives. Thus, the new combinatorial problem can be stated as

$$\min_\Gamma \mathrm{NLE}(\Gamma) = \min_\Gamma \left[\|\Delta_1 \mathbf{A}_\Gamma \Delta_2\|_F^2 + \|\Xi_1 \mathbf{B}_\Gamma \Xi_2\|_F^2 \right]. \tag{21}$$

subject to

$$\mathbf{A}_{\Gamma} = \mathbf{I} - \widetilde{\mathbf{G}}_{s\Gamma}\mathbf{G}_s^{-1} \tag{22}$$

$$\mathbf{B}_{\Gamma} = \widetilde{\mathbf{G}}_{s\Gamma}\mathbf{G}_s^{-1}\mathbf{D}_s \tag{23}$$

$$\mathrm{Re}\left[\lambda_i\left(\mathbf{G}_s\widetilde{\mathbf{G}}_{s\Gamma}^{-1}\right)\right] > 0, i = 1, \ldots, n \tag{24}$$

The conditions stated in (22) and (23) allow to compute both $\mathbf{A}_{\Gamma}$ and $\mathbf{B}_{\Gamma}$ matrices from a particular model parametrization $\mathbf{G}_{s\Gamma}$. The constraint in (24) allows to check the stability condition in steady-state for the current control structure, in the context of IMC [29]. Where $\lambda_i\left(\mathbf{G}_s\widetilde{\mathbf{G}}_{s\Gamma}^{-1}\right)$ is the i eigenvalue of the matrix $\mathbf{G}_s\widetilde{\mathbf{G}}_{s\Gamma}^{-1}$ and Re $[\cdot]$ the real part function. There are several stability and robustness tests in the literature, either for steady-state [26, 30] and dynamic [31, 32] analysis. Note that the final control structure depends on the selected model parametrization, $\mathbf{G}_c(s) = \widetilde{\mathbf{G}}_{s\Gamma}^{-1}(s)\mathbf{F}(s)$.

The optimization problem stated in (21)–(24) is again a combinatorial one and can be solved suitably via GA approach. In this case the number of combinations is $2^{(n\times n)}$ which can grow quickly. Proposing the following chromosome structure

$$\mathbf{C}_{\mathrm{NLE}} = \left[\gamma_{11}\gamma_{12}\cdots\gamma_{1n}\cdots\gamma_{nn}\right] \tag{25}$$

the GA can be used to solve this combinatorial problem (see Appendix).

Furthermore, by considering the base control structure (the decentralized/diagonal) this methodology allows to define the additional interaction loops to be added for net load minimization. In other words, to improve the dynamic response of the process from a diagonal control structure, the following parametrization is proposed

$$\Gamma = \begin{bmatrix} 1 & \cdots & \gamma_{1n} \\ \vdots & \ddots & \vdots \\ \gamma_{n1} & \cdots & 1 \end{bmatrix} \tag{26}$$

$$\mathbf{C}_{\mathrm{NLE}} = \left[\gamma_{11}, \ldots, \gamma_{ij}, \ldots\right], \quad \text{con} \quad i \neq j;\ i, j \leq n; \tag{27}$$

for a decentralized base control structure reorganized in diagonal form. This approach transforms the combinatorial problem dimension into $2^{(n\times(n-1))}$, introducing a considerable reduction compared with (18).

Note that Γ defines which part of the model is taken, $\widetilde{\mathbf{G}}_{s\Gamma}(s)$, according to it the controller is given by $\mathbf{G}_c(s) = \widetilde{\mathbf{G}}_{s\Gamma}^{-1}(s)\mathbf{F}(s)$, taking into account the procedure of IMC design theory. Hence, the controller $\mathbf{G}_c(s)$ design depends on the particular structure of Γ that could be diagonal, block diagonal, triangular or full.

From the combinatorial problem stated in (21) to (24) the following conclusions and properties can be obtained:

1. The minimization of NLE(Γ) reduces the multivariable effects of $\mathbf{y}_s^{net}(s)$ on the CVs (in the transient) produced by set point and disturbance changes.

2. Through the weighting matrices in Eq. 21 the optimal solution is driven by the controller designer to handle the trade-off between servo and regulator behavior accounting the control objectives.

   A full control structure ensures a complete decoupling with respect to set point changes, $\mathbf{A}_\Gamma = 0$, but there is no attenuation for disturbances, $\mathbf{B}_\Gamma = \mathbf{D}_s$.
   A particular control structure $\widetilde{\mathbf{G}}_{s\Gamma}$ could reduce the effect of $\mathbf{B}_\Gamma$, but in this case it will be $\mathbf{A}_\Gamma \neq 0$, necessarily.

3. The minimization of NLE(Γ) does not guarantee the stability of the future control structure, except in the case of $\mathbf{B}_\Gamma = 0$.

   In this context, it is necessary to add the stability criterion of (24) [29].

4. For any plant-wide control approach (heuristic or generalized) it is very useful to have a performance and stability analysis to define the feasibility of the proposed control structure [32]. A typical tool in this case is the structured singular values (SSV), also known as μ-analysis.

2.4 Controller Tuning

According to the IMC design procedure, the TFM of the process can be factorized as a product of two models, $\widetilde{\mathbf{G}}_{s\Gamma}(s) = \widetilde{\mathbf{G}}^-_{s\Gamma}(s)\widetilde{\mathbf{G}}^+_{s\Gamma}(s)$, an invertible and a non-invertible part respectively, and the controller is given by

$$\mathbf{G}_c(s) = \widetilde{\mathbf{G}}^-_{s\Gamma}(s)^{-1}\mathbf{F}(s) \tag{28}$$

where $\mathbf{F}(s) = \text{diag}\left(\left[1/(\tau_{f1}s+1)^{\kappa_1} \dots 1/(\tau_{fn}s+1)^{\kappa_n}\right]\right)$ is the TFM of the low-pass filter, which makes $\mathbf{G}_c(s)$ to be proper (feasible). In case the non-invertible part corresponds to delays,it can be considered that $\widetilde{\mathbf{G}}^+_{s\Gamma}(s) = \text{diag}\left(\left[e^{-\theta_1 s} \dots e^{-\theta_n s}\right]\right)$ is a simplifying assumption.

When the process model has no non-invertible part, $\widetilde{\mathbf{G}}^+_{s\Gamma}(s) = \mathbf{I}$, the IMC approach of Fig. 2 can be converted into a unitary feedback control policy by

$$\mathbf{G}^*_c(s) = \widetilde{\mathbf{G}}^-_{s\Gamma}(s)^{-1}\mathbf{F}(s)\left[\mathbf{I} - \mathbf{F}(s)\right]^{-1} \tag{29}$$

note that the parametrization structure Γ defines the process model selection, $\widetilde{\mathbf{G}}_{s\Gamma}(s)$ and eventually its invertible part, $\widetilde{\mathbf{G}}^-_{s\Gamma}(s)$.

The tuning procedure of this generalized control structure can be made by following the robustness criteria defined in [33, 34] for the SISO case. The time constants of the filters, $\tau_{f1}, \dots, \tau_{fn}$, affect the input channels 1 to n respectively. In case of the existence of time delays in [33, 34] it is suggested that the filter time constant must be greater than the dead time in the control loop, $\tau_{fi} > \theta_i$, when a process model without delay information is used for design. If there is no delay in the control loop the criterion is reduced to $0 < \tau_{fi} < \tau_i^m$, so as to improve the closed-loop time responses. Where τ_i^m is the time constant of the process model for the loop i. Generally, this approach produces good results for decentralized MIMO control structures. In these cases the following tuning criteria can also be applied.

1. Tuning for diagonal controller matrix: the filter time constants are adjusted considering each decentralized loop, $\tau_{fi} = \theta_{ii} + \varphi_i$ or $\tau_{fi} = \tau_{ii}^m/2$ when there is no delay.
2. Tuning for full controller matrix: the filter time constants are adjusted with robustness criteria considering the maximum delay in each input channel, $\tau_{fi} = \max(\theta_{ji}) + \varphi_i$, with $j = 1, \ldots, n$.

   where $\varphi_i > 0$ is a scalar parameter to fulfill the robustness condition $\tau_{fi} > \theta_i$.

3 Plant-wide Control Structure for BPS with PEMFC

In this section the MSD strategy for plant-wide control design is applied to the integrated fuel processor system (BPS) with PEM fuel cell (PEMFC). A complete description about modeling, synthesis, steady-state analysis and working points are given in Chaps. 9 and 10.

The main objectives of the BPS control are to maintain H_2 levels on the anode of the FC, because the starvation can cause permanent damage and overfeeding will lead to hydrogen waste; the CO levels of the anode inlet stream must be low and keeping the temperatures of the reactors set and FC at their operational points prevent damages and finally maintain the system efficiency. The complete set of available variables in the BPS+FC process is given in Table 9 of Chap. 10

According to the algorithm schematically given in Fig. 1 the first step is to achieve the plant stabilization. In this case, the primary control loops needed to stabilize the process are the pressure control of each reactor (ESR, HTS, LTS, CO-PrOx) by manipulating their corresponding exit flow, and the H_2 production to feed the anode of the fuel cell stack, determined by the current demand, with the fresh ethanol flow entering the bio-ethanol steam reforming (ESR). The pressures are assumed under perfect control and the H_2 control loop was adjusted via IMC theory [33, 34]. These five stabilizing control loops are indicated in the Table 9 of Chaps. 10 with (*).

Once the process is stabilized and following the steps of the MSD strategy, the simplified model must be obtained by using system identification (SI) techniques. This procedure allows to obtain both a steady-state and dynamic linearized models of the process for control design purposes. From Table 9 of Chaps. 10, a model with fourteen outputs (y_1 to y_{14}), six potential manipulated inputs (u_1 to u_6) and two disturbances (d_1 and d_2) was identified.

Any SI technique is useful to carry out the linearization. In this particular case the experiments based on subspace state-space SI methods [35] were used and implemented in the Matlab environment [36, 37]. Hence, the specific inputs of the process are excited with random steps, whose amplitudes vary between ±5% (wide-range of linearization) from their nominal values and collect the output data periodically with a suitable sample time, here $T_s = 0.05$ s is adopted. This allows to collect about 100 samples per step change since a period of 5 seconds (the steady-state is reached) is stated for the BPS with PEMFC.

Basically, all the subspace state-space System Identification methods are based on system theory, numerical linear algebra and projections tools. A classical algorithm is the well-known "n4sid" method developed by [35] and implemented in Matlab by [37]. This strategy allows to estimate the system matrices of the space-state linear model, $\hat{\mathbf{A}}$, $\hat{\mathbf{B}}$, $\hat{\mathbf{C}}$, $\hat{\mathbf{D}}$ and the model order using singular value decomposition from the impulse response Hankel matrix. It is obtained from the data by solving a linear least squares problem.

The model structure and the matrix dimension can be observed in (30) and (31). The recorded data base is pre-processed by normalization to zero mean and unit variance for all variables before starting the SI procedure.

$$\begin{aligned} \mathbf{x}(k+1) &= \hat{\mathbf{A}}\mathbf{x}(k) + \hat{\mathbf{B}}\mathbf{u}_t(k) \\ \mathbf{y}(k) &= \hat{\mathbf{C}}\mathbf{x}(k) + \hat{\mathbf{D}}\mathbf{u}_t(k) \end{aligned} \tag{30}$$

where $\mathbf{u}_t(k) = [\mathbf{u}, \mathbf{d}]^T$ is the model input vector containing the manipulated input vector $\mathbf{u} = [u_1, u_2, u_3, u_4, u_5, u_6]^T$ and the disturbance vector $\mathbf{d} = [d_1, d_2]^T$. With

$$\hat{\mathbf{A}} : [15 \times 15];\ \hat{\mathbf{B}} : [15 \times 8];\ \hat{\mathbf{C}} : [14 \times 15];\ \hat{\mathbf{D}} : [14 \times 8], \tag{31}$$

where $\mathbf{x}$ is the state vector of dimension $[15 \times 1]$, $\mathbf{y}$ is the output vector of dimension $[14 \times 1]$ and $\mathbf{u}_t$ is the input vector of dimension $[8 \times 1]$.

The data base obtained from the SI experiment was divided into two groups. One called the estimation data to develop the state-space linear model shown in Eq. 34, and the second one called the validation data to test the accuracy of the model predictions. In Fig. 3 can be observed the validation procedure for some output variables (y_1, y_9 and y_{14}). The order of the model was chosen according to a trade-off between the model size (amount of state) and the mean square prediction error. A very good approximation (either for stationary states and transients) is achieved with the reduced order model.

Following the MSD procedure of Fig. 1, is necessary to define the optimal sensor location (OSL) since some degrees of freedom are available. In this context, 6 controlled variables (CVs) should be selected from 14 available measurements via the SSD approach defined in Sect. 2.2. As it was discussed opportunely, the OSL is a combinatorial problem since it must be chosen among the total amount of $14!/(6!(14-6)!) = 3003$ combinations. In this case an exhaustive search can be implemented without computational problems. When the dimensionality grows some optimization/search tool is required, i.e. genetic algorithms

Solving the problem stated in Eqs. 9 and 10 exhaustively with weight matrices $\Lambda_1 = \mathbf{I}_6$, $\Lambda_2 = \mathbf{I}_8$, $\Theta_1 = \mathbf{I}_2$ and $\Theta_2 = \mathbf{I}_8$, and the steady-state model shown in Table 1 the best sensor location $\mathbf{C}_i$ can be found. In the Table 2 are displayed the best first five solutions, $\mathbf{C}_1 - \mathbf{C}_5$, sorted by the SSD index. Note that $\mathbf{I}_i$ represents the identity matrix with dimension $i \times i$, and the parametrization $\mathbf{C}_i = [c_1, \ldots, c_m]$,

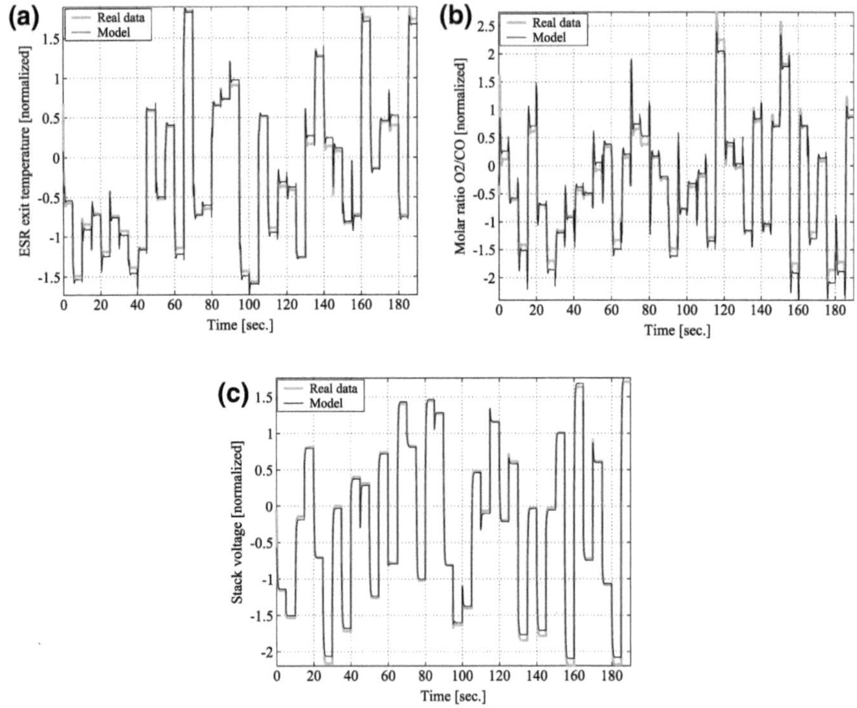

Fig. 3 BPS with FC model validation. **a** y1-ESR exit temperature, **b** y9-Molar ratio O_2/CO, **c** y14-Stack voltage

which belongs to a binary alphabet, represents all the potential measurement points (sensor locations). Thus, $c_j = 1$ indicates that a sensor in the location j is needed and $c_j = 0$ the opposite situation, with $j = 1, \ldots, 14$. Hence, the best solution in Table 2 called $\mathbf{C}_1$ suggests that the controlled variables should be as follows,

y_1 : ESR exit temperature
y_3 : Burner exit temperature
y_8 : CO-PrOx exit temperature
y_9 : CO-PrOx molar ratio O_2/CO
y_{10} : Burner exit molar flow
y_{13} : Oxygen excess in the FC

In this context, a first attempt to propose a decentralized plant-wide control policy is the popular and effective RGA approach. As discussed in the Appendix, the RGA (and its variants) allows to define the input–output pairing by using steady-state tools. In fact, accounting for the OSL proposed by $\mathbf{C}_1$ in Table 2 and the steady-state gains of the process from Table 1 the square process, $\mathbf{G}_s$, can be obtained. Thus, the RGA can be computed easily by $\Lambda = \mathbf{G}_s \otimes \left(\mathbf{G}_s^{-1}\right)^T$ and is shown in Table 3. In this table is highlighted a suitable decentralized input–output pairing with the following

Table 1 Normalized steady-state gains for BPS+FC process

	G						D	
	u_1	u_2	u_3	u_4	u_5	u_6	d_1	d_2
y_1	−0.0012	0.9250	0.0157	0.1658	0.0030	−0.0189	−0.0021	−0.2811
y_2	0.1397	−0.5491	0.0615	0.6641	−0.0040	−0.0015	−0.0036	−0.7037
y_3	−0.2998	−0.2600	0.5627	−0.6554	0.0071	−0.0046	0.3538	0.4096
y_4	0.5028	−0.0852	0.0020	−0.0167	0.0541	−0.0060	−0.0724	0.7594
y_5	0.6500	0.1728	0.0080	0.0236	−0.0058	−0.0040	0.2533	−0.7062
y_6	−0.6530	0.6671	0.0046	0.1194	−0.0009	0.0087	0.0966	−0.1923
y_7	−0.7799	0.0793	0.0004	0.0169	0.0056	0.0070	0.1078	0.8877
y_8	−0.9448	−0.0536	−0.1032	0.1095	0.3593	0.1900	0.3286	−0.0847
y_9	0.7183	0.0184	0.0024	−0.0139	0.1876	−0.0125	−0.1250	−0.9165
y_{10}	0.2080	−0.0347	0.0888	0.9650	0.0023	−0.0013	−0.0048	0.0400
y_{11}	−0.2903	0.1212	0.0856	−0.0709	−0.0680	−0.1415	−0.1304	0.6714
y_{12}	0.0115	0.0167	0.0045	−0.0018	0.0066	−0.0609	0.0087	0.8192
y_{13}	−0.0027	0.0005	0.0001	0.0013	0.0020	0.7526	−0.0010	−0.5717
y_{14}	0.0017	0.0062	0.0017	−0.0026	0.0003	0.4741	0.0031	−0.8188

Table 2 Exhaustive search results

	Sensor locations														SSD
	y_1	y_2	y_3	y_4	y_5	y_6	y_7	y_8	y_9	y_{10}	y_{11}	y_{12}	y_{13}	y_{14}	
C_1	1	0	1	0	0	0	0	1	1	1	0	0	1	0	7.56
C_2	1	1	1	0	0	0	0	1	1	0	0	0	1	0	9.31
C_3	1	0	1	0	0	0	0	1	1	1	0	0	0	1	9.99
C_4	0	0	1	0	0	1	0	1	1	1	0	0	1	0	10.27
C_5	1	1	1	0	0	0	0	1	1	0	0	0	0	1	11.78

Table 3 RGA for the BPS+FC process

	u_1	u_2	u_3	u_4	u_5	u_6
y_1	0.0001	**0.9908**	0.0074	0.0014	0.0004	0.0000
y_3	0.0242	0.0012	**0.8882**	0.0844	0.0020	0.0000
y_8	0.3953	0.0011	0.0143	0.0108	**0.5796**	−0.0011
y_9	**0.5776**	0.0005	0.0004	0.0027	0.4187	0.0000
y_{10}	0.0031	0.0064	0.0897	**0.9008**	0.0000	0.0000
y_{13}	−0.0003	0.0000	0.0000	−0.0000	−0.0007	**1.0011**

control loops: $1 - y_9 - u_1, 2 - y_1 - u_2, 3 - y_3 - u_3, 4 - y_{10} - u_4, 5 - y_8 - u_5$ and $6 - y_{13} - u_6$.

In order to evaluate dynamically the plant-wide control proposed in Table 3, the standard driving cycles given in Chap. 7 are used. These standardized speed profiles represent urban and highway scenarios and were stated for measuring both pollutant emissions and gasoline economy of engines. In this case, the control policy

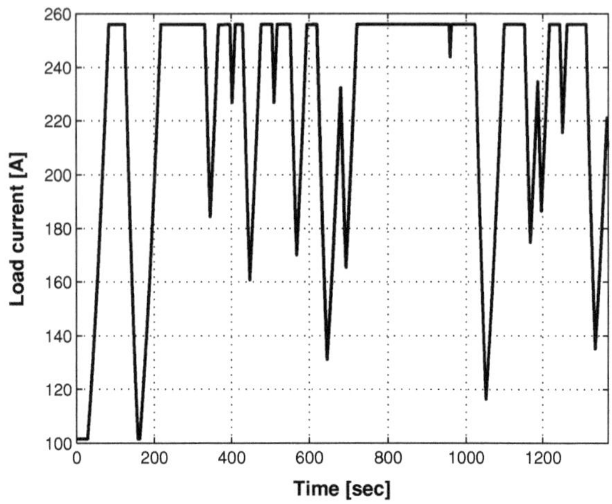

Fig. 4 Load stack current profile-disturbance

Table 4 Controllers tuning

	Loop 1	Loop 2	Loop 3	Loop 4	Loop 5	Loop 6
k_c	0.005	0.039	4.5×10^{-5}	0.001	1×10^{-6}	5.071
τ_i^c	2.671	3.327	0.055	0.047	0.05	0.140

is tested under the urban dynamometer driving schedule (UDDS) cycle because it is considered the most exigent one. For doing this test a connection between the dynamic model, detailed at Chap. 10 and the program ADVISOR (see Chaps. 7 and 11) was performed. According to the aforementioned driving cycle, the current profile demanded to the fuel cell stack is shown in Fig. 4. The energy management strategy presented in Chap. 8, where the power distribution is made according to the power demands and the state of energy of the supercapacitors.

The proposed control loops are implemented via unitary output feedback with PI (proportional/integral) controller type. In this case, the classical internal model control is used for tuning purposes. The Table 4 summarizes the selected parameters for each loop according to (IMC) procedures, recommended in [34, 35]. In Fig. 5 can be seen their localization in the process where the proposed controllers are highlighted with gray background. Note, that in control loops 1 and 6 the controlled variable is the ratio between two quantities. The first loop uses a molar ratio between O_2 and CO and the second uses a relationship between both inlet and used oxygen flow in the PEM–FC. The used oxygen flow can be estimated from the load current stack.

In order to show the effectiveness of the proposed control strategy in the following a complete set of simulations are given under the UDDS cycle. Figures 6, 7, 8, 9, 10 and 11 summarize both the CV and MV dynamic evolutions for each control loop in the process. In addition, Fig. 12 shows the hydrogen production rate profile and the bio-ethanol flow response in this case.

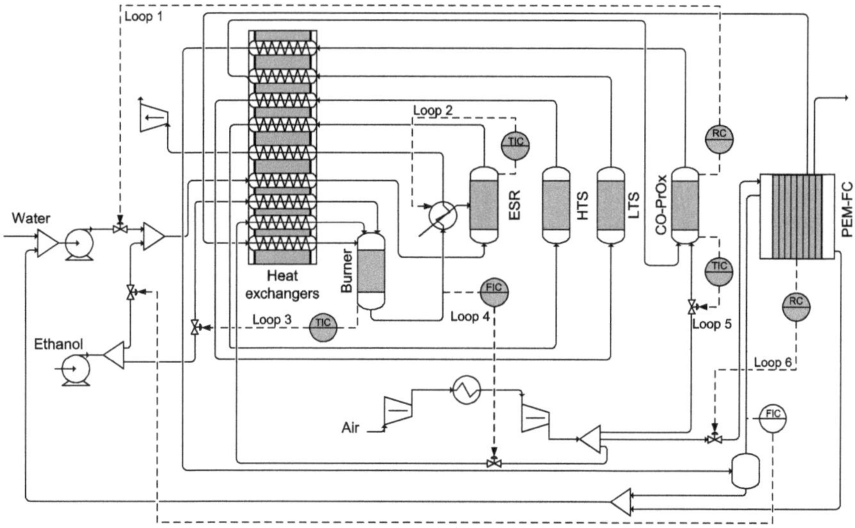

Fig. 5 Plant-wide control of BPS+FC

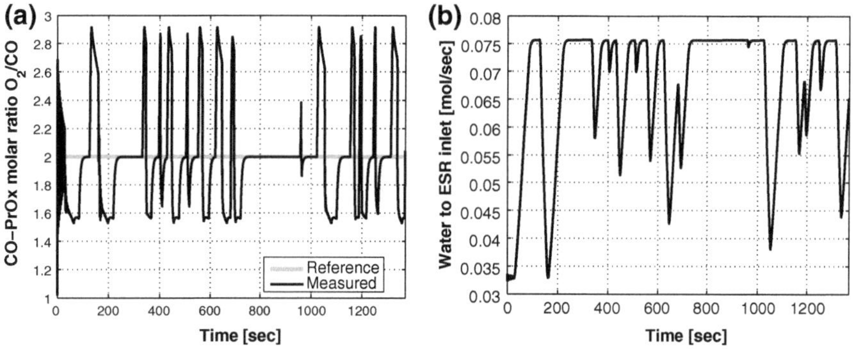

Fig. 6 Loop 1. Dynamic response. **a** y9: CO-PrOx molar ratio O_2/CO, **b** u1: water to ESR inlet flow

Figure 6 shows the dynamic behavior of the control loop no. 1. In this case the CO-PrOx molar ratio O_2/CO (CV) and the water to ESR inlet flow (MV) are shown in Fig. 6a, b respectively. A good regulatory behavior can be observed maintaining the controlled molar ratio next to the reference value of 2. In addition, the water flow presents a suitable evolution without saturations or excessive energy requirements. A similar situation can be observed in Fig. 7 for loop no. 2. In this case the ESR exit temperature as CV and the heat exchanged Q as MV are shown in Fig. 7a, b respectively. Here, it is also observed a good regulator behavior maintaining the ESR exit temperature close to the desired working point, 942.61K, with a variability $\approx \pm 0.3$ K. The MV, heat exchanged Q, presents a good dynamic evolution without saturation.

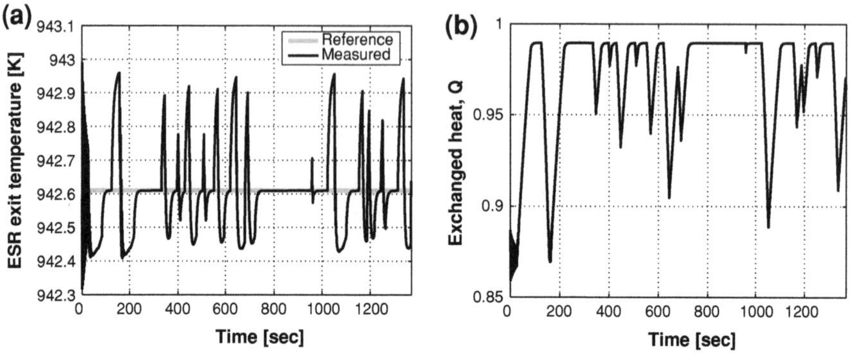

Fig. 7 Loop 2. Dynamic response. **a** y1: ESR exit temperature, **b** u2: exchanged heat Q

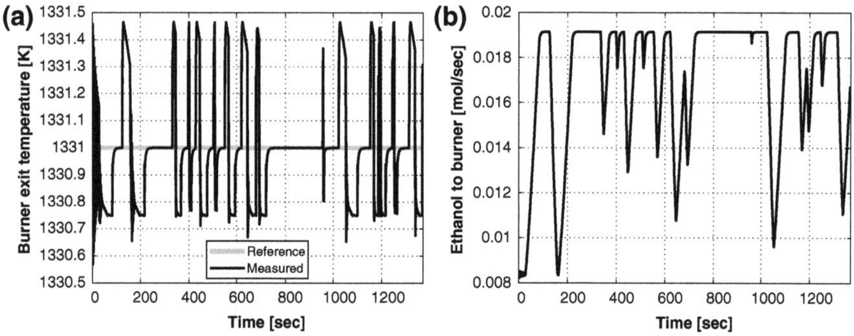

Fig. 8 Loop 3. Dynamic response. **a** y3: Burner exit temperature, **b** u3: ethanol to Burner

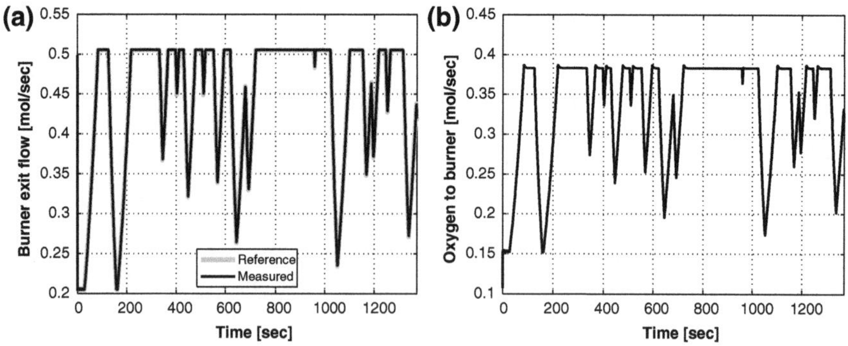

Fig. 9 Loop 4. Dynamic response. **a** y10: Burner exit flow, **b** u4: oxygen to burner

The dynamic response of control loop no. 3 is shown in Fig. 8. Here, the burner exit temperature is the CV and the ethanol flow to Burner the corresponding MV, Figs. 8a, b respectively. This loop presents a very good regulatory behavior maintaining the burner exit temperature around its working point, 1331 K, with a

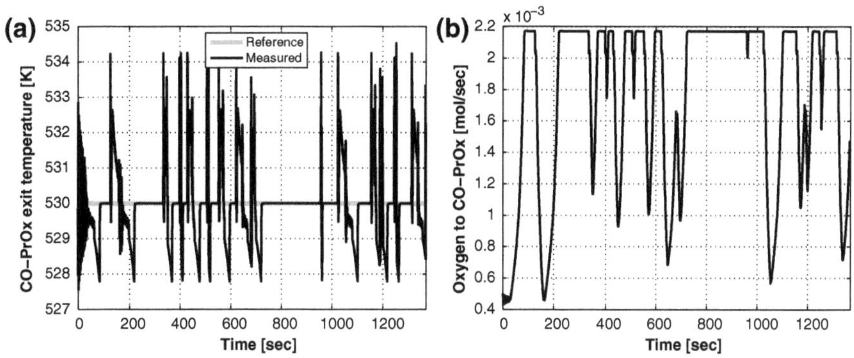

Fig. 10 Loop 5. Dynamic response. **a** y8: CO-PrOx exit temperature, **b** u5: Oxygen to CO-PrOx

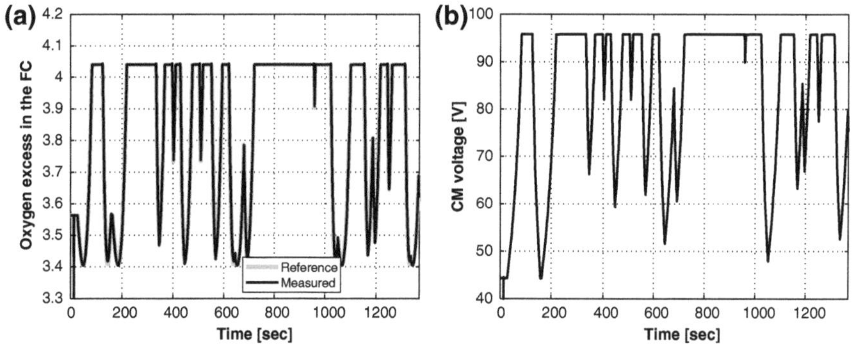

Fig. 11 Loop 6. Dynamic response. **a** y13: Oxygen excess in the FC, **b** u6: CM voltage

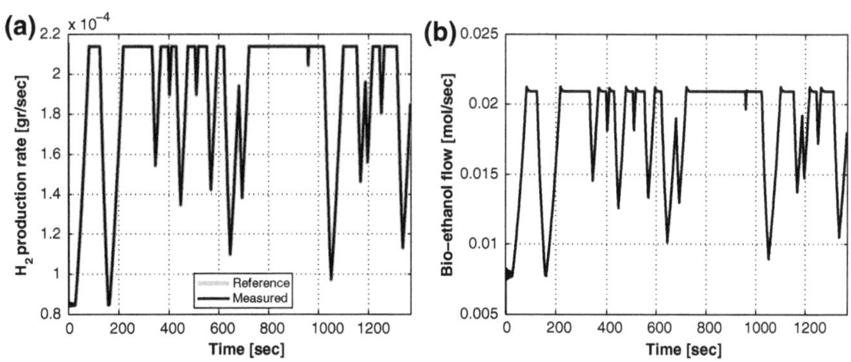

Fig. 12 Production rate loop dynamic response. **a** y19: H2 production rate, **b** u11: bio-ethanol flow

variability $\approx \pm 0.5$ K. In addition, the ethanol flow to Burner presents a suitable evolution within its operation limits.

Figures 9 and 10 summarize the temporal evolution of the control loop numbers 4 and 5 respectively. The former case deserves particular explanation because it has a changing reference (working point). In this case is shown the burner exit flow as CV and the oxygen flow to Burner as MV in the Figs. 9a, b, respectively.

The set point modifications are made according to the disturbance profile. The process was designed for working in a determined current load demand, if that current decreases, the flow through the ESR decreases as a consequence of the lower hydrogen demand. If that flow decreases, and the flow of hot gases passing through the jacket of the reformer are kept constant, then the temperature controller reduces the fraction of heat transferred to the main stream, and it will result as useful heat loss and decaying of the efficiency of the system. That is, to avoid the performance loss by establishing a cascaded set point for the flow of gases going out of the burner, which varies according to the demand corresponding with the following equation:

$$F_{BRN} = 0.001945 \times I_{st} + 0.007867 \quad (32)$$

where F_{BRN} is the reference to the burner exit flow, which has to be followed manipulating the entering flow of air (oxygen).

In the latter case, Fig. 10, the dynamic response of the control loop number 5 is shown where the CO-PrOx exit temperature is the CV and the oxygen flow to CO-PrOx the corresponding MV, Fig. 10a, b, respectively. This loop presents a good regulator behavior maintaining the CO-PrOx exit temperature around its working point, 530 K, with a variability $\approx \pm 4$K. In addition, the oxygen flow to CO-PrOx presents a suitable evolution within its operation limits.

Figure 11 shows the dynamic behavior of loop no. 6, which is the loop that corresponds to the fuel cell. The only variable that could be manipulated is the voltage to the compressor motor, and the considered controlled variable is the oxygen excess in the cathode. The compressor motor voltage controls the amount of air injected into the fuel cell, if it is too little, then the fuel reaction could be incomplete, and that will lead to a low hydrogen utilization. On the other hand, if the amount of oxygen is excessive, a complete reaction will be guaranteed, but the air compressor will require more power, producing a significant decrease in the net efficiency of the system. Then, there is an optimal flow of oxygen for each current load demand, than can be translated into a specific voltage at the inlet of the compressor. According to the recommendations given in [38], the optimal excess of oxygen, according to the current demands, is represented by the following equation, as detailed in Chap. 10:

$$\lambda_{O_2}^{exc} = 5.89873e^{-05} \times I_{st}^2 - 1.79862e^{-02} \times I_{st} + 4.78035 \quad (33)$$

The controlled variable shown in Fig. 11 presents a very good behavior for set point tracking, under a wide variety of demands. It follows closely the reference, both for stationary and transient periods. The measured variable presents a small inverse response, typical of this particular fuel cell when sudden changes in the CM voltage are applied.

Figure 12 shows the production rate control loop. It is one of the main loops of the system, needed to stabilize the plant. Again, it has a changing set point, because the

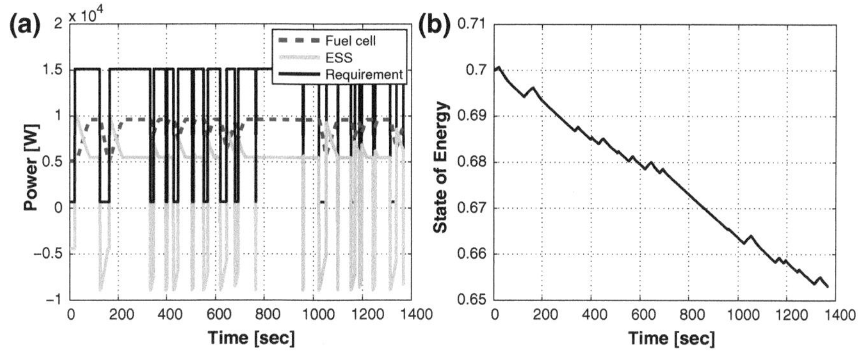

Fig. 13 Production rate loop dynamic response. **a** Energy distribution between the power sources, **b** state of energy of the supercapacitors

hydrogen demand of the fuel cell varies in accordance with the demanded current as described in Chap. 10:

$$W_{H_2} = \frac{n \cdot I_{st}}{2 \cdot F} \tag{34}$$

where W_{H_2} refers to H_2 production (taken as set point), n is the number of cells in the stack and F the Faraday constant. Actual hydrogen production set point should be determined directly from H_2 partial pressure in the anode, but this is a variable that cannot be measured in an FC. This loop is extremely important for the control objectives of the plant. It can be seen that the set point tracking is very good, keeping up to the requirements, and without excessive use of bio-ethanol.

Also, in Fig. 13 is shown the distribution of energy among the power sources and the state of energy of the Energy Storage System (ESS) through the driving cycle. Figure 13a represents the power demand of the hybrid vehicle, and the distribution of energy to fulfill that specific requirement between the fuel cell and the supercapacitors. When the demand is greater than the power that can be obtained from the fuell cell, it works at its maximum point, and the remaining power is supplied by ESS. When the demanded energy is low, the fuel cell goes to its minimum, and the excess of the generated power is used to refill the energy of ESS. In addition, the fuel cell has rate limitations, so when the power change is too sudden, the supercapacitors have to supply the power in order to keep up to the transient demand variation of the system. Figure 13b represents the variation of the State of Energy of the supercapacitors along the driving cycle analyzed here. The different states can be seen, when the supercapacitors are supplying power, the SoE decreases, and when they are recharging, the SoE increases again. A tendency to a low SoE is clear, so a slightly bigger fuel cell could present better performance. Hence, the variability of the energy of the capacitors is very small, indicating that they could be oversized, so less supercapacitors can be enough for the system, decreasing the weight of the vehicle.

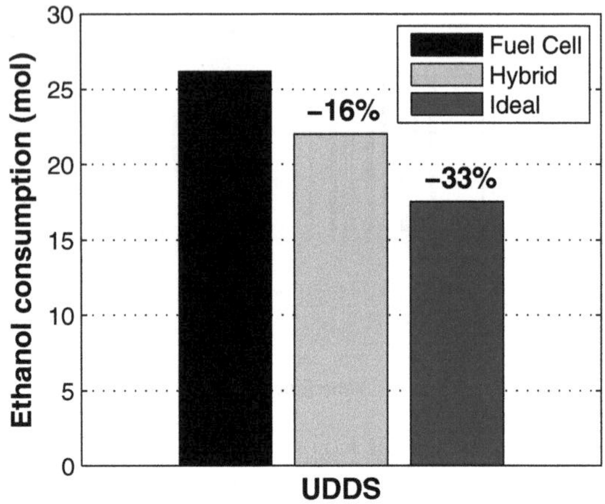

Fig. 14 Fuel consumption for the UDDS

A noticeable result is obtained when ethanol consumption is compared between the different systems considered. Figure 14 represents the moles of fresh ethanol used to fulfill the requirements of BPS within the considered driving cycle. The first bar is the consumption considering a vehicle provided only with the fuel cell. It has to be considered that FC actually cannot reach the maximum power requirement, so power provided is not enough for the cycle during the peaks of the demand and it is not working in the desired condition. The second bar is the consumption of ethanol for the hybrid vehicle as described above, it presents a 16% reduction in fuel needs, and it can provide the necessary power. When the demanded power is higher than the FC capability, the supercapacitors provide the remaining. The third column is the theoretical case when the system is working always in maximum efficiency, and fuel consumption is reduced to its minimum.

In the next part of this section, the final part of the procedure shown in Fig. 1 is followed. Hence, taking into account the strategy defined in Sect. 2.3 the CSD can be analyzed by using the net load effect (NLE) concept. It is done based on the possible contribution that can give the interaction effect between several control loops for rejecting specific disturbances of the process. Thus, considering the optimization problem stated in Eqs. 21–24, the model parametrization in Eq. 27,

$$\Gamma = \begin{bmatrix} 1 & \gamma_{12} & \gamma_{13} & \gamma_{14} & \gamma_{15} & \gamma_{16} \\ \gamma_{21} & 1 & \gamma_{23} & \gamma_{24} & \gamma_{25} & \gamma_{26} \\ \gamma_{31} & \gamma_{32} & 1 & \gamma_{34} & \gamma_{35} & \gamma_{36} \\ \gamma_{41} & \gamma_{42} & \gamma_{43} & 1 & \gamma_{45} & \gamma_{46} \\ \gamma_{51} & \gamma_{52} & \gamma_{53} & \gamma_{54} & 1 & \gamma_{56} \\ \gamma_{61} & \gamma_{62} & \gamma_{63} & \gamma_{64} & \gamma_{65} & 1 \end{bmatrix} \tag{35}$$

and the following chromosome parametrization

$$\mathbf{C}_{\text{NLE}} = \left[\gamma_{12}\gamma_{13}\cdots\gamma_{1n}\cdots\gamma_{ij}\cdots\gamma_{n(n-1)}\right], \ \forall\gamma_{ij} \text{ with } i \neq j \tag{36}$$

Table 5 GA parameters setting

n	N_i	N_c	N_g	P_m	P_{co}	Selection	Crossover	Optimization time
6	6000	30	30	$0.7/N_c$	0.7	Roulette wheel	Double-point	99.31 s

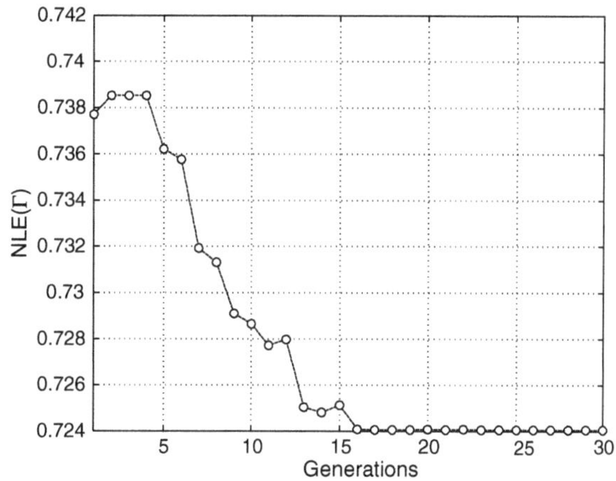

Fig. 15 Fitness profile from GA

the combinatorial size results as $2^{6\times5} \approx 1.0737 \times 10^9$ for which an exhaustive search is impractical and can be efficiently solved by GA. In this case the parameters shown in Table 5 are used to obtain the optimal solution. Here, the amount of decentralized control loops is n=6, the initial population was set to $N_i = 6000 = N_c \times 200$, the chromosome length is 30 due to the off-diagonal elements in Eq. 35, the termination criteria was the amount of generation $N_g = 30$, the crossover and mutation probabilities $P_{co} = 0.7$ and $P_m = 0.7/N_c$ respectively, the selection method based on roulette-wheel and the crossover strategy based on double-point. In this context, the optimal solution, Γ_o, can be observed in Eq. 37 which has a fitness value of NLE $= 0.7241$ and incorporates 17 additional model components to the diagonal control structure stated in the previous step.

The final controller structure depends on the particular model parameterization via Γ_o and the IMC theory (inverse of the selected model). In this context, only the diagonal, full and block-diagonal model parameterization drive to specific controller matrix structure. In several cases, a sparse model selection can drive to control matrix structures with equal or more control loops than the model component parameterizations.

Figure 15 shows the evolution of the fitness profile across generations. The weighing matrices used in this case are $\Delta_1 = \mathbf{I}_6$, $\Delta_2 = \text{diag}([000101])$, $\Xi_1 = \mathbf{I}_6$ and $\Xi_2 = \text{diag}([01])$, so as to assign the same importance to the overall outputs but considering only set point changes in both the CO-PrOx molar ratio O_2/CO and oxygen excess ratio in the FC and the stack current as the most important disturbances.

$$\Gamma_o = \begin{bmatrix} 1\,0\,1\,0\,0\,1 \\ 1\,1\,1\,1\,0\,1 \\ 0\,0\,1\,1\,1\,0 \\ 1\,0\,1\,1\,1\,0 \\ 1\,1\,0\,0\,1\,1 \\ 1\,0\,1\,1\,0\,1 \end{bmatrix} \tag{37}$$

The optimal solution shown in Eq. 42 represents the best model parameterization from both the location and amount of its components point of view. Thus, from a potential search space of 2^{30} combinations the GA selects 17 additional components (control loops) and with the disposition shown in Γ_o. This means that there is no solution with fewer (<17) or more (>17) components that minimize the functional cost. Moreover, a solution with the same amount of loops (=17) but different disposition, to that shown in Γ_o, is a suboptimal one.

In this context it is also useful to know how the NLE index evolves/degrades when the chromosome parameterization, (38), is constrained to take fewer control loops with respect to the optimal solution. Thus, if the best solution to the problem stated in (21)–(24) presents N_{op} additional control loops (model components), now the problem can be modified as,

* for $k = 0$ to $N_{op} - 1$ with step 1

$$\begin{aligned}
&\min_{\Gamma} \mathrm{NLE}(\Gamma) = \min_{\Gamma} \left[\|\Delta_1 \mathbf{A}_{\Gamma} \Delta_2\|_F^2 + \|\Xi_1 \mathbf{B}_{\Gamma} \Xi_2\|_F^2 \right] \\
&\text{subject to} \\
&\mathbf{A}_{\Gamma} = \mathbf{I} - \widetilde{\mathbf{G}}_{s\Gamma} \mathbf{G}_s^{-1} \\
&\mathbf{B}_{\Gamma} = \widetilde{\mathbf{G}}_{s\Gamma} \mathbf{G}_s^{-1} \mathbf{D}_s \\
&\mathrm{Re}\left[\lambda_i \left(\mathbf{G}_s \widetilde{\mathbf{G}}_{s\Gamma}^{-1} \right) \right] > 0,\, i = 1, \ldots, n \\
&\sum_{j=0}^{N_c} \mathbf{C}_{\mathrm{NLE}}(j) \le k
\end{aligned}$$

* end

to evaluate the NLE index degradation. Note that $\mathbf{G}_s$ is the square model selection given by the OSL problem but rearranged to give a diagonal RGA. The same arrangement is applied to $\mathbf{D}_s$. Thus, an effective comparison with several suboptimal control structures can be made as functions of both the amount of control loops and the functional cost value.

The NLE profile containing the optimal, suboptimal, diagonal and full controller structure as solutions is shown in Fig. 16. Clearly, when fewer loops are considered the solutions move toward the diagonal structure degrading their NLE index almost exponentially. Each point in Fig. 16 represents a particular solution with a predefined maximum amount of loops, i.e. $\mathbf{C}_{\mathrm{NLE}}^4$ is the best model parameterization by using up to 4 components, and in this case the fitness value is $\mathrm{NLE}(\Gamma) \approx 0.8$. From this figure can be observed an important valley, from solutions $\mathbf{C}_{\mathrm{NLE}}^4$ to $\mathbf{C}_{\mathrm{NLE}}^{17}$ (optimal),

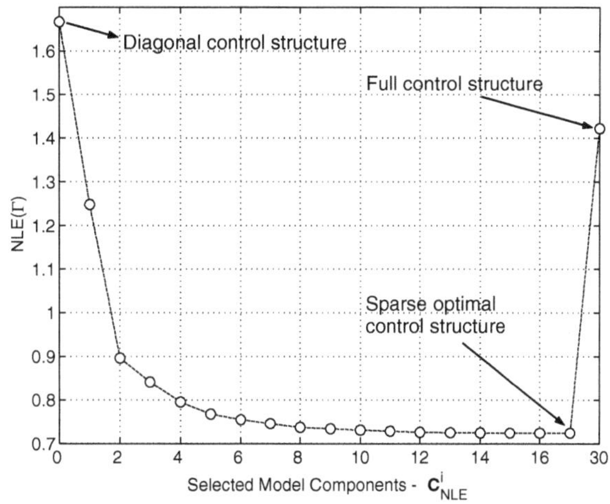

Fig. 16 NLE index profile for suboptimal solutions

Table 6 Optimal and suboptimal solutions from GA

Solutions	Control loops	NLE(Γ)
$\mathbf{C}^{17}_{NLE}$	0 1 0 0 1 1 1 1 0 1 0 0 1 1 0 1 0 1 1 0 1 1 0 0 1 1 0 1 1 0	0.72
$\mathbf{C}^{4}_{NLE}$	0 0 0 0 0 0 0 1 0 0 0 0 1 0 0 0 0 0 0 0 1 0 0 0 1 0 0 0 0 0	0.79
$\mathbf{C}^{30}_{NLE}$	1 1 1 1 1 1 1 1 1 1 1 1 1 1 1 1 1 1 1 1 1 1 1 1 1 1 1 1 1 1	1.42
$\mathbf{C}^{0}_{NLE}$	0 0 0 0 0 0 0 0 0 0 0 0 0 0 0 0 0 0 0 0 0 0 0 0 0 0 0 0 0 0	1.66

this means that there are several solutions with approximately the same NLE index magnitude but using a less number of control loops to implement. It can be observed also that both diagonal and full control structures represent the worst solutions in terms of the NLE index.

In this case the solution $\mathbf{C}^4_{NLE}$ was selected due to a good trade-off between fitness degradation and the number of control loops. Table 6 presents a comparison between this suboptimal solution with the optimal one, full and diagonal. $\mathbf{C}^4_{NLE}$ represents an NLE degradation of $\approx 9.7\%$ with respect to the optimal case, but using only 4 additional control loops. On the other hand, $\mathbf{C}^{30}_{NLE}$ degrades the functional cost in $\approx 97.2\%$ and uses 30 additional control loops. Finally, $\mathbf{C}^0_{NLE}$ degrades the functional cost in $\approx 130.5\%$ with respect to the optimal value but without using additional control loops.

The selected solution, $\mathbf{C}^4_{NLE}$, suggests the following model parameterization

$$\Gamma_4 = \begin{bmatrix} 1\,0\,0\,0\,0\,0 \\ 0\,1\,0\,1\,0\,0 \\ 0\,0\,1\,1\,0\,0 \\ 0\,0\,0\,1\,0\,0 \\ 1\,0\,0\,0\,1\,1 \\ 0\,0\,0\,0\,0\,1 \end{bmatrix}, \quad \widetilde{\mathbf{G}}_{s\Gamma} = \begin{bmatrix} g_{11}(s) & 0 & 0 & 0 & 0 & 0 \\ 0 & g_{22}(s) & 0 & g_{24}(s) & 0 & 0 \\ 0 & 0 & g_{33}(s) & g_{34}(s) & 0 & 0 \\ 0 & 0 & 0 & g_{44}(s) & 0 & 0 \\ g_{51}(s) & 0 & 0 & 0 & g_{55}(s) & g_{56}(s) \\ 0 & 0 & 0 & 0 & 0 & g_{66}(s) \end{bmatrix} \tag{38}$$

where the $g_{ij}(s)$ components are the first order with time delay approximations obtained from the linear model in Eq. 30. Following the MSD approach the next step is the sparse controller structure tuning using the methodology given in Sect. 2.4. Thus, considering the matrix $\widetilde{\mathbf{G}}_{s\Gamma}^{-}$ containing the invertible parts only of $\widetilde{\mathbf{G}}_{s\Gamma}$, and a first order diagonal matrix filter, $\mathbf{F}(s)$, the unitary feedback sparse control policy can be defined as

$$\begin{aligned}
\mathbf{G}_c^*(s) &= \widetilde{\mathbf{G}}_{s\Gamma}^{-}(s)^{-1}\mathbf{F}(s)\,[\mathbf{I}-\mathbf{F}(s)]^{-1} \\
&= \widetilde{\mathbf{G}}_{s\Gamma}^{-}(s)^{-1}\begin{bmatrix} 1/(\tau_{f1}s) & 0 & 0 & 0 \\ 0 & 1/(\tau_{f2}s) & 0 & 0 \\ 0 & 0 & \ddots & 0 \\ 0 & 0 & 0 & 1/(\tau_{f6}s) \end{bmatrix} \\
&= \begin{bmatrix} c_{11}(s) & 0 & 0 & 0 & 0 & 0 \\ 0 & c_{22}(s) & 0 & c_{24}(s) & 0 & 0 \\ 0 & 0 & c_{33}(s) & c_{34}(s) & 0 & 0 \\ 0 & 0 & 0 & c_{44}(s) & 0 & 0 \\ c_{51}(s) & 0 & 0 & 0 & c_{55}(s) & c_{56}(s) \\ 0 & 0 & 0 & 0 & 0 & c_{66}(s) \end{bmatrix}
\end{aligned} \tag{39}$$

where the additional control loops are

$$\begin{aligned}
c_{24}(s) &= \frac{-g_{24}^{-}(s)}{g_{22}^{-}(s)g_{44}^{-}(s)\tau_{f4}s} = -0.4664\left(1+\frac{1}{0.1s}\right) \\
c_{34}(s) &= \frac{-g_{34}^{-}(s)}{g_{33}^{-}(s)g_{44}^{-}(s)\tau_{f4}s} = 0.0542\left(1+\frac{1}{0.1s}\right) \\
c_{51}(s) &= \frac{-g_{51}^{-}(s)}{g_{11}^{-}(s)g_{55}^{-}(s)\tau_{f1}s} = 1.167\times 10^{-4}\left(1+\frac{1}{0.1s}\right) \\
c_{56}(s) &= \frac{-g_{56}^{-}(s)}{g_{55}^{-}(s)g_{66}^{-}(s)\tau_{f6}s} = -1.651\times 10^{-4}\left(1+\frac{1}{0.1s}\right)
\end{aligned} \tag{40}$$

if a diagonal tuning is considered for τ_{f1}, τ_{f4} and τ_{f6}. Note, that the $g_{ij}^{-}(s)$ components are the invertible part of $g_{ij}(s)$. From (40) can be observed that these control loops have a PI structure. In this context, the final control signals given by the sparse controller are as follows:

$$\begin{aligned}
u_1(s) &= c_{11}(s)e_1(s) \\
u_2(s) &= c_{22}(s)e_2(s) + c_{24}(s)e_4(s) \\
u_3(s) &= c_{33}(s)e_3(s) + c_{34}(s)e_4(s) \\
u_4(s) &= c_{44}(s)e_4(s) \\
u_5(s) &= c_{55}(s)e_5(s) + c_{51}(s)e_1(s) + c_{56}(s)e_6(s) \\
u_6(s) &= c_{66}(s)e_6(s)
\end{aligned} \tag{41}$$

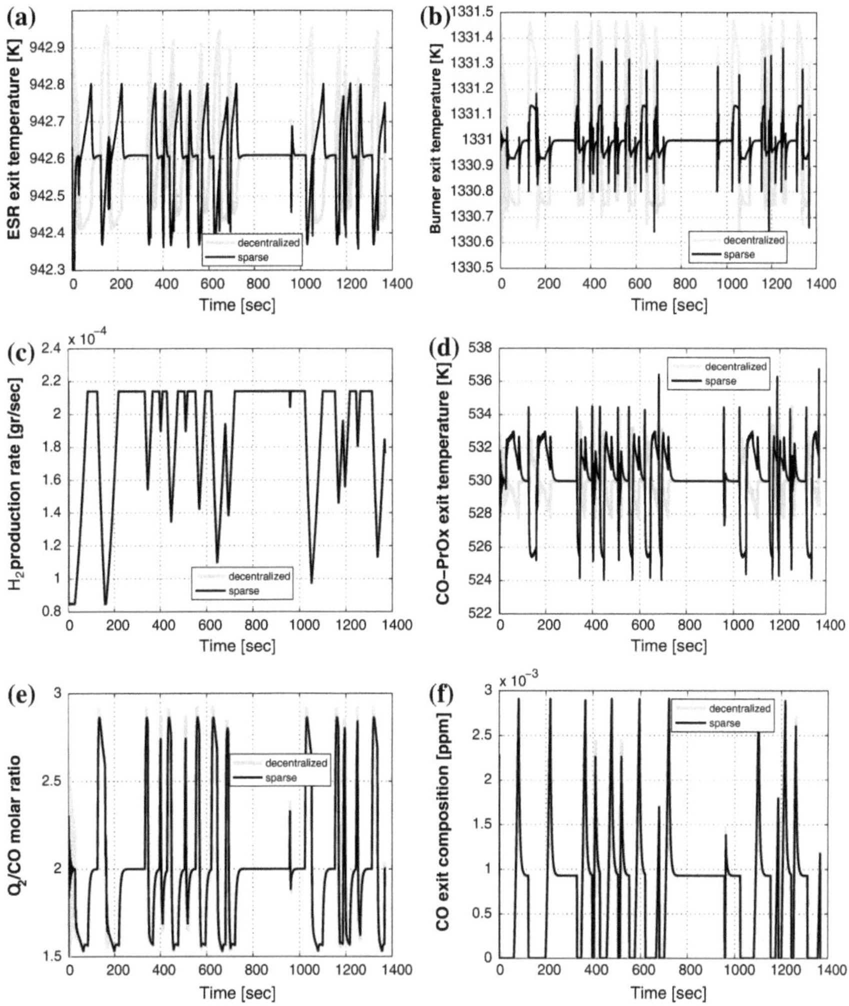

Fig. 17 Decentralized and sparse control structures. **a** ESR exit temperature, **b** burner exit temperature, **c** hydrogen production rate, **d** CO-PrOx exit temperature, **e** CO-PrOx molar ratio O_2/CO, **f** CO-PrOx output CO concentration

where $e_i(s) = y_i^{sp}(s) - y_i(s)$ for $i = 1, \ldots, 6$ is the tracking error for the ith CV, $y_i^{sp}(s)$ the reference signal, and $y_i(s)$ the measured variable. In this case expressed in Laplace domain.

In Fig. 17, the decentralized and sparse (suboptimal) control structures are compared. Tables 7, 8 and 9 present the important parameters and results related to Fig. 17. They are the variability of the controlled temperatures with fixed set point, the Integral Absolute Error (IAE) of the variables with respect to their corresponding references, and the control energy required by the manipulated variables to keep the plant under specifications.

Table 7 Controlled temperatures

Variable	Operating point (K)	Variability (K) Decentralized	Sparse
y_1	942.61	$\approx \pm 0.3$	$\approx \pm 0.2$
y_3	1331.00	$\approx \pm 0.4$	$\approx \pm 0.2$
y_8	530.00	$\approx \pm 2.0$	$\approx \pm 4.0$

Table 8 IAE index-UDDS conditions

Loop no.	CV	Decentralized	Sparse
1	y_9	338.67	324.23
2	y_1	113	89
3	y_3	188.61	58.09
4	y_{10}	1.127	1.126
5	y_6	859.50	1803
6	y_1	6.28	6.03
H_2	y_14	2.38e-4	2.30e-4

Table 9 Control energy - UDDS conditions

MVs	Decentralized	Sparse
u_1	91.78	91.79
u_2	1328.17	1328.19
u_3	23.30	23.29
u_4	460.30	460.29
u_5	2.40	2.40
u_6	114373	114377
Bio-ethanol	24.93	24.93

In Fig. 17a, the dynamic behavior of the ESR exit temperature against the load profile is shown. For the decentralized case it presents a variability of 0.35 K above the set point and 0.2 K under the desired temperature. For the sparse case, the amplitude below the set point is maintained, however, over the reference it decreases to 0.2 K, and also the regulator behavior turns out more aggressive, obtaining a decrease of $\approx$21% in the IAE index. The behavior of the burner exit gases temperature can be seen in Fig. 17b, it improves significantly in the sparse controller structure, decreasing significantly the amplitude of variation, and the deviation from the desired temperature is $\approx$69% lower than the decentralized case. In Fig. 17c, the tracking of the hydrogen production rate is presented, it represents the main product of the reactors set. As can be seen, the new and highly interacted structure is perfectly capable of having a good tracking behavior for the quick changes of the reference value and also improving the IA + E value. Figure 17d represents the exit temperature of the CO-PrOx reactor, which has a slightly higher variability in the sparse case, as a consequence, an increase in the IAE index value is produced.

Figure 17e displays the dynamic evolution of the CV CO-PrOx molar ratio O_2/CO, when both control structures are used. It is clear that the sparse approach produces a variability reduction around its nominal working point. This improvement represents a $\approx$4% of reduction in the integral absolute error (IAE)with respect to the decentralized strategy. On the other hand, Fig. 17f shows the CO-PrOx output CO

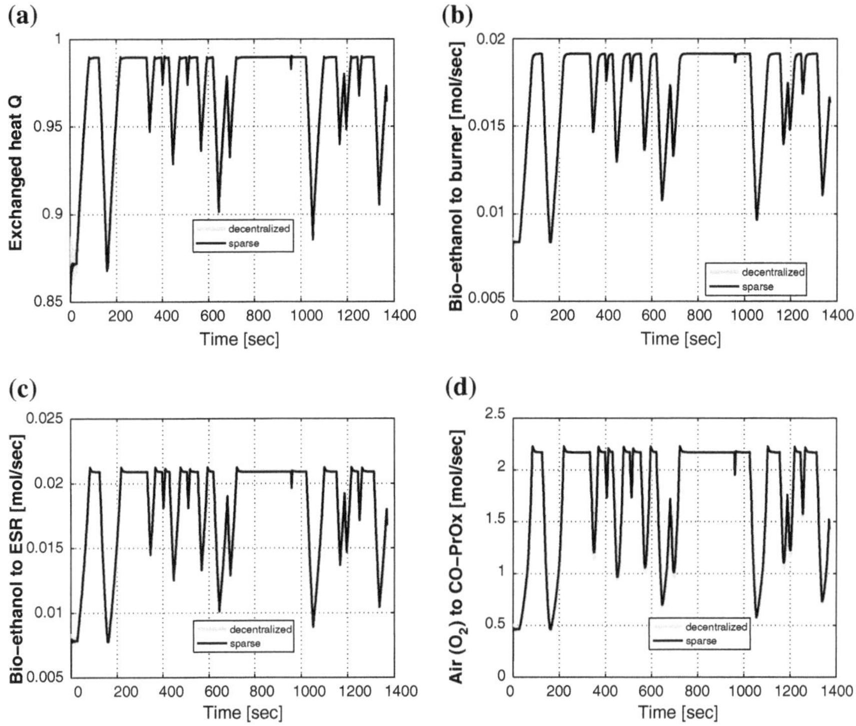

Fig. 18 Decentralized and sparse manipulated variables. **a** Exchanged heat Q, **b** bio-ethanol to burner, **c** Bio-ethanol to ESR, **d** air to CO-PrOx

concentration evolution. This variable is not controlled directly, due to feasibility issues, but it is important because of its influence on the fuel cell. High values with respect to its desired operating point (10^{-6}) in this variable could lead to the poisoning of PEMFC. In this case, by using the sparse control structure an improvement of $\approx$3.4% is obtained with respect to the IAE of the decentralized one. The obtained results are very encouraging, mainly because the sparse controller structure implemented actually represents a suboptimal solution to the control problem. It is a compromise solution that reaches excellent performance without too many new control loops and using the interaction effects for rejecting the disturbances.

In all cases, the required control energy is very similar for both structures. Clearly it denotes that the plant is extremely sensitive, given that slight modifications in the manipulated variables produce noticeable changes in the controlled ones. It can be highlighted that the new sparse control structure does not modify the bio-ethanol consumption at the inlet of the reactors. It represents a satisfactory result, given that the same amount of fuel is consumed, a better set point tracking is achieved, even though it is very little, for the changing reference of the hydrogen requirements from the fuel cell.

In Fig. 18, the time evolutions are shown corresponding to some of the manipulated variables of the two implemented control structures, related to Table 9. Figure 18a shows the heat exchanged Q, which regulates the temperature of ESR. The differences between both control structures are very small, the sparse case is a little more aggressive, presenting small overvalues. However, this tiny difference is enough to produce noticeable changes in the controlled variable, as can be seen in Fig. 17a. Figure 18b presents the bio-ethanol fed to the burner reactor, and the differences are imperceptible but the controlled variable, the burner exit temperature, also presents changes between both strategies. Figure 18c represents the bio-ethanol feed stream at the inlet of the system, it does not present perceivable modifications among structures. Figure 18d presents the air inlet to the CO-PrOx reactor, the differences are clearer in this case, with the sparse control structure showing a more aggressive control with some peaks when changes in the current load demand are produced.

4 Conclusions

In this chapter an integral study of the plant-wide control for the bio-ethanol processor system with fuel cell was presented. The proposed generalized methodology was useful to address effectively the optimal sensor network and the control structure design simultaneously for this challenging industrial plant. Different tools based on steady-state information, system identification, internal model control, genetic algorithms and a new index based on the Frobenius norm have been integrated. As a result a systematic approach, able to be applied even on large dimension systems and minimizing the use of heuristic considerations, is tested in a complex scenario of load requirements. In this context, the optimal sensors network and the plant-wide control structure was defined based on the sum of square deviations (SSD) and the net load effect (NLE) indexes respectively. Mainly, this approach is based on steady-state tools which favor the interconnection/feedback with the process synthesis stage improving the final process design. The new indexes, due to their structures, allow to include information about the principal design objectives with the control structure able to guarantee their achievement. In addition, the inclusion of the proper weighting matrices selection allows to choose the weights according to the inherent importance of those objectives that are more critical for the overall process.

For the particular case addressed here, BPS, the optimal sensor location was solved with 14 chosen CVs and 6 MVs; the selection was performed among 3003 potential solutions,although the problem size could be efficiently addressed by GA or exhaustive search indistinctly. Then, two equally feasible plant-wide control structures were obtained via the application of the proposed methodology. In this context, 6 optimal CVs were selected, able to minimize the SSD index and using the RGA criteria, a decentralized (diagonal) control structure was proposed. It was mainly dedicated to solve the servo problem of the plant. The control strategy was rigorously tested in fuel cell hybrid vehicle within ADVISOR environment, against a challenging driving cycle, named UDDS. It was selected because it presents a wide range of variation, representing one of the most exigent scenarios for the control structure performance evaluation. All the control loops have presented a suitable dynamic behavior, with fast correction and little variation from its set points. Those variables whose references were set as a function of the load demanded, such as references of the hydrogen production and burner exit flow, have presented good servo performance. On the other hand, loops with fixed set points (temperatures, compositions, flow ratios) have shown good regulating responses. All these properties have been evaluated through the integral absolute error and variability percentage. Although they presented a small percentage of variation, a second control structure was considered. It is based on the influence of the interaction between control loops and their potentiality for rejecting some critical disturbances. This control structure was obtained by applying the NLE approach giving as a result a sparse controller design. Initially, a complete analysis was developed considering a combinatorial problem with dimension $2^{6\times 5} \approx 1 \times 10^9$ solved using GA. The optimal solution called $\mathbf{C}_{\mathrm{NLE}}^{17}$ proposes to select 17 model components to be accounted for designing the multivariable controller. However, as another suboptimal sparse control solution, with only four model components was found, it was chosen because it involved a controller structure with less off-diagonal elements ($\mathbf{C}_{\mathrm{NLE}}^{4}$). A detailed analysis was performed from the rigorous comparisons among decentralized, sparse and full controller structures, where the suboptimal sparse controller was selected. This solution presented a suitable trade-off between the hardware requirements associated with the number of extra control loops and NLE degradation with respect to the optimal version. In the last part of this chapter a complete dynamic evaluation was presented comparing this suboptimal sparse control structure and the decentralized one presented previously. The improvements obtained by this sparse approach were quantified using integral absolute errors for tracking reference and the control energy involved in each control policy. The sparse control showed a better dynamic performance than the decentralized one, when the UDDS requirements were considered. This result suggests that the optimal solution $\mathbf{C}_{\mathrm{NLE}}^{17}$ could improve even more the performance given by the option $\mathbf{C}_{\mathrm{NLE}}^{4}$. As future works it will be analyzed whether the implementation of this sparse control structure in the context of model predictive control can improve the dynamic behavior of this process.

Appendix

Relative Gain Array

Considering a multivariable square system represented via a TFM, $\mathbf{G}_s(s)$, a relationship λ_{ij} can be defined to measure how the gain between output y_i and the input u_j is modified when perfect control is considered in the remains outputs respect to the open loop gain. Accounting each process component and their values at steady-state the above relationship can be expressed in matrix form as can be observed in Eq. 42 [13].

$$\Lambda = \mathbf{G}_s \otimes \left(\mathbf{G}_s^{-1}\right)^T = \begin{bmatrix} \lambda_{11} & \dots & \lambda_{1j} \\ \vdots & \ddots & \vdots \\ \lambda_{i1} & \dots & \lambda_{ij} \end{bmatrix} \tag{42}$$

where $\otimes$ is the element-by-element product and,

$$\lambda_{ij} = \frac{\left[\frac{\partial y_i}{\partial u_j}\right]_{u_{k=cte.,k\neq j}}}{\left[\frac{\partial y_i}{\partial u_j}\right]_{y_{k=cte.,k\neq i}}} \tag{43}$$

the Eq. 42 is know as the relative gain array (RGA) and is the most used tool in industrial process control. The elements of this matrix are real numbers. There are some rules to simplify the analysis of Λ for control pairing purposes: positive values of λ_{ij} close to one indicate that the gain of this loop is not affected by closing the control loops remaining. This is a desirable behavior in any control loop. Values of $\lambda_{ij} > 0.5$ indicate that there is some interaction and could influence other control loops. Finally, negative values of λ_{ij} indicate that the sign of the gain could be modified when the remaining loops are closed. This is an unwanted behavior for a potential control loop.

The interaction between control loops is not always harmful. If the control policy is designed suitably this interaction could improve the overall dynamic process behavior. For example, including interaction in specifics parts of the plant.

Relative Normalized Gain Array

From the precursory work of [13], have emerged several variations and complements to the RGA for decades. One of the most recent approach is the proposed by [23]. In this paper the authors suggest the use of dynamic information to normalize the process gains previous to the RGA computing. Thus, the gain of each process element, g_{ij}, is normalized by

$$g_{ij}^n = \frac{g_{ij}}{\tau_{ij}^{rp}} \tag{44}$$

where g_{ij} is the *ij* element of $\mathbf{G}_s$ and τ_{ij}^{rp} is the residence time between the output *i* and the input *j*, in other words, a measurement of the input–output response time. There are different ways to compute this residence time depending on the dynamic model type used. In the case of first order plus delay systems results, $\tau_{ij}^{rp} = \tau_{ij} + \theta_{ij}$, where τ_{ij} is the time constant and θ_{ij} the dead time present in the loop.

Grouping each component g_{ij}^n in a normalized matrix $\mathbf{G}_s^n$ and each residence time τ_{ij}^{rp} in $\mathbf{T}_{rp}$ the relative normalized gain array (RNGA) computing is made analogously to the RGA,

$$\Lambda^n = (\mathbf{G}_s \oslash \mathbf{T}_{rp}) \otimes \left((\mathbf{G}_s \oslash \mathbf{T}_{rp})^{-1}\right)^T = \mathbf{G}_s^n \otimes \left((\mathbf{G}_s^n)^{-1}\right)^T \tag{45}$$

where $\oslash$ is the element-by-element division and $\mathbf{G}_s^n = \mathbf{G}_s \oslash \mathbf{T}_{rp}$.

This RNGA has the properties of the RGA and additionally allows to incorporate dynamic information in an evaluation static purely. Thus, this methodology can suggest a different input–output pairing to the proposed by the RGA. Note that if the time constant and delays are very similar for each process component both matrices, RNGA and RGA, generate the same result. When the process contains dynamics very different between them, generally, the RNGA is preferred to the RGA, because the former selects the fast loops improving the overall process behavior. Anyway, the normalization process accomplished in the RNGA approach may generate numerical problems in the matrices. In fact, the matrix $\mathbf{G}_s^n = \mathbf{G}_s \oslash \mathbf{T}_{rp}$ may result ill-conditioned or singular due to the normalization with $\mathbf{T}_{rp}$. This behavior impacts directly in the RNGA, Λ^n, by producing both erroneous and unfeasible pairings. This problem must be taken into account.

Non-square Relative Gain

The approaches presented above assumed that the process was square, the same number of inputs and outputs. This generally is not true in practice. If the RNGA (or RGA) concepts must be applied, first, the non-square process must be reduced to a square one, $\mathbf{G}_s$, by mean of some methodology (heuristic or generalized). Anyway, other approaches exist addressing the non-square problem directly. In this context, [14] propose the non-square relative (NRG) gain for multivariable non-square systems. Without loss of generality, accounting a non-square process, $\mathbf{G}$, with m outputs and n inputs, and $m > n$, so is not possible maintain all the outputs in the desired working points, there are not degrees of freedom enough. In this sense, the authors suggest and introduce the concept of perfect control and propose the following scenarios:

perfect control in a least square sense, in other words, to minimize the two-norm of the steady-state error for all outputs together.

perfect control on n outputs and to minimize the two-norm of the steady-state error on the (m-n) uncontrolled outputs.

Using the first definition, the IMC theory and considering steady-state environment the following can be stated,

$$\bar{\mathbf{u}} = \mathbf{G}^{\dagger}\mathbf{y}^{sp} \tag{46}$$

where $\mathbf{y}^{sp}$ are the desired values of the m outputs, $\bar{\mathbf{u}}$ is the optimal value of $\mathbf{u}$ under this perfect control condition and $\mathbf{G}^{\dagger}$ is the Moore-Penrose pseudo inverse of $\mathbf{G}$. Note, that the IMC controller is non-square. Thus, by analogy with its square version the NRG is defined as,

$$\Lambda_{NRG} = \mathbf{G} \otimes \left(\mathbf{G}^{\dagger}\right)^T. \tag{47}$$

The row sum in the NRG is not unitary, in contrast with RGA. But this property allows to decide about the controlled variables. In fact, each row sum is interpreted as a measurement of performance in steady-state of the non-square IMC.

In practice, the second definition seems more appropriate, due to the natural tendency to implement decentralized multivariable control structures. In this context, only n outputs from m can be controlled perfectly at steady-state and to minimize the two-norm of the uncontrolled ones.

Relative Disturbance Gain Array

The above sections addressed the input–output pairing with steady-state tools. Sometimes, is necessary to know the control structures behavior when disturbances affect the process. Particularly, in some industrial process is more important the regulator controller behavior than the servo one. Anyway, is essential to know this effects and to select a suitable control policy.

An interesting tool is the proposed by [15, 28] called relative disturbance gain array (RDGA), which allows to quantify the performance of different classic control structures (diagonal, full, block diagonal, triangular) under disturbances. In the following a brief description is given about of this methodology.

Remembering the IMC theory, the definitions in Sect. 2.3 and Fig. 2 the following can be stated,

$$\begin{aligned}\mathbf{y}_s(s) &= \left[\mathbf{I} - \left(\widetilde{\mathbf{G}}_s(s)\mathbf{G}_c(s)\right)\right]\left[\mathbf{I} + \left(\mathbf{G}_s(s) - \widetilde{\mathbf{G}}_s(s)\right)\mathbf{G}_c(s)\right]^{-1}\mathbf{D}_s(s)\mathbf{d}_* \\ &= \left[\mathbf{I} - \left(\widetilde{\mathbf{G}}_s(s)\mathbf{G}_c(s)\right)\right]\mathbf{D}_s^*(s)\mathbf{d}_* \end{aligned} \tag{48}$$

where $\widetilde{\mathbf{G}}_s(s)$ is the square model used from $\mathbf{G}_s(s)$, $\mathbf{G}_c(s)$ is the corresponding IMC controller and $\mathbf{D}_s^*(s) = [\mathbf{I} + (\mathbf{G}_s(s) - \widetilde{\mathbf{G}}_s(s))\mathbf{G}_c(s)]^{-1}\mathbf{D}_s(s)$.

Considering in Eq. 48 that $\mathbf{G}_c(s) = \widetilde{\mathbf{G}}_s(s)^{-1}$ can be observed that, at steady-state, the effect of $\mathbf{d}_*$ is null due to the integrator behavior of $[\mathbf{I} - (\widetilde{\mathbf{G}}_s(s)\mathbf{G}_c(s))]$ by exclusive property of the IMC structure. In addition, the disturbances affect the CVs in the transient (dynamic) proportionally to the gain of $\mathbf{D}_s^*(s)$.

Analyzing at steady-state, can be observed that $\mathbf{D}_s^* = \widetilde{\mathbf{G}}_s\mathbf{G}_s^{-1}\mathbf{D}_s$ and if a full control structure is selected, $\widetilde{\mathbf{G}}_s(s) = \mathbf{G}_s(s)$, then $\mathbf{D}_s^* = \mathbf{D}_s$. This indicates that the full structure is not necessarily the best control policy from a disturbances point of view.

In this context, the RDGA propose by [12, 28] compares several classic control structures with respect to the full version

$$RDGA = \mathbf{D}_s^* \oslash \mathbf{D}_s = \left(\widetilde{\mathbf{G}}_s \cdot \mathbf{G}_s^{-1} \cdot \mathbf{D}_s\right) \oslash \mathbf{D}_s \tag{49}$$

where $\oslash$ is the element-by-element division and $\widetilde{\mathbf{G}}$ is parameterized as stated in Eq. 18.

The control structure (model parametrization) that presents the lowest absolute value in each component of the vector/matrix RDGA, is the preferred control policy. Note that if the process disturbance model, for a controlled plant, has a dimension of $n \times p$ the RDGA will have the same dimension. This indicates that the method loses both clarity and simplicity quickly with the size. In this chapter has been presented an improved approach called MSD.

Genetic Algorithms

In some optimization problems the decision variables belong to an integer set. For example, when the decision is to include or not a control loop, a sensor, etc., in this cases the decision variable will take a binary value. A value of zero if the loop/sensor is not included or one if the opposite happen. If the problem parametrization is correct and a functional cost exists, which measures a particular profit, then the problem can be called a mixed-integer optimization one or combinatorial. In this particular case the combinatorial problem is binary with a dimension of 2^n, with n the potential control loops or sensors locations. In the sensor location case the problem can be augmented with different sensor types. Considering 20 potential sensors locations with three different sensors type, then the problem dimension results $(3+1)^{20} \approx 1 \times 10^{12}$. This shown that the size of the combinatorial problem can increase considerably, making unpractical an exhaustive search of the optimal solution. In this context, is necessary some methodology to solve this kind of problems with the current both software and hardware resources.

An excellent alternative to solve this optimization problems are the genetic algorithms (GA). The GA is a stochastic global search method that mimics the metaphor of natural biological evolution. Mainly, the GA operates on a population of potencial solutions, N_i, applying the principle of survival of the fittest to produce better and better approximation to the solution. In each generation a new set of individuals, $\mathbf{I}_i$, is created based on their level of fitness respect to the specific functional cost for each problem, $FC(\mathbf{I}_i)$, and breeding them together using operators borrowed from natural genetics. Thus, the evolution of the population is developed towards the individuals that are better suited to their environment. Specifically, in this work a code version developed by [39] for Matlab environment is used.

The individuals, $\mathbf{C}_i = [c_1, \ldots, c_{N_c}]$, are encoded as strings (chromosomes) composed over some alphabet, so that the chromosome values, c_j, are uniquely mapped onto the decision variables domain. Decoding the chromosome representation, the decision variables (individuals) can be evaluated considering some performance or fitness function. This function establishes the basis for pairs selection of individuals that will be mated together during reproduction. In this phase, each individual is assigned to a fitness value derived from this objective function and the selection over the population is made with a determined probability according to their relative fitness. Thus, the recombination process is carried out to produce the next generation.

Typically, the initial population is adjusted between 2 and 20% of the problem dimension for a small combinatorial size [39] and about $Nc * 100$ to large scale problems [40, 41]. Generally, the GA termination criteria is the maximum generation number, N_g or a level of fitness. The individuals selection is performed by means of the roulette wheel method according to their fitness measure, thus the best $N_{\text{sel}} = N_i/2$ individuals are retained. The production of new chromosomes is developed by the crossover operator, in this case double-point method with a probability of P_{co}. Analogously to the natural evolution, the mutation produces a new genetic structure and basically is applied with a low probability, P_m. Mutation generally tends to inhibit the possibility of converging to a local optimum.

The following steps summarize the GA procedure to find the optimal set of individuals along generations,

1. *Initialization*: the random initial population, $\mathbf{P}_j = [\mathbf{C}_1, \ldots, \mathbf{C}_{N_i}]^T$ of dimension $N_i \times N_c$, is defined with $j = 0$. In addition, N_i, N_g, N_{sel}, P_{co} and P_m are selected.
2. *Fitness evaluation*: the functional cost is evaluated for each individual, $\mathbf{I}_i U$ from the actual population set, $\mathbf{P}_j$. In addition, the individual with best fitness value is stored in the best population set, $\mathbf{P}_b$.
3. *While*: the termination criteria is false do

   A. *Selection*: the N_{sel} best individuals are selected using their relative fitness values and stored in $\mathbf{P}_s$. The $N_i - N_{\text{sel}}$ remaining individuals are discarded.
   B. *Recombination*: the individuals in $\mathbf{P}_s$ are recombining by the crossover operator and stored in, $\mathbf{P}_r$.
   C. *Mutation*: the recombined individuals in $\mathbf{P}_r$ suffer the mutation process and new genetic structures are obtained and stored in $\mathbf{P}_m$.
   D. *Merging*: both selected and mutated populations are merging together to give the next generation of individuals, $\mathbf{P}_j = [\mathbf{P}_s, \mathbf{P}_m]^T$, with $j = j + 1$ and go to step **B**.

4. *Else* if the termination criteria is true, the set of the best individuals for each generation is obtained from $\mathbf{P}_b$, with a dimension of $j \times N_c$, and the optimization

thus, the better suited individual to their environment is the best individual in the latest generation.

Matrix Properties

Some matrix properties are presented here according to the functional cost used in the optimization problem stated in Eq. 8. Remembering that it was proposed from steady-state analysis to reduce the operating point deviations for the uncontrolled outputs. This functional has some additional/indirect properties.

Minimization of $||\Lambda_2 S_{sp} \Lambda_1||_F^2$: Setpoint effect

The first contribution of the functional cost of the Eq. 8 is $||\Lambda_2 \mathbf{S}_{sp} \Lambda_1||_F^2$. Without loss of generality, the term $||\mathbf{S}_{sp}||_F$ will be analyzed first. Remembering the Frobenius norm definition [32], and the relationships stated in Sect. 2.2,

$$||\mathbf{S}_{sp}||_F = ||\mathbf{G}_r \mathbf{G}_s^{-1}||_F = \sqrt{\sum_{i=1}^{k} \sigma_i^2 \left(\mathbf{G}_r \mathbf{G}_s^{-1}\right)}, \quad \text{con } k = \min(m-n, n) \tag{50}$$

being $\sigma_i \left(\mathbf{G}_r \mathbf{G}_s^{-1}\right)$ the ith singular value of the matrix $\mathbf{G}_r \mathbf{G}_s^{-1}$. In this context and taking into account some singular value properties [42, 43] the following relationships can be written,

$$\underline{\sigma}\left(\mathbf{G}_r\right) \overline{\sigma}\left(\mathbf{G}_s^{-1}\right) = \frac{\underline{\sigma}\left(\mathbf{G}_r\right)}{\underline{\sigma}\left(\mathbf{G}_s\right)} \leq \overline{\sigma}\left(\mathbf{G}_r \mathbf{G}_s^{-1}\right) \leq ||\mathbf{G}_r \mathbf{G}_s^{-1}||_F \tag{51}$$

$$\overline{\sigma}\left(\mathbf{G}_r\right) \underline{\sigma}\left(\mathbf{G}_s^{-1}\right) = \frac{\overline{\sigma}\left(\mathbf{G}_r\right)}{\overline{\sigma}\left(\mathbf{G}_s\right)} \leq \overline{\sigma}\left(\mathbf{G}_r \mathbf{G}_s^{-1}\right) \leq ||\mathbf{G}_r \mathbf{G}_s^{-1}||_F \tag{52}$$

these expressions shown a clear link between the functional cost and the matrix properties of $\mathbf{G}_s$. Here, $\underline{\sigma}$ (or $\sigma_{\min}$) and $\overline{\sigma}$ (or $\sigma_{\max}$) represent the minimum and maximum singular values. Note that the functional cost reduction requires the minimization of the left-hand ratios in Eqs. 51 and 52, so the matrix properties will be modified. Particularly the $\mathbf{G}_s$ properties are important, because this is the final plant to be controlled.

The $\underline{\sigma}\left(\mathbf{G}_s\right)$ value measures how close is the matrix $\mathbf{G}_s$ of the singularity and ill-conditioning. Another important index is the matrix condition number $\gamma\left(\mathbf{G}_s\right) = \overline{\sigma}\left(\mathbf{G}_s\right) / \underline{\sigma}\left(\mathbf{G}_s\right)$, high values in $\gamma\left(\mathbf{G}_s\right)$ indicates ill-conditioning and eventually a poor Λ. All this behaviors indicate that $\mathbf{G}_s$ is very difficult to control and with high sensitivity to modeling errors [26, 27, 32].

Note that the original problem was stated for to select $\mathbf{G}_s$ from the original process $\mathbf{G} = [\mathbf{G}_s^T \ \mathbf{G}_s^T]^T$ so that the cost function is minimized, in other words a rows permutation problem. In this context, and considering the submatrices or interlacing property (corollary 3.1.3 of [42]) the following instances for the Eq. 52 can be stated,

$$0 \leq \overline{\sigma}(\mathbf{G}_r)/\overline{\sigma}(\mathbf{G}_s) \leq 1 \tag{53}$$

$$1 \leq \overline{\sigma}(\mathbf{G}_r)/\overline{\sigma}(\mathbf{G}_s) \leq ||\mathbf{G}_r \mathbf{G}_s^{-1}||_F \tag{54}$$

for $n > m/2$ and $n < m/2$ cases respectively. Equations 53 and 54 show that the functional cost minimization almost has no effect on $\overline{\sigma}(\mathbf{G}_s)$. In the former the ratio is perfectly bounded regardless the functional cost profile, and in the latter case the functional cost produces that $\overline{\sigma}(\mathbf{G}_s)$ and $\overline{\sigma}(\mathbf{G}_r)$ approaching each other. In other words, $\overline{\sigma}(\mathbf{G}_s)$ is bounded and with minimal modifications. This shows that the real

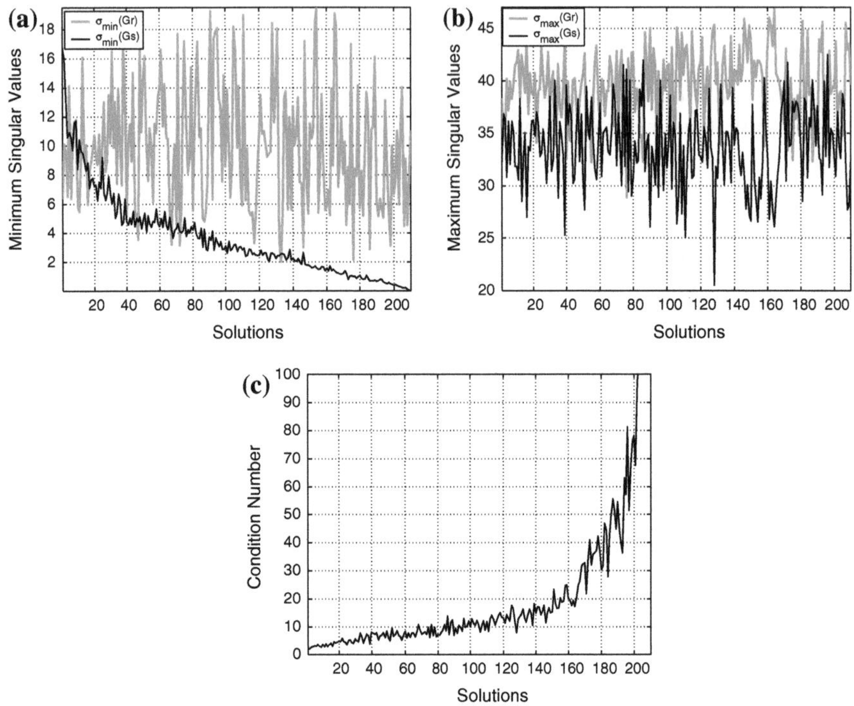

Fig. 19 Matrix properties evolution **a** Minimum singular values, **b** Maximum singular values, **c** Condition number, $\gamma(G_s)$

impact of the minimization is performed on the ratio shown in Eq. 51. This is true because changes in $\underline{\sigma}(\mathbf{G}_s)$ have stronger impact in the functional cost than those in $\overline{\sigma}(\mathbf{G}_s)$.

Summarizing, this minimization selects the best rows of $\mathbf{G}$ into $\mathbf{G}_s$ such that the multivariable transference gain of $\mathbf{G}_r\mathbf{G}_s^{-1}$ is minimized and consequently the matrix properties of $\mathbf{G}_s$ are improved by increasing $\underline{\sigma}(\mathbf{G}_s)$ (better conditioning and RGA)

As an example, Fig. 19 shows the evolution of the matrix properties along the solutions. In this case a 10×4 random process is generated to test the previously discussed properties of minimizing $||\mathbf{S}_{sp}||_F$. The same methodology was used in [26]. Figure 19a displays the profile of the minimum singular value for the matrices $\mathbf{G}_s$ and $\mathbf{G}_r$ when the solutions of the optimization algorithm were ordered from the best (at the left) to the worst (at the right). It can be observed as the $\underline{\sigma}(\mathbf{G}_s)$ increases towards the optimal solution. Analogously, Fig. 19b shows the evolution profile of the maximum singular values in this case. Equation 54 is fulfill here, so $\overline{\sigma}(\mathbf{G}_s)$ and $\overline{\sigma}(\mathbf{G}_r)$ tend to take the same values when $||\mathbf{S}_{sp}||_F$ is minimized. Due to this is expected to have a decreasing condition number of $\mathbf{G}_s$ as is shown in Fig. 19c. The optimal sensor location in this case results $\mathbf{C}_o = [0000111010]$.

Minimization of $||\Theta_2 \mathbf{S}_d \Theta_1||_F^2$: Disturbance effect

Again here without loss of generality the term $||\mathbf{S}_d||_F$ can be analyzed instead $||\Theta_2 \mathbf{S}_d \Theta_1||_F^2$. Remembering that $\mathbf{S}_d = \left[\mathbf{D}_r - \mathbf{G}_r \mathbf{G}_s^{-1} \mathbf{D}_s\right]$ and take into account some singular and eigenvalues properties [42, 43], the following inequality can be stated,

$$\left|\overline{\sigma}(\mathbf{D}_r) - \overline{\sigma}\left(\mathbf{G}_r \mathbf{G}_s^{-1} \mathbf{D}_s\right)\right| \leq \overline{\sigma}\left(\mathbf{D}_r - \mathbf{G}_r \mathbf{G}_s^{-1} \mathbf{D}_s\right) \leq ||\mathbf{D}_r - \mathbf{G}_r \mathbf{G}_s^{-1} \mathbf{D}_s||_F \quad (55)$$

by following the methodology presented in the previous section an upper bound in the minimum singular value of $\mathbf{G}_s$ can be obtained,

$$\underline{\sigma}\left(\mathbf{G}_s\right) \leq \frac{\overline{\sigma}\left(\mathbf{G}_r\right) \overline{\sigma}\left(\mathbf{D}_s\right)}{\overline{\sigma}\left(\mathbf{D}_r\right)} \quad (56)$$

this result shows that in this case (minimization of $||\mathbf{S}_d||_F$) the $\underline{\sigma}\left(\mathbf{G}_s\right)$ cannot be increased freely. In fact, it is bounded above by the process and disturbance transference matrix characteristics. Anyway, in this case also the matrix property $\underline{\sigma}\left(\mathbf{G}_s\right)$ tends to be improved, since it is the main reason of singularity in the left-hand component for the Eq. 55. Note that, if $||\mathbf{D}_r - \mathbf{G}_r \mathbf{G}_s^{-1} \mathbf{D}_s||_F \to 0$ then $\overline{\sigma}\left(\mathbf{G}_r \mathbf{G}_s^{-1} \mathbf{D}_s\right) \to \overline{\sigma}(\mathbf{D}_r)$ and $\mathbf{G}_s$ cannot be ill-conditioned or singular.

So, the minimization of $||\mathbf{S}_d||_F$ can suggests a different solution to the opportunely stated by $||\mathbf{S}_{sp}||_F$, basically due to the constraint in Eq. 56. Obviously, again here appear the classic trade-off control problem between servo and regulator behaviors. In this context, the complemented and weighted functional cost stated in Eq. 8 was proposed.

As an example, in Fig. 20 can be observed the evolution of the matrix properties along the solutions. In this case both a 10×4 random process, $\mathbf{G}$ and 10×2 random process, $\mathbf{D}$, were generated to test the previously discussed properties of minimizing $||\mathbf{S}_d||_F$. The same methodology was used in [26]. Figure 20a displays the profile of the minimum singular value for the matrices $\mathbf{G}_s$ and $\mathbf{G}_r$, and the upper bound $\overline{\sigma}\left(\mathbf{G}_r\right) \overline{\sigma}\left(\mathbf{D}_s\right) / \overline{\sigma}\left(\mathbf{D}_r\right)$ when the solutions of the optimization algorithm was ordered from the best (left) to the worst (right). Can be observed as the $\underline{\sigma}(\mathbf{G}_s)$ increases toward the optimal solution and the Eq. 56 is fulfill. Analogously, Fig. 20b shows the evolution profile of the maximum singular values in this case. Equation 54 is fulfill here, so $\overline{\sigma}(\mathbf{G}_s)$ and $\overline{\sigma}(\mathbf{G}_r)$ tend to take the same values when $||\mathbf{S}_{sp}||_F$ is minimized. Due to this is expected to have a decreasing condition number of $\mathbf{G}_s$ as is shown in Fig. 20c. Remembering the upper limit in Eq. 54 the $\underline{\sigma}(\mathbf{G}_s)$ cannot be increased freely, so the best condition number reached is not the optimal one as in the previous section. The optimal sensor location in this case results $\mathbf{C}_o = [0100011010]$, which is different from the previous case.

A classical problem in control structure design is the servo/regulator behavior trade-off. In fact,it is desirable that the control policy has both a good performance for set point tracking and disturbance rejection simultaneously. The functional in Eq. 8 addresses this problem by accounting for both contributions and the weighting

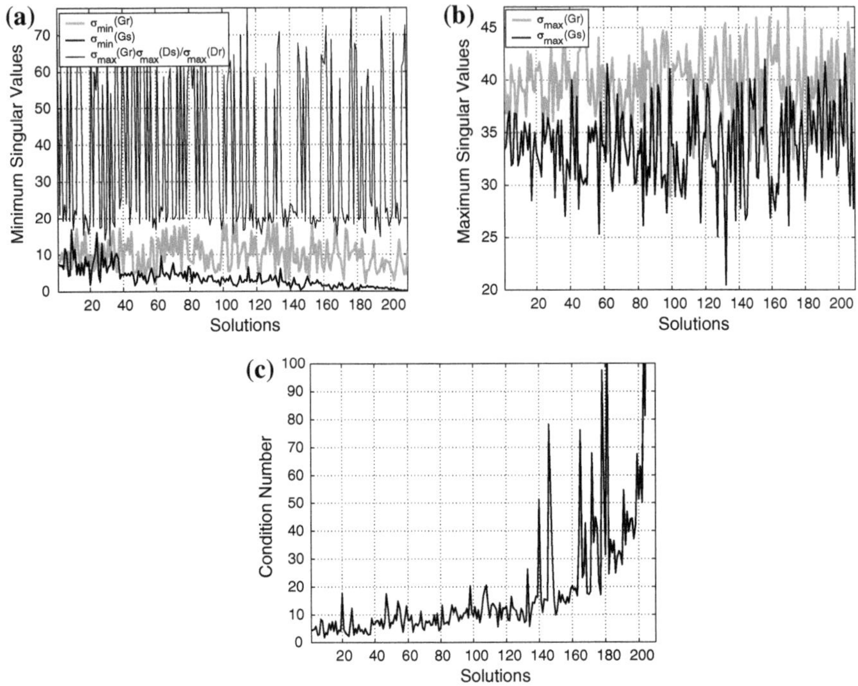

Fig. 20 Matrix properties evolution **a** Minimum singular values, **b** Maximum singular values, **c** Condition number, $\gamma(G_s)$

matrices. So, if these matrices Λ_1, Λ_2, Θ_1 and Θ_2 are used, then the partial control objective from synthesis process and non-normalized models can be accounted. The previous analysis is still valid in this case but now the new matrices become $\mathbf{S}_{sp}^* = \Lambda_2 \mathbf{S}_{sp} \Lambda_1$ and $\mathbf{S}_d^* = \Theta_2 \mathbf{S}_d \Theta_1$, and of course the new optimal solution may differ from the unweighted case.

References

1. Larsson T, Skogestad S (2000) Plantwide control—a review and a new design procedure. Model Indent control 21(4):209–240
2. Buckley PS (1964) Techniques of process control. Willey, New York
3. Luyben WL, Tyreus BD, Luyben ML (1999) Plantwide process control. Chemical Engineering Series. McGraw-Hill, New York
4. Konda M, Rangaiah G, Krishnaswamy P (2005) Plantwide control of industrial processes: An integrated framework of simulation and heuristics. Ind Eng Chem Res 44:8300–8313
5. Cao Y, Saha P (2005) Improved branch and bound method for control structure screening. Chem Eng Sci 60:1555–1564

6. Cao Y, Kariwala V (2008) Bidirectional branch and bound for controlled variable selection. part 1. principles and minimum singular value criterion. Comput Chem Eng 32:2306–2319
7. Robinson D, Chen R, McAvoy T, Schnelle D (2001) An optimal control based approach to designing plantwide control system architectures. J Proc Control 11:223–236
8. Jorgensen J, Jorgensen S (2000) Towards automatic decentralized control structure selection. Comput Chem Eng 24:841–846
9. Skogestad S (2000) Plantwide control: the search for self-optimizing control structure. J Proc Control 10:487–503
10. Skogestad S (2004) Control structure design for complete chemical plants. Comput Chem Eng 28:219–234
11. Alstad V, Skogestad S (2007) Null space method for selecting optimal measurement combinations as controlled variables. Ind Eng Chem Res 46:846–853
12. Suraj Vasudevan S, Rangaiah N, Konda M, Tay W (2009) Application and evaluation of three methodologies for plantwide control of the styrene monomer plant. Ind Eng Chem Res 48:10941–10961
13. Bristol E (1966) On a new measure of interaction for multivariable process control. IEEE Trans Autom Control
14. Chang J, Yu C (1990) The relative gain for non-square multivariable systems. Chem Eng Sci 45:1309/1323
15. Chang J, Yu C (1992) Relative disturbance gain array. AIChE J 38:521–534
16. Mao-Jun He, Wen-Jian Cai, Wei Ni, Li-Hua Xie (2009) Rnga based control system configuration for multivariable processes. J Proc Control 19:1036–1042
17. Molina G, Zumoffen D, Basualdo M (2009) A new systematic approach to find plantwide control structures. Comp Aided Chem Eng 27:1599–1604
18. Zumoffen D, Basualdo M (2009) Optimal sensor location for chemical process accounting the best control configuration. Comp Aided Chem Eng 27:1593–1598
19. Zumoffen D, Basualdo M, Ruiz J (2009) Optimal multivariable control structure design for chemical plants. AIChE annual meeting, Nashville, TN, USA
20. Nieto N, Zumoffen D, Basualdo M, Outbib R (2010) Systematic control structure design of a fuel processor system In International conference on renewable energy. Al-Ain, United Arab Emirates
21. Zumoffen D, Molina G, Basualdo M (2010) Plant-wide control based on minimum square deviation. In: Proceedings of the 9th international symposium on dynamics and control of process systems, Leuven, Belgium, pp 443–448
22. Molina GD, Zumoffen DAR, Basualdo MS (2010) Plant-wide control strategy applied to the tennessee eastman process at two operating points. Comp Chem Eng (in Press)
23. He M, Cai W, Ni W, Xie L (2009) RNGA based control system configuration for multivariable processes. J Process Control 19:1036–1042
24. Arkun Y, Downs J (1990) A general method to calculate input–output gains and the relative gain array for integrating processes. Comp Chem Eng 14:1101–1110
25. McAvoy T (1998) A methodology for screening level control structures in plantwide control systems. Comp Chem Eng 22:1543–1552
26. Morari M, Grosdidier P, Holt B (1985) Closed-loop properties from steady-state gain information. Ind Eng Chem Fundam 24:221–235
27. Skogetad S, Morari M (1987) Implications of large RGA elements on control performance. Ind Eng Chem Res 26:2323–2330
28. Chang J, Yu C (1994) Synthesis of controller structures for robust load performance. Int J Control 60:1353–1369
29. Garcia C, Morari M (1985) Internal model control. 2. design procedure for multivariable systems. Ind Eng Chem Process Des Dev 24:472–484
30. Kariwala V (2004) Multi-loop controller synthesis and performance analysis. Ph.D. thesis, University of Alberta, Canada

31. Goodwin G, Graebe S, Salgado M (2000) Control system design. Prentice Hall, Upper Saddle River
32. Skogestad S, Postlethwaite I (2005) Multivariable feedback control. Analysis and design. Wiley, New York
33. Rivera D (1986) Internal model control. 4. PID controller design. Ind Eng Chem Process Des Dev 25:252–265
34. Rivera D (2007) Una metodología para la identificacion integrada con el diseño de controladores IMC-PID. RIAI (Rev. Iberoamericana de Autom e Inf. Ind), 4(4):5–18
35. Van Overschee P, De Moor B (1994) N4SID: Subspace algorithms for the identification of combined deterministic-stochastic systems. Automatica 30:75–93
36. Ljung L (1999) System Identification (Theory for the user), 2nd edn. Upper Saddle River, Prentice Hall
37. Ljung L (2002) System identification toolbox (user's guide version 5). Technical report, The Math Works Inc.
38. Pukrushpan JT (2003) Modeling and Control of Fuel Cell Systems and Fuel Processors. Ph.D. thesis, University of Michigan
39. Chipperfield A, Fleming P, Pohlheim H, Fonseca C (1994) Genetic algorithm toolbox. for use with matlab. University of Sheffield. Department of automatic control and systems engineering, 44:5645–5659
40. Musulin E, Bagajewicz M, Nougués J, Puigjaner L (2004) Instrumentation design and upgrade for principal components analysis monitoring. Ind Eng Chem Res 43:2150–2159
41. Musulin E, Yélamos I, Puigjaner L (2006) Integration of principal component analysis and fuzzy logic systems for comprehensive process fault detection and diagnosis. Ind Eng Chem Res 45:1739–1750
42. Horn R, Johnson C (1990) Matrix analysis. Cambridge University Press, Cambridge
43. Golub G, Van Loan C (1996) Matrix computations. The Johns Hopkins University Press, 3rd edn. Johns Hopkins University Press, London

Fault Detectability Index for Optimal Monitoring System Design

D. Zumoffen, L. Nieto Degliuomini and M. Basualdo

Abstract In this chapter an approach based on fault detectability study for deciding the optimal measurements selection to build the principal component analysis model using combined statistics is addressed. It is believed that it will improve the monitoring systems design. The methodology applied here to the bioethanol processor with PEMFC is integrated to the previous stage of optimal sensor network and control structure selection detailed in Chap. 12. The problem dimensionality could be important so genetic algorithms for stochastic global search is preferred to solve it. The solution can be driven properly to avoid the inclusion of sensors other than those installed for control purposes. It can be done through a good definition of the functional cost. The capacity of this methodology is demonstrated with a set of typical critical faults that could occur to this complex process.

1 Introduction

In large-scale processes monitoring systems play an important role in supervising the state of the plant. Hundreds and thousands of variables must be handled in each sample time. The increasing demand for obtaining more efficient and rentable process

D. Zumoffen · L. Nieto Degliuomini · M. Basualdo(✉)
CAPEG- CIFASIS-(CONICET-UNR-UPCAM), Rosario, Argentina
e-mail: basualdo@cifasis-conicet.gov.ar

D. Zumoffen
e-mail: zumoffen@cifasis-conicet.gov.ar

D. Zumoffen · M. Basualdo
CAPEG-Chemical Engineering Department, FRRo-UTN, Zeballos 1341, S2000BQA Rosario, Argentina

L. Nieto Degliuomini
e-mail: nieto@cifasis-conicet.gov.ar

M. S. Basualdo et al. (eds.), *PEM Fuel Cells with Bio-Ethanol Processor Systems*, Green Energy and Technology, DOI: 10.1007/978-1-84996-184-4_13,

generates complex control systems and very interconnected plants. In this context, it is extremely difficult to achieve a suitable monitoring without the support of a decision-maker system [1–3].

Principal component analysis (PCA)-based monitoring systems are widely used in industrial processes as well as academic research [4–6] due to their excellent properties for handling noise and large data bases. One of the analyzed problems here is, which is the minimum number of signals to be chosen for developing an efficient PCA monitoring system? This question usually is not addressed because the tendency is to use all the available information sources for developing the PCA model.

In this chapter a new approach to improve the PCA-based monitoring systems performance is presented. Its optimal design is thought to maximize the abnormal events detection and to minimize the investment costs related to the number of sensors to be bought. The detectability indexes are useful for determining which set of variables is enough to achieve the best fault classification. Particularly, it is proposed here a combined statistic based approach [1–3, 7]. Additionally, due to the great problem dimensionality a good solution can be found using genetic algorithms (GA). Thus, the objective function within the GA also considers a cost that penalizes the use of new measurement devices that are not in the current control structure. In other words, this approach tries to avoid the use of additional sensors for monitoring purpose because it will increase the hardware investment cost.

Some previous works in the literature addressed this complex problem in a partial way. For example, Yue and Qin [7] presented the combined statistic using PCA for fault reconstruction tasks on a rapid thermal annealing (RTA) process. Only fault detectability with square prediction error, Q, or hotelling, T^2, statistics were analyzed. Nothing was said about detectability indexes based on combined statistics and the optimal signal selection for PCA implementation. Additionally, in this last area, several works have been published [8–10] based on Kalman filtering in steady-state (final values), integer optimization routines and observability index. It can be remarked that, these approaches were applied in two ways, assuming that the process was working in open loop or with an existing control structure in the process. Musulin et al. [11, 12] presented a signal selection method with PCA-based monitoring using only Q and T^2 statistics. The approach was applied on the Tennessee Eastman (TE) process where the signals selection was done as a trade-off between cost and precision aspects. That strategy was tested in two case studies. First, a chemical plant with recycle is used accounting for faults and disturbances, and second, the TE process by analyzing some of the suggested disturbances without taking into account the potential hardware faults.

The methodology proposed in this chapter presents the philosophy followed in Chap. 12. Therefore, a global analysis, generalized and systematic for solving how to design an optimal sensor network integrated to the plant-wide control structure and the optimal monitoring system design (OMSD) is performed. In Chap. 12 only a steady-state model of the process is accounted and using as less as possible heuristic concepts for reducing the combinatorial problem dimensionality. The methodology presented here complements it by developing an OMSD, based on historical data

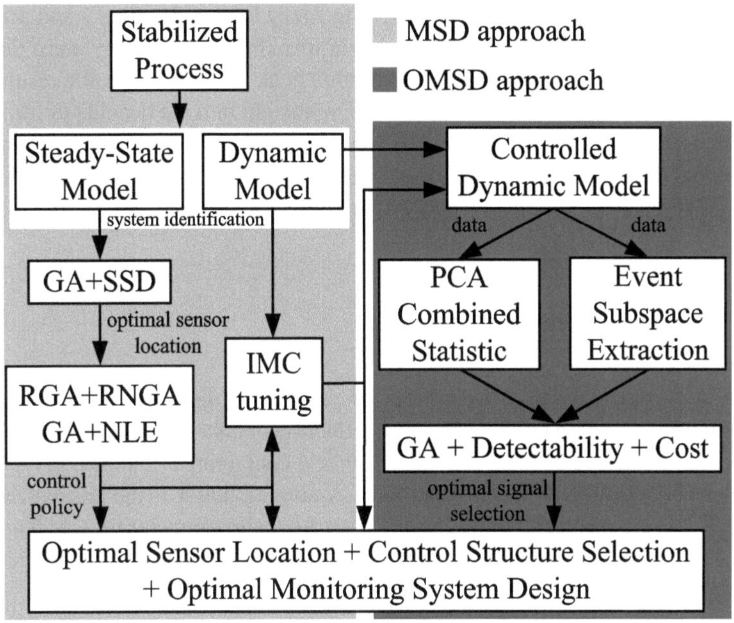

Fig. 1 Proposed strategy

base for both, normal and abnormal behavior of the process. The main steps of this approach are summarized at Fig. 1, showing the involved tools in each stage. It was proposed and tested previously with the benchmark TE in Zumoffen and Basualdo [13]. In Fig. 1, the block diagram with light gray background displays the procedure applied in Chap. 12, the minimum square deviation (MSD) approach. Meanwhile, the dark gray background summarizes the module developed in this chapter and its integration with the OMSD strategy.

Finally, based on the arguments given above, it is important to remark that the approach applied in BPS with FC system is innovative with respect to analogous previous strategies [11, 12]. First, the global treatment of the complete problem, which has demonstrated to be efficient for the TE process [13] could be extended to this challenging and novel case. The problem was solved supported by a well-defined objective function to be optimized through genetic algorithm. Second, the monitoring design based on the detectability index, calculated with the combined statistics, was adopted here because of those successfully obtained when it was applied on complex chemical processes [1–3]. The results presented here considered seven critical faults in sensors of flows, temperatures and ratio between quantities of reactants. The final test about the real capacity of the monitoring-diagnosis-estimation system, accounting for the recommendations given by the detectability index, will be developed with adaptive PCA (APCA) techniques. They present better behavior than conventional ones in plants when the changing operating conditions are normal. The same situation is reported in BPS with PEMFC when they are onboard of vehicles. The APCA, diagnosis and estimation modules, including fuzzy logic (FL) theory and artificial neural networks (ANN), are not discussed in this chapter since they were detailed in Zumoffen and Basualdo [1, 2] and Zumoffen et al. [3]. Therefore, the results presented in this chapter will be useful to get more insight into the benefits of having an efficient system to determine in time if the most critical faults occur. It constitutes a recommendable way for guaranteeing safety and efficiency for the BPS+PEMFC which represents one of the new energy paradigms.

2 PCA-based Monitoring

The PCA is the main tool used here for performing fault detection. PCA is a projection-based method that facilitates a reduction of data dimension. This analysis begins by considering the data matrix $\mathbf{X}$ of $m \times n$ containing m samples of n process variables collected under normal operation. Assuming that $\overline{\mathbf{X}}$ is the normalized version of $\mathbf{X}$, to zero mean and unit variance with the scale parameter vectors $\mathbf{b}$ and $\mathbf{s}$ as the mean and variance vector of the process variables in the data matrix respectively. The normalized data matrix can be represented as

$$\overline{\mathbf{X}} = \mathbf{T}\mathbf{P}^T + \mathbf{E} \quad (1)$$

where $\mathbf{T} \in \Re^{m \times A}$ and $\mathbf{P} \in \Re^{n \times A}$ are the latent and principal component matrices respectively, and A is the number of principal components retained in the model $\mathbf{P}$. The residual matrix $\mathbf{E}$ represents the associated error since only $A \ll n$ principal components were selected.

$\mathbf{P}$ can be obtained by means of singular value decomposition (SVD) from the normalized data correlation matrix as shown in Eq. 2

$$\mathbf{R}_c = \overline{\mathbf{X}} \cdot \overline{\mathbf{X}}^T / (m-1) = \mathbf{U}\mathbf{D}_\lambda \mathbf{U}^T \quad (2)$$

by selecting only the first A columns of $\mathbf{U}$. This factorization produces a diagonal matrix $\mathbf{D}_\lambda = \text{diag}(\lambda_1, \lambda_2, \ldots, \lambda_n)$, where λ_i are the eigenvalues of $\mathbf{R}_c$ sorted in decreasing order ($\lambda_1 > \lambda_2 > \cdots > \lambda_n$) and the corresponding columns of $\mathbf{U}$ are the eigenvectors $\mathbf{p}_i$ and the so-called principal components. Thus, $\mathbf{P} = [\mathbf{p}_1, \ldots, \mathbf{p}_A]$ and $\mathbf{D}_A = \text{diag}(\lambda_1, \ldots, \lambda_A)$.

A reduction of dimensionality is made by projecting every normalized sample vector $\overline{\mathbf{x}}(k)$ (of dimension $n \times 1$) in the principal component space generated by $\mathbf{P}$,

$$\mathbf{t}(k) = \mathbf{P}^T \overline{\mathbf{x}}(k) \quad (3)$$

which is called the principal score vector.

Different approaches for selecting the A principal components retained [14] can be chosen. In this chapter the cumulative percent variance (CPV) is used and displayed in Eq. 4. This index measures the percent variance captured by the first A principal components.

$$\mathrm{CPV}(A) = \frac{\sum_{j=1}^{A} \lambda_j}{\mathrm{trace}(\mathbf{R}_c)} 100\% \tag{4}$$

In this case a search between $1 \le A \le n$ is made to satisfy the condition $\mathrm{CPV}(A) \ge \delta_{\mathrm{cpv}}$ with the minimum A. Where δ_{cpv} is a percentage value, if it achieves lower values it means that only a few principal components retained are needed and if it is close to 100% it means that $A \approx n$.

For generating quality control charts in multivariable online monitoring process with PCA, two statistics are widely used: the Hotelling, $T^2(k)$, and the squared prediction error (SPE), $Q(k)$ [1–3, 13]. Considering the actual process measurements at their normalized version $\overline{\mathbf{x}}(k)$, being k the actual sampling time, these statistics are defined as shown in Eq. 5,

$$T^2(k) = \left|\left|\mathbf{D}_A^{-1/2}\mathbf{P}^T\overline{\mathbf{x}}(k)\right|\right|^2, \qquad Q(k) = \left|\left|\widetilde{\mathbf{C}}\overline{\mathbf{x}}(k)\right|\right|^2 \tag{5}$$

where $\widetilde{\mathbf{C}} = \mathbf{I} - \mathbf{P}\mathbf{P}^T$ and $\Delta\overline{\mathbf{x}}(k) = \widetilde{\mathbf{C}}\overline{\mathbf{x}}(k)$ is the prediction error.

The test consists of declaring as normal operation if $T^2(k) \le \delta_{T^2}$ for Hotelling's statistic and $Q(k) \le \delta_Q$ for SPE statistic, where δ_{T^2} and δ_Q are the control or confidence limits for the above statistics respectively. Supposing a Gaussian distribution the control limits can be approximated by $\delta_i = \mu_i + \nu \cdot \sigma_i$ [15, 16], where μ_i and σ_i are the mean and variance values for the statistic i computed from the normal data matrix ($i = T^2, Q$) and $\nu = 2, 3$ according to the 95 or 99% confidence level respectively.

The model, $\mathbf{P}$, is computed using the normal data matrix which has information about the *common-cause* variations at the surroundings of the process operation point. The Hotelling's statistic, T^2, for a new incoming data sample is a measurement of its distance with respect to the origin of the dominant variation subspace. This origin and its proximities delimit the *in-control zone*. In case an abnormal event happens but the principal score vectors remain in the surroundings of the in-control zone (Hotelling's statistic is under the confidence limit) suggesting that it cannot change enough over the dominant variation subspace to advise about the occurrence of that abnormal event. The use of Q and T^2 statistics, working together, in a combined way [7] could avoid this loss of detection. Hence, in Sect. 5.1 of this chapter a new detectability analysis is done based on the combined statistics. This approach improves the detection properties such as detection times, less false alarms occurrence and missed detections. The combined index $z(k)$ is defined as shown in Eq. 6,

$$z(k) = \frac{T^2(k)}{\delta_{T^2}} + \frac{Q(k)}{\delta_{\mathrm{Q}}} \tag{6}$$

note that $z(k)$, $T^2(k)$ and $Q(k)$ are the computed statistics for the actual measurement $\overline{\mathbf{x}}(k)$. These assumptions can be declared as the normal operation condition if $z(k) \le \delta_z$, where δ_z is the new control limit. A conservative selection could be

$\delta_z = 2$, since false alarms are considered harmful, i.e. robustness is required. A modified version of this approach integrated with adaptive PCA techniques, fault detection, diagnosis and estimation system, which supports fault tolerant control, was deeply tested in [1–3, 13]. Based on the good results obtained in these works it will be applied here for testing the detectability index capacity.

3 Fault Detectability Index Based on T^2 and Q

Fault detectability index can be suitably obtained by using an additive fault model [7] representation of the abnormal process data. This methodology can be applied to both Hotelling and the square prediction error statistics. Yue and Qin [7] presented an approach to develop this index but for T^2 and Q separately. In the following sections an extension of it for computing the fault detectability index based on the combined statistics is proposed.

The normalized process measurement, $\bar{\mathbf{x}}_*(k)$, when a fault is present can be written as,

$$\bar{\mathbf{x}}_*(k) = \bar{\mathbf{x}}_0(k) + \Theta_j \mathbf{f}_j \tag{7}$$

in this fault model two additive effects can be observed: $\bar{\mathbf{x}}_0(k)$ which considers the normal behavior case and $\Theta_j \mathbf{f}_j$ as the fault contribution to the actual measurements. Being Θ_j de fault subspace for the fault j with $j = 1, \ldots, J$ the fault types and $\mathbf{f}_j$ the fault components vector. The fault subspace extraction procedure is addressed properly in Sect. 4

$$\begin{bmatrix} \bar{x}_*^1(k) \\ \bar{x}_*^2(k) \\ \vdots \\ \bar{x}_*^n(k) \end{bmatrix} = \begin{bmatrix} \bar{x}_0^1(k) \\ \bar{x}_0^2(k) \\ \vdots \\ \bar{x}_0^n(k) \end{bmatrix} + \begin{bmatrix} \theta_j^{11} & \theta_j^{12} & \ldots & \theta_j^{1r} \\ \theta_j^{21} & \theta_j^{22} & \ldots & \theta_j^{2r} \\ \vdots & \vdots & \vdots & \vdots \\ \theta_j^{n1} & \theta_j^{n2} & \ldots & \theta_j^{nr} \end{bmatrix} \begin{bmatrix} f_j^1(k) \\ f_j^2(k) \\ \vdots \\ f_j^r(k) \end{bmatrix} \tag{8}$$

where n is the amount of measurement points, and r the fault component vector lengths for the fault type j. The columns of Θ have zero entries except for the measurement affected by the fault, in this case the entry is 1 or -1 depending on the fault direction.

Considering the fault detection process with hotelling statistic, then

$$T^2(k) = \left\| \mathbf{D}_A^{-1/2} \mathbf{P}^T \bar{\mathbf{x}}_*(k) \right\|^2 = \left\| \mathbf{D}_A^{-1/2} \mathbf{P}^T \bar{\mathbf{x}}_0(k) + \mathbf{D}_A^{-1/2} \mathbf{P}^T \Theta_j \mathbf{f}_j \right\|^2 \tag{9}$$

by accounting for the triangle inequality it can be expressed as

$$\left\| \mathbf{D}_A^{-1/2} \mathbf{P}^T \bar{\mathbf{x}}_*(k) \right\| \geq \left| ||\mathbf{D}_A^{-1/2} \mathbf{P}^T \bar{\mathbf{x}}_0(k)|| - ||\mathbf{D}_A^{-1/2} \mathbf{P}^T \Theta_j \mathbf{f}_j|| \right| \tag{10}$$

remembering that for fault detection the condition $T^2(k) \geq \delta_{T^2}$ must be fulfilled and that $||\mathbf{D}_A^{-1/2}\mathbf{P}^T\bar{\mathbf{x}}_0(k)|| \leq \delta_{T^2}$ by construction, the detectability condition using hotelling statistic can be written as

$$\left|\left|\mathbf{D}_A^{-1/2}\mathbf{P}^T\Theta_j\mathbf{f}_j\right|\right| \geq 2\delta_{T^2} \tag{11}$$

and the minimal fault magnitude (MFM) detectable with this statistic can be obtained with a most restrictive condition

$$\left|\left|\mathbf{D}_A^{-1/2}\mathbf{P}^T\Theta_j\right|\right| ||\mathbf{f}_j|| \geq \left|\left|\mathbf{D}_A^{-1/2}\mathbf{P}^T\Theta_j\mathbf{f}_j\right|\right| \geq 2\delta_{T^2} \tag{12}$$

resulting in the following expression for the MFM estimate

$$||\mathbf{f}_j||_{MFM}^{T^2} = \left|\left|\mathbf{D}_A^{-1/2}\mathbf{P}^T\Theta_j\right|\right|^{-1} 2\delta_{T^2} \tag{13}$$

Regarding fault detection with Q statistic it is possible to write

$$Q(k) = \left|\left|\tilde{\mathbf{C}}\bar{\mathbf{x}}_*(k)\right|\right|^2 = \left|\left|\tilde{\mathbf{C}}\bar{\mathbf{x}}_0(k) + \tilde{\mathbf{C}}\Theta_j\mathbf{f}_j\right|\right|^2 \tag{14}$$

using again the triangle inequality it can be expressed that

$$\left|\left|\tilde{\mathbf{C}}\bar{\mathbf{x}}_*(k)\right|\right| \geq \left|||\tilde{\mathbf{C}}\bar{\mathbf{x}}_0(k)|| - ||\tilde{\mathbf{C}}\Theta_j\mathbf{f}_j||\right| \tag{15}$$

where $||\tilde{\mathbf{C}}\bar{\mathbf{x}}_0(k)|| \leq \delta_Q$ by construction. Thus the fault detectability condition with this statistic is summarized as

$$\left|\left|\tilde{\mathbf{C}}\Theta_j\mathbf{f}_j\right|\right| \geq 2\delta_Q \tag{16}$$

and, in this case, the MFM estimation results as

$$||\mathbf{f}_j||_{MFM}^{Q} = \left|\left|\tilde{\mathbf{C}}\Theta_j\right|\right|^{-1} 2\delta_Q \tag{17}$$

4 Fault Subspace Extraction

The fault subspace, Θ, can be extracted from the abnormal data base by processing the overall potential measurements and signals from the process. Initially, due to the typical noise present in the process measurements is applied a smoothed moving average (SMA) filter for consistence. It performs the average (mean) of the original signal $x(k)$ over a specified moving window of dimension $N+1$ samples, as can be observed in Eq. 18, resulting in the filtered version $x_f(k)$.

$$x_f(k) = (N+1)^{-1} \sum_{i=0}^{N} x(k-i) \tag{18}$$

an improved algorithm exists to avoid problems with lagged samples called exponential weighted moving average (EWMA) for online applications [15, 17]. Thus, before the faults subspace extraction the auto-scaled data base is filtered using Eq. 18.

The fault directions are computed using the well-known "3δ *edit rule*" [18] which suggests that if $|x_f(k)| > 3$ the variable is considered to be deviated from its normal state. Analyzing the overall data base for each abnormal case, the fault directions can be computed as

$$F(j,i) = \begin{cases} 1, & \text{if } \bar{x}_f^{ji}(k_*) > 3; \\ 0, & \text{if } -3 \leq \bar{x}_f^{ji}(k_*) \leq 3; \\ -1, & \text{if } \bar{x}_f^{ji}(k_*) < -3. \end{cases} \quad \text{with} \quad \begin{array}{l} j = 1, \ldots, J \\ i = 1, \ldots, N_c \end{array} \tag{19}$$

where $\bar{x}_f^{ji}(k_*)$ is the auto-scaled and filtered version of the variable number i from the abnormal data base (event) j evaluated in the sampled instant k_*. This temporal instant is specified according to the dynamic response of the process trying to avoid the transient behavior due to the fault occurrence $k_* = t_f + N_*$, where t_f is the fault occurrence sample. Thus each row of F represented by $F(j, 1:N_c)$ corresponds to the j fault propagation over the N_c variables analyzed from the abnormal data base. The faults subspace Θ can be computed directly from the fault direction matrix F as can be observed in Eq. 20

$$\Theta_j = \text{nzr}\left\{diag\left[F(j, 1:N_c)\right]\right\}^T \tag{20}$$

where the function $diag(\cdot)$ takes the vector $F(j, 1:N_c)$ as input argument and gives back a diagonal matrix of $N_c \times N_c$ with $F(j, 1:N_c)$ in its diagonal. On the other side, the function $\text{nzr}\{\cdot\}$ takes as input argument the diagonal matrix constructed previously and gives back another matrix Θ_j. This fault subspace matrix contains only the *non zero rows* from the original diagonal matrix. Finally, this fault propagation matrix can be applied as stated in the previous section.

An alternative approach to obtain the fault direction matrix F exists [1–3, 13]. In these works, a strategy based on fuzzy logic tools is applied and the resultant matrix rules represent the fault propagations over the process variables. In this case, the matrix rules evaluation accounts for the mean contribution of the variables within the specified zone of analysis.

5 Optimal Monitoring System Design

In this section, the main steps of the proposed methodology [13], displayed in Fig. 1, are detailed. The approach stated in Chap. 12 (MSD) allows to solve both the optimal sensor location (OSL) and control structure selection (CSS) problems in a systematic and generalized way. These optimal solutions define the framework to perform the monitoring system design stage. Basically, the MSD approach defines,

- the hardware requirements: amount and type of sensors and controllers which are potential fault sources,
- the control structure: how disturbances and abnormal events affect (they are masked by the control) the overall process behavior,

In this context, the OMSD approach can be carried out efficiently. As shown in Fig. 1, the data base from the controlled dynamic model (or eventually from the real plant) is processed and analyzed. The data corresponding to normal behavior are used to develop the PCA model with combined statistic. On the other hand, the abnormal data base is analyzed to build the event/fault subspace, Θ, defined in Sect. 4. This last process is known as event/fault subspace extraction.

In the following subsections the OMSD approach is completed by including the detectability index for combined statistic, investment cost and GA to solve the combinatorial problem and to define the optimal signals selection for monitoring purposes.

5.1 Fault Detectability Index Based on Combined Statistic

In this subsection, the development of the fault detectability index with the combined statistic, $z(k)$, is given. It is done by performing an extension of the concepts given in the previous section. By grouping both hotelling and square prediction error statistics, it can be obtained by improvements in the fault detection performance [1–3, 13]. In this case, the combined statistic can be represented as

$$\begin{aligned} z(k) &= \frac{\bar{\mathbf{x}}_*^T(k)\mathbf{P}\mathbf{D}_A^{-1}\mathbf{P}^T\bar{\mathbf{x}}_*(k)}{\delta_{T^2}} + \frac{\bar{\mathbf{x}}_*^T(k)\tilde{\mathbf{C}}^T\tilde{\mathbf{C}}\bar{\mathbf{x}}_*(k)}{\delta_Q} \\ &= \bar{\mathbf{x}}_*^T(k)\left[\frac{\mathbf{P}\mathbf{D}_A^{-1}\mathbf{P}^T}{\delta_{T^2}} + \frac{\tilde{\mathbf{C}}^T\tilde{\mathbf{C}}}{\delta_Q}\right]\bar{\mathbf{x}}_*(k) \\ &= \bar{\mathbf{x}}_*^T(k)\mathbf{M}\bar{\mathbf{x}}_*(k) \end{aligned} \tag{21}$$

with

$$\mathbf{M} = \left[\frac{\mathbf{P}\mathbf{D}_A^{-1}\mathbf{P}^T}{\delta_{T^2}} + \frac{\tilde{\mathbf{C}}^T\tilde{\mathbf{C}}}{\delta_Q}\right] \tag{22}$$

considering that $\mathbf{M}$ is a symmetric and definite positive matrix [7], and its decomposition is possible by Cholesky factorization to obtain $\mathbf{M} = \mathbf{R}^T\mathbf{R}$. Thus, the combined statistic case presents a similar treatment as the Q statistic in Sect. 3, giving the following fault detectability condition

$$||\mathbf{R}\Theta_j\mathbf{f}_j|| \geq 2\delta_z \tag{23}$$

and its corresponding MFM estimate

$$||\mathbf{f}_j||_{MFM}^z = ||\mathbf{R}\Theta_j||^{-1} 2\delta_z \tag{24}$$

The combined matrix **M** (and eventually **R**) depends on the PCA model developed opportunely. The influences of factors such as sensors location in the process, signals selection to perform PCA model, variance retained and confidence limits are crucial and limit the attainable MFM. Similarly, the fault subspace matrix Θ_j is directly influenced by the sensors network (potential sources of faults) and the selected control structure. In other words, a particular design of these matrices can incur losses or poor quality fault detections. An interesting result can be obtained through the comparison between a PCA model developed using the overall available measurement points with the optimal solution given here [13].

5.2 Optimal Signal Selection Based on Detectability Maximization

The proper signals selection for the PCA model development is performed by focussing on faults detectability maximization. It is based on the existing sensors network in the previously proposed optimal control structure in Chap. 12 (MSD) and the potential additional cost in case new measurement points are required.

In this context, the problem can be defined as follows, considering N_c available signals including controlled as well as manipulated variables in the process and $\mathbf{C}_i = [c_1, c_2, \ldots, c_{N_c}]$ a particular signals selection, where $c_l = \{1; 0\}$ with $l = 1, \ldots, N_c$ represents a binary alphabet indicating the utilization or not of the signal in the l location (see Appendix in Chap. 12). Then, the PCA model construction depends on this particular selection, $\mathbf{P}(\mathbf{C}_i)$ and $\mathbf{D}_A(\mathbf{C}_i)$. In addition, the MFM calculation when combined statistics are used results as

$$\left|\left|\mathbf{f}_j^i\right|\right|_{MFM}^z = ||\mathbf{R}(\mathbf{C}_i)\Theta_j(\mathbf{C}_i)||^{-1} 2\delta_z \tag{25}$$

where i makes reference to the signals selection $\mathbf{C}_i$, with $i = 1, \ldots, 2^{N_c}$ all the possible combinations and $j = 1, \ldots, J$ the considered abnormal event types (disturbances, faulty elements, etc.).

Another important point is considering a cost penalization $\mathbf{c} = [c_1^*, \ldots, c_{N_c}^*] > 0$ each time new measurements are recommended to be included. Note that **c** does not necessarily need to be in monetary units, in fact, its value can be selected accounting a trade-off between detectability and quantity of new extra sensors apart from those installed for the control structure. Thus, lower values of **c** means that the minimization of the objective function in Eq. 26 priories detectability without considers the cost of extra sensors. Otherwise, when **c** has a considerable weight the cost of new sensors is penalized without considering the detectability index. The penalization coefficients must be normalized to contribute in the same order of magnitude as the term $\left\|\mathbf{R}(\mathbf{C}_i)\Theta_j(\mathbf{C}_i)\right\|^{-1} 2\delta_z$. Therefore, the complete problem to be solved can be stated as

$$\min_{\mathbf{C}_i}\left[\sum_{j=1}^{J}\left\|\mathbf{f}_j^i\right\|_{MFM}^z + \mathbf{c}\mathbf{C}_i^T\right] = \min_{\mathbf{C}_i}\left[\sum_{j=1}^{J}\left\|\mathbf{R}(\mathbf{C}_i)\Theta_j(\mathbf{C}_i)\right\|^{-1} 2\delta_z + \mathbf{c}\mathbf{C}_i^T\right] \quad (26)$$

According to the combinatorial characteristic, Eq. 26 has 2^{N_c} potential solutions. In addition, the minimization of the MFM with combined statistics in Eq. 26, drives to the maximization of the faults detectability. In other words, the search is oriented towards finding the optimal signals selection $\mathbf{C}_{op}$ (solution of Eq. 26) that guarantees the best fault detection of the most expected abnormal events at a lowest investment cost.

5.3 Genetic Algorithm Solution

In this subsection, the use of GA (see Appendix in Chap. 12) is discussed for solving the problem displayed in (26) subject to the following restrictions

$$\begin{aligned} &\sum_{l=1}^{N_c}\mathbf{C}_i(l) > A \\ &\left\|\mathbf{R}(\mathbf{C}_i)\Theta_j(\mathbf{C}_i)\right\| > 0 \end{aligned} \quad (27)$$

where the first restriction in Eq. 27 avoids the selection $\mathbf{C}_i$ that does not present dimensional reduction. In addition, if the signal dimension is equal to the principal components retained A, the combined statistic z is reduced to hotelling statistic only ($Q = 0$). The second constraint avoids the individual selection $\mathbf{C}_i$ that produces $\left\|\mathbf{f}_j\right\|_{MFM}^z = \infty$, which means that a specific fault cannot be detected. In other words, the second constraint guarantees the fault detectability condition in Eq. 23.

6 Application Results: BPS with FC

In this chapter the optimal monitoring system design based on detectability maximization is applied on the BPS+FC integrated plant. The optimal plant-wide control for this BPS+FC plant was developed in Chap. 12 and the control structure can be observed in Fig. 5 of Chap. 12. In this context, in Table 1 are summarized the available variables in the BPS+FC plant, with 16 potential measurement points and 7 control signals which are indicated in italic format. In fact, these variables correspond to the seven control loops opportunely proposed (see Fig. 5 of Chap. 12.).

Table 1 Variables in the BPS+FC process

Potential Measured		Manipulated	
y_1	*ESR exit temperature*	u_1	*Water to ESR inlet*
y_2	Jacket exit gases temperature	u_2	*Exchanged heat Q*
y_3	*Burner exit temperature*	u_3	*Ethanol to Burner*
y_4	Burner entering molar flow	u_4	*Oxygen to Burner*
y_5	Molar ratio H_2O/Ethanol	u_5	*Ethanol to ESR*
y_6	HTS exit temperature	u_6	*Oxygen to CO–PrOx*
y_7	*H_2 production rate*	u_7	*CM voltage*
y_8	LTS exit temperature		
y_9	*CO–PrOx exit temperature*		
y_{10}	*Molar ratio O_2/CO*		
y_{11}	*Burner exit molar flow*		
y_{12}	ESR exit molar flow		
y_{13}	CO–PrOx CO exit concentration		
y_{14}	Net Power		
y_{15}	*Oxygen excess*		
y_{16}	Stack voltage		

Table 2 Abnormal events proposed

Faults	Variable	Range [%]	Type
F_1	Molar ratio O_2/CO	±5	Step
F_2	ESR exit temperature	±1	Step
F_3	Burner exit temperature	±3	Step
F_4	Burner exit molar flow	±5	Step
F_5	H_2 production rate	±5	Step
F_6	CO–PrOx exit temperature	±1	Step
F_7	Oxygen excess	±5	Step

The optimal monitoring system design is based on maximizing the abnormal event detectability by accounting for the optimal signal selection for constructing a representative PCA model. In this case, typical problems in the existing process elements [1–3, 13, 19] are considered and summarized in Table 2. Seven potential sensor faults, F_1 to F_7, are accounted for consisting of abrupt bias/offset (step type) within a specific range of magnitude. All the sensors used in the control policy $(y_1, y_3, y_7, y_9, y_{10}, y_{11}, y_{15})$ are considered with a normal noise distribution and magnitude between $[0.5, 5\%]$ with respect to their operating points.

In this section the chosen parameters for the approach implementation are given in Table 3. According to the data given in Table 1 (measurements and manipulated variables) the combinatorial problem has dimension $N_c = 23$ and $2^{23} \approx 8.3 \times 10^6$ possible solutions. Taking into account that a single iteration consumes ≈ 0.1 s, performed in an Intel core duo, 2.53 GHz and 2 GB RAM computer. Clearly, an exhaustive evaluation of the problem is unpractical. In this context, GA is used to solve the combinatorial problem stated in Eq. 26 with the constraints given in Eq. 27. The first step for solving this problem is the fault subspace, Θ, extraction by using the methodology stated in Sect. 4. In this case a modification is introduced to the 3δ-edit rule of Eq. 19 for robustness issues. In fact, the fault directions are computed

Table 3 GA parameters setting

N_i	N_c	J	N_g	P_m	P_{co}	Selection	Crossover	c_i^*	Optimization time
3000	23	7	40	$0.7/N_c$	0.7	Roulette wheel	Double-point	5	0.2 h

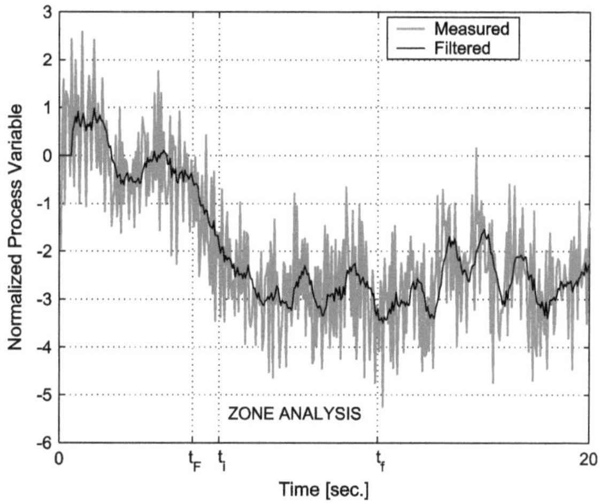

Fig. 2 Zone analysis for 3δ-edit rule

accounting the mean value of the filtered version data in a zone analysis, $[k_i, k_f]$ instead of a single time instant, k_*. Thus, the approach in Eq. 19 becomes,

$$F(j,i) = \begin{cases} 1, & \text{if } \mathrm{mean}[\bar{x}_f^{ji}([k_i, k_f])] > \ \ 2 \\ 0, & \text{if} -2 \le \ \mathrm{mean}[\bar{x}_f^{ji}([k_i, k_f])] \le \ 2 \\ -1, & \text{if } \mathrm{mean}[\bar{x}_f^{ji}([k_i, k_f])] < -2 \end{cases} \tag{28}$$

similar to the fuzzy approach presented in [1–3] the zone analysis allows to obtain a mean contribution of the process variables and to know the trends and effects of abnormal events in a closed-loop context. Obviously, this zone is determined according to the plant dynamic responses and taking into account the slower responses. Figure 2 shows an example of normalized measured process variable, its filtered (smoothed moving average) version, the fault occurrence instant, t_F s, and the zone analysis $[t_i, t_f]$ s. In this case the fault subspace extraction was made using the following parameters: $t_F = 5$ s, $[t_i, t_f] = 6, 12$ s, $N = 10$ samples is the moving window for the filter, $J = 7$ the amount of abnormal events.

Applying the new approach in Eqs. 28 and 20 a complete set of fault subspaces can be obtained from the abnormal data base. In this case, two examples are included corresponding to the faults F_4 and F_5 when a positive offset is considered respectively. Hence, the resulting fault subspaces are shown by Eqs. 29 and 30. The overall process variables are included and obviously the selected amount of them, used to perform the PCA-based monitoring, impact directly the fault detectability conditions. This concept was opportunely stated in Sects. 3 and 5.1.

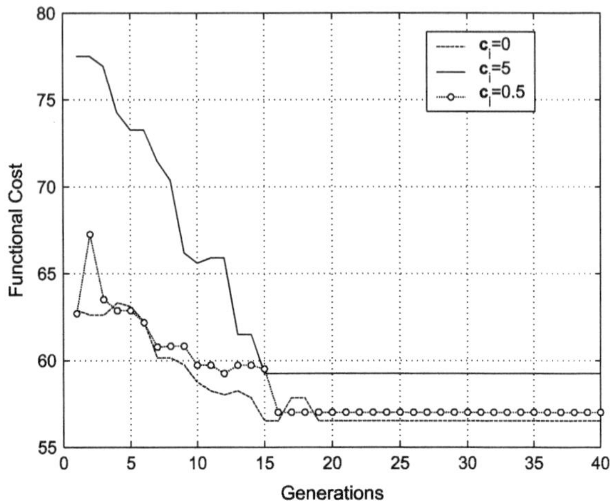

Fig. 3 GA functional cost for different weights c_i^*

$$\Theta_{F_4} = \begin{bmatrix} 0 & -1 & 0 & 0 & 0 & 0 & 0 & 0 & 0 & 0 & 0 & 0 & 0 & 0 & 0 & 0 & 0 & 0 & 0 & 0 & 0 & 0 & 0 \\ 0 & 0 & 0 & 0 & 0 & 0 & 0 & 0 & 0 & 0 & 0 & 0 & 0 & 0 & 0 & 0 & 0 & 0 & -1 & 0 & 0 & 0 & 0 \\ 0 & 0 & 0 & 0 & 0 & 0 & 0 & 0 & 0 & 0 & 0 & 0 & 0 & 0 & 0 & 0 & 0 & 0 & 0 & -1 & 0 & 0 & 0 \end{bmatrix}^T \quad (29)$$

$$\Theta_{F_5} = \begin{bmatrix} 0 & 1 & 0 & 0 & 0 & 0 & 0 & 0 & 0 & 0 & 0 & 0 & 0 & 0 & 0 & 0 & 0 & 0 & 0 & 0 & 0 & 0 & 0 \\ 0 & 0 & 0 & -1 & 0 & 0 & 0 & 0 & 0 & 0 & 0 & 0 & 0 & 0 & 0 & 0 & 0 & 0 & 0 & 0 & 0 & 0 & 0 \\ 0 & 0 & 0 & 0 & 0 & 0 & 0 & 0 & 0 & 0 & 0 & -1 & 0 & 0 & 0 & 0 & 0 & 0 & 0 & 0 & 0 & 0 & 0 \\ 0 & 0 & 0 & 0 & 0 & 0 & 0 & 0 & 0 & 0 & 0 & 0 & 0 & 0 & 0 & -1 & 0 & 0 & 0 & 0 & 0 & 0 & 0 \\ 0 & 0 & 0 & 0 & 0 & 0 & 0 & 0 & 0 & 0 & 0 & 0 & 0 & 0 & 0 & 0 & -1 & 0 & 0 & 0 & 0 & 0 & 0 \\ 0 & 0 & 0 & 0 & 0 & 0 & 0 & 0 & 0 & 0 & 0 & 0 & 0 & 0 & 0 & 0 & 0 & 0 & 0 & 0 & 0 & -1 & 0 \end{bmatrix}^T \quad (30)$$

With the fault subspace, Θ, already obtained, the next step is to solve the optimization problem stated in Eqs. 26 and 27. In this case, the GA approach is selected and the used parameter settings are shown in Table 3. In [13, 20] a complete analysis of how the initial population, N_i, generation number, N_g, mutation probability, P_m, crossover probability, P_{co}, and the weighting parameters, c_i^* affecting the solution of the problem was done. For example, the Fig. 3 shows the functional cost profile when different weights are used to penalize additional sensor utilization than those used for the plant-wide control structure. The unweighted, $c_i^* = 0$, and the slightly weighted, $c_i^* = 0.5$, cases propose the same solutions to the problem. The optimal signal selection to maximize detectability is $[y_1, y_3, y_{14}, u_1, u_2, u_3, u_6, u_7]$. Taking into account Table 1, this optimal solution suggests the use of a new sensor for the variable y_{14} (net power in the FC). On the other hand, the strongly weighted, $c_i^* = 5$, is focussed on penalizing the new hardware utilization more aggressively. In this context, the optimal solution is $[y_1, y_3, y_{15}, u_1, u_2, u_3, u_6, u_7]$, with the benefit that it does not require new measuring points. Thus, this solution only uses the already installed hardware to perform and improve the PCA-based monitoring.

Table 4 Different signals selection

Chromosome	Signals selection
$\mathbf{C}_{op}$	$[y_1, y_3, y_{15}, u_1, u_2, u_3, u_6, u_7]$
$\mathbf{C}_{unf}$	$[y_1, y_3, y_4, y_6, y_9, y_{12}, y_{13}, y_{14}, u_2, u_4, u_7]$
$\mathbf{C}_{full}$	$[y_1, y_2, \ldots, y_{16}, u_1, u_2, \ldots, u_7]$

In the following, a comparison of three different signals selection for PCA-based monitoring is performed (shown in Table 4). The first solution, represented here with the chromosome $\mathbf{C}_{op}$, is the optimal one obtained from the combinatorial problem in Eqs. 26 and 27, using the parameters setting displayed in Table 3. The second signals selection to be compared is the so-called, $\mathbf{C}_{unf}$, it represents an unfeasible solution and does not fulfill with the detectability condition (second constraint in Eq. 27). Finally, the third solution is named $\mathbf{C}_{full}$, it considers all possible measurement points and available signals to build the PCA model. It clearly presents an increase in the cost due to the new measurement points and this sensor network does not guarantee an optimal performance from the monitoring point of view. It seems to indicate that the amount of measurements is not directly related to the quality of fault detection. The last approach is the classical method, but generally not the optimal solution. In fact, we see here that optimal signals selection with the lowest hardware requirements can match and even improve the performance of the full solution from a PCA-based monitoring system point of view. For this comparison the PCA model construction has been made accounting the following common parameters for the three signals selection: cumulative percentage variance, $\delta_{\mathrm{cpv}} = 90\%$; confidence limits for T^2 and Q, 99%; z control limit, $\delta_z = 2$. This setting produces that the following principal components retained are (for each solutions): $A_{\mathrm{op}} = 5$, $A_{\mathrm{unf}} = 7$ and $A_{\mathrm{full}} = 10$.

The first simulation presented in Fig. 4 corresponds to a fault in the molar ratio O_2/CO sensor, called F_1. In this case an offset of +10% with respect to the normal operating point was considered, which occurs at $t = 5$ s. Figure 4a shows the temporal evolution of the combined statistic, $z(t)$ when different sets of measurements and signals are selected according to Table 4. In this figure can be observed that the optimal signal selection, $\mathbf{C}_{op}$, and the full one, $\mathbf{C}_{full}$, have a similar good performance in detecting this abnormal event. In fact, the fault detection is made at ≈ 6 s. On the other hand, the unfeasible solution ($\mathbf{C}_{unf}$) cannot detect this event and clearly maintains its combined statistic below the control limit all the time. This represents a normal state of the process, which is false. In Fig. 4b can be observed the mean contribution of the process variables in the zone of analysis $[t_i, t_f]$ previously defined. This contribution is computed by using the autoscaled/normalized versions of the overall process signals presented in Table 1. This representation allows to know the fault propagation into the process, i.e. is an alternative representation of the fault direction matrices. Clearly, the two variables with the major contribution are the water flow to ESR, u_1, and the oxygen flow to CO–PrOx, u_6, and these signals can be observed in Fig. 4c, d respectively. In fact, fault F_1 affects the local molar ratio O_2/CO control loop, specifically its manipulated variable u_1. In addition, the

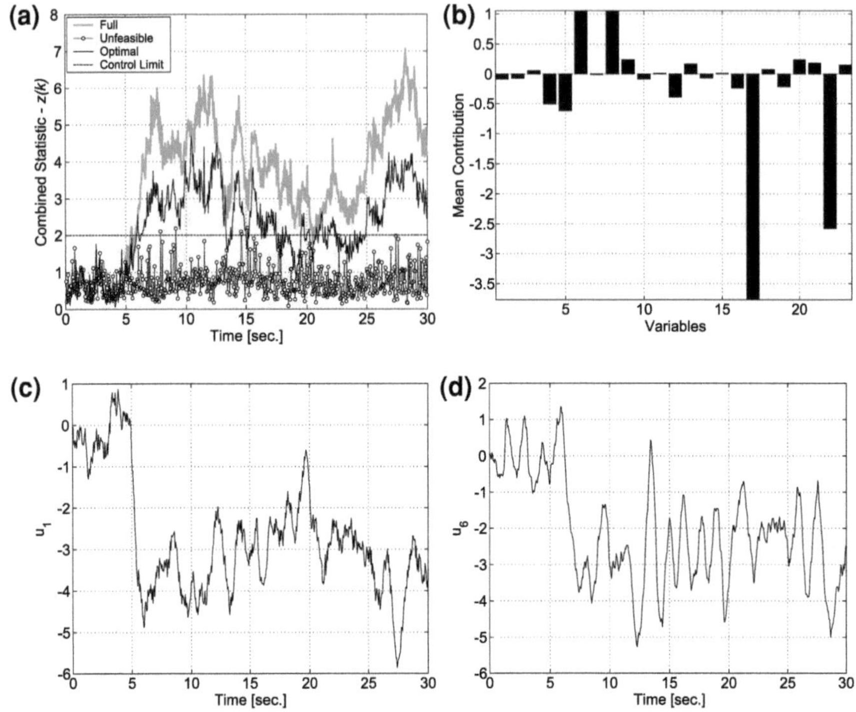

Fig. 4 Fault F_1 (Ratio O_2 /CO offset of +10%) **(a)** Combined statistic **(b)** Mean contribution **(c)** Normalized water to ESR—u_1 **(d)** Normalized oxygen to CO–PrOx—u_6

CO–PrOx temperature is affected too and its corresponding control signal, u_6, must change for rejecting this type of disturbance.

The second simulation case is presented in Fig. 5 and considers the ESR temperature sensor fault, F_2. In this case an offset of +1% with respect to the normal operating point was produced at $t = 5$ s. Figure 5a shows the temporal evolution of the combined statistic, $z(t)$, when different sets of measurements and signals are selected according to Table 4. In fact, in this figure the optimal $\mathbf{C}_{op}$, unfeasible $\mathbf{C}_{unf}$, and full $\mathbf{C}_{full}$ approaches are compared. In all cases the combined statistic profile shows a good behavior from the detectability point of view since the detection at $t \approx 5.2$ s. is performed with high-speed and maintaining its evolution out of the control limit indicating that an abnormal event has occurred. In Fig. 5b can be observed the mean contribution of the process variables in the zone of analysis $[t_i, t_f]$ by using the autoscaled/normalized versions of the overall process signals presented in Table 1. Clearly, the two variables with the major contribution are the HTS temperature, y_6, and the exchanged heat, u_2, and these signals can be observed in Fig. 5c, d respectively. Variable y_6 is not used by the optimal selection $\mathbf{C}_{op}$ but still it ensures a good fault detection performance very similar to the full selection case, $\mathbf{C}_{full}$. This shows that a full selection is not necessary to guarantee abnormal event

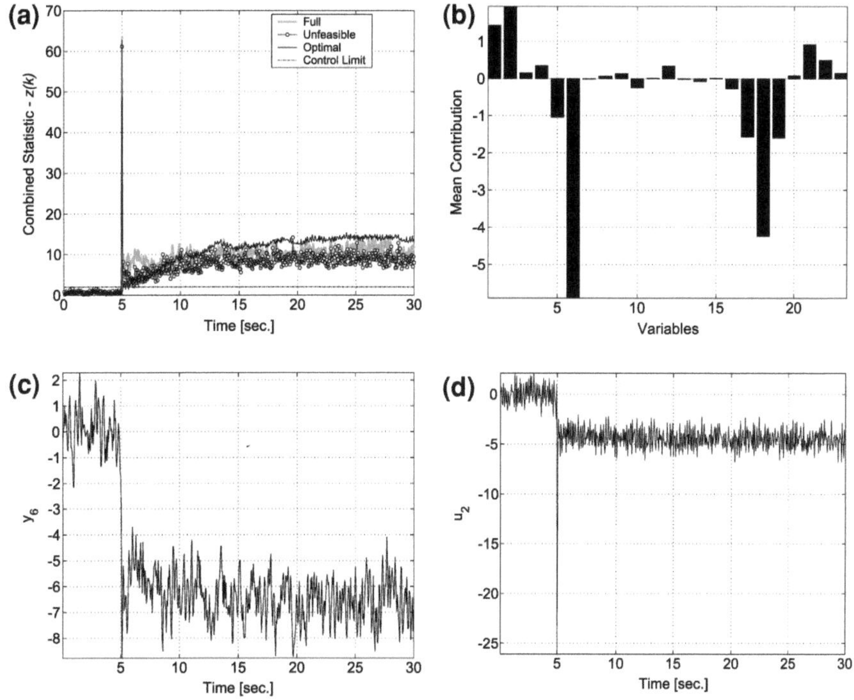

Fig. 5 Fault F_2 (ESR temperature offset of +1%) **(a)** Combined statistic **(b)** Mean contribution **(c)** Normalized HTS temperature—y_6 **(d)** Normalized Exchanged heat—u_2

detectability. Fault F_2 affects the local ESR temperature control loop, specifically its manipulated variable u_2. In addition, the HTS temperature, which is an uncontrolled variable, evolves to another working point due to this abnormal event.

As a third testing scenario in Fig. 6 shows the detectability performance for different signals selection when a burner temperature sensor fault, F_3, occurs at $t = 5$ s with a +3% magnitude. Figure 6a shows the temporal evolution of the combined statistic, $z(t)$, when different sets of measurements and signals are selected according to Table 4. In fact, in this figure the optimal $\mathbf{C}_{op}$, unfeasible $\mathbf{C}_{unf}$, and full $\mathbf{C}_{full}$ approaches are compared. It is clear in this case that the full and unfeasible solution have serious problems to detect this kind of abnormal event. The unfeasible case presents its statistic all the time below the control limit, indicating a concrete miss detection. The combined statistic for the full case displays a behavior very close to the control limit representing both miss and false detection. It is a classic example where a full/complete signal selection does not guarantee detectability from a PCA point of view. In contrast, an appropriate monitoring system design accounting detectability indexes generates an optimal signal selection, $\mathbf{C}_{op}$, that assures fault detection with minimal hardware resources and improving the performance obtained via a full design. The monitoring system based on $\mathbf{C}_{op}$ allows a quick fault detection at

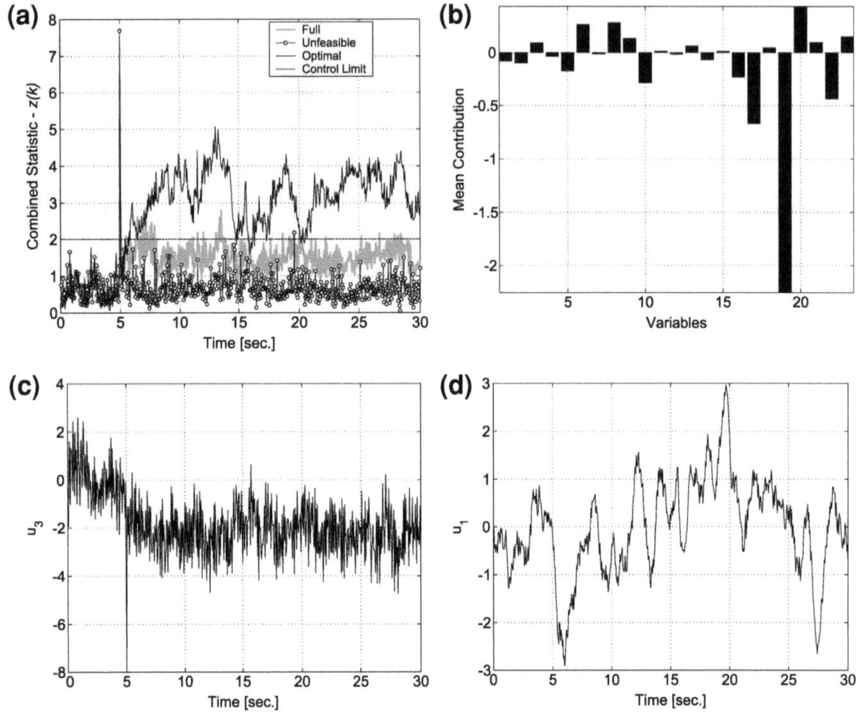

Fig. 6 Fault F_3 (Burner temperature offset of +3%) **(a)** Combined statistic **(b)** Mean contribution **(c)** Normalized ethanol to burner—u_3 **(d)** Normalized water to ESR—u_1

$t \approx 6$ s. In Fig. 6b can be observed the mean contribution of the process variables in the zone of analysis $[t_i, t_f]$ by using the autoscaled/normalized versions of the overall process signals presented in Table 1 of Chap. 13. Clearly, the two variables with the major contribution are the ethanol flow to burner, u_3, and the water flow to ESR, u_1, and these signals can be observed in Fig. 6c, d respectively. Fault F_3 affects the local burner temperature control loop, specifically its manipulated variable u_3. In addition, molar ratio O_2 /CO is affected too so its corresponding control signal, u_1, must vary to reject this kind of disturbance.

Figure 7 shows another faulty scenario, in this case a burner flow sensor fault occurs at $t = 5$ s with a value of +5% with respect to its nominal working point. Figure 7a shows the temporal evolution of the combined statistic, $z(t)$, when different sets of measurements and signals were selected according to Table 4. In fact, the optimal $\mathbf{C}_{op}$, unfeasible $\mathbf{C}_{unf}$, and full $\mathbf{C}_{full}$ approaches are compared. In this case all the solutions have similar performance from a detectability point of view, this fault is quickly detected in about $t \approx 6$ s. Anyway, the system monitoring design based on the optimal solution, $\mathbf{C}_{op}$, shows the most robust behavior so that it is able to maintain its statistic away from the control limit which is very useful for avoiding false alarms. In Fig. 7b can be observed the mean contribution of the process variables in the zone

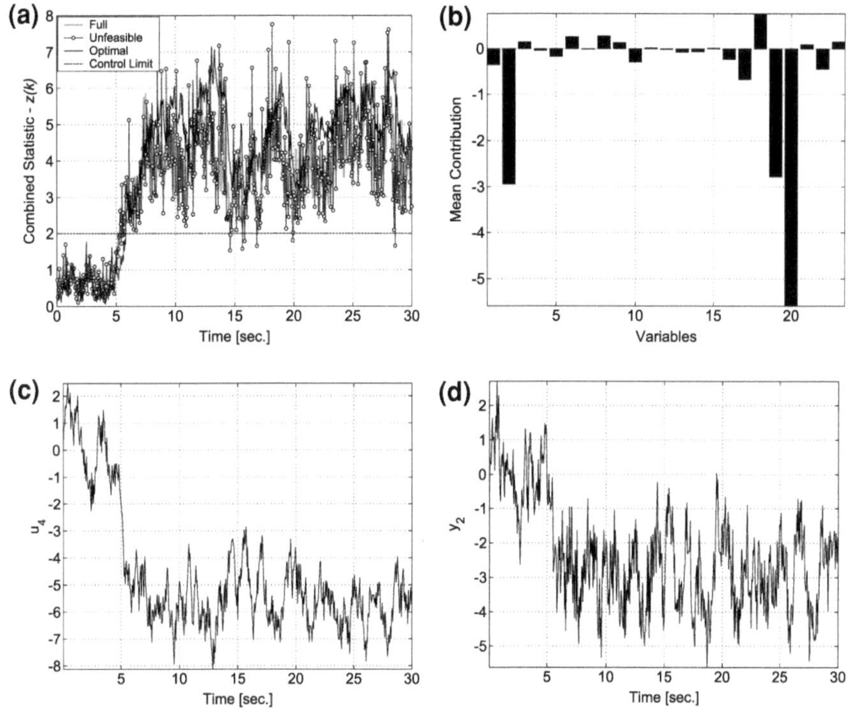

Fig. 7 Fault F_4 (Burner exit flow offset of +5%) **(a)** Combined statistic **(b)** Mean contribution **(c)** Normalized oxygen to burner—u_4 **(d)** Normalized gases temperature—y_2

of analysis by using the autoscaled/normalized versions of the overall process signals presented in Table 1 of Chap. 13. Clearly, two variables with the major contribution are the oxygen flow to burner, u_4, and the gases temperature, y_2, and these signals can be observed in Fig. 7c, d respectively. It is remarkable that, although the variable y_2 was not selected by the $\mathbf{C}_{op}$ configuration, it is capable of ensuring a good fault detection performance. In addition, its behavior is better than that produced by the full selection case, $\mathbf{C}_{full}$. This result is valuable for demonstrating that a full selection is not always enough to guarantee abnormal event detectability. Fault F_4 affects the local burner exit flow control loop, specifically its manipulated variable u_4. In addition, the temperature of the gases that leave the burner is also affected by this fault however, because this variable is uncontrolled it moves toward another operating point.

The fifth testing case is displayed in Fig. 8. Here a H_2 production rate sensor fault occurs at $t = 5$ s with a value of +3% respect to its nominal working point. Figure 8a shows the temporal evolution of the combined statistic, $z(t)$, when different sets of measurements and signals were selected for the three different configurations which have been analyzed in this chapter ($\mathbf{C}_{op}$, $\mathbf{C}_{unf}$, and $\mathbf{C}_{full}$). In this case both, $\mathbf{C}_{full}$ and $\mathbf{C}_{op}$ solutions have similar performance from a detectability point of view

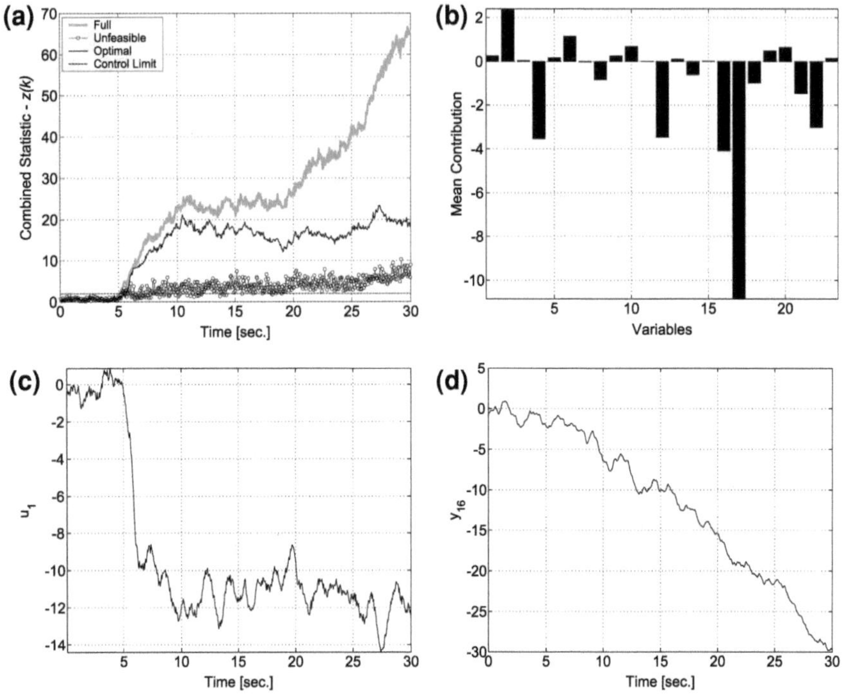

Fig. 8 Fault F_5 (H_2 production rate offset of +3%) **(a)** Combined statistic **(b)** Mean contribution **(c)** Normalized water to ESR—u_1 **(d)** Normalized stack voltage—y_{16}

and provide a well performed fault detection in $t \approx 5.5$ s. On the other hand, the unfeasible solution presents serious problems to detect this kind of an event. In fact, its combined statistic remains near the control limit producing miss detections and false alarms. In Fig. 8b can be observed the mean contribution of the process variables in the zone of analysis by using the autoscaled/normalized versions of the overall process signals presented in Table 1. Clearly, two variables with major contributions are the water flow to ESR, u_1, and the stack voltage, y_{16}, and these signals can be observed in Fig. 8c, d, respectively. Again, $\mathbf{C}_{op}$ and $\mathbf{C}_{full}$ provide a quite similar performance. Fault F_5 affects the local H_2 production rate control loop, but additionally, has a strong effect on the CO–PrOx molar ratio O_2/CO and eventually its manipulated variable, u_1. In this context, F_5 produces an incorrect hydrogen supply to the FC changing its operating conditions, y_{16}. So it is considered a critical fault for this kind of a system and reveals the importance of having an efficient fault detection.

Figure 9 shows another faulty scenario, in this case a CO–PrOx temperature sensor fault occurs at $t = 5$ s with a value of +1% with respect to its nominal working point. Figure 9a shows the temporal evolution of the combined statistic, $z(t)$, when different sets of measurements and signals were selected according to Table 4. In fact, $\mathbf{C}_{op}$,

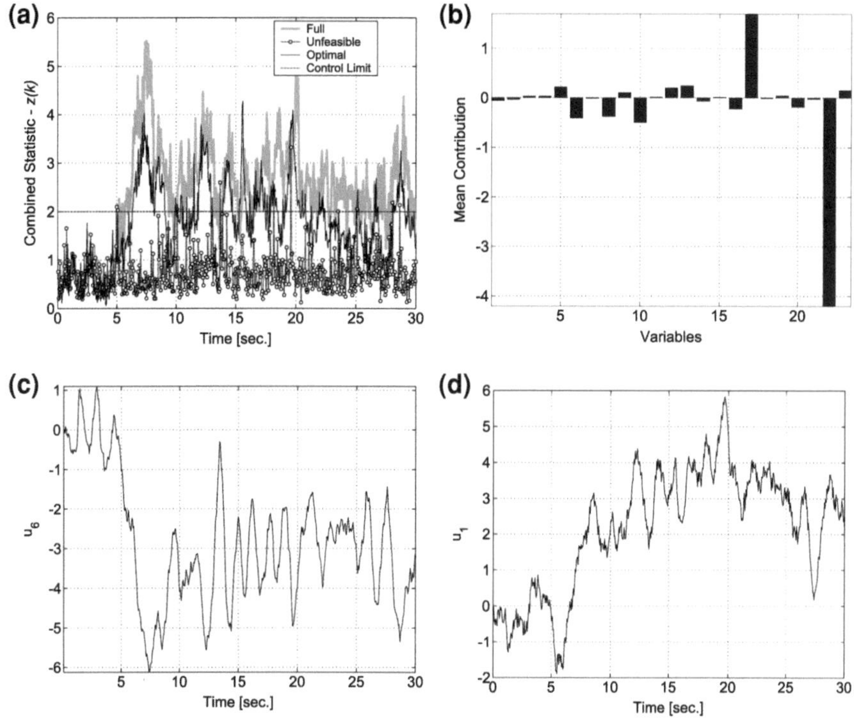

Fig. 9 Fault F_6 (CO–PrOx temperature offset of +1%) **(a)** Combined statistic **(b)** Mean contribution **(c)** Normalized oxygen to CO–PrOx—u_6 **(d)** Normalized water to ESR—u_1

$\mathbf{C}_{unf}$, $\mathbf{C}_{full}$ approaches are compared again. In this case, both full and optimal solutions have similar performance from a detectability point of view, allowing a rapid fault detection in $t \approx 6$ s. On the other hand, the unfeasible solution ($\mathbf{C}_{unf}$) cannot detect this event and clearly maintains its combined statistic below the control limit all the time. It represents a normal state of the process, which is false. In Fig. 9b can be observed the mean contribution of the process variables in the zone of analysis. It can be seen clearly that the major contribution is given by the oxygen flow to CO–PrOx, u_6 and the water flow to ESR, u_1, and these signals are shown in Fig. 9c, d, respectively. Fault F_6 affects the local CO–PrOx temperature control loop, specifically its manipulated variable u_6. In addition, the CO–PrOx molar ratio O_2/CO is also affected, so its corresponding control signal, u_1, must vary so as to compensate this disturbance effect.

Finally, the seventh fault considered in this chapter is presented in Fig. 10 and in this case an oxygen excess sensor fault occurs at $t = 5$ s with a value of +5% a with respect to its nominal working point. Figure 10a shows the temporal evolution of the combined statistic, $z(t)$, for $\mathbf{C}_{op}$, $\mathbf{C}_{unf}$, and $\mathbf{C}_{full}$ configurations. In this case, all the solutions have similar performance from the detectability point of view, allowing a very fast fault detection in $t \approx 5.5$ s. Anyway, the $\mathbf{C}_{op}$, shows the most robust

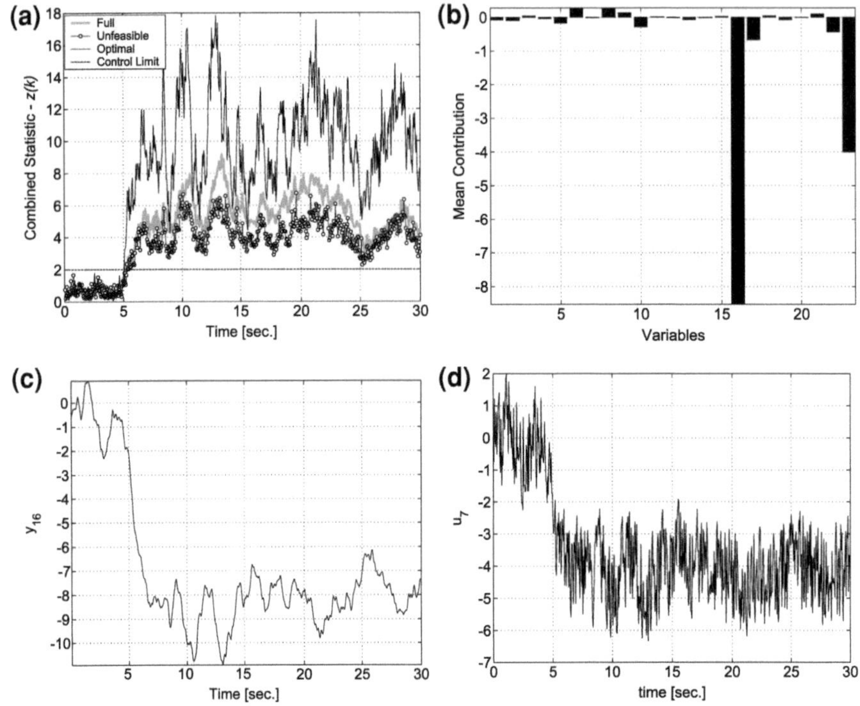

Fig. 10 Fault F_7 (Oxygen excess offset of +5%) **(a)** Combined statistic **(b)** Mean contribution **(c)** Normalized stack voltage—y_{16} **(d)** Normalized CM voltage—u_7

behavior by maintaining its statistic outside the control limit. In Fig. 10b, can be observed the mean contribution of the process variables in the zone of analysis. It can differentiate that two variables present the major contribution which are the stack voltage, y_{16}, and the CM voltage, u_7, as can be seen in Fig. 10c, d, respectively. Fault F_7 affects essentially the local oxygen excess control loop, specifically its manipulated variable u_7. From Fig. 10b can be observed that this fault is practically local to the FC and does not impact BPS.

A useful indicator to evaluate among different signals selection for monitoring system design is named reliability index (RI). This indicator considers the amount of samples that are over the confidence/control limit in percentage mode. Note that in normal operation case the process presents 1% of samples over this limit according to the 99% of confidence. In fact, this index gives information about the degree of robustness for abnormal event detection and its possible isolation and diagnosis [1–3, 11, 13, 20]. Thus, this index is computed and displayed in Table 5 for each abnormal event given in Table 2 and considering different signal selections proposed in Table 4. The unfeasible selection clearly presents loss of detection for faults F_1, F_3 and F_6. The full selection case also presents difficulties to perform a suitable fault detection. For fault F_3 this selection presents lower reliability index (practically loss detection)

Table 5 Reliability index (RI) and hardware requirements (HR)

	RI [%]							HR	
Selection	F_1	F_2	F_3	F_4	F_5	F_6	F_7	Used signals	Extra sensors
Full	96.8	100.0	8.6	99.0	99.6	85.6	99.0	23	9
Unfeasible	1.0	100.0	0.2	96.8	89.2	1.8	99.0	11	5
Optimal	72.4	99.4	94.4	96.8	97.8	43.6	99.8	8	0

and poorer performance than the optimal one. As can be deduced, the optimal signals selection case uses lower amount of variables maintaining a suitable fault detection performance for the overall abnormal events analyzed here. In all cases the reliability index is comparable with the full case and even better for some of them (F_3). In addition, it was considered to include in Table 5 the hardware requirements (HR) so as to have a complete picture of the main subject analyzed in this chapter. In other words, the total amount of signals and new sensors used for each approach to perform the monitoring system design for the same faults. The full case needs the use of 23 signals (all the available information) to develop the PCA model and incorporates 9 extra measurement points besides the 7 sensors already defined for the plant-wide control structure. In spite of the inclusion of extra sensors it is not enough to solve the problem of missed detection and poor detectability performance. The unfeasible case uses 11 signals for monitoring purposes and the methodology suggests to add 5 extra sensors apart from the 7 already defined for control purposes. While this case has less HR than the full one, the detectability and robustness performance presented by the unfeasible solution is worse than the full configuration. In contrast with these previous solutions the optimal one only uses 8 signals to perform the PCA monitoring of the overall process. In addition, no extra sensors are required to carry out a complete and suitable fault detection. These results were clearly reflected in the simulations of Figs. 4, 5, 6, 7, 8, 9, 10 and Table 5. Obviously, lower HR means lower investment cost, and the optimal solution only uses the already installed sensors for the control structure.

7 Conclusions

In this chapter, a systematic and generalized PCA-based monitoring system design for large-scale processes has been presented. The results obtained here show that the use of GA offers an important opportunity to solve large-scale combinatorial problems which cannot be addressed via exhaustive search. Historically, these types of problems are solved with the help of considerably amount of heuristic considerations. The optimal supervisory system is developed using multiple-integrated tools as PCA, combined statistic, detectability analysis, fault subspace extraction and associated investment costs. It was found how to include these elements in the objective function so as to promote the use of the existing control structure, i.e.

already installed hardware. In particular, it must be remarked that the detectability function is extended to be calculated based on the combined statistic which represents a new result in the context of expert fault diagnosis system design. In fact, the optimal solution obtained via detectability maximization allows us to select specific set of signals (inputs–outputs) that guarantees a suitable abnormal event detection performance. It is remarkable the inclusion of the proper selection of the weighting matrices for penalizing the use of extra sensors than those installed for control purpose. This concept provides low-cost alternative solutions which represents a very important issue for industrial implementations. A remarkable result is that the full signal selection does not guarantee the best fault detection performance. In contrast, an optimal selection of these signals, presents comparable and even better detection performance than that obtained by the full selection case. In this chapter was obtained an optimal solution using only the already existing hardware from the optimal control structure. This approach integrated with the previous work of the authors [21–25] drives to a successful, generalized and systematic strategy to solve typical problems in process industries as sensor network , control structure selection and monitoring system design. This methodology was tested for several critical faults, taking into account the overall process requirements given since the synthesis stage. Hence, it was given an integral procedure for an open problem such as the integration of the process synthesis, investment cost analysis, fault detectability and diagnosis in the context of a well driven plant-wide control structure [1–3]. These advances applied on the bio-ethanol processor with the PEMFC constitutes the conceptual engineering to achieve an integral robust design for these kinds of novel systems.

References

1. Zumoffen D, Basualdo M (2008) From large chemical plant data to fault diagnosis integrated to decentralized fault-tolerant control: pulp mill process application. Ind Eng Chem Res 47(4):1201–1220
2. Zumoffen D, Basualdo M (2008) Improvements in fault tolerance characteristics for large chemical plants. Part I: waste water treatment plant with decentralized control. Ind Eng Chem Res 47(15):5464–5481
3. Zumoffen D, Basualdo M, Molina G (2008) Improvements in fault tolerance characteristics for large chemical plants. Part II: pulp mill process with model predictive control. Ind Eng Chem Res 47(15):5482–5500
4. Kourti T, MacGregor J (1995) Process anlysis, monitoring and diagnosis, using multivariable projection methods. Chemom Intell Lab Syst (28):3/21
5. Martin EB, Morris AJ, Lane S (2002) Monitoring process manufacturing performance. IEEE Control Syst Mag 22:26–39
6. Lane S, Martin EB, Morris AJ, Gower P (2003) Application of exponentially weighted principal component analysis for the monitoring of a polymer film manufacturing process. Transac Inst Meas Control 1(25):17–35
7. Yue HH, Qin SJ (2001) Reconstruction-based fault identification using a combined index. Ind Eng Chem Res 40:4403–4414
8. Kadu S, Bhushan M, Gudi R (2008) Optimal sensor network design for multirate system. J Process Control 18:594–609

9. Singh A, Hahn J (2005) Determining optimal sensor location for state and parameter estimation for stable nonlinear systems. Ind Eng Chem Res 44:5645–5659
10. Bhushan M, Narasimhan S, Rengaswamy R (2008) Robust sensor network design for fault diagnosis. Comput Chem Eng 32:1067–1084
11. Musulin E, Bagajewicz M, Nougués J, Puigjaner L (2004) Instrumentation design and upgrade for principal components analysis monitoring. Ind Eng Chem Res 43:2150–2159
12. Musulin E, Yélamos I, Puigjaner L (2006) Integration of principal component analysis and fuzzy logic systems for comprehensive process fault detection and diagnosis. Ind Eng Chem Res (45):1739–1750
13. Zumoffen D, Basualdo M (2009) A systematic approach for the design of optimal monitoring systems for large scale processes. Ind Eng Chem Res, 2010, 49(4), 1749–1761.
14. Li W, Yue HH, Valle-Cervantes S, Qin SJ (2000) Recursive pca for adaptive process monitoring. J Process Control 10:471–486
15. Wold S (1994) Exponentially weighted moving principal components analysis and projection to latent structures. Chemom Intell Lab Syst 23:149–161
16. Ruiz D (2001) Fault Diagnosis In Chemical Plants Integrated To The Information System. PhD thesis, Departament d'Enginyeria Química, Escola Técnica Superior d'Enginyers Industrials de Barcelona, Universitat Politécnica de Catalunya, España, March
17. Musulin E (2005) Process Monitoring and Abnormal Situation Management in Chemical Processes. PhD thesis, Departament d'Enginyeria Química, Escola Tècnica Superior d'Enginyers Industrials de Barcelona, Universitat Politècnica de Catalunya, España, June
18. Chiang L, Pell R, Seasholtz M (2003) Exploring process data with the use of robust outlier detection algorithms. J Process Control 13:437–449
19. ASM. Abnormal situation management consortium. http://www.asmconsortium.com
20. Zumoffen D, Basualdo M (2010) Monitoreo, Detección de Fallas y Control de Procesos Industriales, volume 1. Asociación Argentina de Control automático (AADECA), 1 edition
21. Zumoffen D, Basualdo M (2009) Optimal sensor location for chemical process accounting the best control configuration. Comput Aided Chem Eng 27:1593–1598
22. Molina G, Zumoffen D, Basualdo M (2009) A new systematic approach to find plantwide control structures. Comput Aided Chem Eng 27:1599–1604
23. Zumoffen D, Basualdo M, Ruiz J (2009) Optimal multivariable control structure design for chemical plants. AIChE Annual Meeting, Nashville, TN, USA
24. Zumoffen D, Molina G, Basualdo M (2010) Plant-wide control based on minimum square deviation. Proceedings of the 9th International Symposium on Dynamics and Control of Process Systems, Leuven, Belgium, p 443–448
25. Molina GD, Zumoffen DAR, Basualdo MS (2010) Plant-wide control strategy applied to the tennessee eastman process at two operating points. Comput Chem Eng (In Press), Corrected Proof

Improved Principal Component Analysis Models for Fault Detection Using Delay Adjustment

E. Musulin and M. Basualdo

Abstract Principal component analysis (PCA) has been widely and successfully used with process monitoring purposes. However, it has an important drawback as it does not account for time-delays present in data thus causing an inefficient dimensionality reduction of process variables and the subsequent poor monitoring and disturbance detection performance. In this chapter, a new method, the genetic algorithm based delay adjusted PCA (GA-DAPCA), based on genetic algorithm optimization is proposed to improve the PCA performance in the presence of time delays between process signals. The optimization is performed in two loops. The first loop finds the shift between variables that minimize the number of principal components to be considered as common cause variance, the second loop maximizes the variance contained in the previously selected principal component dimensions. The methodology is demonstrated through motivating case studies in the bio-ethanol processor system (see Chap. 9). The obtained results are presented with the same set of faults considered in Chap. 13.

1 Introduction

Now-a-days the chemical process industry produces a huge amount of information readily available through distributed control systems (DCS) and data acquisition systems. However, if it is not appropriately processed, this information is of limited

E. Musulin · M. Basualdo (✉)
CAPEG-CIFASIS-(CONICET-UNR-UPCAM),
27 de Febrero 210 bis, S2000EZP Rosario, Argentina
e-mail: basualdo@cifasis-conicet.gov.ar

E. Musulin
Control Dept. EIE-FCEIA-UNR, Pellegrini 250, S2000BTP Rosario, Argentina
e-mail: musulin@cifasis-conicet.gov.ar

M. Basualdo
CAPEG-Chemical Engineering Department, FRRo-UTN, Zeballos 1341,
S2000BQA Rosario, Argentina

M. S. Basualdo et al. (eds.), *PEM Fuel Cells with Bio-Ethanol Processor Systems*, Green Energy and Technology, DOI: 10.1007/978-1-84996-184-4_14,

use. Many industrial applications have shown the ability of techniques associated with multivariate statistical process monitoring (MSPM) to monitor chemical processes [1–3]. Among these methods, the principal components analysis (PCA) that builds a linear statistical model of the process using the correlation present in process data is the most widely used [4–6]. PCA can be applied without a wide specific knowledge of the process, in an easy and fast way, and without excessive computational requirements. It performs a dimensionality reduction of the process variables generating a new set of non-correlated latent variables. They are in general no more than three, but include information about all the process variables. Afterwards, some multivariate monitoring charts (e.g. Hotteling statistic T^2, squared prediction error (SPE) can be used online to perform process monitoring.

Although several approaches exist that deal with different aspects of PCA (multi-scale signals and noise [7, 8], nonlinearities [9–11], process changes [12], process dynamics [13], etc.) there are some aspects of PCA that have not been widely studied in the literature. Since PCA only extracts the static linear relations between process signals, the presence of pure time delays between process signals can cause some relations to be delayed and cannot be properly treated by the method, therefore generating a loading matrix of bad quality and the subsequent poor monitoring performance.

Wachs and Lewin [14] proposed an algorithm [and called it delay adjusted PCA (DAPCA)] that consists of applying relative shifts between process signals within a process data matrix in order to find out the position at which the correlation among variables is maximum. The optimal shifts are those that minimize the determinant of the associated correlation matrix.

Although this technique shows improvement in the resolution of PCA, the algorithm assumes that the output variables are correlated among themselves with no delays present, and that the inputs are independent. In real processes it is difficult to discriminate between input and output variables. This distinction is even less realistic if recycles and feedback control systems are present, which is often the case.

This chapter presents a genetic algorithm that manipulates the time shift between signals to increase the normal operation variability while decreasing the permitted variation in the residual subspace, therefore increasing the monitoring tightness. It does not make additional assumptions. The optimization is performed in two loops. The first minimizes the number of principal components (PC) dimensions required to describe the normal variability of the process, while the second maximizes the variation contained in the selected PC dimensions. It is demonstrated that this algorithm improves the projection performance by reducing the dimension of the common variance subspace. In addition, it captures the correlation more accurately, thus improving the detection performance.

In this context, this methodology is tested in the bio-ethanol processor system with PEM fuel cell with the objective of improving the fault diagnosis system specifically designed for this process. Since it is a new plant, there is no other result to be compared for this specific case. Hence, two cases of study are included to capture the benefits of using this technique. The first example is very introductory and the second is the Tennessee Eastman process, a well-known tested benchmark for both plantwide control and fault diagnosis purposes. Finally, the results obtained from BPS with PEMFC accounting the same faults given in Chap. 13 are presented.

The chapter is organized as follows. First, PCA is briefly reviewed. Second, the novel methodology to extract the process delays is presented and explained through two motivating examples. Finally, the methodology is demonstrated in the fuel processor system, where the obtained results allow to extract interesting conclusions.

2 Principal Components Analysis

Consider a matrix $\mathbf{X}$ (of dimension $m \times n$) containing data corresponding to m samples of n process variables. Each column of $\mathbf{X}$ is supposed to follow a normal probability distribution and is normalized with zero mean and unit variance. Let $\overline{\mathbf{X}}$ be this normalized data matrix, and $\mathbf{R}$ its corresponding correlation matrix:

$$\mathbf{R} = \frac{\overline{\mathbf{X}} \cdot \overline{\mathbf{X}}^{\mathrm{T}}}{m - 1} \tag{1}$$

Then, performing singular value decomposition (SVD) on $\mathbf{R}$,

$$\mathbf{R} = \mathbf{U}^T \mathbf{D}_\lambda \mathbf{U} \tag{2}$$

a diagonal matrix $\mathbf{D}_\lambda = \mathrm{diag}(\lambda_1, \lambda_2, \ldots, \lambda_n)$ where λ_i are the eigenvalues of $\mathbf{R}$ sorted in decreasing order is obtained. We note that $\mathbf{R}$ is square and symmetric, so that the generalized SVD, which is applied to non-symmetrical matrices, reduces to the above presented case.

The corresponding eigenvectors $\mathbf{p}_i$ (columns of $\mathbf{U}$) are the principal components and form an orthonormal base in $\Re^n$.

It is possible to divide the principal components into two orthogonal sets, $\mathbf{P} = [\mathbf{p}_1, \mathbf{p}_2, \ldots, \mathbf{p}_A]$ and $\overline{\mathbf{P}} = [\mathbf{p}_{A+1}, \mathbf{p}_{A+2}, \ldots, \mathbf{p}_n]$, the first containing most of the process variance and the second describing the variance due to the noise (called the residual subspace). Methods to select the proper number of PCs are reviewed by Jackson [15] and Himes et al. [16]. A reduction of dimensionality is made by projecting every normalized sample vector $\mathbf{x}'$ in the common variance subspace generated by $\mathbf{P}$ (loading matrix), obtaining $\mathbf{t} = \mathbf{P}^{\mathrm{T}}\mathbf{x}'$, which is called the principal score vector. Because of the normalization of $\mathbf{X}$, when the process is in control, $\mathbf{x}'$ is projected at the origin, or in its proximity. Then, the state of the process can be monitored using the Hotelling's statistic, which is a measure of the Mahalanobis distance to the origin of the new space,

$$T^2 = \left\| \mathbf{D}_\lambda^{-\frac{1}{2}} \mathbf{P}^{\mathrm{T}} \mathbf{x}' \right\|^2 \tag{3}$$

The test consists in declaring normal operation if $T^2 \leq \delta_T^2$, where δ_T^2 is the control confidence limit of T^2. However, monitoring based only on this statistic is not sufficient because it only shows the deviations in the principal components subspace. If new information is not contained in the reference set of data, the projection of the new observation will move to the residual space, and it is possible that the anomaly will not be detected by the T^2 test. To detect this kind of deviation the squared prediction error statistic (*SPE*) is usually used. SPE is defined as the sum of the quadratic error between $\mathbf{x}'$ and the reconstructed signal after the dimensionality reduction,

$$\Delta\mathbf{x}' = \mathbf{x}' - \mathbf{P}\mathbf{P}^{\mathrm{T}}\mathbf{x}' = (\mathbf{I} - \mathbf{P}\mathbf{P}^{\mathrm{T}})\mathbf{x}' = \widetilde{\mathbf{C}}\mathbf{x}' \tag{4}$$

Thus,

$$\mathrm{SPE} = \left\|\widetilde{\mathbf{C}}\mathbf{x}'\right\|^2 \tag{5}$$

If SPE $\leq \delta^2_{\mathrm{SPE}}$, where δ^2_{SPE} is the control limit for SPE, the data vector is declared normal. Different approaches to obtain the threshold limits are presented by Jackson [15].

3 Delay Adjusted Principal Component Analysis

One main aspect of PCA is that it extracts only the static linear relations present in data. However, in general, in chemical processes there exist some degree of autocorrelation due to random noise and uncontrollable disturbances. A dynamic version of PCA (DPCA) has been proposed by Ku et al. [13] that is justified when there exists a substantial amount of auto-correlation in process measurements; pure time delays are not treated, though.

Those delays can limit the performance of PCA (or DPCA) causing the percentage of variance contained in the first few principal components to be low and the difference between the variance contained in the last significant principal component (λ_A) and the next one (λ_{A+1}) to be not accentuated. Therefore, poor dimensionality reduction of process variables is obtained with subsequent poor performance in disturbance detection.

The knowledge of time delays can help to reduce these problems and it is important to improve the projection quality, augmenting the tightness of the detection and obtaining a better filtering of the signals. Wachs and Lewin [14] proposed an algorithm that consists of applying relative shifts between process signals within a process data matrix in order to find out the position at which the correlation among variables is maximum. The optimal shifts are those that minimize the determinant of the associated correlation matrix.

When the number of process variables n is large or the maximum delay considered $d_{\max}$ is high in terms of samples, exhaustive search of the maximum correlation by direct data matrix manipulation is infeasible due to the large size of the problem (it involves $(d_{\max})^n$ determinant calculations). Wachs and Lewin [14] realized this problem and proposed a new algorithm (and called it DAPCA) that reduces the problem size to $d_{\max} \times s \times (n - s)$, where s is the number of inputs to the system. Nevertheless, this algorithm lies on two premises. It assumes that the outputs are correlated among themselves with no delays present, and the independence of the inputs. In most chemical processes it is difficult to discriminate between input and output variables. These assumptions are even more difficult to be realistic if recycles and feedback control systems are present, which is often the case. An important additional drawback of this method is that the solution is not always the optimum. In fact, the condition of minimum determinant is not sufficient to assure that the model obtained is the optimum. As an illustrative example, consider the single-input single-output system introduced by Kramer [9],

$$\begin{cases} u = 0.8\sin(\theta) \\ y_1 = 0.8\cos(\theta) \\ \theta = W[0, 2\pi] \end{cases} \tag{6}$$

where $W[a, b]$ represents the uniform distribution in the range (a, b). Its corresponding data matrix is $\mathbf{X} = [u, y_1]$. Applying PCA, the PCs' variance results in $\lambda_1 = \lambda_2 = 1$, which means that there is no correlation between data, and therefore both principal components describe the same variance. Wachs and Lewin [14] showed that their algorithm found an optimum when the input variable, u, is delayed by $\pi/2$ in relation to the output variable, y_1. Then, performing PCA on the modified array, the eigenvalues $\lambda_1 = 2$ and $\lambda_2 = 0$ are obtained, which indicates that one principal component is enough to model the normal variation present in data without loss of information. Now, we add one output y_2 to the previous example,

$$\begin{cases} u = 0.8\sin(\theta) \\ y_1 = 0.8\cos(\theta) \\ y_2 = -0.8\sin(\theta) \\ \theta = W[0, 2\pi] \end{cases} \tag{7}$$

the corresponding data matrix is $\mathbf{X} = [u, y_1, y_2]$. Since the algorithm considers that the outputs are not delayed, the optimum delay found is $[0, \pi/2, \pi/2]$, that corresponds to the eigenvalues ($\lambda_1 = 2$, $\lambda_2 = 1$, $\lambda_3 = 0$) and $\det(\mathbf{R}) = 0$ (the determinant of the correlation matrix is equal to the product of its eigenvalues). This solution corresponds to the alignment of the input with one output while the other output remains unadjusted. Finally, one can select one principal component to describe 66.6% of the variability or two principal components to model the total variability present in $\mathbf{X}$. Note that the problem is not only the assumption of no-delays between outputs but a wrongly defined objective function. In fact, the correlation between two signals (e.g. a control valve command and the valve flow) will cause that $\det(\mathbf{R}) = 0$ (or $\det(\mathbf{R}) \cong 0$ if we consider measurement noise and fast dynamics), independently of the situation of the remaining process variables. This is an important aspect that is faced in the novel approach presented in this chapter.

4 Genetic Algorithm Based Delay Adjusted Principal Component Analysis

The genetic algorithm-based DAPCA (GA-DAPCA) introduces a conceptual difference with the above presented approach (DAPCA). The objective is now maximizes the sum of the first A eigenvalues (which is equivalent to maximizing the variance contained in the common variance subspace while reducing the variance of the residual subspace as much as possible), instead of the minimization of the determinant of **R**. Therefore, the projection quality can be increased in two aspects:

1. The dimension of the common variance subspace can be reduced without loss of true process variability.
2. The reconstruction results in better quality, allowing a more accurate detection of changes in the measurements correlation (i.e. improving fault detection).

Following with the example of Eq. 7, performing SVD in **R** and searching for the shifts that maximize the first A eigenvalues, with $A = 1$, the solution reached is $[0, \pi/2, \pi]$, corresponding to the eigenvalues $\lambda_1 = 3, \lambda_2 = 0, \lambda_3 = 0$, and $\det(\mathbf{R}) = 0$. Therefore, one principal component is enough to model the total variability of the data matrix instead of the two PCs that would be necessary using the standard DAPCA.

Considering this new approach, if one wants to deal with all the complete adjustment problems without additional assumptions, $(d_{\max})^n$ singular value decompositions of the correlation matrix has to be evaluated. Therefore, a genetic algorithm (GA) [17] has been developed to solve this high combinatorial optimization problem.

In the classical GA formalism a set of N_{ind} candidate solutions (population) are generated randomly. The potential solution in the multidimensional search space (individual) is coded as a vector, called chromosome. The goodness of each individual in the population is evaluated by using a pre-specified fitness criterion. Upon assessing the fitness of all the chromosomes in the population, a new generation of individuals is created from the current population by using crossover and mutation operators.

4.1 Genetic Algorithm-Based Delay Adjusted Principal Component Analysis Technique

In this approach, each chromosome represents a backward shift vector (**DV**) that contains the delays of each process signal with respect to a reference signal. This reference signal can be in general any perturbed process variable, better results will be achieved when the reference signal moves before the remaining variables,

$$\mathbf{DV} = [d_1, d_2, \ldots, d_{n-1}] \quad \forall d_j \in [0, d_{\max}]. \quad j \in [1, \ldots, n-1] \tag{8}$$

Table 1 GA-DAPCA: algorithm outline

1.	Take normal operation conditions data
2.	First Optimization loop: determine the number of latent dimensions
3.	Second Optimization loop: maximize variance in the selected dimensions
4.	Apply optimum shifts to data and generate a PCA model

To speed up convergence, each delay in the chromosome is codified as a binary number of length,

$$L = \text{ceiling}(\log_2(d_{\max})) \tag{9}$$

where $d_{\max}$ is the maximum searched delay, and L is the length of a particular delay in the chromosome. With this modification, the actual $d_{\max}$ results,

$$\text{actual}\, d_{\max} = 2^L \tag{10}$$

and the total chromosome length (i.e. individual length) is,

$$L_{\text{ind}} = L \cdot (n - 1) \tag{11}$$

where n is the number of measured variables.

To generate new populations, individuals are selected from a random initial population and crossed by means of the roulette wheel selection, two point crossover and mutation operators. The algorithm is terminated using one of two stopping criteria: either the number of generations reaches a predefined maximum value, N_G, or the current population does not give sufficient improvement, μ_{stop}, compared with the performance reached G_{stop} generations before. To avoid overtraining, two sets of data can be used. One to train, and the other to stop the optimization early.

The optimization is performed in two loops (see Table 1). The first one is used to set the number of true latent dimensions A, that is used as a parameter in the second loop to maximize the variance contained in the first A principal components.

4.1.1 Genetic Algorithm-Based Delay Adjusted Principal Component Analysis: Determination of the Number of Latent Dimensions

An important decision in PCA is to select the appropriate number of principal components that capture the underlying relationships while eliminating errors and noise. However, when the data matrix is manipulated to adjust the delays, the number of selected principal components will change. A better alignment of process signals will produce a better model and then, the normal relationships will be captured in a smaller number of principal components. Therefore, to obtain a good PCA model, the first optimization loop (Table 2) determines how many true latent dimensions (i.e. true degrees of freedom) has the process. It consists on finding the shift **DV**

Table 2 Determination of the number of latent dimensions (A)

1.	Generate a random initial backward shifts population.
2.	For each individual ($\mathbf{DV}_k$) in the population.
3.	Generate a new delay adjusted data matrix $\overline{\mathbf{X}}_{\mathrm{DA}_k}$ by introducing the backward shifts $\mathbf{DV}_k$ into the normalized data matrix $\overline{\mathbf{X}}$.
4.	Perform SVD on the covariance matrix of $\overline{\mathbf{X}}_{\mathrm{DA}_k}$.
5.	Calculate the fitness of each individual $\mathbf{DV}_k$ as $\Phi_k = A_k$, were A_k is the number of PC selected using parallel analysis or cross correlation on $\overline{\mathbf{X}}_{\mathrm{DA}_k}$.
6.	End.
7.	If the end criteria (minimum improvement μ_{stop}, or number of generations N_{G}) is satisfied then end, else continue.
8.	Generate a new population and go to 2.

Table 3 Maximization of the common cause variance

1.	Generate a random initial backward shifts population.
2.	For each individual $\mathbf{DV}_k$ in the population:
3.	Generate a new delay adjusted data matrix $\overline{\mathbf{X}}_{\mathrm{DA}_k}$ by introducing the backward shifts $\mathbf{DV}_k$ into the normalized data matrix $\overline{\mathbf{X}}$.
4.	Perform SVD on the covariance matrix of $\overline{\mathbf{X}}_{\mathrm{DA}_k}$.
5.	Calculate the fitness of each individual $\mathbf{DV}_k$ as $\Phi_k = \frac{1}{n}\sum_{i=1}^{A}\lambda_i - \alpha_p \cdot p_k$
6.	End
7.	If the end criteria (minimum improvement μ_{stop} or number of generations N_{G}) is satisfied then end, else continue.
8.	Generate a new population and go to 2.

between variables that minimizes the number of latent variables (principal components) describing the common cause of process behavior. Jackson [15] and Himes et al. [16] present a complete review of methods to select the A principal components (e.g. Scree plot, parallel analysis, cross validation, etc.). In this work, A is selected using parallel analysis [18].

4.1.2 Genetic Algorithm-Based Delay Adjusted Principal Component Analysis: Maximization of the Common Cause Variance

The second loop (Table 3) consists of searching the **DV** that maximizes the variance contained in the first A principal components (corresponding to the eigenvalues $\lambda_1, \lambda_2, \ldots, \lambda_A$), to be considered in the description of the normal variability as opposed to the spurious variability contained in the last principal components (corresponding to the eigenvalues $\lambda_{A+1}, \lambda_{A+2}, \ldots, \lambda_n$) that will be discarded. As a consequence, the model will explain the majority of the true process variance in the smallest number of principal components. In addition, spurious variance that lies in the last principal components can be eliminated efficiently, therefore reducing the exclusion of normal variability and the amount of noise embedded into the model.

The displacement of uncorrelated signals have a minimal influence on the eigenvalues. The optimization algorithm will then assign aleatory delays to those signals. To avoid this undesired effect, a penalty term is added to the fitness function,

$$\Phi_k = \frac{1}{n}\sum_{i=1}^{A}\lambda_i - \alpha_p \times p_k \tag{12}$$

where,

$$p_k = \frac{1}{d_{\max} \times n}\sum_{j=1}^{n-1} d_{jk} \tag{13}$$

The penalization term is normalized (divided by $d_{\max} \times n$) so the maximum potential penalization is α_p. Since $\frac{1}{n}\sum_{i=1}^{A}\lambda_i <= 1$ and $p_k <= 1$, α_p allows to set the maximum penalization influence on the individual fitness. For example, $\alpha_p = 0.1$ sets a maximum penalization influence of 10% in relation to the maximum potential value of the fitness function (i.e. $\Phi_k = 1$).

4.2 Genetic Algorithm-Based Delay Adjusted Principal Component Analysis Online Application

Once a PCA model is developed, using the optimum $\overline{\mathbf{X}}_{\text{DA}}$, it is used online in the same way as the standard PCA. $\mathbf{t} = \mathbf{P}^{\text{T}}\mathbf{x}'$, but online signals ($\mathbf{x}$) should be synchronized by the application of **DV** in each sample. Note that due to this synchronization, in the GA-DAPCA the minimum detection time is $max(\mathbf{DV})$. This limitation can imply in some cases the necessity of an alternative detection technique (e.g. standard PCA with conservative limits) to detect abrupt faults. Nevertheless, in the case of small or gradual faults even considering that delay, the possibility of setting lower limits allows a faster and more reliable detection.

The Matlab genetic algorithm toolbox developed by the University of Sheffield [19] has been used to implement the algorithm in the following case studies. All the examples have been solved using an Intel i5-750 CPU with 3 GB RAM running under windows 7.

4.3 Illustrative Example I

To demonstrate the performance of the proposed algorithm an illustrative example is presented in Fig. 1. Figure 2a shows the evolution of three process variables representing: valve command (V_{c}), vapor flow through the valve (F_{v}) and the temperature of the tank (T), heated by means of the vapor.

Two types of delays are present:

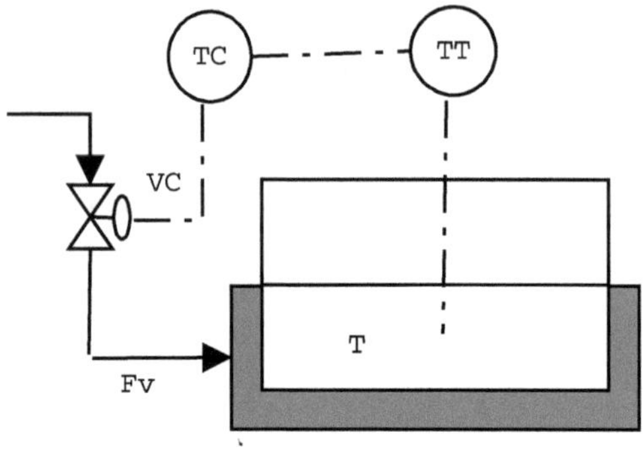

Fig. 1 Illustrative example I: flowsheet

- The valve response is delayed from its command signal.
- The controlled variable T is delayed from the manipulated variable F_v.

The reference signal selected in this case has been the most delayed one (T). The number of generations (N_G), mutation and crossover probabilities (P_m and P_c, respectively) values used in this example are listed in Table 4.

The algorithm converges to the best individual ($\mathbf{DV}_b = [0, 26, 29]$) in three generations, this is the same vector as that obtained by means of an exhaustive search. Nevertheless, a slight difference with the real delay ([0, 20, 24]) has been observed due to the process first order dynamics.

Table 5 presents the eigenvalues of the correlation data matrix of Example 1. It can be seen how after the delay adjustment, the system variance is concentrated on the first eigenvector (in other words, the process variance is explained in a unique direction). Therefore, by maximizing the first eigenvector, the process correlation is also maximized.

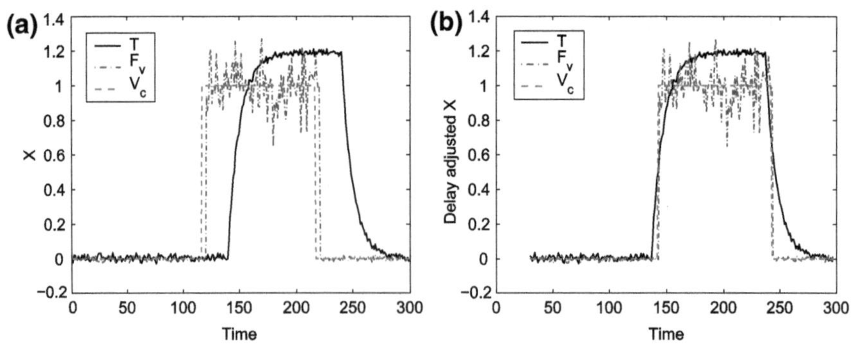

Fig. 2 Illustrative example I. **a** raw process data with delays, **b** adjusted process data

Table 4 Illustrative example I: GA parameters

N_{ind}	N_G	P_c	P_m	d_{max}
50	10	0.7	P_c/L_{ind}	32

Table 5 Illustrative example I: Eigenvalues of the correlation matrix

	First eig.	Second eig.	Third eig.
Raw data	2.39	0.54	0.07
Delay adjusted data	3.08	0.04	0.02

Adjusting the initial data matrix (see Fig. 2a) using the obtained delay vector ($\mathbf{DV}_b$) leads to more consistent data ($\mathbf{Y}_{DA}$) as shown in Fig. 2b.

Thus delays have been correctly identified and their negative effect properly compensated.

4.4 Illustrative Example II

A second application case is the well-known benchmark, the Tennessee Eastman process (TE) (Fig. 3) [20]. It involves the production of two products; G and H, from four reactants, A, C, D and E. Additionally, there are two side reactions that occur and an inert B. All the reactions are irreversible and exothermic. The process has five main units: an exothermic two-phase reactor, a product condenser, a flash separator, a reboiled stripper and a recycle compressor. The process disturbances presented by Downs and Vogel [20] has been simulated in the interval (2 h, 50 h) and measured with a sample time of $T_s = 10$s. The complete set of data includes 41 measured process variables and 12 actuators signals. In this work, the decentralized control system proposed by McAvoy and Ye [21] has been used (Fig. 3). Figure 4a shows the normalized time behavior of the selected signals.

Figure 4b shows the time behavior of the same signals after the implementation of GA-DAPCA with the following parameters (for both loops): $N_{\text{ind}} = 30$, $N_G = 100$ and $d_{\max} = 512$, $Pc = 0.7$, $P_m = P_c/L_{\text{ind}}$. The total computation time to converge was 7.28 s.

Figure 5 shows the percentage of variance contained in each principal component for both models. In PCA, 53.98% of process variance is contained in the first principal component. Applying parallel analysis, two principal components are selected with 90.22% of the process variance. However, as seen from Fig. 6a, none of these situations will be ideal; when selecting one principal component some common cause variance will be discarded in the remaining dimensions. Otherwise, the selection of two or three principal components will lead to a model that contains most of the common cause variability, but including noise.

In GA-DAPCA, 94.53% of the data variance is contained in the first principal component. This is enough to describe the common cause variance of the signals while the remaining dimensions containing spurious variance can be discarded, as is evident from Fig. 6b

Additionally, the goodness of the model can be corroborated from the correlation matrices obtained in each case (see Eqs. 14 and 15). Certainly, the correlation data matrix obtained after the adjustment shows the high correlation existent between the process signals.

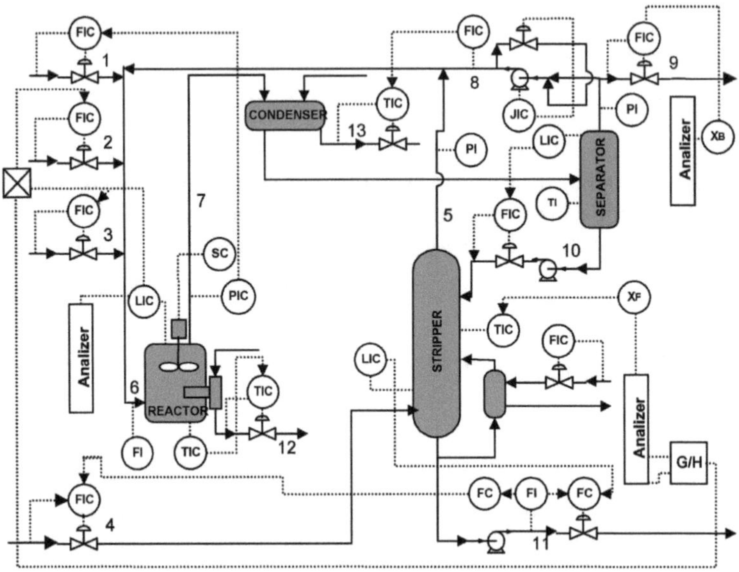

Fig. 3 Tennesee Eastman process flowsheet

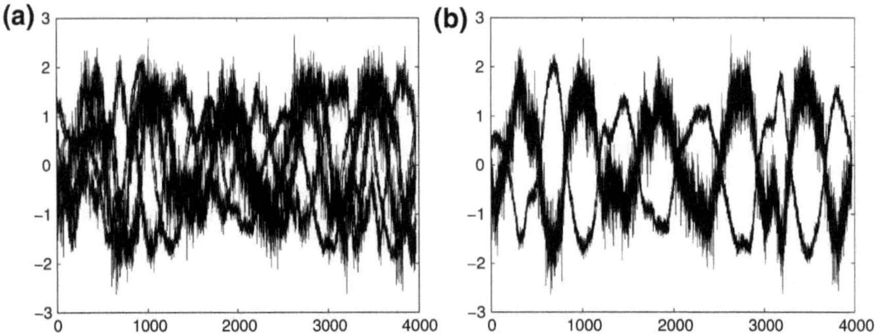

Fig. 4 Normal behavior. a raw measurements, b adjusted measurements

$$\mathbf{R}_{\text{original}} = \begin{bmatrix} 1 & & & & & \\ -0.49 & 1 & & & & \\ 0.28 & 0.37 & 1 & & & \\ 0.43 & -0.77 & -0.46 & 1 & & \\ -0.75 & 0.50 & -0.35 & -0.41 & 1 & \\ 0.36 & 0.28 & 0.98 & -0.36 & -0.45 & 1 \end{bmatrix} \tag{14}$$

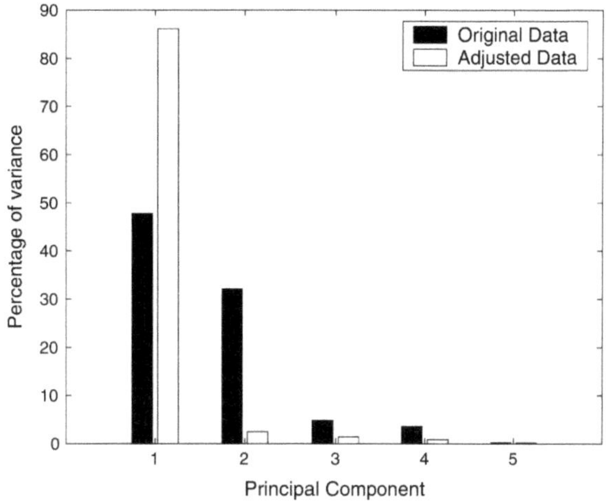

Fig. 5 Percentage of variance contained in each principal component

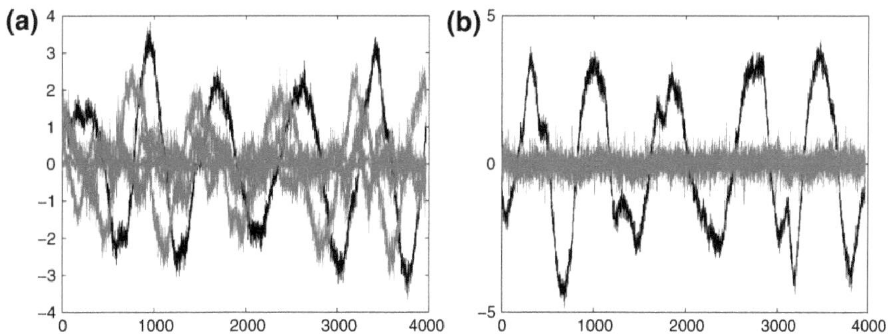

Fig. 6 Latent scores. The black signal corresponds to the first principal component. **a** PCA, **b** GA-DAPCA

$$\mathbf{R}_{\text{adjusted}} = \begin{bmatrix} 1 & & & & & \\ 0.90 & 1 & & & & \\ 0.90 & -0.95 & 1 & & & \\ 0.89 & 0.95 & -0.95 & 1 & & \\ 0.90 & 0.96 & -0.96 & 0.96 & 1 & \\ 0.90 & -0.95 & 0.99 & -0.94 & -0.96 & 1 \end{bmatrix} \quad (15)$$

Then, the monitoring performance of these models is compared. An oscillation of period $T = 1$ min is added to one of the measurements at $t = 1500$ min to simulate an abnormal event. Figure 7 shows the aspect of the data matrix before and after the adjustment.

Detection results are very different in both cases. In Fig. 8, the SPE statistics are presented. In case of GA-DAPCA detection limits are exceeded with more clarity because the residuals are calculated more accurately and detection limits can be set tighter.

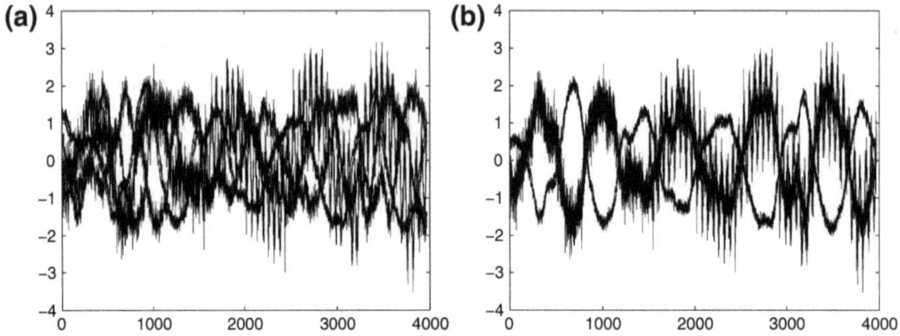

Fig. 7 Process measurements. A disturbance is introduced at $t = 1500$ min. **a** original measurements, **b** adjusted measurements

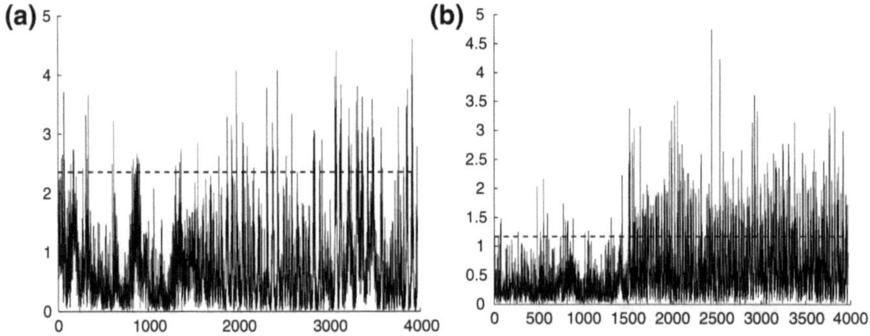

Fig. 8 SPE statistic for different methods. A disturbance is introduced at $t = 1500$ min. **a** PCA, **b** GA-DAPCA

Finally, just to clarify the picture, these statistics are filtered using the exponentially weighted moving average (EWMA) method. Figure 9 displays the filtered version of the SPE statistics obtained by each model. It is clear that the detection performance is highly improved by using GA-DAPCA (Fig. 9b), not only the number of points that exceed the limits increased but also the fault is detected earlier with respect to PCA (Fig. 9a).

5 Application to BPS with PEMFC

Finally, the proposed approach has been evaluated in the bio-ethanol processor system (BPS with PEMFC) described in Chaps. 9 and 10. Table 1 of Chap. 14 summarizes the 16 available measured variables in BPS with PEMFC plant, including seven control signals. In Fig. 5 can be seen the applied control policy where those signals already used in the seven loops are highlighted with a gray background. The process

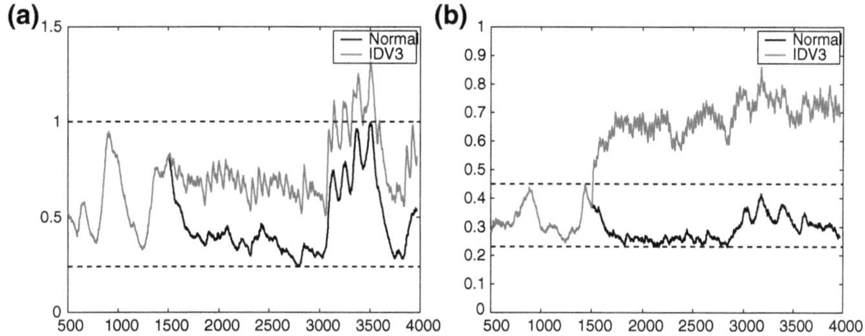

Fig. 9 EWMA of SPE. Black: normal behavior, gray: faulty case. **a** PCA, **b** GA-DAPCA

Table 6 GA parameters Execution to find $d_{\max}$PFS+ FC integrated process

	N_{ind}	N_G(max)	P_c	P_m	α_p	G_{stop}	μ_{stop}
First loop	50	10	0.7	$0.7/L_{\text{ind}}$	–	–	–
Second loop	500	500	0.7	$0.7/L_{\text{ind}}$	0.0001	50	α_p /14

is sampled each 0.5 s, during 30 s the data are recorded to build and test the fault detection algorithms.

5.1 Model Building

The GA-DAPCA model was built using only the control signals (i.e. seven sensors and seven actuator signals), without any additional sensor (see Fig. 5). Data corresponding to normal operation conditions of the process have been used to generate and test the models.

In this case study, $d_{\max}$ and the best reference signals are not considered to be known prior to the algorithm execution.

In order to estimate $d_{\max}$, the algorithm is solved thrice for each reference signal, changing $d_{\max}$ to 16, 32, and 64 samples. The chosen algorithm parameters are in Table 6. A very low penalty ($\alpha_p = 0.0001$) is introduced to avoid spurious delays. The first loop was not configured with early stop because it finalizes the execution in less than one second. The total time for each execution takes a media of 18 s.

Analyzing the delay vectors obtained in each execution (a total of 42), was found that *max*(**DV**) does not go over 32 samples in any case, but surpasses 16 samples in some cases, d_{max} is then set to 32.

To find the best reference signal the algorithm is run again 14 times, once for each reference signal with the parameters of Table 7. In this case, α_p is increased to 0.1 forcing the algorithm to find smaller delays.

Table 7 GA parameters Execution to find the best reference signal PFSα_p FC integrated process

	N_{ind}	N_G(max)	P_c	P_m	d_{max} [samples]	α_p	G_{stop}	μ_{stop}
First loop	50	10	0.7	$0.7/L_{ind}$	32	–	–	–
Second loop	500	500	0.7	$0.7/L_{ind}$	32	0.1	50	α_p /14

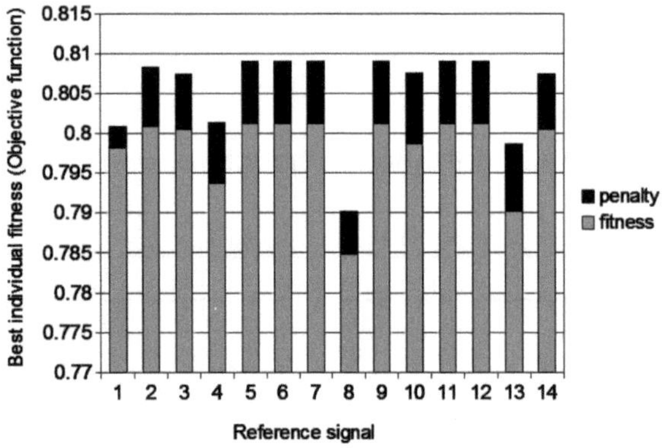

Fig. 10 Fuel processor system. Individual fitness and penalty influence with $d_{max} = 32$

Table 8 Fuel processor system

Delay [in 0.5 s samples]	Variable	Description
21	*Water to ESR inlet*	x_1
0	Exchanged heat Q	x_2
0	Ethanol to Burner	x_3
1	Oxygen to Burner	x_4
0	Ethanol to ESR	x_5
0	Oxygen to CO-PrOx	x_6
0	CM Voltage	x_7
6	*Molar ratio O2/CO*	x_8
0	ESR exit temperature	x_9
0	Burner exit temperature	x_{10}
0	Burner exit molar flow	x_{11}
0	H2 production rate	x_{12}
7	*CO–PrOx exit temperature*	x_{13}
0	Oxygen excess	x_{14}

Delay applied to each process variable

Individual fitness are depicted in Fig. 10, showing also the fitness penalization influence. There are several reference signals that give a similar value to the fitness. From them, the second signal (x_2) was chosen. The resultant delay vector is shown in Table 8.

With x_2 as reference, the first optimization loop selects five principal components. The second loop maximizes the variance in these first five principal components. It

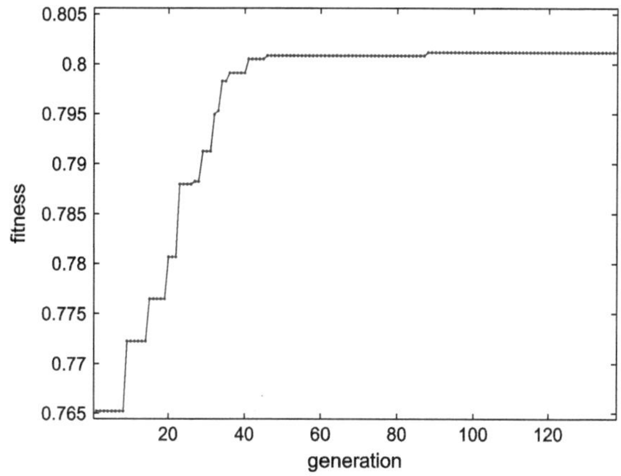

Fig. 11 Fuel processor system. Fitness evolution during the second optimization loop with x_2 as reference

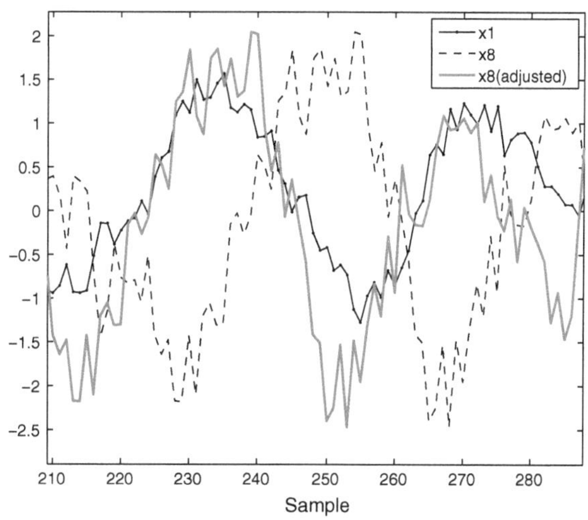

Fig. 12 Fuel processor system. Signal synchronization as a result of the delays introduction.

converges in 94 generations, early stop is reached 50 generations later (fixed by configuration). The total computation time for the 144 generations was 19.79 s. Figure 11 presents the fitness evolution for this particular execution (i.e. x_2 as reference).

Table 8 shows the optimized delay vector that will be introduced into the data matrix before applying PCA.

Figure 12 depicts an example of the introduced changes in the signals synchronization.

To evaluate the model, it has been compared with a standard PCA model built using the same data set.

For PCA, the first five principal components ($A = 5$) are selected, describing 76.5% of the total process variability. On the other hand, GA-DAPCA retain the 81.5% of the total variability in the same number of principal components. In addi-

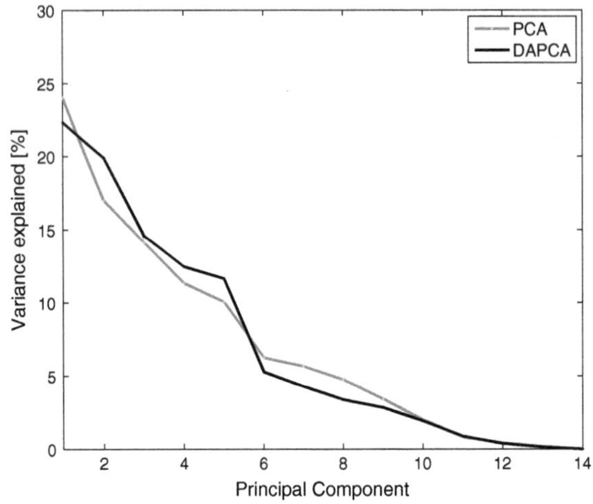

Fig. 13 Fuel processor system. PCA versus DAPCA, variance explained by the first ten latent dimensions

tion, it notability increases the difference between λ_A and λ_{A+1}, making clearer the difference between process and spurious variance (see Fig. 13).

5.2 Monitoring Results

The obtained model was used to monitor the fuel processor system. Detection limits where set to avoid false alarms during the normal operation. Alarms are defined as three consecutive points out of the 99% detection limit. The monitoring performance of GA-DAPCA is compared against that of standard PCA with the same number of selected latent dimensions $(\mathrm{A} = 5)$.

Figure 14 shows the SPE detection charts for PCA and GA-DAPCA during the normal state conditions. Due to the optimization, limits can be set lower in the GA-DAPCA chart without the generation of false alarms.

Table 9 shows the number of points outside the control limits (called reliability) under each disturbance and during the normal state. Disturbances are introduced at the sample 100.

Most faults have no detection problems for GA-DAPCA neither for standard PCA as can be seen for the higher values of reliability. However, the small fault $F3_{1\%}$ presents detection problems for PCA while GA-DAPCA clearly detect it eight samples (4 s) after the fault introduction. Figures 15 and 16 compares those results,

Note that for GA-DAPCA, *max*(**DV**) samples (i.e. 21 samples = 10.5 s) should be added to the online detection time (see Sect. 4.2).

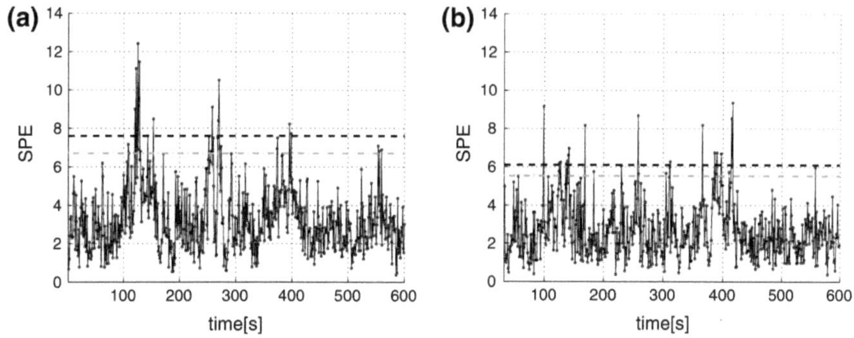

Fig. 14 Fuel processor system. SPE during the normal state **a** PCA, **b** GA-DAPCA

Table 9 Fuel processor system

	PCA		GA-DAPCA	
	T^2	SPE	T^2	SPE
Normal	2.0	2.0	3.9	3.0
F1	2.4	42.3	4.0	51.4
F2	56.1	93.8	17.8	98.2
$F3_{1\%}$	2.2	3.6	1.2	30.2
$F3_{3\%}$	4.7	32.7	5.0	70.2
F4	53.9	99.4	68.8	99.6
F5	51.9	99.4	98.0	99.0
F6	29.1	93.2	21.6	98.6
F7	23.1	96.6	31.0	99.4

Reliability table for PCA and GA-DAPCA

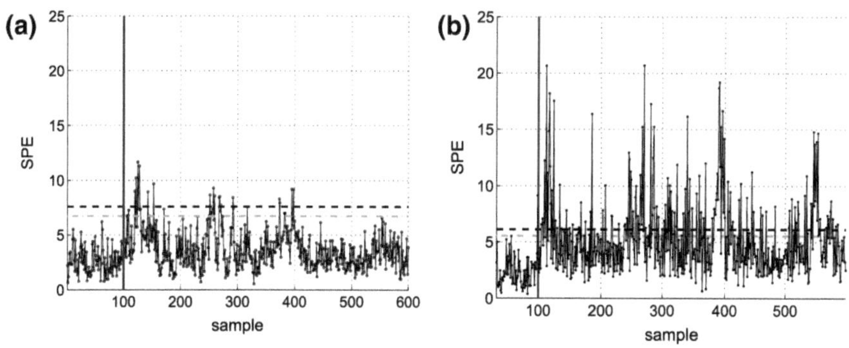

Fig. 15 Fuel processor system. SPE during $F3_{1\%}$. **a** PCA, **b** GA-DAPCA

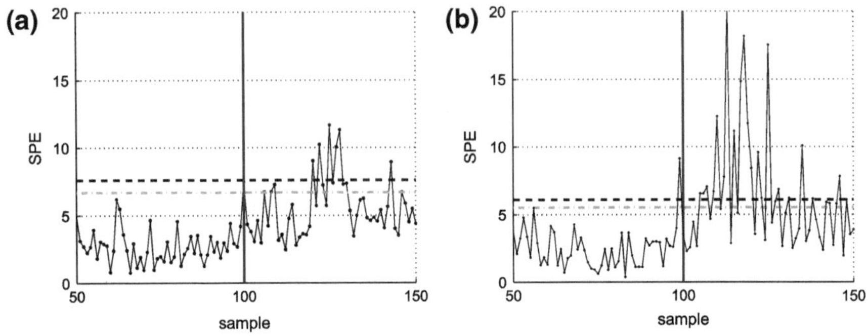

Fig. 16 Fuel processor system. SPE during $F3_{1\%}$. Detail. **a** PCA, **b** GA-DAPCA

6 Conclusions

PCA is a technique that has been widely used for statistical process monitoring for several chemical plants. However, it does not take into account time delays and therefore can lead to the inaccuracy of the correlation model built. A genetic algorithm-based optimization (GA-DAPCA) has been presented that improves the PCA methodology, enabling to deal with time delays in process signals. It helps to build smaller models that better capture the process data correlation.

It is worth noting that the model improvement, and the consequent detection improvement, is reached without considering any faulty case in particular, using only the normal state data. A better detection is expected through a general improvement of the PCA projection.

In all the cases presented here successful application results were obtained. Particularly, for the fuel processor system with PEMFC allowed to detect faults of less magnitude such as the case of $F3_{1\%}$ where the traditional PCA could not detect, GA-DAPCA was able to perceive in only 4 s. Therefore, it can be concluded that the proposed improvements to PCA techniques in Chap. 13 and here are able to give satisfactory fault detection especially for systems where the safety aspects are very important to take into account.

References

1. Kourti T, Lee J, MacGregor JF (1996) Experiences with industrial applications of projection methods for multivariate statistical process control. Comput Chem Eng 20:S745–S750
2. Miletic I, Quinn S, Dudzic M, Vaculik V, Champagne M (2004) An industrial perspective on implementing on-line applications of multivariate statistics. J Process Control 14:821–836
3. Musulin E, Yélamos I, Puigjaner L (2006) Integration of principal component analysis and fuzzy logic systems for comprehensive process fault detection and diagnosis. Ind Eng Chem Res 45(5):1739–1750

4. Nomikos P, MacGregor JF (1995) Multivariate SPC charts for monitoring batch processes. Technometrics 37(1):41–59, February PCA
5. Dunia R, Qin SJ, Edgar TF, McAvoy TJ (1996) Identification of faulty sensors using principal component analysis. AIChE J 42:2797–2812
6. Musulin E, Bagajewicz M, Nougués JM, Puigjaner L (2004) Design and upgrade of sensor networks for principal components analysis monitoring. Ind Eng Chem Res 43:2150–2159
7. Bakshi BR (1998) Multiscale PCA with application to multivariate statistical process monitoring. AIChE J 44(7):1596–1610
8. Misra M, Yue HH, Qin SJ, Ling C (2002) Multivariate process monitoring and fault diagnosis by multi-scale PCA. Comput Chem Eng 26:1281–1293
9. Kramer MA (1991) NonLinear PCA using autoassociative NN. AIChE J 37(2):233–243
10. Dong D, McAvoy J (1996) Nonlinear principal component analysis-based on principal curves and neural networks. Comput Chem Eng 20(1):65–77
11. Fourie SH, de Vaal P (2000) Advanced process monitoring using an on-line non-linear multiscale principal component analysis methodology. Comput Chem Eng 24:755–760
12. Jin HD, Lee YH, Lee G, Han C (2006) Robust recursive principal component analysis modeling for adaptive monitoring. Ind Eng Chem Res 45(2):696–703
13. Ku W, Storer RH, Georgakis C (1995) Disturance detection and isolation by dynamic principal component analysis. Chemom Intell Lab Syst 30:179–196
14. Wachs A, Lewin DR (1999) Improved PCA methods for process disturbance and failure identification. AIChE J 45(8):1688–1700
15. Jackson JE (1991) A user's guide to principal components. Wiley, New York
16. Himes DM, Storer RH, Georgakis C (1994) Determination of the number of principal components for disturbance detection and isolation. In: Proceedings of the Americal Control Conference, IEEE Press, Piscataway, NJ, June 1994
17. Goldberg DE (1989) Genetic algorithms in search, optimization, and machine learning. Addisonwesley, Massachusetts
18. Weiher E, Clarke GDP, Keddy PA (1995) Community assembly rules, morphological dispersion, and the coexistence of plant species. J Veg Sci 6:99–106
19. Chipperfield AJ, Fleming PJ (1995) The matlab genetic algorithm toolbox. In: IEE Colloquium on Applied Control Techniques Using MATLAB, 10/1–10/4 Jan 1995
20. Downs JJ, Vogel EF (1993) A plant-wide industrial process control problem. Comput Chem Eng 17(3):245–255
21. McAvoy TJ, Ye N (1994) Base control problme for the Tennessee Eastman problem. Comput Chem Eng 18(5):383–413

Index

3-D representation, 130

A
Abnormal events, 42, 93, 413, 415–417, 427
Absolute pressures, 158
Activation, 51, 58, 62, 118, 188, 273, 276
Activation loss, 52, 320
Active surface, 162, 173, 175
Adaptive PCA, 407, 410
Adaptive predictive control, 73, 86, 92, 94, 106, 110
Adaptive predictive control with robust filter, 73, 86, 95, 110
Adaptive predictive robust filter control, 94
ADVISOR, 54, 214, 229, 256, 352
Air flow rate, 67, 126
Air humidity, 64, 164
Air supply, 16, 20, 56, 67, 74, 100, 115, 190, 197, 209
Airplanes, 13
Alcohol, 37, 307
Alkaline Fuel Cells, 9
Alkalinity, 9
Anode, 4, 8, 12, 16, 33, 50, 65, 121, 135, 146, 158, 173, 191, 215, 342, 383
APC structure, 93, 96, 98
Artificial neural networks, 408
Asymptotically stable, 75, 117, 144, 148
Atmospheric pressure, 34, 158, 272, 332, 342
Atmospheric temperature, 65
Atomic number, 30
Automobile, 15, 25, 37, 40, 301
Automotive application, 6, 16, 22, 24, 56, 207, 208, 213, 214, 234
Autonomous robots, 14
Auxiliary equipment, 5, 18, 66, 69, 264
Auxiliary power unit, 15

B
Backup power, 13, 15
Ballard, 59, 252
Battery, 4, 10, 15, 18, 24, 29, 210, 356
Battery charger, 13
Behavioral models, 152
Bernoulli's continuity law, 154
Bilinear state model, 118
Bilinear system, 144
Bio-ethanol, 4, 38, 50, 262, 285, 294, 323, 335, 383, 391
Bio-ethanol reforming, 39, 43
Biofuels, 25
Biomass, 25, 30, 37
Bipolar plate, 122, 126, 133, 135
Black-box models, 152
Blower, 66, 288
Boats, 13
Boost converter, 210, 223, 237, 244
Boosting energy flow, 22
Boundary conditions, 124, 137, 176
Boundary conductance, 125
Braking/hydrogen ratio, 229
Buck-Boost converter, 210

C
Capacitance, 60, 155, 160, 209, 213, 218, 236, 254
Capacitive properties, 155
Capacitor voltage, 155

Green Energy and Technology, DOI: 10.1007/978-1-84996-184-4,

C (*cont.*)
Carbon, 6, 9, 12, 30, 37, 270
Carbon dioxide, 9, 11, 31, 36, 38, 274, 303, 310, 314, 326, 331
Carbon monoxide, 11, 31, 272, 302, 310, 315, 326, 333
Carrier, 30, 35, 53, 271
Catalyst, 6, 11, 31, 35, 39, 51, 58, 269, 274, 304, 350
Catalytic partial oxidation, 39, 300, 327
Cathode, 4, 8, 12, 16, 33, 50, 65, 121, 135, 138, 146, 158, 173, 191, 215, 342, 383
Cathode plate, 138
Cell flooding detection, 152
Cell voltage, 52, 59, 66, 77, 153, 166, 171, 177, 265, 287
Channel air, 137
Channel hydrogen, 138
Channel surfaces, 162
Charge, 22, 29, 59, 151, 153, 194, 210, 244, 252, 303, 356
Charging energy flow, 22
Charging mode, 236, 241
Chemical energy, 4, 74, 152, 278, 289, 303, 319
Chemical hydrade, 35
Cholesky factorization, 413
Circuit capacitance, 163
Circuit output, 158
Closed mode, 159
Closed-loop system, 118
Cluster, 129, 131
CO level, 324, 374
Coal, 6, 11, 12, 30, 51
Combinatorial, 361, 364, 368, 371, 375, 385, 393, 397, 406, 415, 419, 427, 436
Combined heat and power, 11, 14, 25
Combined statistic, 43, 406, 413, 419
Combustion, 5, 51, 214, 263, 276, 285, 301, 349
Component, 30, 37, 40, 43, 95, 135, 153, 156, 158, 187, 215, 263, 273, 304, 310, 314, 333, 342, 354, 397, 431, 434, 436, 436, 443
Component nozzle, 155
Compressor, 6, 14, 16, 21, 27, 29, 55, 59, 64, 74, 77, 79, 126, 185, 190, 195, 215, 264, 277, 284, 335, 344, 441
Compressor motor, 18, 65, 77, 104, 190, 321, 382
Concentration, 39, 58, 62, 66, 79, 118, 158, 181, 209, 272, 280, 302, 314, 389, 416
Concentration gradients, 158
Concentration loss, 53, 320
Condensation, 135, 159, 173, 181, 342
Condition number, 369, 399
Conductance, 121, 123, 125
Constrained nonlinear programming, 234, 237, 242, 247, 250, 256
Consumption map, 215, 234, 238, 242
Contamination, 152
Continuous stirred tank reactors, 301, 304, 348
Control horizon, 75, 84, 87, 90, 103
Control law, 91, 98, 117, 145, 201
Control limit, 409, 419, 434
Control objectives, 16, 77, 87, 197, 289, 318, 362, 371
Control strategies, 73, 101, 106, 152, 195, 289
Control strategy, 40, 61, 73, 103, 110, 152, 162, 262, 361, 392
Control structure design, 263, 265, 269, 401
Control structure selection, 361, 366, 412, 429
Control vector, 75
Controllability, 16, 40, 144, 262
Controlled variable, 86, 100, 203, 236, 382, 392, 440
Controller, 17–19, 41, 61, 76, 84, 92–103, 190, 195, 199, 201, 215, 286, 294, 299, 344, 352, 3556, 362, 386, 413
Convection, 122, 125, 138
Convection transfers, 137
Convective exchange, 137
Coolant channel, 135
Cooling fluid, 118
CO-prox, 336, 349, 374, 382, 389, 446
Correlation matrix, 408, 432, 435, 441
Cost function, 76, 87, 197, 235, 242, 399
Cost penalization, 414
Covariance matrix, 96, 112, 438
Cryogenic tank, 34
Cumulative percent variance, 408
Current density, 53, 80, 83, 126, 141, 163, 276, 320
Current source, 26, 156, 163
Cycle life, 35, 213, 254

D
Damage, 16, 17, 21, 73, 98, 100, 152, 173, 178, 185, 209, 236, 286, 324, 374
Degradation processes, 152
Delay, 42, 43, 88, 103, 167, 188, 194, 235, 319, 323, 373, 388, 431, 334–441, 446, 447

Delay adjusted PCA, 43, 431, 436
Delay vector, 436, 441, 446
Density, 4, 8, 15, 34–36, 50, 58, 66, 120, 124, 136, 145, 151, 153, 162, 174, 276, 320
Design variable, 21, 225, 357
Detectability, 42, 43, 405–417, 423
Deviation, 247, 265, 636, 371, 407, 434
Diagnosis, 3, 15, 21, 40, 42, 151, 161, 163, 164, 186, 187
Diagonal matrix, 103, 114, 199, 388, 412, 433
Diffusion layer, 6, 50, 162, 182
Diffusion layer pores, 162
Direct current, 14, 209
Direct hydrogen fuel cell, 232
Direct methanol fuel cell, 6
Discharging mode, 236, 241
Discrete model, 120
Distributed memory, 119, 129
Disturbance, 42, 83, 101, 110, 199, 327, 361, 397, 422, 444
Downstream, 78
Drag coefficient, 65, 160, 162, 214
Dry air flow, 156
Dry gas flow, 158
Drying phenomena, 161
Durability, 12, 15, 110, 118
Dynamic matrix control, 73, 84, 110, 287
Dynamic PCA, 434
Dynamic programming, 228
Dysfunction, 152, 167, 172, 177

E

Edit rule, 412, 416, 417
Efficiency, 3, 7, 9, 11, 12, 14, 15, 17, 19, 21, 26, 28, 29, 33, 35, 37, 39, 49, 51, 53, 54, 55, 59, 65, 67, 69, 73, 98, 190, 213, 221, 230, 261, 263, 278, 286
Efficiency map, 215, 216, 234, 238, 240, 247, 257, 355
Eigenvalues, 401, 433, 435, 436, 438
Eigenvectors, 408, 433
Electric circuit, 153
Electric current, 4
Electric environment, 155
Electrical current, 153, 276
Electrical energy, 4, 10, 54, 117, 265, 278, 289, 303, 319
Electrical environment, 153, 154
Electrical equivalence models, 153
Electrical equivalent, 151, 153, 155, 160, 182, 217
Electrical network analysis, 161
Electrical resistivity, 161, 163
Electrical topology, 26, 210, 231, 237
Electrical-thermal analogy, 122
Electricity, 4, 9, 10, 14, 24, 25, 33, 36, 302, 303
Electric-pneumatic analogy, 153
Electro chemical reaction, 4, 9, 10, 12, 15, 17, 33, 36, 51, 60, 122, 209
Electro osmotic, 65, 157, 159, 160, 162, 163
Electrochemical cells, 121, 122
Electrochemical energy, 4
Electrochemical impedance spectroscopy, 71
Electrochemical reaction, 4, 9, 15, 17, 33, 36, 51, 60, 122, 209
Electrode, 4, 7, 8, 11, 50, 53, 58, 59, 117, 121, 122, 127, 133, 135, 138, 152, 172, 209, 213, 319
Electrolysis, 4, 30, 33, 52
Electrolyte, 4, 6, 8, 9, 11, 33, 49, 50, 51, 57, 121, 276
Electrons, 6, 10, 33, 52, 54, 121, 276
Electro-osmosis, 64
Embedded systems, 152
Emissions, 3, 5, 16, 25, 33, 38, 51, 214, 220, 302, 303, 353, 355, 359, 377
Empirical models, 57
Endothermic reaction, 31, 308
Energy balance, 30, 120, 261, 262, 266, 277, 278, 305, 307, 309, 313, 317, 325, 342
Energy carrier, 30, 53
Energy conservation laws, 153
Energy density, 3, 8, 14, 52
Energy flow, 22, 209, 210, 223, 233, 237
Energy management strategy, 24, 27, 222, 237, 358, 378
Energy storage system, 24, 207, 208, 211, 212, 216, 223, 230, 244, 356, 359
Energy vector, 3, 36
Entropy, 127
ESR, 217, 262, 265, 279, 286–288, 302, 305, 308, 316, 322, 334, 339, 374, 382, 416, 422
Estimated parameter, 111, 112
Ethanol, 31, 37–39, 263, 269, 271, 285, 305, 322, 327, 331, 384, 432
Ethanol Steam Reforming, 39, 268, 271, 322, 339, 374
Ethernet LAN network, 131
Euclidean distance, 189, 193
Evaporation, 127, 156, 159
Execution time, 118, 119, 148
Exothermic reaction, 117, 137, 138, 272

F
Factorization, 97, 114, 116, 408, 413
Failure modes, 152, 161
Failure origin, 153
Failures, 116, 152, 161
Fall rate, 235, 237, 239, 242, 244
Fan, 5, 14, 15, 24, 56, 66, 278
Faraday number, 18, 54, 81
Faraday's constant, 64, 126, 320, 383
Fault components vector, 410
Fault detection, 21, 45, 115, 186, 187, 193, 194, 196, 205, 295, 404, 408, 410, 411, 413, 415, 419, 421, 424, 429, 431, 436, 450
Fault diagnosis, 3, 40, 43, 48, 115, 151, 153, 161, 182, 185, 187, 189, 191, 193, 195, 196, 204, 205, 428, 429, 432, 433, 451
Fault isolation, 186, 188
Fault magnitude, 21, 186, 189
Fault signature, 187, 190, 192, 194
Fault simulator, 190, 192, 193
Fault tolerance, 48, 196, 197, 201, 203, 205, 428
Fault tolerant control, 16, 45, 48, 93, 96, 185, 186, 205, 410, 428
FDI module, 197, 201
Federal Test Procedure, 219, 221, 243
Feedback controller, 18, 102
Feedback stabilization, 117, 118, 133, 139, 140, 143, 144
Feedforward controller, 18
Feeding channels, 6, 50
Filter, 21, 73, 74, 86, 91, 94, 95, 97, 98, 105, 110, 161, 186, 196, 356, 370, 373, 374, 388, 406, 411, 412, 417, 434, 444
Finite impulse response, 88
FIR coefficients, 88, 96
FIR model, 88, 90, 92, 97, 102, 103, 105, 106, 109
Fitness, 385, 387, 397, 398, 436, 438, 439, 446, 447
Flooding, 59, 71, 151, 152, 161, 166, 171, 174, 179, 181, 183, 190
Flooding failure mode, 164
Flow rate, 16, 60, 62, 64, 67, 74, 77, 78, 100, 118, 125, 127, 256, 264, 265, 283, 286, 288, 289, 326
Fluid circulation process, 134
Fluid mass, 154
Forklifts, 13
Fossil sources, 4, 31
FPGA, 253, 254
Free dynamics, 140
Free response, 75
Frobenius, 361, 368, 392, 399
Fuel cell advantages, 5
Fuel cell applications, 13, 270, 295, 297, 327
Fuel cell description, 4
Fuel cell efficiency, 53, 54, 66, 233, 237
Fuel cell energy flow, 22
Fuel cell hybrid system, 3, 22, 24, 26, 46, 57, 207, 209, 211, 214, 223, 230, 233, 234
Fuel cell hybrid vehicle, 24, 25, 45, 70, 207, 208, 213, 231, 233, 257, 359, 360, 392
Fuel cell life, 12
Fuel cell model, 49, 69, 70, 155, 190, 358
Fuel cell physical structure, 4, 50
Fuel cell principle of operation, 4, 6
Fuel cell stack, 14, 16, 21, 22, 24, 39, 43, 51, 55, 58, 62, 65, 69, 71, 74, 85, 115, 118, 121, 152, 156, 169, 194, 207, 208, 215, 232, 263, 276, 319, 322, 327, 374, 378
Fuel cell system, 3, 5, 7, 14, 17, 19, 22, 44, 47, 55, 57, 60, 61, 66, 67, 69, 71, 73, 74, 77, 83, 84, 110, 116, 136, 148, 149, 151, 152, 155, 177, 183, 185, 186, 190, 192, 194, 195, 197, 201, 204, 205, 207, 208, 215, 223, 231, 233, 243, 244, 261, 278, 283, 285, 286, 295, 297, 300, 327, 328, 404
Fuel cell technology, 12, 14, 44, 148, 231
Fuel cell types, 4, 8, 10, 11
Fuel cell voltage, 52, 53, 69, 73, 77, 287
Fuel crossover, 52, 276
Fuel processor system, 37, 264, 301, 374, 446
Fuel system control, 151
Full, 23, 28, 57, 127, 145, 370, 372, 419, 427
Fusion data, 152
Fuzzy logic, 152, 234, 404, 408, 412, 429, 450

G
GA-DAPCA, 43, 431, 436
Gain matrix, 145, 363
Gas channels, 58, 123
Gas circuit, 156, 158
Gas distribution, 158
Gasoline, 31, 36, 51, 219, 377
Generator, 10, 14, 47, 182, 186, 205
Genetic algorithm, 361, 363, 368, 375, 397, 404, 405, 415, 431, 436, 451
Genetic Algorithm Based Delay Adjusted PCA, 43, 436–439

Golf carts, 13
Greenhouse, 33, 302

H

Hardware Requirements, 393, 413, 427
Heat conduction, 120
Heat equation, 118
Heat exchanger, 38, 134, 139, 263, 272, 299, 309, 318, 335, 342
Heat losses, 135
Heat rejection, 118
Heat source, 137, 277, 317
High energy density, 3, 15
High temperature water gas shift, 264, 272, 310, 318, 322, 340, 351, 368, 374, 416, 420
Higher Heating Value, 278, 289
High-power, 126
Highway Fuel Economy Test, 219, 224, 229, 243, 246
Histogram, 219, 222
Host computer, 252
Hotelling, 406, 409, 433
HTS, 264, 266, 272, 280, 284, 310, 312, 322, 340, 374, 421
Humidifier, 14, 57, 99, 157
Humidifying system, 156
Hybrid system, 3, 10, 18, 22, 43, 46, 57, 196, 207, 209, 211, 214, 225, 230, 233, 243
Hybridization degree, 26, 212, 215, 225
Hydraulic, 153
Hydrogen carrier, 35
Hydrogen circuit, 158
Hydrogen consumption, 17, 28, 207, 215, 223, 226, 233, 241, 249, 251
Hydrogen economy, 22, 25, 44, 208, 221, 227, 234, 247
Hydrogen Fuel Cell Vehicle, 25, 46
Hydrogen oxidation, 121, 279, 333
Hydrogen production, 3, 15, 19, 30, 33, 37, 43, 47, 261, 267, 282, 295, 301, 308, 320, 325, 378, 383, 389, 393
Hydrogen security, 36
Hydrogen starvation, 19
Hydrogen supply, 16, 56, 74, 197, 424
Hydrogen uses, 36
Hydroxide, 9
Hygrometry, 179

I

Ideal gas law, 35, 154, 157, 288
Ideal pipe, 154
Identification, 48, 92, 102, 111, 116, 152, 161, 178, 182, 186, 301, 322, 327, 365, 374, 392, 404, 428, 451
Incompressible liquid, 154
Indirect adaptive control, 92
Indirect controlled variable, 105, 111
Industrial processes, 36, 42, 186, 365, 402, 406
Initialization, 97, 99, 103, 114, 127, 347, 398
Inlet gas, 58, 165, 342
Input gas, 58, 165, 342
Input gas system, 156
Input nozzle, 155
Integral absolute error, 109, 383, 393
Interaction, 45, 71, 115, 158, 258, 262, 300, 320, 327, 361, 372, 384, 391, 403
Internal combustion engine, 15, 25, 51, 54, 214, 301
Internal currents, 52, 276
Internal model control, 41, 47, 91, 301, 324, 328, 363, 366, 378, 382, 385, 395, 403
Internal plant model, 197
Internal temperature, 118, 308
Isolation, 21, 186, 193, 205, 307, 310, 315, 426, 451
Isomorfism, 155
Isotherm volume, 127
Isothermal, 123, 135, 139, 266, 176, 196, 307, 328, 340
Isothermal fluid, 139

K

Kalman filter, 21, 161, 406
Kalman filtering, 161, 406
Kirchoff's node, 153
Kolmogorov-Smirnov, 171, 180

L

Labview, 252, 258
Laminar flow, 126, 154
Laplacian, 120
Large-scale, 8, 14, 42, 118, 127, 361, 398, 405, 427
Latent, 127, 342, 408, 429, 432, 437, 443, 448
Least square, 76, 111, 367, 385, 395
Linear approximation, 144
Liquid cooling, 126
Liquid phase, 173, 182
Liquid state, 52, 153, 162, 173
Lithium hydrade, 36
Load current, 28, 60, 68, 84, 102, 253, 378
Load voltage, 167
Load-following mode, 209, 239
Location, 303, 368, 376, 386, 397, 414

L (*cont.*)
Locomotive, 13
Loop, 16, 41, 74, 88, 94, 100, 118, 126, 149, 153, 172, 194, 204, 294, 299, 301, 322, 330, 349, 359, 361, 370, 403, 406, 415, 422, 431, 437, 441, 444
Loop laws, 153
Low Temperature water gas shift, 264, 272, 279, 311, 318, 322, 340, 351, 374, 416
Lower Heating Value, 54, 229, 247, 278, 289

M
Manifold, 135, 208, 320
Manipulated variables, 41, 73, 76, 84, 110, 286, 294, 326, 351, 366, 389, 414
Mass, 18, 30, 37, 53, 58, 71, 81, 125, 148, 153, 154, 166, 177, 181, 192, 208, 214, 219, 252, 255, 261, 273, 294, 303, 313, 316, 336, 342, 350, 354
Mass flow, 17, 61, 78, 100, 118, 154, 252, 255, 336
Mass flow rate, 61, 64, 100, 118
Mass motion, 125
MATLAB, 197, 214, 299, 322, 329, 343, 350, 360, 374, 397, 404, 439, 451
Matrix, 11, 73, 84, 89, 96, 103, 110, 124, 128, 142, 149, 188, 192, 199, 201, 204, 301, 323, 357, 363, 368, 385, 388, 393, 408, 412, 432, 447
Maxwell Technologies, 213, 231, 252, 258
MEA, 50, 121, 122, 138
Mean, 65, 155, 170, 350, 375, 408, 433
Measured, 38, 77, 86, 104, 107, 167, 187, 190, 199, 252, 262, 286, 319, 323, 353, 379, 416, 441
Measured variables, 18, 253, 382, 389, 437, 444
Measurements, 41, 152, 161, 187, 190, 206, 363, 368, 375, 405, 409, 434, 442
Mechanical losses, 154
Mechanistic models, 57
Membrane, 6, 15, 33, 49, 57, 69, 73, 79, 98, 118, 127, 133, 142, 146, 151, 157, 160, 169, 173, 191, 209, 266, 273, 276, 296, 302, 319
Membrane conductivity, 63
Membrane dehydration, 59, 162
Membrane deterioration, 152, 161
Membrane electrode assembly, 50, 121, 138, 152
Message Passing Interface, 129
Methane, 4, 11, 31, 38, 263, 267, 282, 300, 303, 308, 326, 330, 334, 349
Micro fuel cell, 15
Micropower, 4
Minimal fault magnitude, 411, 414
Minimum square deviation, 364, 374, 388, 397, 406, 412
Model Predictive Control, 17, 45, 48, 71, 74, 86, 92, 110, 115, 185, 194, 363, 393, 428
Model-based fault diagnosis, 186, 190, 193, 204
Model-based method, 21, 45, 185
Molar flow, 153, 304, 309, 315, 323, 339, 349, 376, 416, 446
Molar mass, 18, 64, 126, 192
Molten Carbonate Fuel Cells, 6, 11, 31, 36, 295
Monitoring, 42, 48, 96, 206, 253, 404, 405, 413, 417, 427, 431, 443, 448
MPI, 117, 128, 148
Multiple-Input/Multiple-Output, 374
Multiprocessors, 119, 127

N
Nafion, 50, 69
National instruments, 252
Natural gas, 6, 10, 13, 19, 25, 30, 36, 300, 319, 327
Net load evaluation, 365, 370, 384
Network nodal modeling, 122
New European Driving Cycle, 16, 28, 219, 229, 243
NEXA, 252, 254
Nitrogen, 32, 56, 61, 64, 158, 192, 264, 330
Nitrogen diffusion, 158, 159
Nodal network, 123, 128, 137, 148
Nodal principle, 123
Node, 120, 123, 146, 161
Nodes, 121, 130, 138, 146
Nonlinear model, vii, 19, 41, 84, 135, 197
Nonlinear system, 61, 116, 117, 144, 215, 429
Non-parametric identification, 152
Norm, 177, 361, 368, 392, 395, 399
Normal distribution, 170
Normal operation, 97, 172, 408, 426, 432, 437, 445
Nozzle, 64, 78, 155
Number of moles, 154, 304, 350
Number of Reynolds, 156, 307

O
Objective function, 43, 361, 397, 406, 414, 427, 435

Observer, 21, 161, 187, 193, 300, 327
Ohmic loss, 53, 161, 320
Ohmic voltage, 63, 276
Ohm's law, 154, 276
Open circuit voltage, 52, 58, 62, 191, 319
Open loop mode, 172
Open mode, 158
Operating conditions, 19, 28, 39, 45, 57, 67, 152, 155, 167, 175, 178, 261, 271, 280, 284, 307, 320, 324, 330, 407, 424
Operating range, 11, 19, 118, 140, 199, 208
Operating temperature, 4, 21, 31, 36, 118, 135, 229, 254, 275, 278, 284, 342
Operation modes, 29, 151, 182
Optimal monitoring system design, 405, 415
Optimal operating point, 16, 21, 185, 325
Optimal sensor location, 42, 48, 327, 361, 364, 366, 382, 400, 429
Optimal signal selection, 406, 414, 416, 419, 421
Optimization, 18, 20, 28, 43, 70, 74, 149, 196, 232, 234, 242, 258, 295, 307, 320, 357, 361, 384, 397, 406, 417, 431, 436, 446
Outlet valve area, 78, 199, 205
Output nozzle, 155, 162
Oxidant, 4, 118, 319
Oxidation process, 121
Oxygen circuit, 158
Oxygen excess ratio, 16, 58, 68, 77, 110, 192, 195, 287
Oxygen starvation, 17, 45, 79, 195, 205, 209, 231

P
Pairing, 364, 370, 376, 394, 395
Parallel algorithm, 119, 130, 131
Parallel computers, 119
Parallel computing, 118, 119, 129, 148
Parallel programming, 119
Parallel Virtual Machine, 129
Parameter setting, 481
Parasitic compressor power, 68
Partial differential equations, 120, 148
Partial oxidation, 30, 39, 300, 319, 327
Partial pressure, 15, 67, 100, 156, 159, 191, 199, 265, 287, 314, 383
PEM fuel cell, 15, 16, 20, 21, 36, 42, 49, 58, 69, 129, 135, 185, 190, 201, 214, 325, 374, 432
Penalization, 88, 90, 414, 439, 446
Penalty, 88, 414, 439, 446
Perfluorosulfonic acid copolymer, 50
Phosphoric acid, 6, 10, 14
Phosphoric Acid Fuel Cell, 6, 10, 14
Physico-chemical phenomena, 152
PI, 18, 100, 378, 388
PID, 17, 74, 352
Pipes, 154
Plant-wide control, 327, 361, 418, 428
Plates, 121, 135
Platinum, 6, 11, 51, 269, 275
Pneumatic capacitance, 155
Pneumatic conductor, 154
Pneumatic environment, 153
Pneumatic resistance, 154
Poiseuille's law, 154
Polarization curve, 20, 53, 118, 166, 177, 215
Polarization zone, 177
Pollution, 5, 15, 25
Portable electronics, 6, 8, 13, 15
Portable power, 13
Power conditioning subsystem, 57
Power conservation principle, 64
Power converter, 210, 237, 239
Power source, 15, 18, 22, 27, 208, 221
Power split, 29, 43, 233, 243
Power/Energy storage system ratio, 212
Predicted outputs, 75, 86
Prediction horizon, 74, 96, 103, 197, 201
Preferential oxidation, 261, 264, 288, 303, 312, 334, 340
Pressure gradient, 162, 170
Pressure losses, 162
Pressure vessel, 34
Preventive maintenance, 152
Primary source, 14
Principal component, 40, 43, 405, 408, 431, 437, 438, 441
Principal component analysis, 405, 431
Probability density, 151, 166
Process response, 61, 110, 319
Programming paradigms, 119, 127
Propane, 31, 32
Proton Exchange Membrane Fuel Cell, 3, 6, 151, 264, 276, 299, 300, 319, 341
Protonic conductivity, 7
Protons, 6, 7, 33, 65, 121, 276
Pure fuel cell vehicle, 26, 212
PVM, 117, 118

Q
Q, 76, 84, 110, 120, 142, 145, 199, 265, 323, 350, 379, 392, 406, 409, 413
Quadratic function, 111, 321, 359
Quadripole model, 155
Quasi-load-following, 257

R

R, 76, 84, 92, 98, 127, 143, 154, 167, 174, 177, 279, 280, 284, 414, 433, 436
Random variable, 167, 170
RC circuit, 155
Reactant, 4, 9, 17, 52, 66, 130, 288, 305, 307, 341, 402, 441
Reactant flow subsystem, 56
Real time applications, 153
Recursive least square, 112
Reference trajectory, 86, 90, 107
Reformer, 11, 24, 51, 60, 194, 261, 271, 274, 279, 282, 294, 300, 334, 342, 382
Reforming, 12, 19, 30, 31, 39, 43, 52, 261, 266, 268, 269, 272, 278, 280, 302, 305, 308, 334
Regenerated energy flow, 22
Regenerative breaking, 22, 24, 29, 208, 210, 223, 229
Regulation valves, 156
Relative disturbance gain array, 370, 396
Relative gain array, 19, 363, 393, 394
Relative humidity, 68, 156, 191, 342
Relative normalized gain array, 365, 394
Reliability, 8, 9, 151, 152, 426, 448
Reliability index, 426
Renewable energy sources, 4, 30
Requirements, 16, 37, 42, 58, 110, 208, 213, 219, 223, 263, 282, 302, 322, 338, 419, 427
Residual, 21, 161, 182, 186, 192, 272, 408, 432
Residual generation, 21, 161, 187, 192
Resistance, 7, 53, 58, 63, 138, 154, 190, 209, 214, 217, 236, 254, 270, 276, 340
Return manifold, 63, 65, 78, 190
Reversible open circuit voltage, 319
Rise rate, 236, 42
Robust, 15, 20, 28, 73, 86, 94, 187, 214, 363
RTOS computer, 252, 254

S

S, 121, 127, 130, 136, 139, 146, 188
Safety, 35
Sample time, 42, 86, 90, 113, 323, 405, 441
Saturation pressure, 160, 181, 192, 342
Saturation voltage, 160
Scooters, 13
Sensitivity, 187, 192, 204, 261, 287, 324, 354, 399
Sensor, 42, 96, 166, 254, 263, 327, 361, 375, 392, 397, 400, 405, 412, 428, 445
Sensor bias, 175
Separately excited direct current, 100, 107
Sequential time, 130
Shared memory, 119, 127, 133
Signal selection, 406, 414, 421, 428
Signature matrix, 189, 192, 204
Simulink, 214, 301, 348, 358
Single-Input/Single-Output, 86, 435
Singular value, 323, 375, 408, 433, 436
Singular value decomposition, 323, 375, 408, 433
Smoothed moving average, 417
Sodium borohydride, 35
Solar power, 25
Solid Oxide Fuel Cells, 14
Source term, 120, 138
Space partition, 129
Sparse matrix, 128
Specific energy, 25, 52, 211, 226, 244, 254
Specific heat, 64, 67, 78, 120, 124, 136, 305
Specific heat capacity of air, 67
Specific power, 209, 212, 226, 254
Speedup, 117, 130, 133, 289
Square predictive error, 375, 413, 432
Stability, 93, 98, 103, 106, 109, 142, 271, 364, 371
Stabilization, 117, 133, 139, 143, 167, 322, 365, 370, 374
Stack current, 68, 83, 98, 107, 199, 286, 320, 385
Stack integrity, 152
Stack life time, 254
Stack model, 155, 276
Stack reliability, 152
Stack voltage, 60, 83, 101, 160, 164, 197, 253, 323, 424
Stand-alone application, 24
Standard driving cycle, 219, 229, 377
State model, 139, 264, 294, 324, 330, 367, 375, 406
State of charge, 29, 213, 218, 244, 353, 356
State of energy, 217, 221, 229, 236, 240, 378, 383
State-of-health, 152
Static model, 166, 215
Stationary power generation, 6, 10, 13
Statistical analysis, 153, 170
Statistical approach, 166, 169, 172, 176, 177, 182
Statistical methods, 21, 40, 180
Statistics, 219, 406, 409, 413, 443
Steady-state, 25, 40, 68, 84, 103, 110, 186, 199, 262, 264, 286, 289, 294, 300, 330, 353, 361, 367, 392
Steam, 11, 30, 39, 135, 156, 181, 272, 302, 337
Steam-methane reforming, 31

Step response, 61, 74, 94, 110
Stochastic approach, 166
Stochastic system, 170
Stoichiometry, 20, 68, 79, 126, 274, 286, 310, 331
Storage device, 27, 57, 154, 209, 213, 230, 300
Subspace extraction, 410, 417, 427
Sulfur, 12
Sum of square deviation, 364, 371, 392
Supercapacitor, 18, 43, 57, 211, 236, 252, 378
Supply manifold, 61, 78, 190, 320
Synchronization, 28, 97, 106, 119, 439, 447
Synthesis, 19, 38, 42, 261, 268, 270, 289, 294, 303, 308, 325, 361, 374, 392, 402
System identification, 111, 323, 327, 366, 374, 392

T
Tafel's equation, 161
Taylor series, 120
Technological instrumentation, 40, 152
Telecommunications, 13
Temperature distribution, 124, 152
Temperature drop, 162, 180
Temperature gradient, 51, 173
Tennessee Eastman, 406, 432, 441
Thank-to-wheel, 14
Thermal behavior, 117, 133, 148
Thermal capacity, 121
Thermal conduction, 120, 138
Thermal conductivity, 120, 125, 133, 307
Thermal energy, 117
Thermal transfer, 123, 136
Thermocouples, 152
Threads, 129, 133
Three-dimensional models, 58
Threshold, 10, 96, 187, 434
Throttle, 60, 78
Traction Control System, 101
Transfer Function Matrix, 301
Transient response, 22, 56, 60, 83, 103, 110, 208, 215, 234, 287, 370
Transportation, 4, 25, 38, 302
Tube model, 154
Tuning, 41, 74, 103, 299, 324, 366, 371, 373, 378, 388
Two-dimensional models, 58

U
Uncontrolled Variable, 366, 367
Underwater vehicles, 13
Unified Parallel C, 129
UPC, 117–119, 128–133, 148
Upstream, 78
Urban Dynamometer, 219, 220, 243, 378
Urban Dynamometer Driving Schedule, 220, 378
Utility vehicles, 13

V
Valve, 19, 56, 60, 73, 78–84, 85, 110, 195, 197, 199, 203, 205, 286, 287, 289, 435, 439, 440
Vapor phase, 160, 331, 332
Variance, 43, 174, 176, 179, 180, 375, 408409, 414, 419, 431–441, 443, 446, 448
Variance decomposition, 176
Vehicle, 5, 9, 14, 17, 23–29, 31, 51, 55, 98, 185, 186, 209, 210, 212, 214, 219–230, 233, 234, 252, 303, 330, 352–360, 383, 384, 392
Vessel, 32, 35, 303
Virtual sensors, 152
Voltage losses, 10, 52, 161
Voltage source, 158
Voltage variance, 176
Volumetric density, 34

W
Wastewater Treatment Plants, 13
Water accumulation, 175
Water activity, 160
Water balance, 7, 57, 58
Water diffusion, 66, 160, 191, 192
Water diffusion coefficient, 66, 191, 192
Water distribution, 174–176
Water electrolysis, 30
Water Gas Shift, 31, 32, 262, 268, 270–274, 276, 280, 288, 300, 302, 308, 310, 311, 332, 334, 340, 349, 350
Water management, 7, 9, 51, 57, 58, 162, 215, 264
Water vapor, 25, 33, 36, 54, 64, 162, 172, 173
Water vaporization rate, 156
Water-gas shift reaction, 31, 32
Wave power, 25
Weight matrix, 84
Well-to-tank, 14
Wind power, 25
Workstations, 119, 131, 132

Z
Zone, 58, 73, 98, 101, 102, 107, 167, 169, 177, 179, 208, 219, 221, 257, 358, 409, 417, 419, 424

MIX
Papier aus verantwortungsvollen Quellen
Paper from responsible sources
FSC® C105338

If you have any concerns about our products,
you can contact us on
ProductSafety@springernature.com

In case Publisher is established outside the EU,
the EU authorized representative is:
Springer Nature Customer Service Center GmbH
Europaplatz 3, 69115 Heidelberg, Germany

Printed by Libri Plureos GmbH
in Hamburg, Germany